Human-Computer Interaction and Cybersecurity Handbook

Human Factors and Ergonomics

Series Editors

Waldemar Karwowski *University of Central Florida, Orlando, USA*
Pamela McCauley *University of Central Florida, Orlando, USA*

PUBLISHED TITLES

Around the Patient Bed: Human Factors and Safety in Health Care
Y. Donchin and D. Gopher

Cognitive Neuroscience of Human Systems: Work and Everyday Life
C. Forsythe, H. Liao, M. Trumbo, and R. E. Cardona-Rivera

Conceptual Foundations of Human Factors Measurement **D. Meister**

Content Preparation Guidelines for the Web and Information Appliances: Cross-Cultural Comparisons *H. Liao, Y. Guo, A. Savoy, and G. Salvendy*
Cross-Cultural Design for IT Products and Services *P. Rau, T. Plocher and Y. Choong*

Data Mining: Theories, Algorithms, and Examples **Nong Ye**
Designing for Accessibility: A Business Guide to Countering Design Exclusion **S. Keates**
Ergonomics in Design: Methods and Techniques **Marcelo M. Soares and Francisco Rebelo**
Ergonomic Workplace Design for Health, Wellness, and Productivity **Alan Hedge**
Fall Prevention and Protection: Principles, Guidelines, and Practices **Hongwei Hsiao**
Handbook of Cognitive Task Design **E. Hollnagel**

The Handbook of Data Mining *N. Ye*
Handbook of Digital Human Modeling: Research for Applied Ergonomics and Human Factors Engineering *V. G. Duffy*
Handbook of Human Factors and Ergonomics in Health Care and Patient Safety Second Edition *P. Carayon*
Handbook of Human Factors in Air Transportation Systems *Steven James Landry*
Handbook of Human Factors in Web Design, Second Edition *K. Vu and R. Proctor*
Handbook of Occupational Safety and Health *D. Koradecka*
Handbook of Standards and Guidelines in Ergonomics and Human Factors *W. Karwowski*
Handbook of Virtual Environments: Design, Implementation,and Applications, Second Edition, *K. S. Hale and K M. Stanney*
Human-Computer Interaction and Cybersecurity Handbook Abbas Moallem

Human-Computer Interaction and Cybersecurity Handbook

Edited by
Abbas Moallem

CRC Press is an imprint of the
Taylor & Francis Group, an **informa** business

First published in paperback 2024

First published 2019 by CRC Press
2385 NW Executive Center Drive, Suite 320, Boca Raton FL 33431

and by CRC Press
4 Park Square, Milton Park, Abingdon, Oxon, OX14 4RN

CRC Press is an imprint of Taylor & Francis Group, LLC

Publisher's Note
The publisher has gone to great lengths to ensure the quality of this reprint but points out that some imperfections in the original copies may be apparent.

ISBN: 978-1-138-73916-1 (hbk)
ISBN: 978-1-03-291968-3 (pbk)
ISBN: 978-1-315-18431-9 (ebk)

DOI: 10.1201/b22142

Visit the Taylor & Francis Web site at
http://www.taylorandfrancis.com

and the CRC Press Web site at
http://www.crcpress.com

To the loving memory of my father, mother, and brother

Contents

Section VI: Perspective

List of figures

List of tables

Preface

The recently discovered vulnerability flaw of ubiquitous microprocessors* is only one example of the spreading security risks which emerge with the exponential growth of Internet interconnections and the development of the Internet of Things. Cyberattacks endanger individuals and companies, as well as vital public services and infrastructures.

A survey conducted in 2016 found that over 60% of respondent businesses had an information technology security breach in 2015, and 42% of them reported that the breach resulted in insignificant negative impact.† Confronted with spreading and evolving cyber-threats, organizations and individuals are falling behind in defending their systems and networks, and they often fail to implement and effectively use basic cybersecurity practices and technologies, leading to questions about why security seems to be so difficult.‡ Successful security depends on companies and governments collaborating to identify threats, weaknesses, and solutions. However, many initiatives today focus on systems and technology, without addressing well-known user-related issues. In fact, users have been identified as one of the major security weaknesses in today's technologies, as they may be unaware that their behavior while interacting may have security consequences. However, if users are to be considered one of the greatest risks to system security, they are also one of the greatest hopes for system security.§ In this perspective, human–computer interaction (HCI) becomes a fundamental pillar for designing more secure systems. By considering the user—what they know, how they use the system, and what their needs are—designers will be better positioned to empower them in their digital security role and increase the usability of security solutions.

This new handbook on HCI in cybersecurity represents the current state of the art in this young but rapidly developing and maturing scientific domain, which is composed of HCI principles, methods, and tools in order to address the numerous and complex threats that put computer-mediated human activities at risk in today's society. This is progressively becoming more and more intertwined with and dependent on interactive technologies.

* Wired. 2018. *A Critical Intel Flaw Breaks Basic Security for Most Computers*, https://www.wired.com/story/critical-intel-flaw-breaks-basic-security-for-most-computers/.

† AT&T. 2016. The CEO's guide to cyberbreach response: What to do before, during, and after a cyberbreach. AT&T *Cybersecurity Insights*, Volume 3. Retrieved January 9, 2018, https://www.business.att.com/cybersecurity/docs/cyberbreachresponse.pdf.

‡ J. M. Haney and W. G. Lutters. 2017. The Work of Cybersecurity Advocates. In *Proceedings of the 2017 CHI Conference Extended Abstracts on Human Factors in Computing Systems* (CHI EA '17). Association for Computing Machinery, New York, 1663–1670. DOI: https://doi.org/10.1145/3027063.3053134.

§ J. D. Still. 2016. Cybersecurity needs you! *Interactions* 23, 3 (April 2016), 54–58. DOI: https://doi.org/10.1145/2899383.

The scope of investigation is broad, covering authentication and access management, trust and privacy in HCI and cybersecurity, insider threats, social engineering, and smart networks and devices, as well as governance law and regulation.

This pioneering handbook, the first of its kind in this field, fills a significant gap in an important domain of investigation by reflecting recent developments, consolidating present knowledge, and opening new perspectives for the future. It provides a structured guide for beginners, a reference collection for more experienced practitioners and researchers in the field, as well as an important educational tool for undergraduate and postgraduate students.

The book is structured in six parts and contains 18 chapters written by 32 authors from ten countries, coming from academia, research institutions, industry, and public policy institutions.

In summary, this handbook represents a great contribution toward further advancing the concepts and principles of HCI in cybersecurity, for the benefit of all citizens in the rapidly expanding information society. For this, I express my appreciation and extend my congratulations to the editor and the authors of this handbook.

Constantine Stephanidis

Acknowledgments

I would like to thank Professor Louis Freund, for without his suggestions and help, this book would not be possible. I am also grateful to Professor Sigurd Meldal for generously sharing his time and expertise in editing this book.

Introduction

Cyberspace has become a new site of crime and illegal behavior. While a wide range of acts of crime and criminality—including robbery, identity theft, ransom, spying, subterfuge, deception, and black markets—have been part of the experience of social life, globalization and the expansion of new media technologies have presented us with new changes and challenges. With the expansion of digital media, these activities have taken unique forms requiring specific, and sometimes fundamentally distinct, ways of understanding.

Some scholars have referred to our global and postindustrialized societies as risk societies [1]. Exposure to vulnerability and risk seems to be an integral component of globalization especially because of the expansion of techlands, mediascapes, and digital spaces. The concept of a mediascape includes the ways in which electronic capabilities produce and disseminate information through media. The spatial notion of scape refers to the fluid and fragmentary nature of media and global flows [2]. The notion of cybersecurity and security studies have entered the scholarly literature to describe the specificity of our digital age. Cybersecurity, or information technology (IT) security, includes a focus on protecting computers from criminal behavior, but nothing has changed regarding the types of criminal activity. What has changed is its scale, facility to conduct the crime, and its volume. Cybercriminals are increasingly hacking into homes, offices, hospitals, and government establishments and stealing information without breaking any physical doors or being identified. They attack hundreds of individuals and major organizations and ask for ransoms to give them back their own data [3]. Bank accounts of masses of people are being compromised without any trace [4]. Stolen identities of individuals are used to obtain loans, purchase products, or even seek medical treatments [5]. Communications and movements of people—even heads of state—are listened to and spied on [6]. A huge black market, publicly available on the Internet, exists and provides the opportunity to purchase any illegal product from illicit and nonillicit drugs to stolen credit card information [7]. Some financial activities are conducted with minimal tracking using Bitcoins [8]. These are just a few of the many types of criminal activities investigated by spying agencies, security agencies, or private investigation firms.

Human societies have been transformed through the expansion of digital technologies, and along with this comes the increase in opportunities for criminal activity, which continues to dramatically accelerate every day. These activities are affecting not only the public sphere and institutions but also the private sphere of families and communities. For example, residential units are being equipped with more security devices including surveillance cameras, temperature controls, medical tracking devices, and smart refrigerators to just name a few. At the same time, human communication is largely being carried out digitally through mobile devices and social networking. Furthermore, all manners of financial transactions—from banking to purchasing products and services—are being performed online. All types of organizations, both public and private enterprises, are being thoroughly digitalized. Every aspect of human activity, from when you enter a building, to communicating with customers, clients, and coworkers, to when you design and create new products, is managed, recorded, and tracked in the digital realm. Even "in-person" meetings are oftentimes recorded or executed via video conferencing. All countries, rich and poor, are being fully restructured through digital information. This form of reorganizing may include anything from managing citizen information and requests, to monitoring and surveillance and, possibly soon, even voting systems. Last but not least, countries all over the world are competing for who can first get their hands on information, which can be used for anything from military defense/offense, to commercial, financial, and technological advantages.

New problems, old solutions

All this transformation is happening within the old frameworks of human society. Our societies are still not fully equipped to face these transformations and deal with this new paradigm of criminal activity. This is partially caused by those in power: government agencies, various public and private institutions, and transnational corporations with significant power to control the world. Ulrich Beck calls it a "risk society" [1] where people are increasingly living on a high technological frontier in a society where no one understands or can predict the future.

It is possible, if not likely, that cyberwars among nations, organizations, and people will soon become everyday events. Consequently, a scholarly focus on cybersecurity is imperative for every person, group, organization, or entity. According to Gartner, global spending on IT security reached $77 billion in 2015, an increase of 4.7% over 2014. Estimates are that global spending on information security will hit $101 billion in 2018 [9].

Human actors or agents are at the center of all security systems and are at the center of all security research. Some people refer to human actors or users as the "weakest link in the security chain." According to IBM Security, 95% of all security incidents involve human error [10]. Thus, the understanding of human performance, capability, and behavior must be one of the leading areas that experts in cybersecurity should focus on, both from a human–computer interaction (HCI) point of view and a pure human factors perspective.

The vulnerability of human agents depends on their performance and what limits them in this process. For decades, human factors professionals have advocated for the incorporation of human factors into design and for consideration of human behavior when evaluating risks, accidents, or preventive measures for people's protection. Despite the considerable vulnerability of human agents, we observe that many companies just invest in security technology by way of firewalls, encryption, secure access devices, and hard passwords, rather than considering the human factors in cyberattacks and cybersecurity. As one famous hacker once testified to the US Senate Committee on Governmental Affairs, he had obtained more passwords by tricking users than by cracking [11]. Besides the security risks related to technology—i.e., hacking computers, bad firewalls, encryption and data protection, etc.—there is a variety of areas identified as risky specifically due to user behavior or errors.

In terms of users' behavior, social agents may mainly interact with computers in the following roles: end user, admin agent, security officer, hacker, and group or community. To understand each of these roles, we can take a close look at a few scenarios. For instance, as users, we may log into our bank account, order merchandise through e-commerce sites, check our e-mail or Facebook, view our medical test reports on our healthcare provider sites, and see or manage our security. How do we know which site is safe to use?

When should users trust their computer, instant messaging, e-mail, smartphone, or cloud services? Understanding user behavior is fundamental in helping them make a sound judgment and create technology that helps them make a good decision. Unless users learn about secure and safe behavior, they will be vulnerable to cyberattacks. According to a Pew Research Center report [12], 69% of online adults say they do not worry about how secure their online passwords are—more than double those (30%) who admit to having worries about their personal password security, and Americans who have personally experienced a major data breach are generally no more likely than average to take additional means to secure their passwords (such as using password management software). 54% of online adults report that they utilize potentially unsecure public Wi-Fi networks—with around one in five of these users reporting that they use these networks to perform sensitive activities such as e-commerce or online banking. Potential hackers are also aware of these data, and this puts a lot of people at risk.*

Human agents are needed to administer and manage a computer system, and this would be the way to access systems in order to change the configuration or problem-solve issues.

Administrators might have different levels of access to a system to manage specific areas. For example, a banking system admin trying to problem-solve a user issue in accessing his or her account might not see the bank account of the customers, but a higher-level agent might have greater access. Former National Security Agency subcontractor Edward Snowden is one among many voices who illustrate how sensitive admin behavior is with regard to cybersecurity.† A variety of other scenarios also shows the sensitive roles of human agents with more access capability to their systems. (See the footnote for scenarios.)‡

Human agents can also be ill-intentioned hackers who try to break into a system and access information for a variety of reasons such as financials, information spying, or investigation.

There are numerous scenarios of hacking government data or individuals, and these are frequently reported on everyday news. Hacking cases are now focused not only on financial gain but also on influencing public opinion and population beliefs through fake news and election systems.§

* "A hacker calls company employees and tells them that he works for IT support and needs their password to update some programs on their machine. Since no admin accounts have been set up on many personal computers in this company, IT support staff often need to ask users for their passwords when they want to get into these machines. This is a contextual issue—if systems are set up—so users are regularly asked to disclose their passwords; it is difficult for them to distinguish in which context disclosure is safe, and when it is not" [11].

† In 2007, TJX disclosed that hackers had been inside its network stealing data for at least 18 months before they were discovered. An investigation revealed that the hackers obtained access by sitting in the parking lot of two Marshall's stores in Miami, aiming a powerful antenna at its wireless network and gaining access. TJX was found to have used a weak and outdated encryption standard to protect its data, among other things [12].

‡ "A contractor working with the hospital had downloaded patients' files onto a personal laptop, which was stolen from the contractor's car.

The data on the laptop were password protected but unencrypted, which means anyone who guessed the password could have accessed the patient files without a randomly generated key.

According to a hospital press release, those files included names, addresses, and social security numbers—and, in a few cases, "diagnosis-related information" [13,14].

§ "From the 1990s, private investigator Jonathan Rees reportedly bought information from former and serving police officers, customs officers, a VAT (value added tax) inspector, bank employees, and burglars, as well as bloggers who would telephone the Inland Revenue, the Driver and Vehicle Licensing Agency, banks, and phone companies, deceiving them into providing confidential information. He then sold that information to the major press. News of the World paid £150,000 a year to a man who obtained information from corrupt police and illegal sources" [15].

All these cases reveal that the human agent is the main weak point in breached security. In many user scenarios, human error is often found to be the origin of the security failure that allows criminals to obtain credentials. However, too large a portion of the investment in cybersecurity goes to different technologies and not to human factors.

To understand the amount of investment that significant companies put on the human side of their cybersecurity, I tried to obtain information from the management of several companies. While they never wanted to give me a dollar amount, they admitted that the percentage of their cybersecurity budget that was committed to the human side was very small. By looking at the actions undertaken by companies and institutions on the human side of cybersecurity, we see that one of the most common actions is requiring employees to take an online training on cybersecurity. Recently some bigger companies with more financial resources have also launched a phishing e-mail exercise to give more awareness to employees to be more careful when they click on links.

While any training is helpful to people, I have not yet found any research that shows the efficiency of cybersecurity online employee training sessions. It seems that these types of trainings are offered simply for the legal purpose of illustrating that the company has taken action to train employees and can thus transfer the blame from the company to employees. This is a reminder of the early stages of the human factors and ergonomics field where it was common to blame human operators for bad design instead of incorporating human characteristics into the design in the first place in order to avoid errors.

Need for an HCI in cybersecurity handbook

A few years ago, when working on the design for a course on HCI in cybersecurity, I tried to see if a course of this nature was already available. After a close review of 25 educational programs [6] in the cybersecurity area, I discovered that educational programs are only offered to postgraduates in the United States. Among the 70 topics identified, only the following were offered by multiple institutions: forensics (12), cryptography (8), information assurance (9), information security (10), IT (6), network security (10), and security management (4) [16].

After looking at the content of the programs, only one institution was offering a course on human factors and managing risk. Even this course was not an all-encompassing human factors course. Even though there is a significant number of publications and conference papers about cybersecurity, the teaching of this topic falls behind the research and scholarship [17].

As an effort to fill the gap, I decided to assemble this handbook and create space for a resource that covers major themes and topics related to HCI and cybersecurity. This collection came out of the contributions of a group of prominent experts in the field of cybersecurity.

This collection has six parts. Part One covers authentication and access management with three chapters: "User Authentication: Alternatives, Effectiveness, and Usability" (Chapter One); "Biometrics" (Chapter Two); and "Machine Identities: Foundational to Cybersecurity" (Chapter Three). User authentication is a fundamental aspect of security, representing the frontline defense for many systems at point of entry and beyond. Biometric technology is fast becoming an essential cybersecurity tool, as its use for authentication and identity proofing is becoming routine in everyday life, and we will cover digital certificates that are foundational to the security of the digital economy while remaining among the least understood concepts by users.

Part Two is dedicated to the important question of trust and privacy in HCI and cybersecurity. Chapter Four, "New Challenges for User Privacy in Cyberspace," covers the growing impact of emerging ITs—such as the Internet of things (IoT), augmented reality, biometrics, cloud computing, and big data—on persons' privacy and the analysis of several specific privacy-destroying technologies. It will also adopt a holistic view to take into account the mutual dependencies among presented technologies. Chapter Five, "Trust," introduces and discusses several trust constructs and factors that predict appropriate or inappropriate trust outcomes. Further, the chapter will explore the application of the trust construct to two user populations: cyberdefenders and consumer users of the Internet.

Part Three includes three chapters covering a few significant areas of research consisting of insider threat, social engineering, and money laundering. Chapter Six ("Insider Threat") addresses questions of insider threats and best practices regarding effective insider threat mitigation, followed by the outcomes of the challenges of detecting insider threats. It will become clear that any potential solution cannot merely rely upon technology but will need to adopt a holistic approach. Chapter Seven ("Social Engineering") reviews a broad set of distinctly nefarious activities in cybersecurity, along with techniques that social engineers use to acquire sensitive information from legitimate sources. Chapter Eight ("Money Laundering and Black Markets") highlights the critical nexus between cyber-enabled and money laundering crimes. It offers an overview of current popular methods and avenues that illicit funds (an increasing portion of which are generated through cyber-enabled crimes) that are purposefully concealed and moved around the globe.

Part Four covers smart networks and devices with five chapters: "Smart Home Network and Devices" (Chapter Nine), "Trusted IoT in Ambient Assisted Living Scenarios" (Chapter Ten), "Smart Cities under Attack" (Chapter Eleven), "Securing Supervisory Control and Data Acquisition Control Systems" (Chapter Twelve), and "Healthcare Information Security and Assurance" (Chapter Thirteen). Chapter Nine reviews the characteristics of home networking before discussing related research in this area and usability and security challenges, followed by potential solutions. Chapter Ten analyzes the risks of IoT-based ambient assisted living applications, focusing on the elements that need to be protected, types of attacks that can be launched by malicious attackers, and how such attacks might be mitigated. Chapter Eleven explains the threat landscape by defining security requirements and known threats of smart city infrastructures. Then, it investigates cybercrimes in smart cities by covering use cases, known vulnerabilities, attack scenarios, and real-world cyberattacks already experienced. Chapter Twelve reviews the user interface design that promotes cybersituation awareness of the interactions between the physical and cyber/control processes and teamwork that enhances coordinated response to intrusions among personnel of different disciplines and analyzes the risk that justifies the effective appropriation of security investments. Chapter Thirteen covers healthcare information security and assurance. It examines how healthcare has progressed and increased its use of IT. It also sheds light on how emerging technologies could impact stakeholders and reviews the usability of mobile devices and applications that make the access and exploitation of medical records easier.

Part Five is dedicated to governance and gives readers an overall review of law and regulation, security agencies, and how those might be applied in an organization. There are five chapters in this part: "US Cybersecurity and Privacy Regulations" (Chapter Fourteen), "Impact of Recent Legislative Developments in the European Union on Information Security" (Chapter Fifteen), "Privacy and Security in the IoT: Legal Issues" (Chapter Sixteen), "US Government and Law Enforcement" (Chapter Seventeen), and "Enterprise Solutions and Technologies" (Chapter Eighteen). Chapter Fourteen focuses on corporate compliance with US cybersecurity and privacy regulations. Chapter Fifteen gives an overview of recent legislative developments in the European Union (EU) that are expected to have a significant impact on information security. Chapter Sixteen discusses the government and law enforcement agencies involved in cybersecurity and their roles, governing structures, and coordinating entities. Chapter Seventeen discusses the US government agencies involved in cybersecurity threats facing the United States, their role, governing structures, and coordinating entities.

Chapter Eighteen provides an understanding of how enterprises employ a myriad of technologies, policies, practices, and programs to influence behavior to guard their intellectual property and confidential data and reduce risks associated with disclosure.

Part Six includes a chapter about the perspective of the future. It summarizes how experts in cybersecurity, including the contributors of this book, see the future of human factors in cybersecurity. The intention is to give a sense of how experts might think about the future with respect to the issues presented in this book.

As for the final part, I have included a section on the entertainment industry where I have incorporated my reviews of a number of major movies and documentaries related to cybersecurity. Most of these films are Hollywood-style entertainment films with a focus on cybersecurity plots. With the influence of films and media on people's awareness of security, watching films even when they are not very realistic can help bring consciousness to the public. Of course, they can also be effective resources for discussions in cybersecurity training sessions. Several educational materials, such as TED-ED, TED Talks, and Podcast media are also reviewed and presented.

At the conclusion of the book is an extensive glossary that provides easy-to-understand definitions to common terms, abbreviations, and acronyms to help the reader contextualize or further understand any concept or idea presented in the book.

Preparing organizations and people for cyberattacks is going to be a serious concern confronting our societies in coming years. The active approach to improve security cannot be achieved without considering human behavior, responsibilities, and capabilities. It is urgent that all aspects of society—businesses, organizations, and individuals—organize, confront, and deal with these new security issues and that they take measures to educate people to manage their way through the cyberworld of today and tomorrow.

Abbas Moallem

References

1. Giddens A. (2017): Risk and responsibility, source: *Modern Law Review*, Vol. 62, No. 1 (Jan. 1999), pp. 1–10, Wiley on behalf of the *Modern Law Review*, http://www.jstor.org/stable/1097071, Accessed: November 27, 2017, 20:41 UTC.
2. Appadurai A. (1990): Disjuncture and difference in the global cultural economy: *Theory Culture Society*, Vol. 1990, No. 7, p. 295, http://journals.sagepub.com/home/tcs.
3. Winton R. (2016): $17,000 bitcoin ransom paid by hospital to hackers sparks outrage, *Los Angeles Times*, February 19, 2016, http://www.latimes.com/local/lanow/la-me-ln-17000-bitcoin-ransom-hospital-outrage-20160219-story.html.
4. Wallace G. (2015): Hackers stole from 100 banks and rigged ATMs to spew cash, *CNN Tech*, February 16, 2015, http://money.cnn.com/2015/02/15/technology/security/kaspersky-bank-hacking/.
5. Armour S. (2015): How identity theft sticks you with hospital bills, *Wall Street Journal*, August 7, 2015, http://www.wsj.com/articles/how-identity-theft-sticks-you-with-hospital-bills-1438966007.

6. Ball J. (2013): NSA monitored calls of 35 world leaders after US official handed over contacts, *The Guardian*, October 25, 2013, http://www.theguardian.com/world/2013/oct/24/nsa-surveillance-world-leaders-calls.
7. Nelson S. (2015): Buying drugs online remains easy, 2 years after FBI killed Silk Road, *US News* October 2, 2015, http://www.usnews.com/news/articles/2015/10/02/buying-drugs-online-remains-easy-2-years-after-fbi-killed-silk-road.
8. Mihm S. (2013): Are bitcoins the criminal's best friend? *Bloomberg* November 19, 2013, http://www.bloombergview.com/articles/2013-11-18/are-bitcoins-the-criminal-s-best-friend.
9. Gartner (2015): *Gartner Says Worldwide Information Security Spending Will Grow Almost 4.7 Percent to Reach $75.4 Billion in 2015*, Gartner, Stamford, CT, September 23, 2015, http://www.gartner.com/newsroom/id/3135617.
10. Howarth F. (2014): The role of human error in successful security attacks, *SecurityIntelligence*, September 2, 2014, https://securityintelligence.com/the-role-of-human-error-in-successful-security-attacks/.
11. Schwartz J. (2016): Hacker gives a hill how-to: Mitnick tells panel loose lips sink systems, *Washington Post*, March 3, 2016, http://www.washingtonpost.com/wp-srv/WPcap/2000-03/03/044r-030300-idx.
12. Olmstead K. and Smith A. (2017): *Americans and Cybersecurity*, Pew Research Center, Washington, DC, January 26, 2017, http://www.pewinternet.org/2017/01/26/americans-and-cybersecurity/.
13. Mitnick K. D. and Simon W. L. (2002): *The Art of Deception*, Wiley, Hoboken, NJ.
14. Schultz D. (2012): As patients' records go digital, theft and hacking problems grow, *Kaiser Health News*, June 3, 2012, http://khn.org/news/electronic-health-records-theft-hacking/.
15. Zetter K. (2009): 4 years after TJX hack, payment industry sets security standards, *Wired*, July 17, 2009, http://www.wired.com/threatlevel/2009/07/pci.
16. Davies N. (2011): Jonathan Rees: Private investigator who ran empire of tabloid corruption, *The Guardian*, March 11, 2011, http://www.theguardian.com/media/2011/mar/11/jonathan-rees-private-investigator-tabloid.
17. Moallem A. (2016): *Applied Human Factors and Ergonomics International News*, February 2016, http://www.ahfe2018.org/news/2016/February.html.

Editor

Abbas Moallem is an executive director of UX Experts, LLC, and an adjunct professor at San Jose State University and California State University, East Bay, where he teaches human–computer interaction (HCI), human factors, and cybersecurity. Dr. Moallem holds a PhD in human factors and ergonomics from the University of Paris (Paris XIII) and has over 20 years of experience in the fields of human factors, ergonomics, HCI, and usability. He has worked or consulted with numerous companies that include PeopleSoft, Oracle Corporation, Tumbleweed, Axway, NETGEAR, Sears HC, Polycom, Cisco Systems, HID Global, Lam Research, and Applied Materials.

Contributors

Jeffrey G. Baltezegar is a security analyst with Liberty Healthcare Management and an information security consultant and an adjunct faculty member at the University of North Carolina at Wilmington. He holds a BS degree in Industrial Technology and an MS degree in Network Technology from East Carolina University, as well as CEHv8 and CSA+ certifications.

Jill Bronfman is an affiliate scholar of the Privacy and Technology Project at the Institute for Innovation Law and has taught Data Privacy and Privacy Compliance Law as an adjunct professor of law at the University of California Hastings College of the Law. She was named to *The Recorder's* 2014 list of the 50 "Women Leaders in Tech Law." Also, Professor Bronfman was selected as a 2014–2016 USC. Annenberg alumni ambassador. Professor Bronfman was formerly an assistant general counsel and network security subject matter expert for Verizon in the San Francisco office. At Verizon, she designed and moderated several in-house training programs in data security, compliance, and intellectual property. She also teaches at San Francisco State University, including an advanced seminar in Mobile Media and a graduate seminar in Media Ethics. Professor Bronfman was selected to workshop her technology-driven fiction at a juried literary conference, and she has read her work at Litquake, San Francisco's renowned celebration of writers and writing.

Daniel Calvo obtained his master's degree in Electronics and Telecommunications Engineering from the University of Cantabria in 2009. From then, he was involved in several European and national research projects such as FP7 Pharaon or Catrene Cobra. He performed research activities related to design and verification of advanced hardware/software embedded systems based on multicore and many-core platforms. In 2011, he also cofounded TedCas, a start-up that was partially funded by Telefónica through the Wayra initiative. In 2015, he joined the Internet of Everything Lab of Atos Research & Innovation. His main research interests are focused on IoT technologies applied to smart cities, connected vehicles, and energy.

Wojciech Cellary is a professor of computer engineering working at the Poznań University of Economics and Business, in Poland, serving as the head of the Department of Information Technology. He has been a visiting professor at different universities in France, Italy, Macau, and Portugal. His research interests are focused on e-business, e-government, IoT, and three-dimensional multimedia. He is the author of ten books and over 150 papers. He served as a consultant to the European Commission and many Polish ministries. He was the chair of over 40 national and international scientific conferences and a leader of many national and international projects.

Michael Cook is a lifelong IT professional. He began consulting at the age of 12 and has spent the entirety of his adult career dedicated to cybersecurity within the California State University System. He is currently the information security officer for and authored the first information security program at San Jose State University, the oldest public university in the state of California. He currently holds a bachelor's degree in Computer Science from the California Polytechnic State University, in San Luis Obispo, master's degrees in Business Administration and Information Security from California State University, Dominguez Hills, and numerous information security certifications.

Francisco Corella is the cofounder and chief technology officer of Pomcor, a US company, where he conducts research on applications of biometrics and cryptography to Internet identity. He has a PhD in Computer Science from the University of Cambridge, an MS in Computer Engineering from Stanford University, and an Engineering degree from ParisTech. Prior to Pomcor, he worked as a manager at Symantec, as a researcher at Schlumberger Palo Alto Research and IBM, and as an engineer–scientist at Hewlett–Packard. He is an inventor with 14 patents granted and several patents pending.

Alexandros Fragkiadakis is a researcher at the Telecommunications and Networks Laboratory of the Institute of Computer Science at Foundation for Research and Technology–Hellas. He received his PhD in Computer Networks from the Department of Electronic and Electrical Engineering of Loughborough University, in the United Kingdom. His research interests include wireless networks, intrusion detection and security in wireless networks, IoT, reprogrammable devices, open source architectures, cognitive radio networks, wireless sensor networks, and blockchain. Alexandros is highly involved in projects related to IoT with several publications in book chapters, journals, and conferences.

Steven Furnell is a professor of information technology (IT) security at the University of Plymouth and an adjunct professor at Edith Cowan University and an honorary professor at Nelson Mandela University. His research interests include user authentication, security culture and awareness, usable security, and cybercrime. He has authored over 280 refereed journal/conference papers, as well as various books, chapters, and other published outputs. He currently chairs Technical Committee 11 (security and privacy) of the International Federation for Information Processing and is a board member of the Institute of Information Security Professionals.

Ryan Gerdes, PhD, is an assistant professor at the Bradley Department of Electrical and Computer Engineering at Virginia Polytechnic Institute and State University, Virginia, United States. He received his BS, MS, and PhD degrees from the Department of Electrical and Computer Engineering at the Iowa State University in 2004, 2006, and 2011, respectively. His primary research interests are signal and data authentication, hardware and device security, computer and network security, transportation security, and applied electromagnetics.

Aqeel Kazmi is a postdoctoral researcher at the Insight Centre for Data Analytics, NUI Galway. He holds a BSc degree in Computer Science from GIFT University, Pakistan. He holds MSc and PhD degrees from the University College Dublin, Ireland. Currently, Dr. Kazmi is actively involved in multiple IoT research projects that are funded by the European Commission under the FP7-ICT and H2020 frameworks. His research interests are in the area of ubiquitous computing, cloud computing, data analysis, information visualization, wireless sensor networks, IoT, and smart city applications.

Nathan Lau, PhD, is an assistant professor at the Grado Department of Industrial and Systems Engineering at Virginia Polytechnic Institute and State University, Virginia, United States. He received his BASc and PhD degrees from the Department of Mechanical and Industrial Engineering at the University of Toronto in 2004 and 2012, respectively. His primary research interests are cognitive engineering in safety critical systems, particularly on situation awareness, cybersecurity, cognitive work analysis, and user interface design.

Karen Lewison is the cofounder and chief executive officer of Pomcor, a US company, where she conducts research on applications of biometrics and cryptography to Internet identity in addition to leading the company and managing government-funded research projects. Prior to founding Pomcor, she worked as a radiologist and nuclear medicine physician. She has a BS degree in Biology from Cornell University, an MD degree from New York Medical College, and a masters degree in Clinical Research from University of California, San Diego, and is a member of the Alpha Omega Alpha Medical Honor Society. She is an inventor with four patents granted and several patents pending.

Athanasios Lioumpas has been working at Cyta Hellas since 2013, being the scientific lead in research projects of the company in the telecoms sector, while also being a principal engineer in the mobile virtual network of the company. He holds a diploma in Electrical and Computer Engineering (2005) and a PhD degree in Telecommunications Engineering (2009), both from Aristotle University of Thessaloniki. He has published more than 50 scientific articles in international journals and conferences in the field of machine to machine (M2M), IoT, and mobile communications, while he has been granted five patents.

George Moldovan graduated from the University Transilvania of Brasov in 2006 and finished his MSc in Computer Science in 2009 at the RheinMain University of Applied Sciences, Wiesbaden, where he worked in distributed systems and machine learning laboratories. Until 2014, he worked as a PhD researcher (with the Goethe University of Frankfurt). Since 2010, he was involved in security, privacy, and performance research problems for ad hoc sensor networks and has been, since 2014, active in Siemen's internal and EU-supported security-oriented projects.

Hari Nair is responsible for product strategy, vision, and roadmap at Venafi, a cybersecurity start-up that is focused on the security and management of machine identities. Hari has spent all his career with public key infrastructure, developing and delivering strong authentication solutions for the largest, most security-conscious organizations in the world. Hari has a master's degree in Management Science from Stanford University and is a padawan in the chief information security officer program at Carnegie Mellon University.

Maria Papadaki is an associate professor at the Centre for Security, Communications and Networks Research and the head of Cyber Security at Dartmouth Centre for Sea Power and Strategy. Her research interests include insider threats, incident response, maritime cybersecurity, security assessment, social engineering, security usability, and security education. Her research outputs include 24 journals and 31 conference papers. Dr. Papadaki is active in a variety of professional bodies, a fellow of the Higher Education Academy, and a member of the BCS, the Institute of Information Security Professionals (IISP), ISACA, and the Global Information Assurance Certification Advisory Board (GIAC). Maria holds GIAC Intrusion Analyst, and GIAC Penetration Tester (GPEN) certifications.

Gerald Quirchmayr holds doctorate degrees in computer science and law from Johannes Kepler University Linz (Austria). He is currently a professor in the Faculty of Computer Science at the University of Vienna and an adjunct professor in the School of School of Information Technology and Mathematical Sciences at the University of South Australia. His major fields of interest are information security and legal aspects of information security. He has contributed over 200 articles to books, journals, conference and workshop proceedings, and technical reports. His research activities are composed of national and European projects, as well as work for the United Nations and close cooperation with researchers in Australia, North America, and Asia.

Santiago Reinhard Suppan was awarded the grade of master of science with honors in the area of business informatics by the University of Regensburg and the Bavarian Elite Network in 2012. In 2013, he was granted a doctorate scholarship by Siemens AG, where he accompanied several funded research projects (NESSoS, ICeWater, and Rerum) and researched security in industrial systems, the smart grid, data protection, and privacy-enhancing technologies in the IoT (focus area), holding academic publications and industry patents. Today, he advises, trains, and coaches business units worldwide as a security consultant at Siemens AG.

Greg A. Ruppert is the chief of the Financial Crimes Risk Management at the Charles Schwab Corporation, which includes fraud investigations and cybercrime investigations. Prior to joining Schwab, Greg spent over 17 years with the Federal Bureau of Investigation (FBI) working complex financial investigations, as well as terrorist financing, terrorism, and cyberthreats. Greg served as the section chief of the Threat Section of the FBI Cyber Division, covering complex cyberintrusions emanating from Asia, the Middle East, Africa, and Eastern Europe. Greg is also clinical professor at the University of the Pacific (UOP), School of Engineering and Computer Sciences' Data Analytics Master's Program, and an industry advisory board member for the UOP's Cybersecurity Program. Greg is also a member of the State Bar of California, United States.

Brita Sands Bayatmakou is the managing director of the Financial Intelligence Unit at Charles Schwab. In this rapidly changing environment, she works to develop data-driven solutions, fusing emerging threat data with external research related to a multitude of crimes, including fraud, cyber-related crimes, and money laundering, to provide actionable intelligence to the firm. Her research interests include the use of social network analysis and visualization to identify and investigate organized criminal activity. Prior to joining Schwab, she worked at Wells Fargo Bank within the Global Financial Crimes Intelligence Group where she focused on anti-money laundering compliance and financial crimes associated with emerging technology. Brita holds an MA in International Policy Studies from the Monterey Institute of International Studies and is a certified financial crime specialist and a certified antimoney laundering specialist.

Mark Schertler is the Chief Information Security Officer of TIBCO Software, Inc. where he is responsible for the security of all things that involve bits. He received his MS Computer Science from George Washington University and BS Computer Science from the University of Illinois Urbana-Champaign. During his career Mark has tackled security challenges for the US government, large enterprises, and SAAS providers, as well as a few startups and smallish companies. Mark has edited and contribute to IETF RFCs, ITU standards, and the NATO Adhoc Working Group on Security.

David Schuster is an assistant professor of psychology at San Jose State University. He holds a PhD in Psychology specializing in applied experimental and human factors psychology from the University of Central Florida. Dr. Schuster's research centers on understanding individual and shared cognition in complex environments. He has conducted research in domains such as cybersecurity, aviation, transportation security training, and military human–robot interaction. In 2016, Dr. Schuster was awarded a National Science Foundation grant to investigate the cognitive factors of computer network defense.

Alexander Scott is a master's student of Human Factors and Ergonomics at San Jose State University and a user experience research intern at McAfee. He holds a bachelor's degree in Behavioral Science from San Jose State University. Alexander Scott's research involves investigating trust in automation and its antecedents, leveraging findings into automated system designs. His primary areas of interest are cybersecurity, automated vehicles, virtual and augmented realities, and emerging technology.

Martin Serrano is a recognized research data scientist with extensive experience in the industry and applied research with a track record of successful European FP6, FP7, H2020 RIA/IA projects, Irish, and Enterprise Ireland innovation projects. Dr. Serrano is a continuous contributor to the Strategic Research and Innovation Agenda (SRIA) for Europe. He is actively investigating semantic-based cloud infrastructures and big data analytics, cyberphysical systems, privacy, and cybersecurity. Dr. Serrano is an active member of the Institute of Electrical and Electronics Engineers (IEEE) and the Association for Computing Machinery (ACM) with more than 100 peer-reviewed publications and an author of four academic books.

Stavros Shiaeles is a lecturer in computer and information security at Plymouth University and a member of Centre for Security, Communications and Network Research. He is a certified ethical hacker, certified EC-council instructor, certified advanced penetration tester, and cyberoam certified network and security professional. He is actively involved in cybersecurity professional training as well as consulting. He is a fellow of Higher Education Academy and fellow of BCS. Dr. Shiaeles' research interest is machine learning applied in cybersecurity and specifically in the areas of distributed denial-of-service attacks, malware, insider threats, open-source intelligence, social engineering, digital forensics, and blockchain.

Ralf C. Staudemeyer is Professor for IT-Security at the Faculty of Computer Science of the Schmalkalden University of Applied Sciences in Germany. He holds a PhD in Computer Science and a German Diploma (Diplom-Informatiker). He is an expert in IT security & privacy, computer networks, and deep learning. He lectured at several academic institutions in Germany, South Africa, and the Fiji Islands. He was recipient of various research fellowships and supported EU research projects as a coordinator, work-package leader, and researcher. He is author of more than 30 publications.

Chee-Wooi Ten, PhD, is an associate professor at the Department of Electrical and Computer Engineering in Michigan Technological University, Michigan, United States. He received BS and MS degrees in Electrical Engineering from Iowa State University, in Ames, in 1999 and 2001, respectively. Later he obtained his PhD in 2009 from the University College Dublin, National University of Ireland. His primary research interests are cybersecurity for power grids and software prototype and power automation applications on supervisory control and data acquisition systems. Chee-Wooi was the editor for the *IEEE Transactions on Smart Grid* and the *Elsevier Journal Sustainable Energy, Grids and Networks.*

Elias Z. Tragos is a research manager at the Insight Centre for Data Analytics, UCD, Ireland. Dr. Tragos holds a PhD in Wireless Communications and a master's degree in Business Administration in techno-economics. He has been actively involved in many EU and national research projects as researcher, technical manager, and project coordinator. His research interests lie in the areas of wireless and mobile communications, Internet of things, cognitive radios, network architectures, fog computing, security, and privacy. Dr. Tragos has published more than 70 peer-reviewed conference and journal papers, receiving more than 1500 citations.

Artemios G. Voyiatzis is key researcher with SBA Research, in Austria, and a principal researcher with Athena Research and Innovation Center, in Greece. His research focuses on networked systems security with emphasis on IoT security and privacy. He holds a PhD degree in Electrical and Computer Engineering, MSc and BSc degrees in Computer Science, and a second BSc degree in Mathematics. He has worked in IT, enterprise resource planning, and NOC environments designing and securing information and communications technology services at a scale of thousands of users. He is a senior member of both IEEE and ACM.

Hao Wang, MS, is a PhD candidate at the Grado Department of Industrial and Systems Engineering in Virginia Polytechnic Institute and State University, Virginia, United States. He received his MS degree in Statistics from Virginia Polytechnic Institute and State University in 2017 and BS degree in Electrical Engineering and Industrial and Operations Engineering from the University of Michigan in 2012. His research interests are cyber-situation awareness, work domain analysis, human factors, and statistical modeling in cybersecurity.

Rolf H. Weber is a professor emeritus of business law at Zurich University, the chairman of the executive board of the Center for Information Technology, Society, and Law, the codirector of the University Priority Research Program "Financial Market Regulation," and a practicing attorney-at-law in Zurich. His main fields of research and practice are IT and the Internet, international business, competition, and international financial law. Rolf publishes and speaks regularly on Internet-related legal issues.

Adam Wójtowicz, PhD, is an assistant professor working at the Department of Information Technology, Poznan University of Economics and Business, in Poland. His research interests are focused on information, system, and user security, particularly on new access control methods for multimodal and context-aware systems, on security in the Internet of things (IoT) and in virtual reality/augmented reality systems, and on privacy-preserving systems. He is the author of 25 publications and many reviews, and he has been involved in a dozen research projects. He lectures on IT system security, information security in organizations, e-business security, and computer programming.

Ulku Yaylacicegi Clark's research interests span ICTs, telecom policy, information security, IT productivity; healthcare IT, quality management, and innovative education. She holds an MS degree in Information Technology and Management and a PhD degree in Management Science with management information systems concentration from the University of Texas at Dallas. Her publications appeared in various academic journals, such as *Journal of Management Information Systems, IEEE Technology and Engineering Management,* and *International Material Data System.*

section one

Authentication and access management

chapter one

User authentication

Alternatives, effectiveness, and usability

Steven Furnell

Contents

1.1 Introduction

Authentication is the most readily recognized aspect of cybersecurity for most users, being encountered and used multiple times per day across various devices and services. It is fundamentally about ensuring that the right people have access, by checking that they are who they are claiming to be. Consequently, authentication represents a key aspect of security, as it typically represents the frontline protection standing between a system and a would-be impostor.

While users authenticate themselves on a regular and frequent basis, they are likely to find that these encounters deliver varying experiences. Authentication can appear in multiple guises and with different demands and expectations involved. So whether users ultimately judge it to be a *positive* encounter is likely to reflect the choice of technology and the effort of using it.

At the highest level, the multiple guises that authentication can take will fall into one or more of the following three categories, depending on what is required from the user in order to prove their identity:

- Something the user *knows* (e.g., a secret, such as a password or a personal identification number [PIN])
- Something the user *has* (e.g., a physical token/possession, such as a card)
- Something the user *is* (i.e., a biometric characteristic, based upon an aspect of their physiology or behavior)

Methods falling within these categories are often used as distinct, stand-alone approaches. However, they can also be used in combination—thus requiring multiple things of the user and potentially introducing an immediate impact in relation to the resulting usability. Indeed, the usability is likely to be an important issue for the legitimate user. After all, they are already going to be perfectly confident of their own identity and may have a limited tolerance in terms of trying to prove this to the system. While a small overhead may be seen as an acceptable price to pay for protection, encountering an involved process or facing repeated challenges and doubts from the system are likely to engender resistance and dissatisfaction. So in short, it is important to get it right, because authentication is the thing that regular users will see front and center on a daily basis.

The actual choice of what to deploy in practice will typically depend on several factors and associated questions, including the following:

- *Security*—What level of protection are we looking to provide?
- *Users*—Who is expected to use it, and what level of effort can reasonably be expected of them?
- *Device*—What approach(es) can the intended access device(s) naturally and easily support?
- *Context*—Where will users be authenticating themselves and under what conditions (e.g., will it be in a fixed location or on the move)?

These decisions will clearly have an impact upon the resulting usability. For example, a mismatch between the chosen method and the target device will make it more difficult to operate and therefore represent an ongoing challenge to those expected to use it. Even the most common methods can fall foul of this, as will be recognizable to anyone who has tried to use password-based approaches on a device such as a smartphone and then been frustrated by the small size of the on-screen keyboard that they are required to use and the need to alternate between different versions of it in order to get access to all the characters that they might wish to enter (which, if following good password practice, would be a mix of upper- and lowercase letters, numbers, and punctuation symbols, thus likely necessitating at least three versions of the virtual keyboard in order to access them all).

Although we can refer to it via a single word, the usability of an authentication technique or process is typically not tied to just one aspect. Various factors are likely to influence the usability perceived by the user, which may include elements such as the following:

- *Mental effort*—e.g., the extent to which the technique relies upon the user's ability to memorize and recall things and how precise this must be
- *Convenience*—e.g., the speed with which the user is able to log in and the effort/engagement required to do so
- *Applicability*—e.g., whether the technique will work effectively on desktop, mobile, and handheld devices, with differing input mechanisms and screen sizes/resolutions
- *Flexibility*—e.g., the ease with which the user can change their authentication credentials in the event of compromise

As will become apparent as different methods are introduced in the discussion, there can be some significant variations in how they stack up against these factors. For example, the secret-based approaches can significantly depend upon mental effort, but can be widely applicable and flexible to change. Meanwhile, biometrics take the mental effort away, but are often less applicable and offer very limited potential for changing them. Meanwhile, in all cases, the convenience factor is very often tied to the exact way in which the authentication has been implemented and integrated within the wider system or the device that is using it.

Usability is a long-established consideration for authentication and is often an implicit driver behind many of the techniques that we see. Indeed, as with security more widely, it must be recognized that anything we ask users to do is essentially eating into their so-called compliance budget [1], reflecting that there is only so much that they will have the time for and inclination to do in service of security when there are primary tasks demanding attention. Given that many users seem to budget very little in the first place, it is important to ensure that we get them to spend it effectively! Viewed from a slightly different perspective, this has also been referred to as *security fatigue* [2,3], and flagging that excessive burden from security processes and procedures increases the potential that users will tire of it and stop using or following it.

At this stage, it is probably worth setting expectations appropriately (and spoiling the ending!) by saying that the chapter is not going to end up highlighting any single method as the best answer for all circumstances. In short, there is no panacea. However, the various choices, the context in which they are used, and the way in which they are deployed can all have implications for the usability of the authentication experience. The prominence of passwords means that they naturally receive a fair chunk of the discussion. Another key reason is that usability issues encountered with passwords are often why some of the other alternatives get considered in the first place (the other reason unsurprisingly being the greater level of security that the alternatives are considered to offer). As such, it is relevant to understand the baseline method before turning our attention to the others.

1.2 Passwords: The unpopular favorite

If authentication is the most familiar form of security, then it is safe to say that passwords are the most familiar form of authentication. However, they are often less than well loved and often with more reason than simply because familiarity breeds contempt. Part of the problem is that while it is easy to get the idea of passwords (i.e., chose one and keep it secret!), it is rather more difficult to follow the associated good practice for using them securely. Indeed, passwords can present usability challenges in terms of how we *select* them and how we *manage* them. The typical style of a standard advice that tends to be issued in this respect is summarized in Table 1.1, and while each point probably seems simple enough in isolation, when combined, they can represent something of a tall order for many users to follow, at least without resorting to some form of compromise and potential weakening of the protection as a result.

One of the frequent problems with such guidance is that it is often issued without any attempt to explain the reason behind it (which in turn can have an influence on whether or not users understand and accept it). In order to avoid falling into that trap, the next few paragraphs provide some of the accompanying background.

Table 1.1 Traditional guidelines for selecting and managing passwords

Selection guidance	Management guidance
Length—e.g., use at least *x* characters	Avoid writing them down
Composition—e.g., use alphanumeric and punctuation characters	Do not share with other people
	Change them regularly
Avoid reusing passwords from other systems	
Avoid dictionary words and personal information	

Firstly, the guidance around the length and composition of the password aims to safeguard against attempts to discover it by brute force (i.e., trying all the character permutations until the correct one is found). For those unfamiliar with how such an attack would be mounted, it does not involve the attacker sitting at the keyboard physically attempting all the possibilities, but rather makes use of an automated tool that encrypts successive strings of characters and compares them to an encrypted password that has already been acquired until a match is found. This process is considerably faster than manual testing—offering the potential to try thousands of attempts per second—but can still take a long time to perform when a long and complex enough password has been used (given the exponential increase in password strength as the length increases).

Avoidance of dictionary words is also linked to the use of automated tools—in the sense that they will already have a preencrypted set of these words and so can just compare the captured password via a look-up table, yielding an instant match without the need to resort to the brute force approach.

Avoiding password reuse is based on the simple premise of not putting all eggs into one basket. It is recognized that users *do* reuse passwords in multiple places, so having discovered the password for an account on one system, it would be fairly natural for an attacker to then test to see if the same thing works in other places.

Finally, avoiding personal information is based on the premise that such details could also be known to someone who knows the target user or could be discovered/acquired by means of social engineering (or by seeking out information online, via social media sites or even general web search).

Meanwhile, in terms of the management advice, the idea of not writing passwords down and not sharing them is clearly to prevent them being inadvertently discovered or intentionally given away. Basically, once the password (or indeed any other secret used for authentication purposes) is disclosed, it becomes shared knowledge. As such, it is essentially no longer within the control of the legitimate user and ceases to be a reliable basis for verifying that an individual presenting the information is actually them.

Lastly, the advice to change passwords regularly seeks to provide a safeguard in the situation where the current password has already been compromised. The longer it remains unchanged, the longer an impostor user may proceed to gain unauthorized access. As soon as the password is changed, their access is cut off, and they must go back to square one if they want regain entry to the system.

The selection points all add to the immediate cognitive burden of performing the task (i.e., actually working out a password that satisfies the criteria), as well as the subsequent challenge of actually committing the resulting choice to memory. Indeed, concerns over the usability of traditional password guidance motivated the UK's National Cyber Security Centre to advise against one of the long-standing elements of advice:

> *Most administrators will force users to change their password at regular intervals, typically every 30, 60 or 90 days. This imposes burdens on the user (who is likely to choose new passwords that are only minor variations of the old) and carries no real benefits as stolen passwords are generally exploited immediately. . . . Regular password changing harms rather than improves security, so avoid placing this burden on users* [4].

As an aside, some have questioned this advice, particularly the claim that stolen passwords are exploited immediately, given that various long-standing breaches have come to light and have shown that unchanged passwords would consequently have left users vulnerable for longer [5].

Table 1.2 Most popular password choices of 2012–2016

Rank	2013	2014	2015	2016
1	123456	123456	123456	123456
2	password	password	password	password
3	12345678	12345	12345678	12345
4	qwerty	12345678	qwerty	12345678
5	abc123	qwerty	12345	football
6	123456789	123456789	123456789	qwerty
7	111111	1234	football	1234567890
8	1234567	baseball	1234	1234567
9	iloveyou	dragon	1234567	princess
10	adobe123	football	baseball	1234

Given the various challenges of following good practice, many users simply do not bother and resort to obvious choices. Indeed, according to successive findings from SplashData [6], users have a pretty appalling track record of choosing passwords sensibly. Table 1.2 illustrates this with the top 10 choices observed over a four-year period and clearly suggests that attackers might expect to get some success just by trying a small set of predictable choices.

So from a usability perspective, passwords are clearly less than ideal and always have been. Indeed, back in 1979, the abstract of the seminal paper on password security by Morris and Thompson [7] concluded by referring to the "compromise between extreme security and ease of use." This being the case, it is worth considering what can be done to compensate for it and whether such steps are actually being taken as often as they should be.

1.3 *Improving password usability: Compensating or compromising?*

Unfortunately, the usability challenge of passwords does not end once the user has chosen a password—they must then of course remember it (which, quite frankly, may end up being harder now that they have been encouraged to make it longer and put a wider variety of characters into it!). With this in mind, and with the ever-wider dependency upon passwords across numerous websites, technology has attempted to assist the user by taking some of the effort out of it. This is illustrated in Figure 1.1, which shows Apple's Safari web browser offering to remember a password for future use, such that when the user visits the site again the browser automatically populates the field for them. While this is fine from the perspective of providing a convenience for the legitimate user, it introduces obvious weaknesses if they elect to store the password on a device that other people also use or could gain access to.

Figure 1.1 Web browser offering to remember a website password.

New password

Use Safari suggested password:
NXJ-JfM-PeU-YQ8

This password will be saved in your iCloud Keychain so it is available for AutoFill on all your devices.

show password

Figure 1.2 Web browser autosuggesting and saving a password.

Browser-based automation also extends to creating passwords for new sites and services, as illustrated by the further screenshot in Figure 1.2. Furthermore, as shown in the accompanying text, in this case, the password will also be stored in the user's iCloud Keychain—such that any of their other devices connected through iCloud will also be able to provide the password automatically as well (noting that other non-Apple platforms also offer similar features for both password generation and cloud-based storage).

So let us consider the pros and cons of this approach from a usability and security perspective. For the user, it certainly does serve to make several things easier:

- They are relieved of the effort of having to devise a new password.
- They no longer need to type it in.
- They are not required to remember it.

However, the last point has a clear flipside—by letting the system generate and store the password for them, the user themselves could quite literally never know what the password actually is in the first place. So having been remembered on some devices, it could end up being very usable in *those* contexts, but impossible for the user to provide on any system that does not have it prestored. On the positive side, this does at least have the ironic security advantage that if the user does not know it, they cannot compromise protection by sharing it with anyone else!

Meanwhile, from a security perspective, the storage and automatic provision of the password is good only up to the point where the device(s) concerned remain solely accessible by the original user. This may become particularly challenging when dealing with multiple devices synchronized through the cloud. While *some* of the devices will be used by a single user only, others might be used by a wider group. A relatable example here is often found in the domestic context, with the contrast between the personal/individual nature of smartphones and laptops, versus the frequently shared/communal nature of tablets and desktop systems.

An alternative solution for offsetting the memorability problem is to use password manager (also known as password safe) tools. These may be hardware, software or online (e.g., cloud- or web-based), and well-known examples as of 2017 include 1Password, KeePass, and LastPass. The general principle is that the user protects their passwords for other systems and services via a master password required to access the manager tool. Depending on the specifics of how the tool has been implemented, this is often essentially trading the usability challenge of *remembering* the password to the extra effort involved in *retrieving* it. In some cases, they can also provide usability boosts by automatically populating the login fields, thus also providing a security safeguard against keylogging in addition to reducing the user effort required. However, while password manager tools have been available as an option for some time, many people still choose not to use them (with a 2016 study suggesting that only 16% of users chose to adopt—or were already using—password management tools from among a group of over 280 participants [8]).

The efforts required to make passwords work serve to shine a light on different levels of usability that can be encountered. Indeed, we can talk in terms of whether technologies can offer *innate* or *assisted* usability. Passwords fall firmly into the second category—they need a lot of support in order for people to select them appropriately and then further measures to make them easier to manage—and even then, they may still not be considered truly usable. However, it is still worth considering the extent to which providing some assistance can improve the situation. As such, we will return to this theme later in the chapter. Prior to this, the discussion turns toward some other modes of authentication, in order to see what they can offer in the usability context.

1.4 Stepping beyond passwords

For all the attention they receive, passwords are just one form of secret-based approach. Another frequently encountered option is of course the PIN, as commonly used for scenarios such as accessing mobile devices and card payment transactions. However, for the purposes of this discussion, PINs can essentially be regarded as numeric-only passwords and can share several of the same weaknesses (indeed, several of the options from Table 1.2 are essentially PIN strings). Looking more widely, various other forms of secrets can also be used, and two that can be readily found in widespread use are challenge questions [9] and graphical/image-based secrets [10].

As the name suggests, the idea of challenge questions is that users must prove their legitimacy by providing the correct response to presented questions. While this would typically be too cumbersome and time-consuming to use as a frontline, point-of-entry authentication method, it can often be found as a fallback in a "forgotten password" or account recovery process (particularly on web-based services). As an example, let us consider the various challenge questions that users can select from on a popular online auction site, as listed in Figure 1.3. As can be seen, the site lets the user choose and set responses to three of these, which are then used should they need to recover their account later.

At first glance this seems a reasonable approach and certainly offers the user a fair number of questions to choose from. However, looking through the list, it is easy to spot some potential issues in terms of both usability and security. On the usability side, one possible problem is several of these questions will not work for all users. For example, what if they have never had a car, a pet, or an other half? Additionally, various questions are framed around asking for favorites, which means that the information is arguably subject to change between the time that the user sets their response and a future point when they may need to answer the question (e.g., up until now, your favorite book may have been *Cybercrime: Vandalizing the Information Society*, but it could now be this book instead!). As such, while these questions may be usable enough in the first instance, they could end up causing problems later (although to be fair, when setting the responses, the site does at least advise that it should not be an answer that would frequently change). Looking at the example in Figure 1.3a, we can also note another usability constraint—it is not permissible to use some characters that would otherwise be a natural part of your answer (so, *Cybercrime: Vandalizing the Information Society* cannot be your favorite after all; not to worry, it is not anyone else's favorite either!). This constraint could conceivably affect the ability to answer several questions (particularly those involving names, where people may have double-barreled names separated by hyphens) and potentially force the user into having to adapt their preferred answer to something they may then not remember when needed.

Meanwhile, on the security side, various questions would appear to be susceptible to compromise by someone who knows the user well or is prepared to do a bit of background

PICK YOUR SECRET QUESTIONS

Give yourself another way to recover your account securely in case your information becomes outdated.

Question 1:

Name of favourite book?

Cybercrime: Vandalizing the Information Society

Answers are not case-sensitive, must not include symbols and should not be an answer that would frequently change.
Looks like you're using a symbol we don't allow.

Question 2:

Select an option

Answer

Question 3:

Select an option

Answer

Confirm

Cancel

(a)

First company you worked for?
Name of favourite book?
Model of your first car?
Name of your first pet?
Last name of favourite actor?
Name of your favourite band or singer?
Dream job as a child?
City where you met your other half?
City/country where you want to retire?
First name of your best friend?
Name of the street you grew up on?
Name of favourite film or series?
Your childhood nickname?
First name of your oldest cousin?
First name of best man or maid of honour?
First three words of favourite quote?
First name of your favourite boss?
Choose your own phrase (at least 2 words)

(b)

Figure 1.3 Setting challenge questions on an online auction site: (a) guidance received when setting and (b) available options.

research. These criticisms are by no means unique to the implementation of this particular website, and many sites where challenge questions are used can end up suffering from the same problems when presenting preset questions to choose from. To overcome this, the final option lets users set their own phrase, for which the would-be impostor does not get an on-screen clue. What is notably not offered is the option for users to set their own questions. The likely reason is that users could then end up choosing questions that others would know or guess the answers to. After all, given that many users cannot select good passwords, there is little reason to believe that they would be naturally better at creating security questions.

So as with passwords, challenge questions can have a role to play, but must be approached carefully in order to get usability and security aspects right. Turning to graphical approaches, these can also be found making occasional appearances for website authentication, but they have become particularly prominent in the context of mobile devices such as smartphones and tablets. Two common examples are presented in Figure 1.4, namely, Android's Pattern Unlock and Windows' Picture Password. The Android approach enables users to create a secret pattern using a series of on-screen dots, while the Windows technique requires them to associate three gestures (which can be a mixture of drawing circles, straight lines, and taps) with particular parts of a chosen image (with the regions and gestures then becoming the secret used for authentication).

In both cases, the approaches take advantage of the devices they run on by allowing the user to interact via the touchscreen, thus making them appear more usable and naturally

(a)

(b)

Figure 1.4 Graphical secrets, illustrations of (a) Android's Pattern Unlock and (b) Windows's Picture Password.

suited to the devices concerned (although it is notable that the Picture Password approach can also be used via a mouse on Windows devices that do not have touchscreen, where it feels noticeably less natural as a result). Unfortunately, however, the ease of use again comes with potential downsides. Firstly, both approaches still run the risk of weak or obvious secrets being selected (e.g., just drawing a simple square or a predictable Z shape as an unlock pattern or tapping on three corners or other obvious "hotspot" points in a picture password). Meanwhile, more complicated choices may be harder to commit to memory, particularly with the Picture Password if a combination of circles, lines, or taps has been used.

Meanwhile, on the security side, a potential concern is the discoverability of the secrets. Both methods can be considerably more observable than passwords or PINs (this is particularly the case for the Android pattern if a relatively simple one has been used or the device is configured to show it on-screen as it is being entered). Stand next to someone, or sit behind them, and you can often spot the exact pattern being entered. Moreover, touchscreens themselves can retain clues for a potential impostor, with smear marks from fingers potentially offering a residual pattern that can be retraced. This problem can be overcome by choosing patterns that overlap or double-back on themselves (or indeed via more careful device hygiene), but this then becomes something else that the user has to explicitly consider in their choice of secret and use of devices.

1.5 Usability versus tolerability

Usable authentication is typically perceived to be about speed and lack of friction in the process, in order to allow the legitimate user access as quickly as possible. However, these factors are not always natural bedfellows alongside the level of protection that needs to be provided. As such, the question must often change from "how easy can we make it?" to "how much complexity will the user tolerate before finding it unusable and rejecting it?" What is tolerable depends upon what we are trying to protect and the effort we think is appropriate to safeguard it (and indeed, the level of protection that we take comfort in seeing as standing in the way of an impostor). As a result, the types of authentication technology and/or process that we may be willing to accept will often vary depending upon the context involved.

The widespread acknowledgment of password weaknesses, and the recognition of related breaches, has driven a move toward two-factor authentication (complete with its own trendy and industry-/product-friendly acronym—2FA). So rather than simply relying upon a single form of authentication—such as knowing a secret—it is combined with another mode in order to make the task of an impostor more difficult.

Online banking sites are a good example here. The login and authentication process tends to be far more involved than we typically encounter on standard websites and services. Whereas most sites are traditionally satisfied with performing authentication via a combination of user identification (ID) (often, the user's e-mail address) and password, online banking logins are typically a multistage process involving customer numbers, secret codes, and challenge questions—and very often involving nonstandard input methods in order to reduce the risk of threats such as keyloggers (e.g., asking for selected characters from codes rather than the full secret and/or providing input using on-screen features rather than via the keyboard). However, at this stage, it could still just be a question or asking for multiple secrets and so essentially a two-*step* rather than two-*factor.* Many sites go even further than this, requiring not only the aforementioned secrets, but also an interaction with a separate hardware device to generate a one-time code. In some cases, these devices take the form of card readers, requiring the user to insert their ATM card and then authenticate using the accompanying PIN, whereas in others, the devices are self-contained and have their own PIN that must be provided before an access code is generated (and another variation is to use a smartphone app to generate the code, thus making the possession of the smartphone a second factor of the authentication).

Generalizing away from the specifics of any individual banking site, the resultant process for the user then becomes as follows:

1. Go to the website and perform some initial identification/authentication.
2. Authenticate to the secondary device and obtain a one-time code.
3. Provide the one-time code to the website.

As an example of how this looks in practice, Figure 1.5 illustrates the type of on-screen interfaces that a user will see during the first and last phases of this process, in this case, using screenshots taken from the online banking service provided by Barclays Bank. Figure 1.5b also depicts the card reader (which, in Barclays' parlance, is known as the PINsentry) and the user interactions involved during the second phase.

The login and authentication process can easily end up taking upward of 30–60 s, compared to the 5–10 s it might take to type in a username and password on other sites. So, it is in no sense more *usable,* but if we appreciate the reason (and therefore see the value in being expected to do it), then it remains *tolerable.* In the case of online banking, the user does not necessarily need much convincing about the reason and value aspects, because most will readily recognize their finances as something worth protecting. However, if they are being asked to jump through hoops in other contexts, where they do not feel they have as much at stake, then the basis for doing so might not be so obvious.

1.6 A token gesture?

When it comes to the usability of approaches based around something the user has, there is arguably not so much to say; it basically comes down to the need to possess an item of some sort, the likelihood of having it with you when needed, and the ease of then presenting it to the system to achieve the authentication. While there are lots of underlying technicalities in terms of how different approaches work under the surface (e.g., including

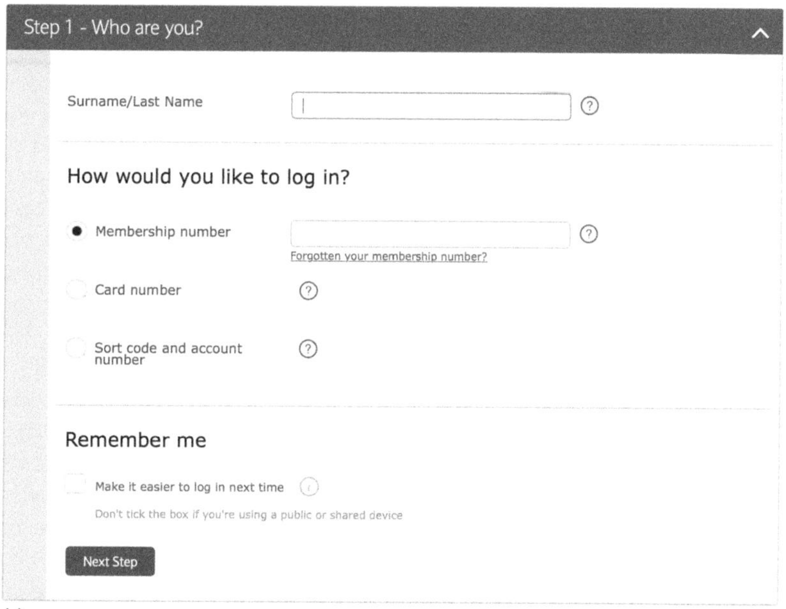

(a)

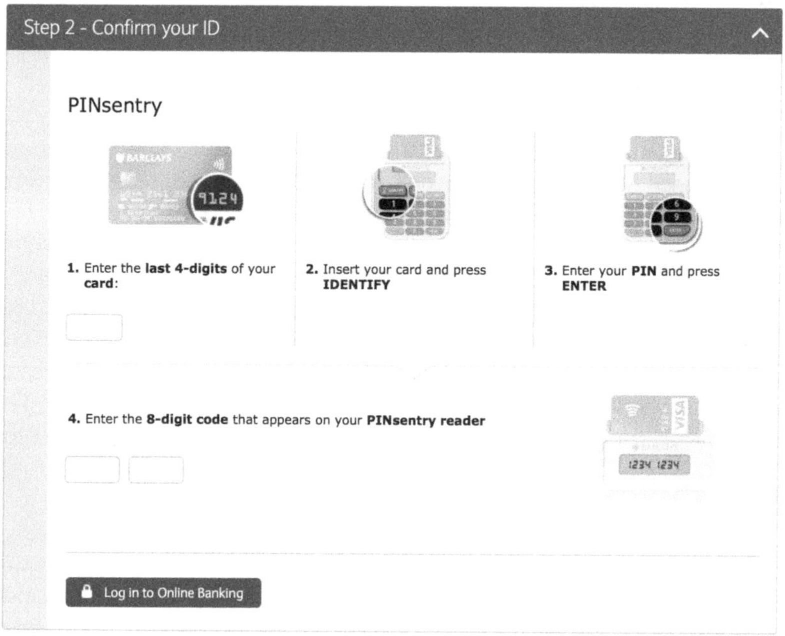

(b)

Figure 1.5 Online banking example from Barclays Bank: (a) first stage of banking login and (b) second stage of banking login.

aspects such as smartcard technologies, cryptographic protocols, and near-field communications), these details are essentially outside the scope of this discussion. What is of interest here is what a possession-based approach demands of the user. In some cases, it is simply having the device and presenting it to be seen/detected by the system—an example here would typically be in an area such as physical access control, such as presenting or swiping a card for door entry, or similar. By contrast, the online banking example from the previous section requires the user to have both their bank card *and* the card reader. A similar principle applies to other approaches, such as the RSA SecurID token (which generates time-limited one-time codes for accessing other systems)—users are not required to present the token itself, but they must have it with them and interact with it to get further information needed (depending upon the device concerned, this may simply involve reading a current code off the display or firstly authenticating on the device itself [i.e., the user needs to authenticate themselves on the device that generates the one-time code before the code will be generated] and only then generating the one-time code required).

While this clearly introduces a further level of security (particularly in variants that combine the physical possession with the need to know some secret knowledge), it can end up impacting the level of flexibility perceived by the user. For example, consider the following quote from an online banking customer, offered in response to HSBC's introduction of its Secure Key token back in 2011, which meant that users now needed to use a small device to generate a one-time code each time they wished to access online banking:

> *It annoys me intensely as I cannot now access my online banking at work, as the device is too big to take in my wallet.*

While the same user might have no objection at all to taking an access card to work, because it would feel like a usual requirement of the environment concerned, carrying the online banking token around just in case it is needed is rather less natural and raises objection as a result. In the HSBC case, this was also the only choice available at the time, as there was initially no longer an option to access online banking without the Secure Key. This was ultimately changed in 2015, when the service was modified to offer the option to login to a restricted set of banking functionality without needing the Secure Key. However, in the intervening period, it was a nonnegotiable requirement and represented a usability restriction as a result.

This exemplifies how token-based authentication can undermine usability. The approach arguably works best if the token is something that you would always expect to have with you at times requiring access. Examples here could be a mobile phone, smartwatch, or other wearable technology—and in the future, perhaps some sort of implant for those who really want to be at one with their technology! Staying with existing techniques, a common example now is to replace the issuance of tokens such as SecureID or the Secure Key with smartphone apps that can generate the codes instead. So the possession of the smartphone becomes the key to being able to gain access (and should add a further layer of security by virtue of the user having enabled some level of authentication on the device itself).

Another current example, provided within the Apple ecosystem, is the ability for users to unlock their computer via their smartwatch. In this particular case, the user must firstly enable an option on their computer to permit automated unlocking to occur (Figure 1.6a), which (if they have not already done so) will also necessitate enabling 2FA for their Apple ID. Once enabled, they no longer need to have an explicit interaction to reauthenticate themselves on the computer if their account is already logged in, as the presence of their (preauthenticated) watch is sufficient to let them in (Figure 1.6b). Meanwhile, any other user attempting to wake the computer would be confronted with the password prompt

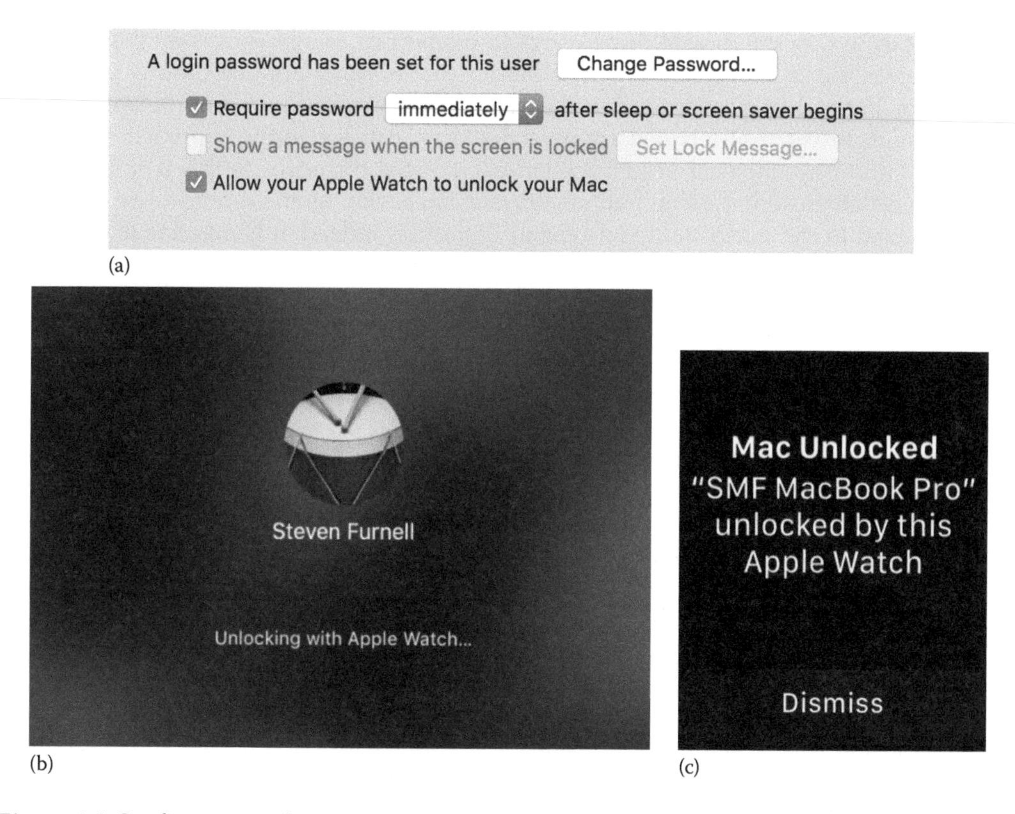

Figure 1.6 Configuring and using automatic unlock from Apple Watch to Mac: (a) configurating automatic unlock, (b) what is seen on the computer, and (c) what is seen on the watch.

as normal. The watch displays a notification to confirm that the process has happened (Figure 1.6c), thus giving a useful alert just in case the legitimate user was simply in close proximity and their presence had inadvertently allowed someone else to gain access instead.

The approach has clear advantages in terms of usability, as it removes an element of the friction involved in waking the device and carrying on with activities. At the same time, it arguably serves to accent those occasions when the password *is* still required (e.g., when first logging in or during restarts) and potentially makes even that small interaction now feel like more of an overhead than it previously did.

1.7 Biometrics: A factor of who you are

Biometrics represent the final category of authentication, based upon something the user *is*. Going beyond this summary assessment, there is no single definition that perfectly encapsulates the term and the ways in which it can be used, so let us use two relevant examples taken from *Newbold's Biometric Dictionary: For Military and Industry* [11]:

- "A measurable biological (anatomical and physiological) and behavioral characteristic that can be used for automated recognition"
- "The automated use of physiological or behavioral characteristics to determine or verify identity"

It is notable that the first of these refers to biometrics as a *characteristic* that is being used, whereas the second refers to it as a *process* that is being applied (noting that it can be used for both identification and authentication purposes). And even between them, the definitions are not entirely consistent in how they categorize the approaches (i.e., is it anatomical and physiological or just the latter?). For the purposes of this discussion, we will use physiological and behavioral as key subcategories, as these tend to be the more accepted norm in the wider field and related literature. Indeed, it is possible to group all the mainstream techniques (and some less mainstream ones) into one of these two groupings, as shown in Table 1.3. While the details of these are not explored as part of the current discussion, it should already be apparent that their applicability and ease of use is not going to be equivalent in all contexts (e.g., while measuring hand shape/geometry may work well enough for physical access control scenarios, it is less likely to suit login situations on a desktop or laptop computer).

Biometrics receive more extensive attention in a later chapter, so the treatment here is necessarily brief and is primarily intended to provide some contrast against the approaches that we have already considered. In this context, biometrics should theoretically offer the best potential for delivering usable authentication, as they are based on inherent characteristics of the user and therefore require nothing to be remembered, and (other than in rather extreme circumstances) there is nothing that can be accidentally lost or left behind. Therefore, if implemented correctly, the demands on the user can be relatively minimal. At the same time, there are advantages from the security perspective, in the sense that users cannot share their biometrics with others or be tricked into giving them away, as they could with a secret or a token.

Having said this, it cannot be assumed that biometrics are a perfect and fool-proof solution. For example, the aforementioned "extreme circumstances" in which they could be lost would include situations such as the user suffering physical harm that rendered their biometric unusable (e.g., loss or injury to a finger or an eye could prevent the use of previously successful fingerprint or retina-/iris-based methods). There is also a clear concern to be raised if a biometric is lost or otherwise compromised, insofar as the user has no option to change or replace it as they would with a password or a token.

The suitability of potential biometric authentication techniques can be assessed based on a number of associated traits, as proposed by Jain et al. [12]:

- *Uniqueness*—the ability to successfully discriminate people
- *Universal*—the ability for a technique to be applied to a whole population of users
- *Permanence*—the ability for the characteristics not to change with time
- *Collectable*—the ease with which a sensor is able to collect the sample
- *Acceptable*—the degree to which the technique is found to be acceptable by a person
- *Circumventable*—the ability not to duplicate or copy a sample

Table 1.3 Physiological and behavioral biometrics

Physiological	Behavioral
Face print	Gait
Facial thermogram	Keystroke dynamics
Fingerprints	Linguistic style (stylometry)
Hand geometry	Mouse dynamics
Iris pattern	Signature recognition
Retinal pattern	Touch dynamics
Vein pattern	Voiceprint

These clearly vary across the set of measures previously listed in Table 1.3. For example, something such as keystroke dynamics, while effective for some users, does not allow the degree of granularity to accurately discriminate across a large population. Similarly, its universality is constrained by the fact that it needs users to have at least a baseline ability to type in order for a characteristic rhythm to be determined, and its permanence is variable as the typing ability—and hence style—of some users will evolve over time. However, it is easily collectable, on the basis that most devices provide a keyboard-based input as standard, and the acceptability is reasonable from perspective that it does not require the user to perform any additional actions (although they may be uncomfortable to feel that their typing is being monitored if it is happening throughout the session). Finally, the ease of circumvention is likely to depend upon the maturity of the user's typing style and whether a more skilled typist can impersonate it. Meanwhile, switching to a different biometric, we would see a different set of assessments. Fingerprints tend to do well in relation to the first three characteristics, but require a specific sensor to be present in order to collect the information. While in the early days the technique was perhaps stigmatized by association with criminology, it now appears to be generally acceptable to most users provided that it is implemented in a usable manner. It can be circumvented, but unless the system has simply been poorly implemented, it requires a nontrivial effort to do so (including access to the legitimate user's prints).

The point about implementation in a usable manner is worth exploring further, and fingerprint recognition represents a good case in point because the traditional implementation on laptop and mobile devices has been via a distinct sensor whose only purpose is to capture the print. Examples are illustrated in Figure 1.7, depicting an HP iPAQ from 2002 (which illustrates an early example of biometrics making an appearance on mobile

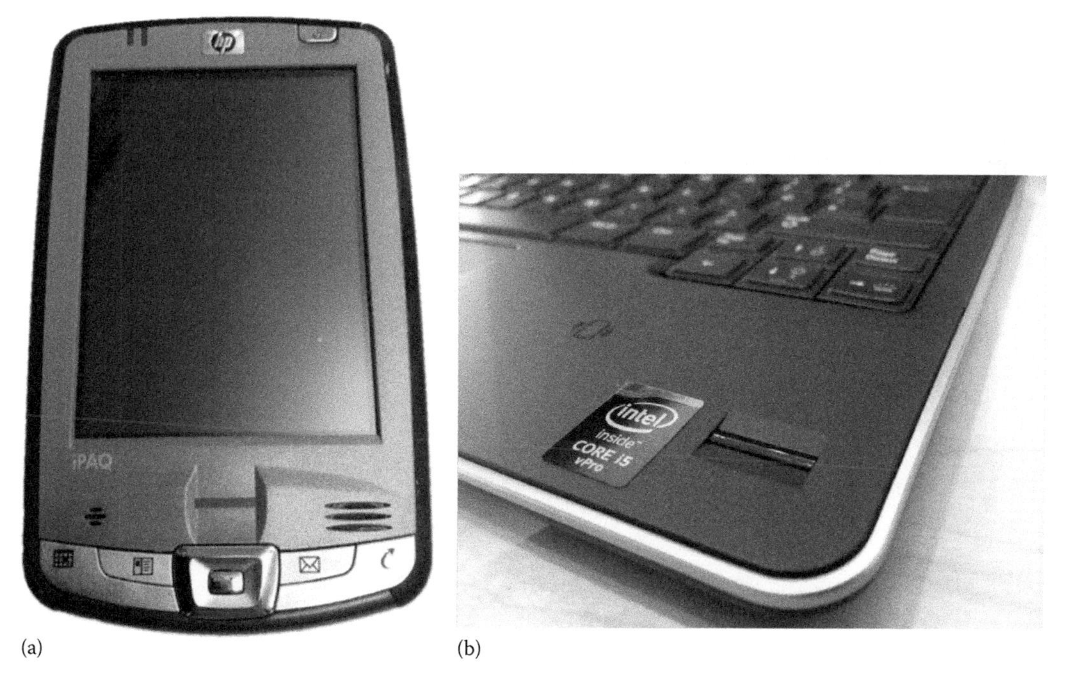

Figure 1.7 Fingerprint sensors provided as separate features on (a) a personal digital assistant and (b) a laptop personal computer.

devices) and a typical laptop-based deployment of today (in addition, many sensors have previously been made available as separate universal serial bus devices, enabling them to be plugged into desktop and laptop systems as add-ons). Cases such as these necessitate the manufacturer having to explicitly integrate a distinct sensor into the physical design of the device and require the user to interact with something that they would not otherwise use for any nonsecurity purposes.

By contrast, more recent deployments of fingerprint recognition have given more explicit consideration to how the sensor can become a more natural part of both the parent device and the resulting user experience. As an example, looking at how Apple has integrated biometrics within its Touch ID solution, there is clear support for innate usability by making the interaction a transparent aspect within an activity that the user would be performing anyway. In each case, the Touch ID sensor is located within a button that would normally be used to activate/wake the device (see Figure 1.8), and the speed of response means that the action required is just a momentary press—certainly no more than would normally be required on a device without the fingerprint sensor present. As such, its use under normal circumstances is frictionless from the user perspective and serves to make fingerprint authentication a natural part of their experience (although a caveat exists—particularly in the smartphone context—where the device may be used outdoors in wet conditions or where the user has dirty, damaged, or covered fingertips, all of which could prevent the sensor from capturing an effective sample).

Much of the discussion to this point has implicitly assumed that authentication is taking place in a traditional manner, at the outset of activity, before any tangible level of access has been granted. However, this in itself creates a situation where the authentication activity can be viewed as a barrier between a legitimate user and what they are actually trying to do. In addition, it creates a potential weakness, insofar as once this stage is passed, the user's legitimacy may not be verified again—thereby enabling an impostor to proceed unchallenged if they have managed to spoof the initial authentication or taken the place of the legitimate user at a subsequent point. However, today's technology offers an opportunity to deliver authentication beyond point of entry, with the ability to obtain a continuous (or periodic) measure of the user's legitimacy. Moreover, this can be done by leveraging natural user interactions as a source for collecting authentication data. This is one of the principles underlying so-called active authentication approaches advocated by

(a)

(b)

Figure 1.8 Fingerprint sensors embedded within other buttons on (a) an iPhone and (b) a MacBook Pro.

the Defense Advanced Research Projects Agency, aiming to overcome the weaknesses of traditional technologies. To quote from their call for research in early 2012,

> *The current standard method for validating a user's identity for authentication on an information system requires humans to do something that is inherently unnatural: create, remember, and manage long, complex passwords.... The Active Authentication program seeks to address this problem by developing novel ways of validating the identity of the person at the console that focus on the unique aspects of the individual through the use of software based biometrics.... This program focuses on the behavioral traits that can be observed through how we interact with the world* [13].

Biometrics offer significant potential to contribute in this context, insofar as there are several approaches that have the potential to be implemented and applied in a transparent and nonintrusive manner. For example, facial recognition can be operated as a background service in order to detect whether the person currently sitting in front of the computer is recognized as the expected user, keystroke dynamics can be used to detect whether their typing rhythms match, and voice verification can be applied if the user is speaking. While these may not all be available to measure at the same time, at least one feed is likely to be on offer regardless of what the user is currently doing. As such, there is a basis for making authentication an ongoing process that is implicitly more usable because (as with the aforementioned integration of fingerprint recognition into Touch ID) it only requires the user to be doing the things they are naturally doing anyway. There is, however, a clear potential to worry or alienate the user if they feel that their every move on the system is now being monitored and assessed. This comes down to how much they are willing to trust the information not to be used for nonsecurity purposes, and this in turn may be linked to how well the situation is explained to them. If they are told that the system is only monitoring *how* things are done (i.e., to verify their identity and prevent impostor use), it is different from leaving them to assuming that it is looking at the details of *what* they are doing and making judgments about things such as their work productivity.

The point about explaining things to users brings us back to a more general issue, namely, that even the more overtly usable technologies should not be deployed without providing some awareness of what they are doing and why. Ultimately, whatever technology is chosen is going to need users to interact with it and use it effectively, so their understanding of what is going on can affect the success of the approach. To illustrate this, we will now leave biometrics and return to the realm of passwords in order to consider whether users have been fairly supported in their past interactions with them.

1.8 Supporting the user

The earlier discussion has already presented some fairly significant criticism of passwords, and to be honest, they are a flawed technology for today's uses, because we need to pick and manage so many of them. However, given that we are still confronted with a requirement to use them in many places, it is worth recognizing that they can be used *better* than they have been and that there are at least some steps that could be taken to improve matters.

A fundamental point, which many readers may recognize from experience, is that passwords are very easy to use badly. Moreover, the poor use of passwords is typically the natural behavior that users can settle into, because they do not fully appreciate the ways in which passwords are vulnerable and lack any instinctive sense of what makes one password stronger than another. This is not meant as a criticism of the users, and indeed, it is not surprising or unexpected that their natural inclinations would be this way (after all, when asked to choose a password, very few people would naturally think that it would be important to choose a long string with multiple character types so as to withstand a machine-based brute force attack!). However, this realization ought to be a prompt that further action is required and that getting passwords to be used well (or at least better) is not likely to happen without a degree of additional effort. Unfortunately, however, this often fails to be the case. Many readers may again be able to reflect on this from contexts such as their workplace, where it is commonplace for staff to be expected to *use* passwords, but far less common for them to receive clear guidance in *how* to do so effectively. This problem is also apparent in a much wider context. For example, a series of prior studies assessed the degree to which leading websites had taken steps to (a) promote password guidance to users and (b) enforce good practice when users are making password choices. These studies were conducted in 2007 [14], 2011 [15], and 2014 [16] and thus provide a means of tracking how aspects of password usage have evolved over time.

Looking at the results from the final version of the study, Table 1.4 summarizes the extent to which users were provided with support during the initial sign-up/registration process. Several aspects were notable. Firstly, the provision of *any* sort of support to users was clearly a minority pursuit, with 6 out of 10 sites providing neither guidance nor password meter feedback. One must wonder about the likely quality of the passwords that users would be likely to make in these contexts, particularly given the extent to which sources such as the SplashData surveys have shown poor choices to be made.

Further examination revealed that the level of guidance differed at other stages of the password life cycle. For example, if the user elected to change their password at some point, then Google still provided selection guidelines, but WordPress no longer did. Meanwhile, LinkedIn, which had not provided guidance at sign-up, was found to do so at the password change stage. Furthermore, if the user was to use the password reset facility, then the availability of guidance was far more commonplace (with only Microsoft Live, Pinterest, Wikipedia, and Yahoo! sites now not doing so). From a certain perspective, this difference makes sense; if the user has found himself/herself needing to use the reset option, then it implies that they have probably forgotten their original password and may benefit from guidance in how to select a replacement. Although from another perspective, the difference makes little sense at all; why not simply offer guidance at *all* stages so that the user is appropriately and consistently supported throughout their use of passwords within the site? One may conjecture that from a service provider perspective, a key objective when new users wish to sign up is not to put obstacles in their way. As such, the provision of password rules and too many enforcements could serve as a disincentive.

Reflecting the fact that some sites come and go in terms of popularity, the 10 sites sampled in the final study were not the same as those in the 2007 version. However, five sites were consistently featured across all three studies, and it is interesting to observe the extent to which their provision evolved. As Table 1.5 shows, even by 2014, only 1 of the 5 (Yahoo!) had progressed to a stage where all the suggested baseline checks were being made, and even then, it was not demonstrating good practice in other aspects (referring back to the lack of password guidelines at any stage of selection). Meanwhile, it was more than a little surprising (and disappointing) to find market-leading retail (Amazon) and social media (Facebook) sites having advanced their practices so little in the intervening period.

Table 1.4 Password guidance and enforcement provisions at sign-up on leading websites

	Guidance to users		Restrictions enforced					
Site	Selection guidelines	Password meter	Enforces min length	Prevents surname	Prevents user ID	Prevents *password*	Enforces composition	Prevents dictionary words
Amazon	✗	✗	6	✗	✗	✗	✗	✗
Facebook	✗	Ratings	6	✓	✗	✗	✗	✗
Google	✓	✓	8	✓	✓	✓	✗	✓
LinkedIn	✗	✗	6	✗	✓	✓	✗	~
Microsoft Live	✗	✗	8	✓	✓	✓	✓	✗
Pinterest	✗	✗	6	✗	✓	✓	✗	✗
Twitter	✗	✓	6	✗	✓	✓	✗	~
Wikipedia	✗	✗	✗	✗	✗	✗	✗	✗
WordPress	✓	✗	6	✗	✓	✓	~	✓
Yahoo!	✗	✗	8	✓	✓	✓	✓	✓

Table 1.5 Evolution of password choice restrictions on leading websites

		Restrictions enforced					
Site	Year	Enforces min length	Prevents surname	Prevents user ID	Prevents *password*	Enforces composition	Prevents dictionary words
Amazon	2007	✗	✗	✗	✗	✗	✗
	2014	6	✗	✗	✗	✗	✗
Facebook	2007	6	✓	✗	✓	✗	✗
	2014	6	✓	✗	✗	✗	✗
Google	2007	8	✗	✓	✓	✗	✗
	2014	8	✓	✓	✓	✗	✓
Microsoft Live	2007	6	✗	✓	✗	✗	✗
	2014	8	✓	✓	✓	✓	✗
Yahoo!	2007	6	✗	✓	✗	✗	✗
	2014	8	✓	✓	✓	✓	✓

Assessing the situation again in early 2017, during the writing of this chapter, revealed that there were still significant shortcomings to be observed. While the full study was not repeated, Amazon, Facebook, and Twitter were all checked to see whether the situation had improved in terms of guiding and supporting users at sign-up. However, 3 1/2 years on from the prior study, none of them was managing to provide any upfront guidance to support password selection. Moreover, their enforcement of password restrictions was still rather questionable. For example, the Amazon site still accepted the word *password* as a valid choice without any word of warning or complaint. Meanwhile, the Facebook site had evolved a little, with attempts to use *password* being prevented and met with an error message (with the same warning being issued in relation to attempts to use several other poorly chosen strings such as *password1*, *Password*, *Password1*, *qwerty*, *apples*, and *123456789*). Unfortunately, however, the attempt to use variations involving the user's name (e.g., *jones1* and *fredjones*) still met with (preventable) success. Finally, the Twitter site, while retaining the support and restrictions previously shown in Table 1.4, still managed to permit the use of *1234567890* and rated it as acceptable on the password meter.

Meanwhile, the Yahoo! website (which Table 1.4 has previously shown to be well rated in terms of enforcing good practice) was still found to perform well and prevented all dubious choices listed earlier. However, it still did so without presenting any upfront guidance on what a good password should look like. As a result, attempts to choose passwords that did not comply with expectations could find themselves met with a variety of responses (a series of which are shown in Figure 1.9). From a usability perspective, one could argue that it is surely better to provide users with some guidance upfront rather than forcing them to encounter feedback in a piecemeal fashion through a process of potential trial and error. The latter may not only cause them to get progressively more frustrated at their inability to select an acceptable password, but could also lead them to make a rash choice that they then fail to remember (noting that in common with Facebook and Twitter cases, the Yahoo! sign-up page did not ask the user to enter their password twice to verify that they can remember it and had not mistype it).

At this point, one could possibly take the view that this does not matter, because any users prone to choosing poor passwords would arguably still do so regardless of any attempt that sites could make to guide them otherwise. With this in mind, some further research was conducted to determine the *effect* that password guidance and feedback

Your password isn't strong enough. Please try making it longer.
Please create a stronger password, the one that you submitted is too easy to guess.
Your password cannot include your name.

Figure 1.9 A series of warning messages resulting from weak password choices on the Yahoo! website.

could have on users' practices. In the first instance, some initial work was conducted with a small sample group of 27 users [17]. These participants were not informed that they were involved in a study relating to password selection, and from their perspective, the main focus of the requested task was to conduct a usability evaluation of a website (noting that this use of mild deception was agreed via ethical approval before the study was conducted). However, prior to performing their evaluations, all users were required to create an account and a password. This enabled the quality of their resulting choices, made under realistic conditions, to then be scored based on their compliance with five basic criteria:

- Use of at least eight characters
- Use of alphabetic and numeric characters
- Use of other characters (e.g., punctuation symbols)
- Avoidance of dictionary words
- Avoidance of personal information

During the study, participants were allocated to two alternative groups. In one group, the registration page included a small set of password tips based upon the preceding list, plus a password meter. In the other group, users were left to select their passwords in an unguided manner. Neither group involved any enforcement of password rules, and users were still permitted to proceed with whatever they chose. Allocating a point to each of the items of good practice listed earlier, 13 users allocated to the guided group averaged 3.8, whereas the remaining 14 in the unguided averaged just 1.8 (with aspects such as password length, use of punctuation characters, and avoidance of personal information being notably less well adhered to in the latter cases).

The difference observed in this initial study encouraged a more substantial investigation in follow-on work, this time involving a more sizeable sample of 300 participants [18]. This study was mounted on the pretext of asking users to register and participate in a survey about social media practices, but again the underlying and ethically approved aim was merely to use this as a basis for getting the participants to create password-protected accounts. While the preliminary study had simply assessed the difference between guided and unguided contexts, the extended experiment also sought to investigate whether the format and style of guidance had any additional effect. As a result, the overall participant group was divided into five experimental subgroups, each with 60 participants, and exposed to one of the following scenarios:

1. An unguided password selection task
2. Password selection supported by basic guidance, as shown in Figure 1.10a
3. Baseline guidance plus a standard password meter, with three single-word ratings prompts (weak, medium, strong), as shown in Figure 1.10b

Figure 1.10 Interface examples from password guidance and feedback study: (a) the basic guidance to users, (b) standard meter-based feedback, and (c) emoji-based feedback.

4. Baseline guidance plus an image-based feedback approach, based on sad, neutral, and smiling emojis (which were also color coded as red, yellow, and green)
5. As previously stated but with the emoji also accompanied by one of three messages ("This is not good enough!" and "Ok, but you could do better!" and "Well done!"), as illustrated in Figure 1.10c

The emoji-based approach was a new idea investigated in order to determine whether appealing to the user at a more emotional level would have any effect compared to traditional measures (i.e., they may be motivated by wanting to please the system). This was similarly the basis for using more emotive language in the textual prompts for the final scenario, as opposed to the impersonal ratings used alongside the standard password meter.

As with the preliminary study, it should be noted that while scenarios 2–5 from the previous list all *presented* guidance, none of them actually *enforced* any good practice, so the intention was once again to determine whether password selections were tangibly affected by the presence of information and feedback. The resulting password choices were scored out of 100 points using a utility obtained from GitHub [19], based upon the rules shown in Table 1.6 (noting that if passwords scored over 100, the value was then capped). As shown in the table, three scoring ranges were then used to denote weak, medium, and strong

Table 1.6 Password scoring rules, rating ranges, and examples

Rules	Rating	Examples	
		Password	Score (pts)
5 pts—Unique character	0–40	1234567	35
2 pts—Repeated character (one already used anywhere else in the password)	Weak	iloveyou	37
		luke33	37
15 pts—Each time a new character type is included (uppercase letter, lowercase letter, number, or symbol) after the first type used in the password	41–70	Luke23	50
	Medium	BROK3R-	52
		foL34p!	65
	71–100	maggie9876543	72
	Strong	neBemvor1893	77
		Lafe@9856!e	82

password choices, thereby enabling a broad comparison between the effectiveness of the different guidance/feedback scenarios.

The resulting performance is depicted in Figure 1.11. While none of the methods got anywhere near perfect password selection behavior, there are clear differences to be observed between the unguided scenario and the effect when alternative levels of guidance and feedback are provided. The starkest contrast is between the first and final scenarios, where the prevalence of weak passwords has been more than halved and the extent of medium-rated passwords is more than doubled. Moreover, strongly rated passwords—which were entirely absent from the choices made without guidance—now account for more than a 10th of the resulting selections. Meanwhile, the average password lengths observed rose from 6.7 characters in the unguided scenario to 8.8 characters in the second of the emoji-based scenarios.

On the basis of the preceding statement, key observations arising from the study were the following:

- Providing any form of guidance and feedback has a clear impact on the resulting user behavior.
- Emoji-based feedback methods gave tangibly better performance than standard guidance and/or password meters.

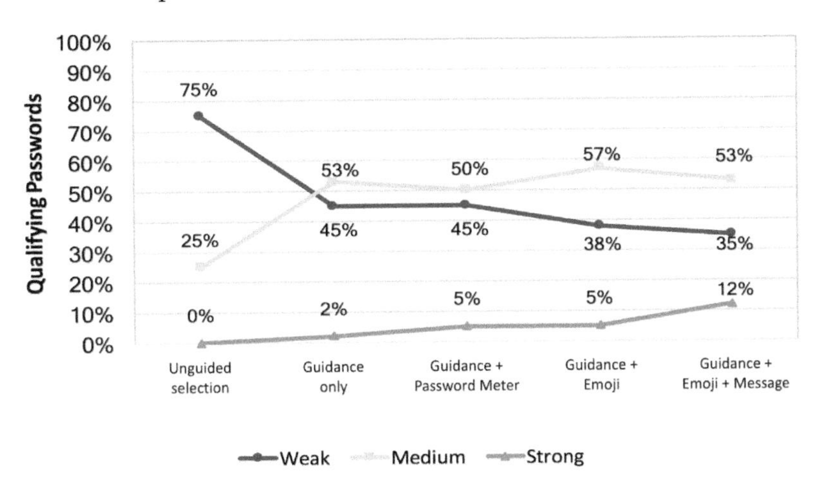

Figure 1.11 Comparing the strength of password choices achieved using different levels of guidance and feedback.

The second point potentially needs to be offset against the perceived novelty of the method (i.e., some of the positive impact may have been because participants were not used to seeing feedback of this nature, whereas they may have become somewhat desensitized to the more familiar password meter approach). However, the validity of the first point should be in less dispute—particularly given the alignment of these findings with those of the earlier initial study.

These findings offer us a wider lesson in the value of supporting other authentication technologies, and indeed other aspects of security, more effectively. To be honest, the fact that guiding and supporting users enable better performance should not be an unexpected or surprising revelation! Nonetheless, it has been an area in which users have been demonstrably undersupported, and so for many system designers and providers, it remains a lesson that it is clearly useful to learn.

1.9 Conclusions

As promised at the outset, we have not settled upon any single method as the perfect choice to ensure usability. Even passwords, which are often dismissed entirely [20], have still been shown to offer the potential to be better used if some consideration and effort are directed toward doing so. However, even then, it does not match the strength of some other methods, and it cannot provide the transparent, ongoing measure of authentication that alternative approaches could offer. Meanwhile, all other approaches present drawbacks too, particularly in terms of their applicability across the range of devices that an individual may be using. So while specific biometrics are only feasible if the device concerned has the necessary sensor to capture them, passwords (and PINs) simply require a keyboard or some other character-entry system—which has been proven to be achievable on anything from traditional desktop and laptop systems through to set-top boxes and smartwatches. However, as anyone who has tried this will realize, these devices offer varying degrees of usability when actually trying to enter the information. Indeed, a key point is that usability is not universal; what works well in one context may not be so effective in others.

It is also worth recognizing that in some cases, usability is coming at the expense of strength. Are we satisfied with the compromise (i.e., having some security is better than none), or should we swallow hard and accept that the protection we need demands extra effort to achieve it? Again, there is no one-size-fits-all response, and the right answer depends on an honest assessment of each situation.

One thing we can be sure about moving forward is that the authentication landscape will no longer be as dominated by passwords as it has been in the past. The use of alternative approaches (particularly biometrics) has already become commonplace, so users can expect a far more varied experience than they have previously been offered. Moreover, if the technology choices are appropriately judged and correctly matched to their needs, then users should be able to expect authentication to feel less of a barrier to legitimate use. The options and opportunities are there—but they still require manufacturers and service providers to take them.

References

1. Beaumont, A., Sasse, A.M., and Wonham, M. 2008. The compliance budget: Managing security behaviour in organisations, *Proceedings of the 2008 Workshop on New Security Paradigms (NSPW '08)*. Association for Computing Machinery, New York, 47–58.

2. Furnell, S., and Thomson, K. 2009, November. Recognising and addressing "security fatigue," *Computer Fraud & Security*, 2009, 7–11.
3. Stanton, B., Theofanos, M.F., Prettyman, S.S., and Furman, S. 2016, September–October. Security fatigue, *IT Professional*, 18, 26–32.
4. National Cyber Security Centre. 2016, January. *Password Guidance: Simplifying Your Approach*, National Cyber Security Centre, London. https://www.ncsc.gov.uk/guidance/password-guidance-simplifying-your-approach (accessed October 25, 2016).
5. Winder, D. 2016. As Amazon uncovers login credential list online, does controversial GCHQ password advice still stand? *SC Magazine*, 13 October.
6. Morgan. 2017. *Announcing our Worst Passwords of 2016*. https://www.teamsid.com/worst-passwords-2016/.
7. Morris, R., and Thompson, K. 1979. Password security: A case History, *Communications of the ACM*, 22(11), 594–597.
8. Aurigemma, S., Mattson, T., and Leonard, L. 2017, January. So much promise, so little use: What is stopping home end-esers from using password manager applications? *Proceedings of the 50th Hawaii International Conference on System Sciences (HICSS-50), Hilton Waikoloa Village, Hawaii*, 4061–4070.
9. Haga, W.J., and Zviran, M. 1991. Question-and-answer passwords: An empirical evaluation, *Information Systems*, 16(3), 335–343.
10. Biddle, R., Chiasson, S., and Van Oorschot, P. 2012. Graphical passwords: Learning from the first twelve years. *ACM Computing Surveys*, 44(4), 19.
11. Newbold, R.D. 2007. *Newbold's Biometric Dictionary: For Military and Industry*. AuthorHouse, Bloomington, IN.
12. Jain, A.K., Bolle, R., and Pankanti, S. 1999. Introduction to biometrics, in Jain, A.K., Bolle, R., and Pankanti, S. (eds.), *Biometrics: Personal Identification in Networked Society*. Kluwer Academic Publishers, Norwell, MA.
13. Defense Advanced Research Projects Agency. 2012, January. *Broad Agency Announcement—Active Authentication*, DARPA-BAA-12-06. Defense Advanced Research Projects Agency, Arlington County, VA.
14. Furnell, S.M. 2007. An assessment of website password practices, *Computers & Security*, 26(7–8), 445–451.
15. Furnell, S. 2011, December. Assessing password guidance and enforcement on leading websites, *Computer Fraud & Security*, 2011, 10–18.
16. Furnell, S. 2014, 2014. Password practices on leading websites—Revisited, *Computer Fraud & Security*, 2014, 5–11.
17. Furnell, S., and Bär, N. 2013, July. Essential lessons still not learned? Examining the password practices of end-users and service providers, in L. Marinos and I. Askoxylakis (Eds), *Human Aspects of Information Security, Privacy, and Trust*, LNCS 8030, Springer-Verlag, Heidelberg, pp. 217–225.
18. Furnell, S., and Esmael, R. 2017, January. Evaluating the effect of guidance and feedback upon password compliance, *Computer Fraud & Security*, 2017, 5–10.
19. Mohamed, K. 2014, September. *Password Meter Tutorial*. https://github.com/lifeentity/password-meter-tutorial.
20. Fido Alliance. 2013, February. Lenovo, *Nok Nok Labs, PayPal, and Validity Lead an Open Industry Alliance to Revolutionize Online Authentication*, Press Release. https://fidoalliance.org/lenovo-nok-nok-labs-paypal-and-validity-lead-an-open-industry-alliance-to-revolutionize-online-authentication/.

chapter two

Biometrics

Francisco Corella and Karen Lewison

Contents

2.1 Introduction

After more than a century of biometric usage for forensics and physical identification, and decades of evolution of biometric technology at a measured pace, we are now in the middle of a biometric revolution, fueled by the availability of biometric sensors in smartphones, tablets, and laptops and, more recently, by breakthroughs in biometric technology. The usage of biometrics is becoming routine in everyday life, as fingerprints are used for unlocking smartphones, selfies are used for online identity proofing, and banks use speaker verification for customer authentication; and machines are now claimed to be better than humans at identifying faces (Lu and Tang 2015).

At the same time, biometric technology is facing severe security and privacy challenges. Security challenges come from spoofing techniques, which are improving just as quickly as biometric accuracy. It is possible to spoof a fingerprint reader with an artifact constructed after photographing a finger with a high resolution camera from several meters away (Khandelwal 2014); a highly accurate deep neural network for face recognition may be deceived by colored eyeglass frames (Sharif et al. 2016); real-time voice morphing may allow an impersonator to fool not only speaker recognition software, but also even human verifiers (Mukhopadhyay et al. 2015). A privacy challenge arises from the fact that biometric characteristics used for biometric verification may be used to link the activities of the subject across both cyberspace and the physical world. If, for example, a selfie is used for authentication at a web site, the site operator or an adversary who breaches the database of users of the site can use the selfie to link each user's account to the user's activities on social networks and the user's visits to physical stores equipped with customer identification cameras.

Biometric usage for human–computer interaction is mostly concerned with *biometric recognition*, or *matching*. Biometric recognition can be divided into two categories, one-to-many matching, also called *identification*, and one-to-one matching, also called *verification*. In identification, the biometric task is to assign a biometric sample as belonging to one out of a large collection of individuals, while in verification, the task is to decide whether or not a sample belongs to a given individual. Use cases of biometric recognition include forensics, surveillance, photo tagging, and identification of customers who walk into a store. Use cases of biometric verification include physical access control, authentication of automated teller machine customers, phone unlocking, remote identity proofing, authentication, and privilege escalation. Of the two categories, biometric verification is the one most relevant to human–computer interaction, and is therefore the focus of this chapter.

The rest of the chapter is organized as follows: Section 2.2 introduces basic concepts related to biometric verification. Section 2.3 describes four paradigms commonly used today for biometric verification, which make use of biometric templates, statistical models, deep neural networks, and biometric cryptosystems (a.k.a. revocable biometrics or biometric key generation). Section 2.4 discusses biometric security, including zero-effort attacks, presentation attacks and their mitigation, and security architectures. Section 2.5 describes in some detail biometric modalities most commonly used today and other modalities in less detail; it also discusses the fusion of biometric modalities. Section 2.6 concludes by suggesting possible avenues for future research in biometric verification.

2.2 *Biometric verification concepts*

A *biometric characteristic*, or *trait*, is a measurable aspect of the human body that can be used to distinguish individuals from each other, such as a fingerprint, an iris image, a facial image, or acoustic features of the human voice. A *biometric sample* is a sample of a biometric characteristic. A *biometric modality* is a class of biometric systems that deal with a particular biometric characteristic.

In biometric verification, a verifier compares two biometric samples and decides whether they come from the same individual. Biometric verification is a two-phase protocol. In an *enrollment phase*, an enrollment sample is acquired from a subject. In a subsequent *verification phase*, the subject provides biometric input comprising a verification sample to a verifier, which compares it to the enrollment sample or to data derived from the enrollment sample. In addition to the verification sample, the biometric input may provide clues that the verifier can use for *presentation attack detection*, as described in Section 2.4.2.3.

There are two kinds of biometric verification, which are often called *authentication* and *identity proofing*. In authentication, the subject presents the enrollment sample to the verifier. If the verification sample later matches the enrollment sample (or a template or other enrollment data derived from the enrollment sample), the verifier learns that the individual presenting the verification sample is the same subject who provided the enrollment sample, but nothing else. In identity proofing, the verifier, which may have no prior relationship with the subject, obtains the enrollment sample or other enrollment data from an *identification authority* with which the subject has enrolled earlier, together with a binding of the enrollment data to attributes of the subject. If the verification sample matches the enrollment data, the verifier learns that the attributes belong to the subject presenting the verification sample and uses those attributes to identify the subject.

The verification sample is said to be *genuine* if it comes from the same subject as the enrollment sample. For a given configuration of the verifier, the accuracy of the verification process may be defined by two probabilities: the probability that the presented sample is accepted as genuine when it is not genuine, called the *false accept rate* (FAR) or the *false match rate* (FMR), and the probability that it is rejected when it is in fact genuine, called the *false reject rate* (FRR) or *false nonmatch rate* (FNMR).

The verifier may be configured to be more or less forgiving of differences between the presented sample and the enrollment sample. At one extreme, FAR = 0, while FRR = 1. At the other extreme, FAR = 1, while FRR = 0. As FAR increases, FRR decreases; therefore, if the FAR and the FRR are modeled as continuous functions of a real-valued configuration parameter, there is a value of the parameter for which the FAR and the FRR have the same value, which is called the *equal error rate* (ERR). If the FAR is further modeled as a strictly monotonic function of the configuration parameter, each FAR determines a value of the parameter, which in turns determines a value of the FRR. The function that thus maps each FAR to the corresponding FRR is called the *receiver operating characteristic* (ROC). (The term *accuracy rate* is typically used in the context of identification rather than verification, to refer to the complement of the classification error rate.) Accuracy metrics are highly dependent on the sample space and can only be assigned precise numeric values when the sample space has been defined, for example, with reference to a database of biometric samples used for benchmarking, such as the Labeled Faces in the Wild (LFW) database (University of Massachusetts 2017).

2.3 *Biometric matching paradigms*

There is a great variety of techniques for comparing biometric samples, across biometric modalities as well as within each modality. They can be roughly classified into four different paradigms, based on whether they use biometric templates, statistical models, deep neural networks, or biometric cryptosystems. For some modalities, techniques pertaining to multiple paradigms are available.

2.3.1 *Biometric matching using biometric templates*

In this paradigm, a *biometric template* is derived from the enrollment sample and matched against the verification sample or against a template derived from the verification sample.

A biometric template is an encoding of characteristic features of a biometric sample. The order of the features encoded in the template may or may not be significant. If it is significant, the template is a *feature vector*, or an encoding of a feature vector. If not, it is a *feature set*, or an encoding of a feature set. An example of a feature set is a fingerprint template consisting of a set of minutiae. A fingerprint minutia is either the end of a friction ridge or a bifurcation of a friction ridge. Each minutia is described in the template by its type (end or bifurcation), its position, and its orientation. An example of a feature vector is an iris code (Daugman 2003) described in Section 2.5.2.

2.3.2 *Biometric matching using statistical models*

In this paradigm, multiple enrollment samples are used to construct a statistical model of the subject's biometric characteristic. A general model of the biometric characteristic is also constructed using samples from a large number of individuals, and a statistical test is used to estimate the likelihood that the verification sample comes from the subject rather than a random individual. An example of this paradigm is the Gaussian mixture model–universal background model (GMM-UBM) verification method (Reynolds et al. 2000) often used in voice biometrics.

2.3.3 *Biometric matching using a deep neural network*

Deep neural networks, further described in Section 2.5.3, are multilayer artificial neural networks that are being used very successfully in applications such as facial and speech recognitions. When a deep neural network is used for face verification, the network is trained with millions of labeled faces belonging to thousands of people, but there is no need to specifically train the network with enrollment samples of the subject. The enrollment and verification samples are separately input to the network, which produces a mathematical output for each sample. The outputs are then compared according to some similarity metric and deemed to belong to the same person if their similarity metric is above a certain threshold. In the case of Google's FaceNet (Schroff et al. 2015), the output is a vector with 128 coordinates, each of which is a single byte, and the similarity metric used to compare the vectors derived from enrollments and verification samples is the Euclidean distance between the two vectors.

2.3.4 *Biometric matching using a biometric cryptosystem*

As illustrated in Figure 2.1, in a biometric cryptosystem (ISO/IEC 2011, Rathgeb and Uhl 2011), error correction techniques are used to consistently generate a *biometric key* from varying but genuine biometric samples. At enrollment time, an enrollment biometric template is derived from an enrollment sample, and a random biometric key and *helper data* are generated from the enrollment template and random bits produced by a random or pseudorandom bit generator (NIST 2016). At verification time, a verification biometric template is derived from a verification sample, and an error correction algorithm attempts to recover the biometric key from the verification template and the helper data. If the verification sample is genuine, the error correction algorithm is able to recover the key with a probability equal to the complement of the FRR, 1 – FRR.

Even though the helper data are derived from the enrollment template, randomization makes it computationally unfeasible to derive any useful biometric information from it. Thus, the confidentiality of the subject's biometric information is preserved even if an adversary captures the helper data. By contrast, traditional biometric templates reveal biometric information (Cappelli et al. 2007, Ross et al. 2007).

Different kinds of biometric key generation techniques are used with different kinds of biometric templates. Techniques based on the concept of a *fuzzy commitment* (Juels and Wattenberg 1999) may be used with feature vectors, where the order of the features matters, while techniques based on the concept of a *fuzzy vault* (Juels and Sudan 2006) may be used with feature sets, where the order does not matter.

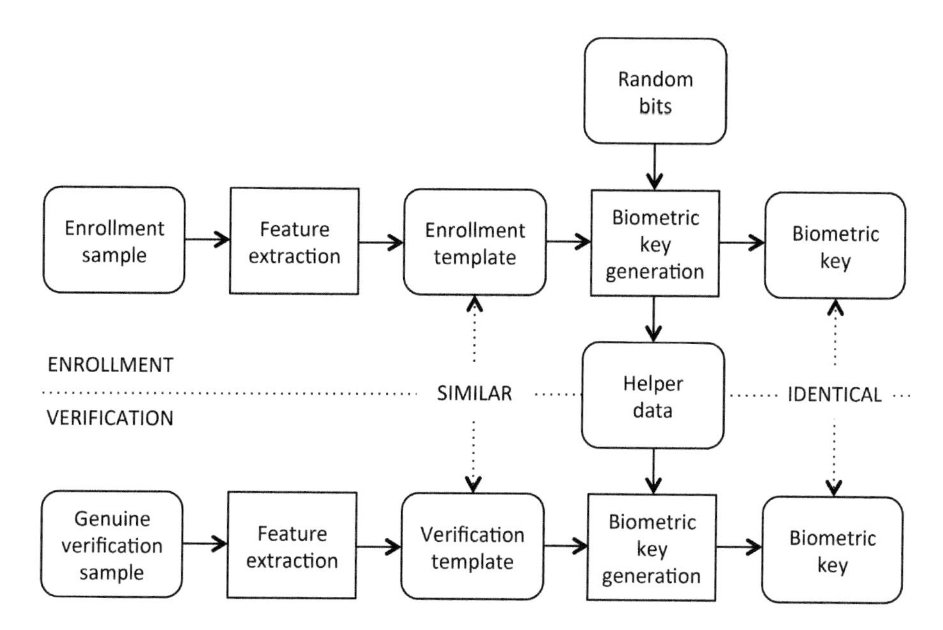

Figure 2.1 Biometric cryptosystem.

A biometric cryptosystem can be used for a variety of purposes. For example, the biometric key can be used to encrypt data. In such a use case, if the biometric key is compromised, it can be replaced with a different random key generated from the same biometric characteristic of the subject, and the data can be encrypted anew with the replacement key. The biometric key is said to be revocable, and this motivates referring to biometric cryptosystem technology as *revocable biometrics*, or *cancelable biometrics*.

On the other hand, when a biometric cryptosystem is used for biometric matching as discussed here, the biometric key is not used for encryption or any other cryptographic use. It is used to check whether the verification sample is genuine, by verifying that the error correction algorithm is able to produce the same key that was generated at enrollment time. The biometric key cannot be stored along with the helper data for that purpose, because the helper data and the biometric key together do reveal biometric information. But a cryptographic hash of the biometric key can be stored together with the helper data, and the biometric key produced at verification time can be verified by hashing it and comparing the resulting hash to the stored hash.

The use of a biometric cryptosystem for biometric matching raises a practical difficulty. Neither the enrollment sample nor the enrollment template is available at verification time. Therefore, it is not possible to perform any geometric alignment of the enrollment and verification samples or templates, in modalities that require such alignment. Methods for solving this difficulty are discussed in Section 2.5 in connection with fingerprint and iris modalities.

2.4 *Biometric security*

The goal of biometric verification is to prevent the impersonation of the subject by an adversary, and that requires protecting against a variety of attacks that may be carried out by the adversary. The adversary may carry out a *zero-effort attack* by presenting a biometric sample from his or her own body and hoping that it will be accepted as genuine.

Biometric accuracy mitigates zero-effort attacks. However, the sample presented by the adversary may not come straight from the adversary's body. It may come from an artifact, or the adversary may wear a disguise, or the sample may be a digital transformation of a sample originating from the adversary, or it may be a digital copy of a genuine sample coming from the subject's body. Attacks with such samples are *presentation attacks*, informally known as *spoofing attacks*.

2.4.1 Presentation attacks

To understand presentation attacks and how to provide protection against them, it helps to classify them along the following dimensions:

- A presentation attack may be *physical* or *digital*, according to whether it is performed before or after a sensor has digitized the biometric sample.
- The presented sample may be *artificial*, if it is produced by a physical artifact or is digitally generated; *disguised*, if it comes from a physically or digitally disguised adversary; or *genuine*, if it originates from the impersonation victim.
- The target, i.e., the biometric characteristic of the subject that the adversary wants to impersonate, may be *known* or *unknown*.

These classification facets are illustrated by the following examples.

When a fake finger is used to hack the fingerprint sensor of a smartphone (Chaos Computer Club 2013), or a photo of the impersonation victim is presented to a smartphone camera in an attack against face verification, the attack is *physical*, the sample is *artificial*, and the target is *known*.

When a MasterPrint (Roy et al. 2017) is used to make a fake finger that is presented to a partial fingerprint sensor, the attack is *physical*, the sample is *artificial*, and the target is *unknown*.

When a wig, a fake nose, and makeup are used to disguise an adversary against face verification (Pavlidis and Symosek 2000), the attack is *physical*, the sample is *disguised*, and the target is *known*.

When colored eyeglass frames are used in a perturbation attack against a deep neural network (Sharif et al. 2016), the attack may be *physical* or *digital*, the sample is *disguised*, and the target is *known*.

When voice morphing is used to disguise the voice of an adversary reading prompted text in an attack against speaker verification (Mukhopadhyay et al. 2015), the attack is *digital*, the sample is *disguised*, and the target is *known*.

When a video of the impersonation victim is obtained by a malicious verifier and replayed to another verifier, the attack is *digital*, the sample is *genuine*, and the target is *known*.

2.4.2 Protection against presentation attacks

2.4.2.1 Biometric confidentiality

Biometric characteristics of an individual are not secrets, and hoping that the adversary does not know the target characteristic should not be the only defense against impersonation. However, the biometric characteristics used in some biometrics modalities, such as iris or retina verification, may be difficult for an adversary to acquire without the subject's consent. Furthermore, the adversary may not know the target characteristic because the adversary does not know the identity of the target subject. Therefore, efforts to protect the confidentiality of a biometric characteristic are a useful mitigation against attacks that require the target characteristic to be known, besides being motivated by privacy considerations.

Biometric confidentiality can be protected by presenting samples only to trusted verifiers over secure connections and by protecting databases containing biometric enrollment samples or templates against security breaches. It can also be protected by using a biometric cryptosystem as described in Section 2.3.4.

2.4.2.2 *Combination with a password*

A biometric sample can acquire secrecy by combining it with a password or passphrase. In text-dependent speaker verification, a short sample text becomes a passphrase simply by treating it as a shared secret between the subject and the verifier (Novoselov et al. 2014). A password has also been used in combination with lip reading (Cheung 2017) and was used, two decades ago, in combination with behavioral biometrics based on keystroke dynamics (Monrose et al. 1999).

2.4.2.3 *Presentation attack detection*

While the confidentiality protection and combination with a password are useful mitigations, the best defense against presentation attacks is presentation attack detection.

Different techniques have been proposed for detecting different kinds of attacks against different modalities. In some modalities such as fingerprint verification, presentation attack detection is performed by the sensor. In other modalities, such as iris, face, or speaker verification, presentation attack detection is performed on the digital output of the sensor.

Challenge–response is a technique available against attacks where the adversary presents a genuine sample obtained from the subject. In the face verification modality, for example, the verifier may ask the subject to stream a video of him/herself reading a challenge sequence of digits, rather than a static selfie. To detect a possible replay attack, the verifier may then use a lip reading technique to check that the digits being read are those in the challenge sequence (Kollreider et al. 2007). In any challenge–response interaction, the challenge should be chosen by the verifier at random with high entropy.

To detect attacks that use an artificial sample coming from a physical artifact, the verifier may check for the presence or absence of signs indicating that the sample comes from a live human body. The absence of such signs indicates a presentation attack. For example, a fingerprint sensor may look for indications of perspiration coming from the pores on the friction ridges (Schuckers and Johnson 2014), or the contraction of the pupil in response to brighter light may indicate liveness in iris scanning.

To detect a disguise worn by the adversary, the verifier may look for specific kinds of disguise, which may require using additional sensors. For example, disguises, such as makeup, a fake nose, or even a wig made out of human hair, are revealed by imaging in the upper near-infrared (IR) spectrum (0.8–1.4 μm) (Pavlidis and Symosek 2000). Some smartphones have near-IR cameras, which are used for iris scanning (Mayhew 2016) but could be used for other purposes in the future.

Protection against presentation attacks with digitally generated or modified samples is an open area of research.

Multiple presentation attack detection techniques may be used together for protection against different kinds of attacks on the same modality. For example, face verification may be exposed both to replay attacks and disguise attacks. The preceding lip reading challenge–response technique may be used together with imaging from a near-IR camera to protect against both attacks.

2.4.2.4 Remark: Liveness detection

The term *liveness detection* is sometimes used as a synonym of presentation attack detection. Strictly speaking, however, liveness detection should refer to presentation attacks with samples that are not live. The word *live* may refer to the real-time presentation of a sample or to a sample that comes from a live body. Hence, liveness detection may refer to the detection of replay attacks with genuine samples or the detection of samples that come from artifacts.

2.4.3 Security and privacy implications of biometric verification architectures

Biometric verification for human–computer interaction involves components, such as a sensor, enrollment data, and biometric matching software, and devices such as a smartphone, a personal computer, a smartcard, or a server. A biometric verification architecture determines what components reside on what device. There is a wide variety of possible architectures, ranging from an old-fashioned one where a fingerprint is obtained by a sensor attached to a desktop and compared to a template stored in a smartcard plugged into a card reader also attached to the desktop to more recent ones such as where a credit card is equipped with a fingerprint sensor (Mastercard 2017). This section examines the security and privacy implications of four architectures commonly used today, as illustrated in Figures 2.2 through 2.5.

In Figures 2.2 and 2.3, the subject locally authenticates to a personal device, such as a smartphone or a laptop, by presenting a biometric sample to a sensor located on the device. The purpose of the authentication may be to unlock the device or to authorize a secondary nonbiometric authentication to a remote server. The latter purpose is the goal of the Fast IDentity Online (FIDO) Universal Authentication Framework (FIDO Alliance 2016), where the secondary authentication to the remote server is by means of an uncertified key pair.

Figures 2.2 and 2.3 differ by the kind of presentation attack detection that is performed, if any, which depends on the kind of modality and sensor that are used. In Figure 2.2,

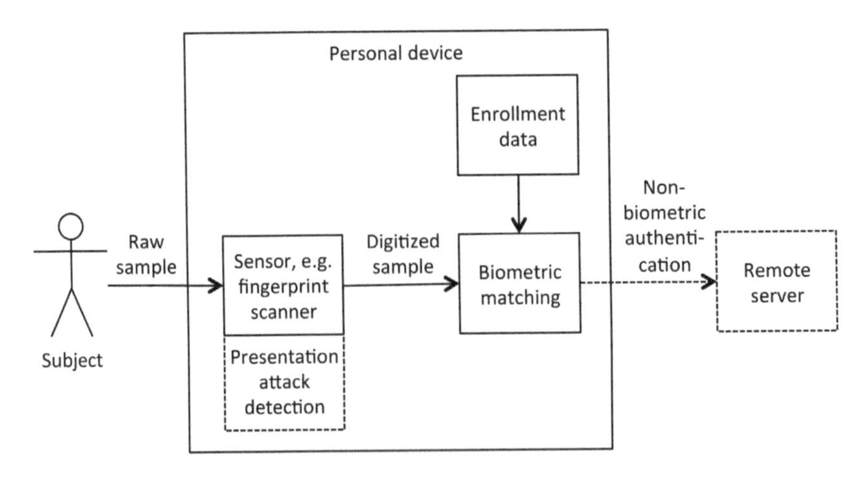

Figure 2.2 Local authentication with presentation attack detection by sensor.

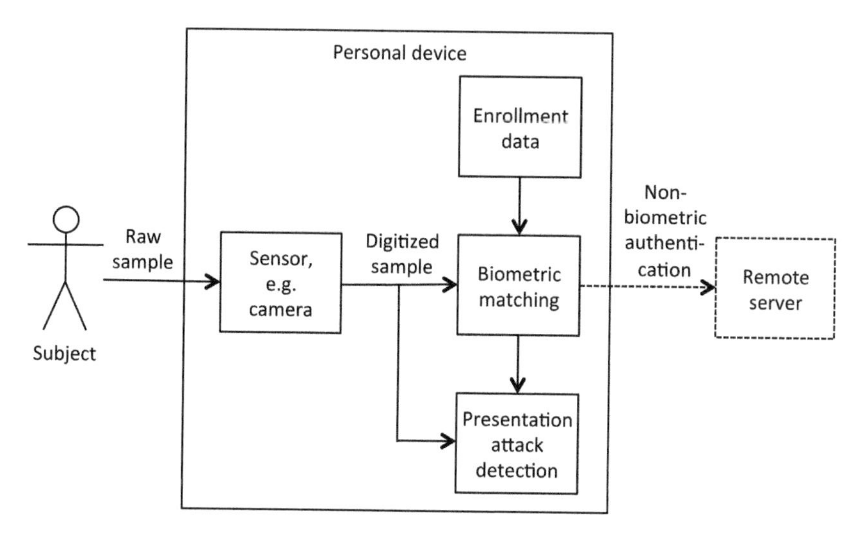

Figure 2.3 Local authentication with presentation attack detection performed on a digitized sample.

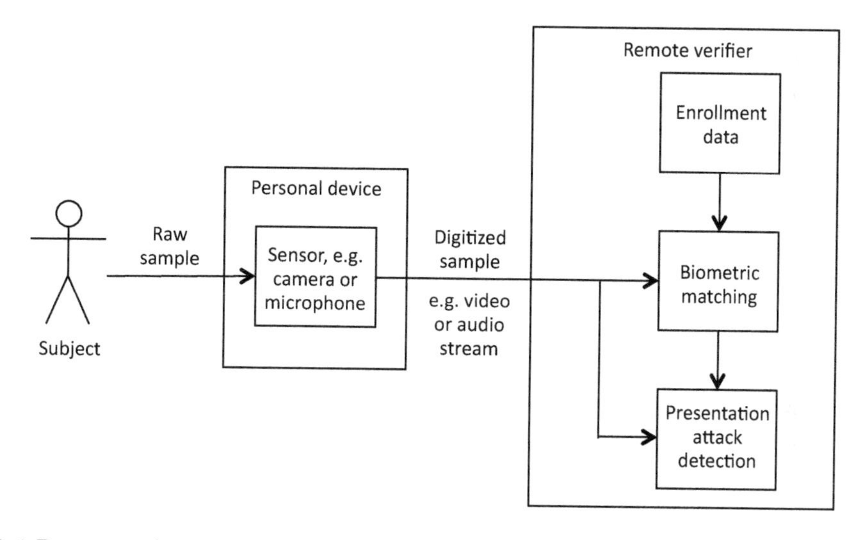

Figure 2.4 Remote authentication or identity proofing against a database.

presentation attack detection is carried out by a sensor such as a fingerprint scanner, or is omitted. Today, fingerprint scanners found on smartphones do not perform any presentation attack detection, but they may do so in the future, as discussed in Section 2.5.1. In Figure 2.3, presentation attack detection is carried out on the digitized output of the sensor, which may be, for example, a video captured by a camera.

The architectures in Figures 2.2 and 2.3 provide strong biometric privacy protection because biometric samples are never sent outside the personal device. The protection is even stronger if the enrollment data are stored in tamper-resistant hardware such as a secure element or a Trusted Platform Module (TPM), as is sometimes the case for the fingerprint template in some smartphones equipped with a fingerprint sensor. In that case, the subject's biometric information is protected against an adversary who physically captures the subject's device. Equivalent protection can be achieved by using a biometric

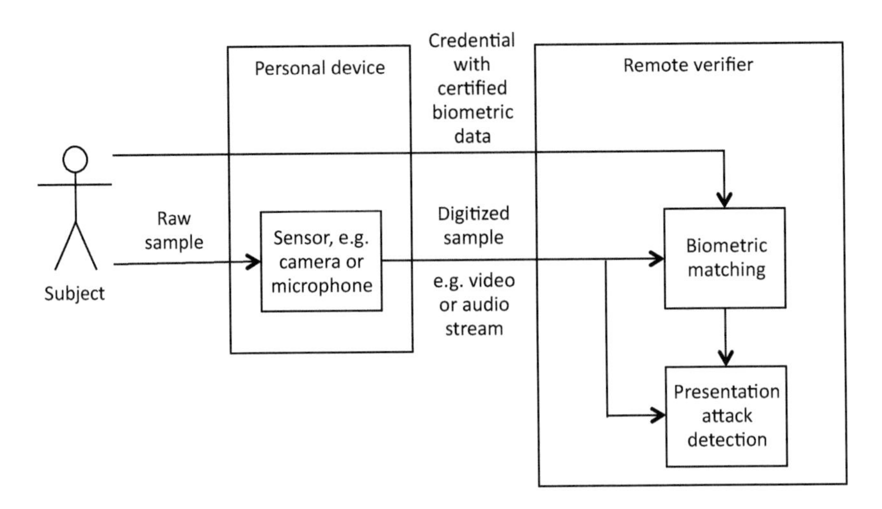

Figure 2.5 Remote authentication or identity proofing against certified enrollment data.

cryptosystem and using helper data together with a hash of a biometric key to verify the authentication sample, as described in Section 2.3.4.

The security provided by the architectures in Figures 2.2 and 2.3 depends on the modality, the sensor, and the efficacy of the presentation attack detection. It is poor today when a fingerprint sensor is used without presentation attack detection, as discussed in Section 5.1.

In Figures 2.4 and 2.5, the subject presents a biometric sample to a sensor located in a personal device, which digitizes the sample and forwards it to a remote verifier. The sensor may be, for example, a camera that streams a video of the subject's face to the verifier for face verification or a microphone that streams speech uttered by the subject for speaker verification. Presentation attack detection is performed on the digital output of the sensor transmitted to the remote verifier. These architectures may be used both for recurring authentication against an enrolled sample and for identity proofing to a verifier that has no prior relationship with the subject.

Figures 2.4 and 2.5 differ in the location of the biometric enrollment data.

In Figure 2.4, the enrollment data are stored in a database that is accessed by the remote verifier. In the authentication case, the database is built by the verifier as subjects enroll biometric samples. For example, the subjects may be users of an online service, and the database may be the user database of the service; each user record may then contain enrollment data used for authenticating the user. In the identity proofing case, the database is not specific to the verifier. It is provided instead by an identification authority such as, in the United States, a department of motor vehicles of a state (Gamlin 2016).

In Figure 2.5, the enrollment data are included in a credential and certified by a digital signature on credential data. The credential is submitted by the subject together with the biometric verification sample. It may be a physical token, such as a passport with an embedded near-field communication (NFC) chip (International Civil Aviation Organization 2015), a Personal Identity Verification (PIV) card carried by employees of the US Federal Government (NIST 2013), or a national identity card of a country that issues such cards. It may also be a purely digital credential, which may provide multifactor authentication, such as a biometric certificate (Dulude and Musgrave 2001) or a rich credential (Lewison and Corella 2016).

By contrast with the architectures in Figures 2.2 and 2.3, those of Figures 2.4 and 2.5 are not vulnerable to the physical capture of the subject's personal device.

The biometric architecture in Figure 2.4 has a serious privacy drawback, because, given the current state of cybersecurity, the database containing the enrollment data may be breached, and if so, the adversary may capture at once biometric information pertaining to a large number of subjects. This drawback may be eliminated by using a biometric cryptosystem for biometric matching and using helper data for verification together with a hash of a biometric key, as described in Section 2.3.4. In Figure 2.5, by contrast, an adversary can only attempt to capture biometric information of one subject at a time.

2.5 *Biometric modalities*

2.5.1 *Fingerprint verification*

The biometric characteristic that is measured in fingerprint verification is the pattern of the friction ridges of a finger, which can be observed by different kinds of sensors at different resolutions. Most sensors used today, for example, those on smartphones, are capacitance-based sensors that measure the variations in electrical capacitance between the finger and an array of microelectrodes embedded in the sensor, the capacitance measured under ridges being higher than the capacitance measured under valleys.

Different features can be extracted from the friction ridge pattern at different resolutions. At 500 pixels per inch (PPI), which is the resolution of today's smartphone sensors, the extracted features are the minutiae already described in Section 2.3.1. At lower resolutions, the extracted features are ridge shapes known as arches, loops, and whorls. At higher resolutions, it is possible to measure the thickness of the ridges and to count the perspiration pores located on the ridges. It is even possible to detect pores that are open and emitting perspiration onto adjacent valleys, which can be used for liveness detection (Schuckers and Johnson 2014).

Higher resolutions may be available in future smartphones by using optical rather than capacitance sensors. US patent 9,570,002 (Sakariya and Nauta 2017) refers to inserting IR light-emitting diodes and sensing IR diodes between the subpixels of the phone screen in order to image the ridges of a fingerprint placed on the screen, potentially achieving higher resolution.

At 500 PPI, a fingerprint template may be a list of minutiae, each described by its type, its x and y coordinates, and the angular orientation θ of the ridge that ends or bifurcates at the minutia. Matching two templates requires finding a correspondence between some of the minutiae in one template and some of the minutiae in the other. A matching score can then be computed from parameters including, among others, the number of minutiae that have been matched.

Minutiae can be viewed as an unordered set of features, and a biometric cryptosystem based on the concept of a fuzzy vault can be used for biometric matching in the fingerprint modality. However, this requires corresponding minutiae to be mapped to elements of a finite field. If the mapping is based on the x and y coordinates and orientation θ of the minutiae, measured at some granularities, enrollment and verification samples must be digitally aligned precisely enough for x, y, and θ to have the same values at those granularities. However, as pointed out in Section 2.3.4, the enrollment sample and template are not available at verification time. The first biometric cryptosystem proposed for fingerprint matching (Clancy et al. 2003) avoided this problem by requiring the physical alignment of the samples, but this is not practical. A subsequent proposal (Uludag and Jain 2006) addressed the problem by adding ridge shape information to the helper data. But this goes counter to the essential tenet that helper data should not reveal any useful biometric information. More recently, there have been proposals to make the mapping of minutiae to elements of the finite field independent of geometric alignment (Wencheng et al. 2014, Li and Hu 2016).

The fingerprint sensors used today to unlock smartphones provide little security, for two reasons. The first reason is that due to the absence of any presentation attack detection, the sensor can be hacked with an artifact constructed from a latent print lifted from the phone itself. The iPhone TouchID was thus hacked shortly after it was introduced (Chaos Computer Club 2013, Rogers 2013). A fingerprint can also be obtained for that purpose from a photograph of a finger (Khandelwal 2014). The sensor on the Samsung Galaxy S5 was also hacked, four days after being introduced (Storm 2014). An improved version of TouchID was introduced for the iPhone 6, but it was also hacked in the same way as that for the iPhone 5S (Bort 2014).

The second reason is that fingerprint sensors on today's smartphones capture only partial fingerprints. A partial fingerprint template has less entropy and is more likely to be matched by a zero-effort attack than a full template. Furthermore, in order to avoid false rejection when the partial verification fingerprint does not match the partial enrollment fingerprint, users are allowed to record multiple partial fingerprints, further reducing entropy and increasing exposure to zero-effort attacks. Also, some partial templates have been shown to occur with higher probability than others. An adversary can further increase his/her chances by constructing an artifact that produces such a template and applying it to the sensor (Roy et al. 2017). In the future, it should be possible to scan full rolled fingerprints with a sensor embedded in the smartphone screen as described in the aforementioned US patent 9,570,002.

Fingerprint verification is also used on smartphones for purposes other than unlocking the phone. It is used in particular to secure payments both online and in stores equipped with NFC payment terminals. Mastercard has recently announced an alternative way of using a fingerprint to secure payments in stores, by means of a sensor located on a Europay, Mastercard, and Visa (EMV) chip card (Mastercard 2017). The fingerprint is matched against an "encrypted" template stored in the chip. However, if the template is encrypted, it is not clear what key can be used to decrypt it that would not also be available to an adversary who tampers with the card to obtain the template.

2.5.2 *Iris verification*

In the iris verification modality, the biometric characteristic that is used is the texture of the iris, as imaged in the near-IR spectrum where it exhibits a richer structure than in the visible wavelength spectrum, particularly for brown eyes. Iris verification is a very accurate modality. Based on 200 billion cross comparisons performed on a database of 632,500 iris images acquired at border crossings (Daugman 2006), the FMR was estimated to be 10^{-15} when setting a threshold of 0.225 for a matching score (best of seven Hamming distances of iris codes at different relative rotations, as described in the following) for which the FNMR was estimated to be less than 1%. Until recently, the most common use of iris verification was for traveler identification at border crossings, but some phones now feature a near-IR camera that is used for unlocking the phone by means of iris verification (Mayhew 2016).

Iris verification uses algorithms for computing a 2048-bit biometric template, called an *iris code*, invented and patented by John Daugman in the 1990s (Daugman 1994) and still in use today. An iris code is computed in three steps. First, the pixels of the iris are located in the near-IR image of the eye by identifying the boundaries between the iris and the pupil on one hand and the iris and the limbus on the other hand. Second, areas where the view of the iris is obstructed by eyelids, eyelashes, or specular reflections are identified. Third, the 2048-bit iris code is constructed by demodulating the iris pattern with pairs of quadrature two-dimensional Gabor wavelets, extracting spatial phase information, but discarding amplitude information that depends on imaging contrast, illumination, and camera gain (Daugman 2003). Each bit of the iris code corresponds to a relative position in the iris defined by pseudopolar coordinates. The pseudopolar coordinates eliminate the dependency on the dilation and contraction of the pupil, but not on its rotation, which depends on head tilt, torsional eye rotation within its socket, camera angle, etc. An iris image is verified against an enrollment image by comparing an iris code computed from the enrollment image to seven iris codes computed from the verification image using seven angular origins for the pseudopolar coordinates. A modified Hamming distance is computed between the enrollment iris code and each verification iris code by counting the number of bits that differ between the codes, ignoring bits that belong to areas where the view of the iris is obstructed in either of the images. The smallest of the seven modified Hamming distances is used as a matching score.

A biometric cryptosystem for iris verification has been described (Hao et al. 2006). Since an iris code is a feature vector where order matters, it is based on the concept of a fuzzy commitment rather than a fuzzy vault. At enrollment time, a random biometric key K is generated; a commitment C to K is computed using a cryptographic hash function, $C = \text{hash}(K)$; a 2048-bit error correction codeword W is obtained by adding redundancy to K; an enrollment iris code E is computed from the enrollment iris image; and the helper data H are computed by x-oring E and W, $H = E \text{ xor } W$. At verification time, seven iris codes are generated from the verification image for seven angular origins of the polar coordinates, and each verification code V is xored with the helper data to compute $W' = V \text{ xor } H = V \text{ xor } (E \text{ xor } W) = (V \text{ xor } E) \text{ xor } W$. If the verification image is genuine and the angular origin of V results in good alignment, then the iris codes V and E differ only in a few bits, W and W' differ only in those same bits, and the error correction system may be able to recover W from W', compute K by removing the redundancy from W, and compute $C = \text{hash}(K)$. Verification succeeds if this process successfully produces C for one of the seven iris codes.

This verification method is not able to ignore the bits that correspond to areas where the view of the iris is obstructed in one of the iris images, because the enrollment image is not available at verification time. Hao et al. compensate for this by combining two error correction techniques, one for random errors and one for burst errors caused by the presence of such areas of obstruction.

While iris verification has a low FMR, it is only secure against an adversary who attempts to impersonate the subject if the adversary does not have an iris image of the subject or effective presentation attack detection is implemented. The detection of presentation attacks against iris verification is a nascent area of research (Raghavendra and Busch 2015, Czajka 2016).

2.5.3 *Face verification*

Following three decades of the steady development of traditional techniques in academia, the field of face verification has been disrupted in the past five years by two independent phenomena: the advent of deep learning and the appearance of commercial face verification systems. Deep learning systems and commercial systems have both furthered the state of the art of face verification, each in its own way, but both have been lacking in protection against presentation attacks.

Traditional techniques for face verification rely on methods for face detection such as Viola–Jones (Viola and Jones 2004) or Histograms Oriented Gradients (HOG) (Dalal and Triggs 2005) and a broad variety of methods for face matching. Face matching methods may be based on linear subspace analysis, where the facial images to be compared are projected onto a linear subspace determined by training images, and a similarity score is computed from the resulting coefficients (Yambor et al. 2002); or on comparing the texture of the facial images as described by Local Binary Patterns (LBP) (Ojala et al. 1996); or on identifying landmark points on each face, linking them into graphs that model the faces and comparing the models (Wiskott et al. 1997). Traditional techniques are challenged by pose, illumination, and expression variations. Three-dimensional (3D) imaging techniques have been proposed to help with those challenges, but some of them require complex imaging setups, such as using multiple cameras, which are not practical for most use cases.

Deep learning refers to machine learning using a deep neural network. Machine learning using artificial neural networks goes back to the 1950s, but its performance has improved spectacularly over the last few years for applications including face and speech recognition. These improvements are based on the use of deep neural networks (Zeiler and Fergus 2013, Goodfellow et al. 2016), composed of many layers of neurons (e.g., eight layers in Facebook's DeepFace or 22 layers in Google's FaceNet) and millions of parameters (120 million in DeepFace, 140 million in FaceNet) that are adjusted by training on millions of inputs. The training of such networks has been made possible by the use of arrays of graphics processing units (GPUs) with thousands of cores each. Deep neural networks are used by social networks and search engines for photo tagging, where they match or surpass human-level performance for facial image classification (Taigman et al. 2014, Lu and Tang 2015, Schroff et al. 2015). Photo tagging is an identification task rather than a verification task, but FaceNet is explicitly intended for use in both identification and verification.

While deep neural networks provide surprisingly accurate face recognition when used in a nonadversarial setting, they seem to be surprisingly weak when under attack. It is possible to compute quasi-imperceptible perturbations that cause natural images to be misclassified with high probability by state-of-the-art deep neural networks (Moosavi-Dezfooli et al. 2017). Independently, it has been shown that such a perturbation can be inconspicuously achieved with colored eyeglass frames and can be targeted for the impersonation of specific individuals (Sharif et al. 2016).

Face verification is now used commercially for many purposes, including identity proofing (Brinded 2016, Gamlin 2016), authentication (http://zoomlogin.com, http://keylemon.com), unlocking of phones (Mayhew 2016), and unlocking of password managers (http://biomids.com, http://1uapps.com). Most commercial face verification systems now use some form of presentation attack detection, but even advanced detection techniques were defeated by displaying a 3D model of a face reconstructed from photographs on a virtual reality system (Xu et al. 2016).

2.5.4 *Speaker verification*

In speaker verification, or voice verification, the biometric characteristic being measured consists of the acoustic features of the human voice. Speaker verification may be text dependent or text independent. Text-dependent verification has the advantage that, as noted earlier in Section 2.4.2.2, the text that is spoken may be a password, injecting secrecy into the biometric characteristic. Text-independent verification, on the other hand, has the advantage that the text to be spoken may be prompted by the verifier for challenge–response protection against replay attack.

Since the year 2000 and to this day, most techniques used for text-independent speaker verification have been based on the GMM-UBM verification method (Reynolds et al. 2000). In this method, a statistical GMM is constructed for a hypothesized speaker based on cepstral analysis of training speech, and a UBM is constructed from speech samples obtained from a large number of speakers representative of the expected range of possible speakers. At verification time, a likelihood ratio test is applied to the verification speech sample to compare how much better it fits the hypothesized speaker model than the UBM. The resulting statistic is then compared to a configured threshold to decide whether to accept or reject the sample as genuine.

Recently, deep neural networks have been very successful at speech recognition, and this has motivated attempts at using them for speaker recognition as well. Today, deep neural networks are being used in combination with GMM-UBM techniques to improve performance (Richardson et al. 2015).

Text-independent speaker verification is being used by financial institutions and call centers to authenticate speakers during phone calls, using previously recorded calls as training speech (Barclays 2016, Citigroup 2017, Pindrop 2017a). Text-dependent speaker verification is being used as well (OCBC 2016).

The speaker verification modality has two drawbacks. One is practical: the acoustics of human voice change with age much faster than a fingerprint or an iris code. Coping with this may require adjusting the statistical speaker model after each successful verification.

The other is a serious security vulnerability. Voice morphing may be used to change the sound of human voice in real time. This is done for fun or game playing, using widely available and inexpensive commercial software to make the voice sound as coming from a younger or older person or from a person of a different gender. However, it can also be done to mount an impersonation attack against a specific individual. This can be done by building a model of the victim's voice from a very limited number of speech samples using acoustic-to-articulatory inversion mapping, according to a study (Mukhopadhyay et al. 2015). In the study, voice morphing was used to attack several speaker verification systems, which were only able to reject the fake voices at a rate of 10–20%. Even humans had difficulty identifying fake voices, which they rejected at a rate of 50%. The authors assert that the voice morphing technique that they used can be used in real-time communications, but this was not part of the study. Whether victim-specific voice morphing can be done in real time is important, because if so, voice morphing could be used against speaker verification systems currently used by financial institutions and call centers. Anecdotal evidence that voice morphing attacks are already being carried out in the wild can be found in a call center fraud report that attributes a 113% fraud rate increase from 2015 to 2016 at least in part to "voice distortion software" (Pindrop 2017b).

2.5.5 *Other biometric modalities*

A variety of biometric characteristics are used by other biometric modalities:

- *Retinal scanning* is based on the pattern of blood vessels in the retina, observed using a beam of IR light that scans the retina as the subject looks into the scanner.
- *Eye vasculature biometrics* is based on the pattern of veins in the sclera (the white part of the eye).

- *Finger vein recognition* is based on the pattern of surface veins in the finger, imaged with IR light while the finger is inside a scanner.
- *Electrocardiogram biometrics* is based on the cardiac rhythm, which may be measured by an NFC-enabled wristband.
- *Behavioral biometrics* is based on a pattern of human activity detected by a collection of signals such as keystroke dynamics, mouse movements, movements of the hand that holds a smartphone, details of touchscreen gestures, etc.
- *Gait* as observed by a camera is not a practical biometric for human–computer interaction, but it becomes practical if observed by the accelerometer and gyroscope in a smartphone carried by the subject while walking. Gait may then be a component of a broader collection of behavioral biometric signals.

2.5.6 Biometric fusion

Biometric fusion refers to the observation of multiple biometric characteristics, resulting in multiple biometric samples, for biometric verification. There are many ways of combining the multiple samples to reach a decision as to whether they are genuine or not. The samples may be different instances of the same biometric modality, such as fingerprints from multiple fingers or images of both irises, or pertain to different modalities. They may be acquired by one sensor or multiple sensors. If biometric templates are used for matching, the samples may be processed together to produce a joint template or separately to produce multiple templates. The decision may be based on a joint matching score or on separate matching scores. If multiple matching scores are computed, they may be used together to reach the decision, or they may be separately compared to thresholds to reach separate decisions and then combine the decisions, for example, by requiring all the matching scores to exceed their thresholds or only some of them. The following are notable examples of biometric fusion systems:

- Veridium's (2017) product 4 Fingers Touchless ID combines four fingerprints captured together by a camera.
- Derakhshani (2012) describes a biometric cryptosystem that combines eye vasculature with micro features found in the tear duct and below the lower eyelid.

2.6 Conclusion

The preceding sections have hopefully shed light on the security and privacy challenges mentioned in the introduction and possible ways of addressing them.

The security challenge is the need to provide protection against presentation attacks. There are two aspects to this challenge. One is the need to spread awareness of the threat of presentation attacks among implementers and users of biometric verification systems. Progress has been made in that respect over the last few years, but work remains to be done. The other is the need to address presentation attacks that are particularly difficult to cope with.

One class of such attacks is illustrated by virtual reality and voice morphing attacks in Sections 2.5.3 and 2.5.4. In those attacks, the adversary digitally constructs an alternate reality where the verifier sees what it needs to see in order to accept the biometric evidence as genuine. If the subject's biometric characteristic is known to the adversary, as it is prudent to assume, the adversary may have all the information needed to construct a digital sample indistinguishable from a genuine one. If the alternate reality is constructed in real time, the adversary may also have all the information needed to respond to a challenge presented by the verifier. Protection against this class of attacks is an open area of research.

Another class of challenging attacks is illustrated by the universal perturbation and the eyeglass frame attacks in Section 2.5.3. Here the difficulty seems to come from a weakness in the current deep neural network technology. This calls for research in understanding the weakness and finding ways of eliminating or mitigating it. Such research is already under way.

In biometric verification, there will always be an arms race between verifiers and impersonators. Therefore, biometric verification should be used in combination with other methods of identity proofing and authentication, so that the emergence of an unforeseen method of attack is not catastrophic for verifiers.

As noted in the introduction, a privacy challenge arises from the linkability of a biometric characteristic across cyberspace and the physical world. When biometric matching is used for authentication, biometric information can be kept from the verifier by using the architectures in Figure 2.2 or 2.3. When biometric matching is used for identity proofing, a credential containing certified biometric data can be used to obviate the need for storing biometric information in a database vulnerable to hacking, as illustrated in Figure 2.5. If a database is used for identity proofing as illustrated in Figure 2.4, a biometric cryptosystem would make it possible to store in the database helper data that would reveal no useful biometric information to a hacker, instead of a traditional biometric template that leaks such data.

This suggests several areas of research that may lead to mitigations of the privacy challenge.

Digital credentials with certified biometric data protect biometric information against hackers by enabling the identity proofing architecture in Figure 2.5 and may further enhance privacy if they accommodate biometric cryptosystems and provide selective disclosure of attributes and selective presentation of verification factors, as rich credentials do (Lewison and Corella 2016). However, the widespread deployment of such credentials would require an ecosystem of issuers and verifiers who agree on standard protocols for issuance, presentation, and validation of the credentials. Research is needed into such protocols.

Biometric cryptosystems can be used to protect biometric information against hackers in the architecture in Figure 2.4, but little work has been done on biometric cryptosystems for face or voice verification. More research is needed in those areas.

Traditional biometric templates were once thought to hide biometric information, but, as noted in Section 2.3.4, this was shown not to be the case. That led to the development of biometric cryptosystems, where helper data are deemed not to leak biometric information. Research needs to be done in determining the biometric information that may be leaked to an adversary by the output of a deep neural network on a facial image, assuming that the adversary has access to the trained network.

References

Barclays. 2016. Barclays launches voice security technology to all customers. Press release. Barclays, London. http://www.newsroom.barclays.com/r/3383/barclays_launches_voice_security_technology_to_all_customers (accessed April 29, 2017).

J. Bort. 2014. The guy who just hacked Touch ID in the iPhone 6 says it's safe . . . for now. *Business Insider.* http://www.businessinsider.com/iphone-6-touchid-is-safe-for-now-2014-9 (accessed April 26, 2017).

L. Brinded. 2016. HSBC is letting customers verify their bank accounts like Airbnb does with selfies. *Business Insider.* http://www.businessinsider.com/hsbc-implements-selfie-security-verification-technology-for-banking-2016-9 (accessed April 28, 2017).

R. Cappelli, A. Lumini, D. Maio, and D. Maltoni. 2007. Fingerprint image reconstruction from standard templates. *IEEE Transactions on Pattern Analysis and Machine Intelligence*, 29(9):1489–1503.

Chaos Computer Club. 2013. Chaos Computer Club breaks Apple Touch ID. Chaos Computer Club, Germany. https://www.ccc.de/en/updates/2013/ccc-breaks-apple-touchid (accessed April 10, 2017).

Y. Cheung. 2017. HKBU scholar invents world's first "lip password." Hong Kong Baptist University News, Hongkong. http://hkbuenews.hkbu.edu.hk/?t=enews_details/1758&acm=50_726 (accessed April 20, 2017).

Citigroup. 2017. Citi tops 1 million mark for voice biometrics authentication for Asia Pacific consumer banking clients. Citigroup, New York. http://www.citigroup.com/citi/news/2017/170321b.htm (accessed April 29, 2017).

T. C. Clancy, N. Kiyavash, and D. J. Lin. 2003. Secure smartcard-based fingerprint authentication. In *Proceedings of the ACM SIGMM Multimedia, Biometrics Methods and Applications Workshop*, 45–52.

A. Czajka. 2016. Iris liveness detection by modeling dynamic pupil features. In *Handbook of Iris Recognition*, K. W. Bowyer and M. J. Burge (eds), 439–467. Springer, London.

N. Dalal and B. Triggs. 2005. Histograms of oriented gradients for human detection. In *Proceedings of the 2005 IEEE Computer Society Conference on Computer Vision and Pattern Recognition*, 886–893. IEEE Computer Society, Washington, DC.

J. Daugman. 1994. Biometric personal identification system based on iris analysis. US Patent 5,291,560. US Patent and Trademark Office, Alexandria, VA.

J. Daugman. 2003. The importance of being random: Statistical principles of iris recognition. *Pattern Recognition*, 36:279–291.

J. Daugman. 2006. Probing the Uniqueness and Randomness of IrisCodes: Results From 200 Billion Iris Pair Comparisons. *Proceedings of the IEEE*, 94(11):1927–1935.

R. Derakhshani. 2016. Biometric template security and key generation. US Patent 9,495,588. US Patent and Trademark Office, Alexandria, VA.

R. S. Dulude and C. Musgrave. 2001. Biometric certificates. US Patent 6,310,966. US Patent and Trademark Office, Alexandria, VA.

FIDO (Fast IDentity Online) Alliance. 2016. Fast IDentity Online (FIDO) Universal Authentication Framework (UAF). FIDO Alliance, Wakefield, MA. https://fidoalliance.org/download/ (accessed April 25, 2017).

R. Gamlin. 2016. Alabama's new weapon in the war against tax refund fraud? The selfie. *Birmingham Business Journal*, Birmingham, AL. http://www.bizjournals.com/birmingham/news/2016/05/27/alabamas-new-weapon-in-the-war-against-tax-refund.html (accessed April 25, 2017).

I. Goodfellow, Y. Bengio, and A. Courville. 2016. *Deep Learning.* MIT Press, Cambridge, MA.

F. Hao, R. Anderson, and J. Daugman. 2006. Combining crypto with biometrics effectively. *IEEE Transactions on Computers*, 55(9):1081–1088.

International Civil Aviation Organization. 2015. Doc 9303—Machine readable travel documents, seventh edition. International Civil Aviation Organization, Montreal, QB http://www.icao.int/publications/pages/publication.aspx?docnum=9303 (accessed April 25, 2017).

ISO (International Organization for Standardization)/IEC (International Electrotechnical Commission). 2011. ISO/IEC 24745: Information technology—Security techniques—Biometric information protection. ISO, Geneva; IEC, Geneva.

A. Juels and M. Sudan. 2006. A fuzzy vault scheme. *Designs, Codes and Cryptography*, 38:237–257.

A. Juels and M. Wattenberg. 1999. A fuzzy commitment scheme. In *Proceedings of the 6th ACM Conference on Computer and Communications Security*, 28–36. ACM, New York.

S. Khandelwal. 2014. Hacker clones German defense minister's fingerprint using just her photos. *The Hacker News.* http://thehackernews.com/2014/12/hacker-clone-fingerprint-scanner.html (accessed April 18, 2017).

K. Kollreider, H. Fronthaler, M. I. Faraj, and J. Bigun. 2007. Real-time face detection and motion analysis with application in "liveness" assessment, *IEEE Transactions on Information Forensics and Security*, 2(3–2):548–558.

K. Lewison and F. Corella. 2016. Rich credentials for remote identity proofing. Pomcor, Carmichael, CA. https://pomcor.com/techreports/RichCredentials.pdf (accessed April 25, 2017).

C. Li and J. Hu. 2016. A security-enhanced alignment-free fuzzy vault-based fingerprint cryptosystem using pair-polar minutiae structures. *IEEE Transactions on Information Forensics and Security*, 11(3):543–555.

C. Lu and K. Tang. 2015. Surpassing human-level face verification performance on LFW with GaussianFace. *AAAI Conference on Artificial Intelligence, North America*. https://www.aaai.org/ocs/index.php/AAAI/AAAI15/paper/view/9845 (accessed April 23, 2017).

Mastercard. 2017. Thumbs up: Mastercard unveils next generation biometric card. Johannesburg and Purchase, New York. https://newsroom.mastercard.com/press-releases/thumbs-up-mastercard-unveils-next-generation-biometric-card/ (accessed April 24, 2017).

S. Mayhew. 2016. Princeton Identity to license its patented iris recognition technology to Samsung. Biometrics Research Group, Inc, Toronto. http://www.biometricupdate.com/201609/princeton-identity-to-license-its-patented-iris-recognition-technology-to-samsung (accessed April 2017).

F. Monrose, M. K. Reiter, and S. Wetzel. 1999. Password hardening based on keystroke dynamics. In *Proceedings of the 6th ACM Conference on Computer and Communications Security*, 73–82. ACM, New York.

S.-M. Moosavi-Dezfooli, A. Fawzi, O. Fawzi, and P. Frossard. 2017. Universal adversarial perturbations. In *IEEE Conference on Computer Vision and Pattern Recognition*, 86–94. Available at https://arxiv.org/pdf/1610.08401.pdf.

D. Mukhopadhyay, M. Shirvanian, and N. Saxena. 2015. All your voices are belong to us: Stealing voices to fool humans and machines. In *Computer Security—ESORICS 2015*, G. Pernul, Y. A. P. Ryan, and E. Weippl (eds). Springer International Publishing, Cham.

NIST (National Institute of Standards and Technology). 2013. Federal Information Processing Standard (FIPS) 201-2. NIST, Gaithersburg, MD. http://csrc.nist.gov/publications/PubsFIPS.html (accessed April 25, 2017).

NIST. 2016. Special Publication (SP) 800-90A-C. 2016. NIST, Gaithersburg, MD. http://csrc.nist.gov/publications/PubsSPs.html#SP%20800 (accessed April 24, 2017).

S. Novoselov, T. Pekhovsky, A. Shulipa, and A. Sholokhov. 2014. Text-dependent GMM-JFA system for password based speaker verification. In *IEEE International Conference on Acoustics, Speech and Signal Processing*, 729–737.

OCBC (Oversea-Chinese Banking Corporation). 2016. OCBC Bank is first in Singapore to use voice recognition technology to enhance customer experience. Press release. OCBC, Singapore https://www.ocbc.com/assets/pdf/media/2016/may/ocbc%20media%20release%20-%20ocbc%20rolls%20out%20voice%20recognition.pdf (accessed April 29 2017).

T. Ojala, M. Pietikäinen, and D. Harwood. 1996. A comparative study of texture measures with classification based on feature distributions. *Pattern Recognition*, 29:51–59.

I. Pavlidis and P. Symosek. 2000. The imaging issue in an automatic face/disguise detection system. In *Proceedings of the IEEE Workshop on Computer Vision Beyond the Visible Spectrum: Methods and Applications*, 15–24. IEEE Computer Society, Washington, DC.

Pindrop. 2017a. Phoneprinting technology explained—How the largest global contact centers stop fraud and protect customers. Pindrop. https://www.pindrop.com/phoneprinting-webinar/ (accessed April 29, 2017).

Pindrop. 2017b. Call center fraud report. Pindrop. https://www.pindrop.com/wp-content/uploads/2017/04/Fraud-Report-Global-4-24-17-FINAL.pdf (accessed April 29, 2017).

R. Raghavendra and C. Busch. 2015. Robust scheme for iris presentation attack detection using multiscale binarized statistical image features. *IEEE Transactions on Information Forensics and Security* 10(4):703–715.

C. Rathgeb and A. Uhl. 2011. A survey on biometric cryptosystems and cancelable biometrics. *EURASIP Journal on Information Security* 2011:3.

D. A. Reynolds, T. F. Quatieri, and R. B. Dunn. 2000. Speaker verification using adapted Gaussian mixture models. *Digital Signal Processing*, 10(1):19–41.

M. Rogers. 2013. Why I hacked Apple's TouchID, and still think it is awesome. https://blog.lookout.com/blog/2013/09/23/why-i-hacked-apples-touchid-and-still-think-it-is-awesome/ (accessed April 26, 2017).

F. Richardson, D. Reynolds, and N. Dehak. 2015. Deep neural network approaches to speaker and language recognition. *IEEE Signal Processing Letters*, 22(10):1671–1675.

A. Ross, J. Shah, and A. K. Jain. 2007. From template to image: Reconstructing fingerprints from minutiae points. *IEEE Transactions on Pattern Analysis and Machine Intelligence*, 29(4):544–560.

A. Roy, R. Memon, and A. Ross. 2017. MasterPrint: Exploring the vulnerability of partial fingerprint-based authentication systems. *IEEE Transactions on Information Forensics and Security*, 12(9): 2013–2025.

K. V. Sakariya and T. Nauta. 2017. Interactive display panel with IR diodes. US Patent 9,570,002. US Patent and Trademark Office, Alexandria, VA.

F. Schroff, D. Kalenichenko, and J. Philbin. 2015. FaceNet: A unified embedding for face recognition and clustering. In *Proceedings of the IEEE Conference on Computer Vision and Pattern Recognition*, 815–823.

S. Schuckers and P. Johnson. 2014. Fingerprint pore analysis for liveness detection. US patent application 14/243,420. US Patent and Trademark Office, Alexandria, VA.

M. Sharif, S. Bhagavatula, L. Bauer, and M. K. Reiter. 2016. Accessorize to a Crime: Real and stealthy attacks on state-of-the-art face recognition. In *Proceedings of the 2016 ACM SIGSAC Conference on Computer and Communications Security*, 1528–1540. ACM New York.

D. Storm. 2014. Researchers spoof fingerprint, bypass Samsung Galaxy S5 security, access PayPal. Computerworld, IDG Communications Inc, Boston. http://www.computerworld.com/article/2476130/cybercrime-hacking/researchers-spoof-fingerprint—bypass-samsung-galaxy-s5-security—access-paypal.html (accessed April 26, 2017).

Y. Taigman, M. Yang, M. A. Ranzato, and L. Wolf. 2014. DeepFace: Closing the gap to human-level performance in face verification. In *Proceedings of the IEEE Conference on Computer Vision and Pattern Recognition*, 1701–1708. IEEE Computer Society, Washington, DC.

U. Uludag and A. K. Jain. 2006. Securing fingerprint template: Fuzzy vault with helper data. In *Proceedings of the IEEE Workshop on Privacy Research in Vision*, 163–163. Full paper available at http://biometrics.cse.msu.edu/Publications/SecureBiometrics/UludagJain_FFVHelper_PRIV06.pdf.

Veridium. 2017. https://info.veridiumid.com/hubfs/Content/Datasheets/datasheet-4-fingers-touchless-id.pdf (accessed April 29, 2017).

University of Massachusetts. 2017. Labeled Faces in the Wild. http://vis-www.cs.umass.edu/lfw/ (accessed April 28, 2017).

P. Viola and M. Jones. 2004. Robust real-time object detection. *International Journal of Computer Vision*, 57(2):137–154.

Y. Wencheng, J. Hu, S. Wang, and M. Stojmenovic. 2014. An alignment-free fingerprint bio-cryptosystem based on modified Voronoi neighbor structures. *Pattern Recognition*, 47(3):1309–1320.

L. Wiskott, J.-M. Fellous, N. Krüger, and C. von der Malsburg. 1997. Face Recognition by Elastic Bunch Graph Matching. *IEEE Transactions on Pattern Analysis and Machine Intelligence*, 19(7):775–779.

Y. Xu, T. Price, J.-M. Frahm, and F. Monrose. 2016. Virtual U: Defeating face liveness Detection by building virtual models from your public photos. In *Proceedings of the 25th USENIX Security Symposium*, 497–512.

W. S. Yambor, B. A. Draper and J. R. Beveridge. 2002. Analyzing PCA-based face recognition algorithm: Eigenvector' selection and distance measures. In *Empirical Evaluation Methods in Computer Vision*, ed. H. I. Christensen, P. J. Phillips, 39–60. World Scientific.

M. D. Zeiler and R. Fergus. 2013. Visualizing and understanding convolutional networks. In *Proceedings of the 13th European Conference on Computer Vision*, 818–833. Springer International Publishing, Cham.

chapter three

Machine identities

Foundational to cybersecurity

Hari Nair

Contents

3.1 Trust in the digital world

The *Oxford Dictionary* defines *trust* as the "firm belief in the reliability, truth, ability, or strength of someone or something." A wonderful article on ChangingMinds.org offers multiple interpretations for trust. When considering trust in the digital world, one in particular is especially relevant:

> Trust means making an exchange with someone when you do not have full knowledge about them, their intent and the things they are offering to you. (Trust—changingminds 2017)

In the physical world, trust is established based on identity or context, built on familiarity (the frequency of our interactions), and ultimately dependent on experience. Just as importantly, trust is nuanced: We do not trust everyone equally. For example, we generally find that people trust friends and family more than neighbors or casual acquaintances. There are exceptions, of course, but in general, the notion of trust in an entity is built over time and dependent on the frequency and the nature of our interactions with that entity.

Yet frequently, we must trust people that we do not see often, such as doctors, mechanics, and tax accountants. What gives us the confidence to depend on these people? Specifically, how do we know that a doctor is a doctor? The fact that we can "see" that the doctor is accredited or affiliated with a well-known hospital certainly helps, as do physical attributes such as the doctor's office, location, and the notion of reviews by other people.

Now let us contrast how we build trust in the physical world with how we establish trust in the digital world, where we cannot see anything. For example, I connect to my bank, my e-mail provider, and a variety of e-commerce sites—each of which requires me to provide personally identifiable information (PII) and, in some cases, credit card data. I can identify the websites I frequently visit based on the logo, colors, and layout, but attacks such as phishing have long since rendered my ability to recognize the "look and feel" of an (online) entity practically useless. Without a tangible identity, there is no way I can build familiarity and, hence, trust.

How then do I know I am connecting to the online service provider that I want to use? This is where digital keys and certificates have such an important role to play. A certificate, much like a credit card or a passport, is issued by a "trusted" authority (an enterprise or a financial or government institution in the real world) and has an associated validity and purpose.

The similarities end there, however. Although we can "view" certificates, the attributes that make them unique (and, hence, linked irrevocably to a physical entity) can only be "verified" by applications such as a web browser or an e-mail client. Theoretically, it should then be possible for an application to identify and, over time, trust an entity, should it not?

Not so fast. There are a couple of reasons that this is not practical yet:

- *Unlike a physical attribute (such as a face, voice, or fingerprint), a digital attribute is inherently transient in nature.*
  The digital "key" that serves as the unique attribute to identify an online entity is valid for a specified period of time and must then be replaced. Periodically replacing a digital key mitigates the risk of the key being duplicated. The more a key is used, the greater the chance it will be compromised. As such, best practices recommend the periodic regeneration of keys. The familiarity with a specific key (sometimes referred to as "certificate pinning") is not particularly useful—especially when the keys themselves get replaced as often as every 90 days.

- *Unlike the physical world, on the Internet, the concept of trust is not as tangible.* We cannot establish identity based on sight, and digital keys—the mechanism designed to verify identity and establish trust—are frequently updated, rendering familiarity impractical. Trust must be established every time and cannot be based on frequent interactions.

3.2 *Human and machine identities*

As the lines between the physical and the digital world blur, the role of identity in establishing trust that translates across both worlds becomes critical. As we look at the interactions in our daily lives, we can see the increasing dependence on smart "things" to enable us in our functions and responsibilities—phones, cars, homes, utilities, health, finance, and education. As the famed futurist Larry Kurzweil notes, in the next 10–15 years, the following will take place:

- Self-driving cars will start to take over the roads.
- "Nanobots" will become smarter than current medical technology.
- The Turing test will start to become passable, rendering humans incapable of distinguishing between human and machine.

One thing that will not change, however, is the necessity to identify (and eventually, trust) who we interact with, regardless if they are humans or machines. The notion of human identity, even in the digital world, is relatively well understood—whether it is for e-mail, social networking, or logging into travel portals, we normally identify ourselves using a "username" (typically an e-mail address) and a "password." We normally set our own passwords and end up using some form of mnemonic device that is easy enough for us to remember, instead of having to write it down. However, since these passwords are generated by humans, they are inherently easy to "crack" as well—today, there are automated password-cracking tools that can run more than a million variations of a password in about 24 hours. As such, we are very sensitive to password security:

- We have encouraged users to move from passwords to longer "passphrases."
- Regulations such as the Sarbanes–Oxley act have required organizations of all types to force users to not just use long and strong passwords but also change them frequently.
- There is a plethora of "password manager" tools that users can use to store, change, and use passwords.

Organizations such as National Institute of Standards and Technology have mapped out the different types of identities through documents such as SP 800-63 and the risk/assurance they provide in establishing trust online (SP800-63 2017). Passwords/passphrases are generally regarded as having the lowest level of "assurance." On the other end of the scale are cryptographic keys, generally not only associated with high value systems—think of these as "machine identities"—but also used by security-focused organizations such as the government and the military to protect humans through the use of devices such as smart cards.

Unlike humans, machines have the capability to generate long and complex passwords to identify themselves, both to humans and to other machines. Machines, however, do not (yet) have the cognitive capability to *remember* passwords without having to store them—power down a machine, and it loses its ability to recollect the password/passphrase it is required to use for its interactions. It is for these reasons that cryptographic keys have been widely adopted and accepted as the best practice to identify machines. Cryptographic keys are generally much longer and stronger than passwords, rendering them comparatively immune to cracking attempts.

While we as an industry have spent billions of dollars trying to protect passwords, there has been very little thought and effort put into the security of machine identities. This gap becomes all the more glaring when we consider the rate at which encryption has become ubiquitous, while the pace of innovation reaches dizzying heights. In 2017, the volume of encrypted Internet traffic surpassed traffic that is unencrypted, hitting a seminal milestone that signals the increase in the usage of keys and certificates (Encrypted Web 2017). This will only continue to grow as browsers such as Google Chrome have started to penalize web sites that are *not* encrypted by default. In addition, as organizations start to embrace the cloud and development and operations (DevOps) practices, infrastructures will become a lot more ephemeral, necessitating more identities to be generated *and trusted* for short periods. Indeed, banks such as Monzo (https://monzo.com) have shown the way to become viable financial institutions while maintaining literally no physical presence.

On a related, but different note, every connected device in the Internet of things (IoT) space will need keys to identify itself. Autonomous vehicles, such as self-driving cars, will have millions of certificates. We expect that *34 billion* IoT devices will be connected by 2020, and considering that these devices will all need identities in order to be able to communicate and trust each other, the implications of the protection of machine identities assume truly staggering proportions (IOT stats 2015).

3.3 Keys and certificates: Foundations of cybersecurity

All organizations rely on cryptographic keys and digital certificates to secure their business. These software devices were designed to solve the original Internet security problem—accurately identifying servers and browsers so that they could safely communicate back and forth independently. Machines rely exclusively on keys and certificates to know what to trust and what not to trust in our digital world. Any time data are being transferred, whether it be your business or personal information, there is a key or certificate that is being used to protect it. If the communication channel is not trusted, your data are not secure.

Secure sockets layer (SSL) was first introduced in the 1990s by Netscape to protect digital communication just as the Internet, and e-commerce, was taking off and was implemented in their Navigator browser. It served as the de facto standard for encryption for a number of years and has been updated over the years (the last, SSLv3 was only deprecated in 2015) to address vulnerabilities and account for stronger security standards. It is for this reason that cryptographic protocols are often referred to as "SSL."

At a high level—there are two primary benefits that keys and certificates afford in the context of digital interactions:

1. *They identify (and potentially authenticate) the participants* in a transaction (depending on the nature of the transaction, the participants may be referred to as clients, servers, or peers).
2. *They protect the data* that get transferred between the participants.

There are a number of common applications of keys and certificates that readers may be familiar with.

- *Identity*—This enables users to authenticate themselves to access sensitive content. This is typically limited to government and military personnel, typically using smart cards initiatives such as the common access card or the personal identity verification program. The higher security associated with these credentials has also made it attractive for e-government applications. Countries such as Canada (using concepts such as *meaningless but unique numbers* or MBUN), Estonia, and New Zealand, among others, have rolled out government-to-citizen programs built on the bedrock of security that cryptographic keys and certificates provide (MBUN 2017). Elsewhere, countries around the world have implemented "national identification" initiatives to enable access to electronic voting, utility, transportation, and medical programs.
- *Document signing*—European entities, both government and business, have adopted standards such as XML DSig, XAdES, PAdES, and CAdES to standardize the document signing process and allow for interoperability within and between organizations. Tax returns are a common application of document signing, and countries around the world have standardized on keys and certificates as a means to not only digitize the process, but also secure it from the perspective of both nonrepudiation (guaranteeing the source and integrity of the content) and auditability. While these practices are being adopted in the United States as well, we have a fair way to go before we catch up with our European counterparts.
- *Code signing*—The notion of signing an application code to prove its integrity and trustworthiness has assumed particular significance in the last 20 years because of the explosive growth of malicious software (commonly referred to as "malware"). To date, there have been more than 120 *million* instances of known malicious software, a number that will continue to grow exponentially as a direct result of the adoption of mobile applications, devices, and IoT. Partly in response to this threat, operating system vendors, including Microsoft, Apple, and Google flag unsigned code, requiring users to consciously acknowledge and accept the risks of running unsigned code. Sadly, though, the practice of requiring and then validating signed code is not as prevalent. The WannaCry malware exploit in 2017 had a particularly catastrophic impact on more than 100 countries around the world and could have been thwarted if strong code validation processes were in place. Controls are much stricter for mobile operating systems, where the plethora of mobile application developers (as of March 2017, there were more than five *million* applications on Apple iOS and Google Android) has increased the risk of infecting devices that are increasingly used for personal and business transactions (App statistics 2017). Apple, for example, insists on self-signing all code that can run on its mobile devices.
- *Encryption*—In addition to guaranteeing the "integrity" of the date, cryptographic keys and certificates can also be used to protect data from being accessed by malicious entities. This is true for data at rest (data being stored on user, server, and cloud systems) *and* data in motion (data being transferred from one location to another). While the technology that is used to protect the data in each case is slightly different, cryptographic keys are used in both cases to minimize the risk of data being compromised. For data at rest, encrypted storage/backups are particularly important today, when sensitive data can be accessed on devices that are inherently insecure—most organizations have a "bring your own device" initiative to allow users to access data on their personal devices, including mobile phone and tablets. Unfettered access to their devices or, worse yet, loss of these devices puts user's personal and professional

data at risk, and encryption provides a necessary safeguard in the worst-case scenario of a device getting in the hands of a malicious entity.

For data in motion, the process is relatively simpler—most common data transfer applications (browsers, file, and e-mail transfer clients) have encryption controls built into them, and the standards that govern the implementation of these controls are fairly mature and robust.

Another common application of encryption is to protect e-mail communications. While the advent of text messaging and social networks has given users other options, e-mails remain the most commonly used means for communication for business and personal use. There are many sensitive data that are typically accessible in users' e-mail communications, which makes e-mail providers a prime target for malicious actors. While the data themselves are stored and transmitted securely between e-mail senders and recipients, it is possible to intercept this traffic while it is in transit, and this is where encryption comes in handy to provide an additional layer of security. Essentially, e-mail traffic is protected such that even if the communication channel (sometimes referred to as a tunnel) is compromised, only the recipient(s) have the capability to read the actual contents of the e-mail. While e-mail encryption is essentially a user-specific task and as such is challenging to roll out to individuals (training users to implement e-mail encryption is an onerous process, especially for nontechnical users), organizations have frequently deployed encryption at the enterprise gateway in an attempt to minimize the risk of compromise when data are in transit between the sender and the recipient.

- *Network/Wi-Fi access*—Accessing the Internet, whether over wired or wireless networks, within and outside the enterprise, is another critical application of digital keys and certificates, independent of whether the access happens at work, at home, or in public spots. While the process is essentially transparent to the user, in the background, both the client device (desktops, laptops, tablets, and phones) and the network essentially need to authenticate each other before they attempt to provide or use access. Kerberos, remote authentication dial-in user service, Network Access Control, and 802.11 are all standards related to secure network access that are built on a foundation of keys and certificates.

In summary, keys and certificates are in play any time you are moving data via SSL/transport layer security (TLS); secure shell (SSH); and mobile, cloud, and IoT connections. While these are common applications of keys and certificates, in truth, *any* application that is used today, whether it is for personal, professional, or public access, needs to be secured, and cryptography is the foundation of cybersecurity. At the very minimum, every interface of applications, whether it be for administrative interfaces [for example, information technology (IT) administrators configuring their web application firewalls] or user interfaces (citizens checking their private e-mail), needs to be protected using digital keys and certificates. By extension, a compromise of the underlying cryptolayer can end up compromising the entire security fabric of the organization's application.

3.4 Encryption fundamentals

Now that we have identified some of the reasons why keys and certificates are foundational to security for digital transactions, let us consider some of the concepts that will be useful to become familiar with. Public key cryptography (sometimes referred to as asymmetric cryptography) uses pairs of keys, which are derived from two random long prime numbers. The *private* key is, as the name indicates, meant to keep secret and is only known to the owner of the key. The *public* key, on the other hand, is meant to be shared with relying entities in digital transactions. Typically, the public key is shared in a standard format—a digital certificate—that provides data about the key owner and the intended usage for the key pair (which includes the private key). The private and public keys are cryptographically related by the fact that operations performed by one of the keys can be verified by, and only by, the other key. Which key is used depends on the function being performed.

For public key *encryption,* the sender of the information encrypts data to be transmitted with the public key of the recipient (that is accessible using the recipient's digital certificate). Once encrypted, only someone with access to the private key (which should be the recipient—hence the reason to keep it private) can decrypt data that were sent. This ensures the *confidentiality* of data for the intended recipient—all the sender needs is the recipient's certificate, which is shareable.

For public key *signatures,* the sender of the information attaches a hash of the data that are being sent, encrypts this hash with their private key, and appends it to the data. To verify the integrity of the data (that is, to ensure that what was sent by the sender has not been modified in transit by a malicious actor or corrupted by other means), the recipient decrypts the hash using the public key from the sender's certificate and compares it a hash of raw data that the recipient generates on their own. If the two hashes are identical, it proves that only someone with access to the private key (which should be the sender) could have sent the message. This ensures both the origin of the data and its *integrity.*

Figures 3.1 and 3.2 serve to illustrate the two processes:

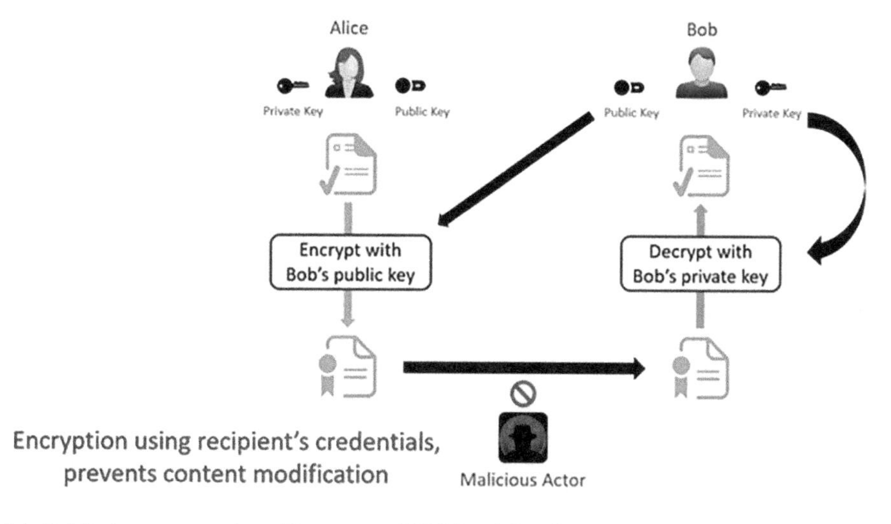

Figure 3.1 Public key encryption. (Courtesy of US Naval Academy, Annapolis, MD, http://usna.edu.)

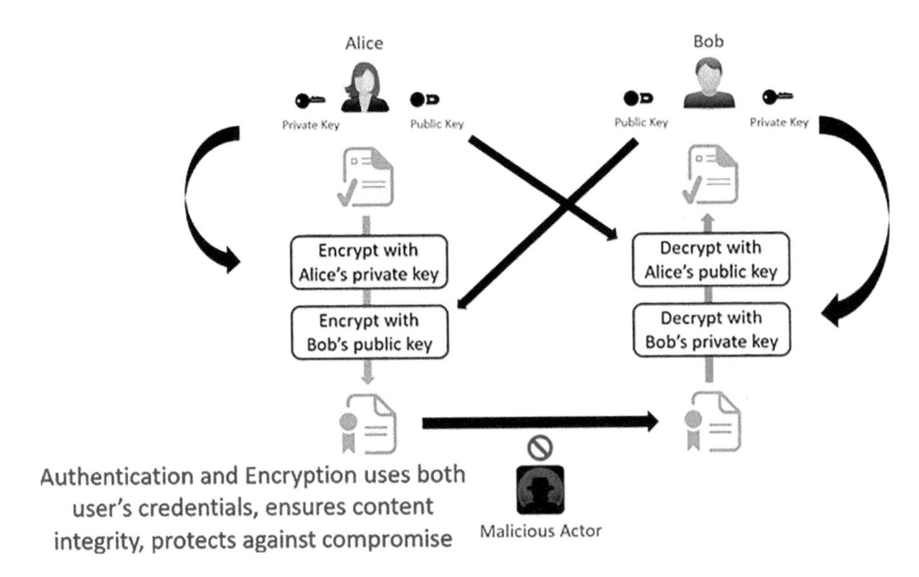

Figure 3.2 Public key signatures and encryption. (Courtesy of US Naval Academy, Annapolis, MD, http://usna.edu.)

3.4.1 *Key and certificate properties*

Now that we have talked about how keys and certificates can be used to guarantee the integrity and confidentiality of data that are transmitted electronically, let us look into different types of keys and certificates that are commonly used.

Keys are used to encrypt data, and there are different types of algorithms that are used to generate these keys. The most commonly used key algorithm is RSA, originally pioneered by Ron Rivest, Adi Shamir, and Leonard Adleman in 1977, out of the Massachusetts Institute of Technology. While RSA keys are still the de facto standard, there are a number of other algorithms that have also been used over the years: Digital signature algorithm and elliptic curve cryptography (ECC). Of these, ECC offers the advantage of smaller key sizes (for the same amount of computational complexity) that makes it ideal for resource-constrained environments such as IoT devices.

Certificates are vehicles that make it possible to share the public key component of an entity's key pair (Figure 3.3). X509 is a standard that is widely deployed and understood when it comes to defining the format of digital certificates. A typical certificate, much like a physical passport, has a number of properties that can be used to identify the owner of the key pair and its intended usage. Some of these are as follows:

- *Validity*: governs when a certificate was issued and when it expires to ensure that keys are periodically regenerated (much like passwords) to ensure that they do not become susceptible to cracking attempts
- *Subject distinguished name*: who the certificate was issued to—this could be a human or a machine identity
- *Issuer distinguished name*: who issued the certificate—depending on whether the issuer is a known or unknown entity; this can be used to determine the level of trust placed in the owner of the certificate

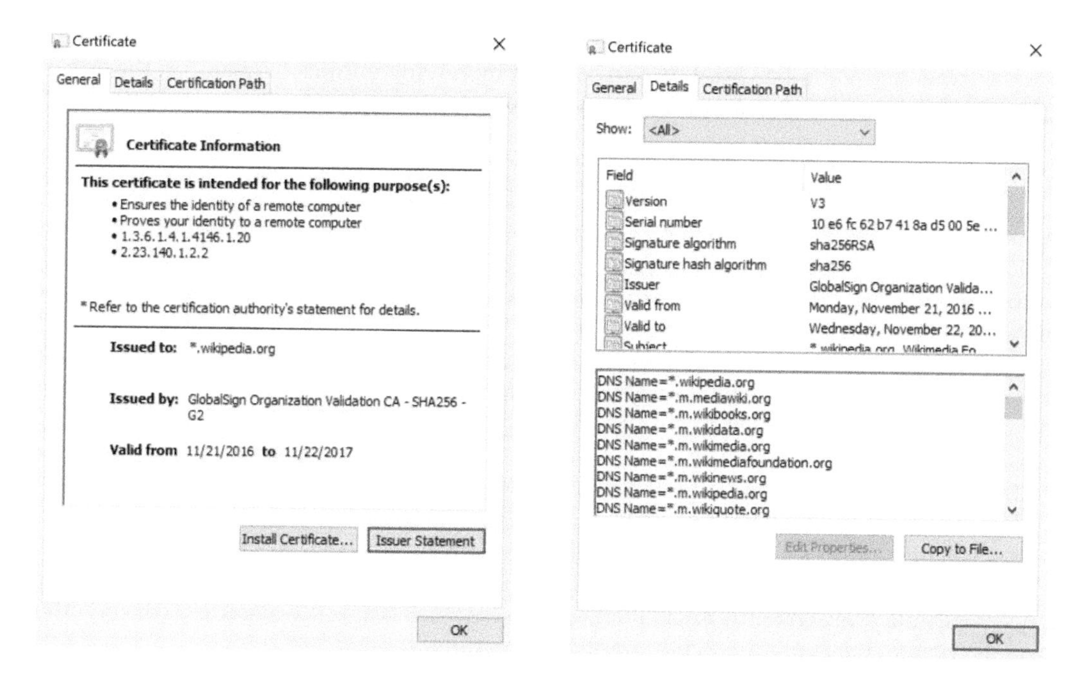

Figure 3.3 Digital certificate properties.

- *Key usage and extended key usage*: controls what the certificate can be used for (to authenticate digital identities, to encrypt messages, for smart card authentication, etc.)
- *Public key*: records the public key part of the entity's key pair that can be used to send encrypted messages to the entity

3.4.2 *Public key infrastructure*

Public key infrastructure (PKI) is used to refer to the ecosystem that controls the issuance, storage, and distribution of digital certificates and includes the following components:

- *Certification authority (CA)*—This issues digital certificates. CAs can be public (trusted by anyone on the Internet) or private [trusted only by specific organization(s) for the purposes of internal transactions] and are the root of trust.
- *Registration authority (RA)*—This is responsible for the verification of identities prior to the issuance of certificates.
- *Certificate database*—This maintains a record of certificates that have been issued or revoked for audit purposes.
- *Key escrow/archival server*—This is used to store copies of private keys corresponding to entities to audit/inspect communications between human and machine entities or for disaster recovery purposes.
- *Certificate management system*—This uses centrally defined policies that govern the issuance, distribution, and life cycle management of certificates.
- *Certificate revocation lists (CRLs)*—Certificates that have been issued but do not need to be trusted any longer (for a variety of reasons such as key compromise and entities

that have left the organization) are revoked by the issuing authority (CA) and put on a "blacklist" called a CRL that can be used by relying entities to check on the status of known/unknown parties in a transaction.

3.4.3 *Cryptographic protocols (SSL and TLS)*

As described earlier, public key cryptography provides two distinct benefits—authentication and encryption. Cryptographic protocols utilize both of these benefits to secure communications over computer networks. In any such interaction, there is a "client" (for example, a browser) that initiates the transaction to a "server" (say, a website). To secure the data that are being transmitted (say, credit card information in an e-commerce transaction), and to ensure that the data are being sent to the right server (for example, a retail website), an elaborate exchange takes place between the two parties in the transaction:

1. The client initiates the transaction by sending a "Client Hello."
2. The server responds with a "Server Hello" and its (public key) certificate.
3. The client creates a (symmetric) session key.
4. The client encrypts the session key with the public key extracted from the server certificate and sends the encrypted session key to server—this ensures that only the server can decrypt and access the client generated session key.
5. The server decrypts the session key.
6. The session key is used for the remainder of the SSL session.

We introduced SSL earlier in this section. TLS was proposed as a stronger alternative to SSL in 1999 and is now the required by most modern applications for encryption. Like with SSL, TLS has gone through multiple versions with TLS 1.3 being the latest version of the protocol.

While SSL/TLS are most often associated with websites, where the client application is typically a web browser, these protocols can also be used by other applications such as e-mail, instant messaging, voice over Internet protocol (VoIP) and printing/faxing.

3.4.4 *Symmetric keys*

We introduced the notion of "symmetric keys" as part of the discussion on cryptographic protocols in the previous section. Unlike asymmetric key encryption where there are separate public and private keys, symmetric key algorithms use the same keys for both encryption and decryption. Symmetric keys are preferred over asymmetric keys as they offer better encryption performance, yet have the requirement of both parties in a transaction needing access to the (same) symmetric key. It is for this reason that symmetric keys are typically used to secure data at rest—disk encryption, file encryption, database encryption, etc. To secure data in motion, symmetric key encryption is used (to encrypt the data) in conjunction with asymmetric key pairs (to authenticate the participants and encrypt the symmetric key used for a session).

3.4.5 *SSH*

SSH is another example of a cryptographic protocol that is best known for accessing and administering remote servers, mostly for non-Windows platforms, although Microsoft has recently added support for SSH in Windows 2016. SSH can be used to establish secure communication channels over unsecured networks, typically for remote login scenarios.

If SSL/TLS is how applications interact with one another, SSH is how users can administer these applications. SSH can also be used by machines to move data between machines in different networks and hence serves as an identification *and* access protocol. While SSH also depends on asymmetric key cryptography, there are important differences—there is not a notion of certificates with SSH, only keys. As such, there is no identifying meta-data about SSH keys, beyond where they are located, that can be used to identify the owner of the SSH key pair (be it a human or a machine identity). There is also no validity associated with an SSH key pair—SSH keys can live forever. Most importantly, SSH keys are typically self-generated—there is no notion of an issuing authority such as a CA for certificates. This last distinction becomes important in the context of how trust is established for SSH sessions, a concept that will be described in detail in the next section.

3.5 Encryption key and certificate risks

There are a number of risks associated with encryption keys and certificates:

- Downtime and system outages—Certificates that are not renewed and replaced before they expire can cause serious downtime and outages.
- Key compromise—Direct access to keys by administrators, weak access controls, administrative turnover, and inability to regularly replace keys.
- Compliance violations—Auditors are increasingly scrutinizing encryption key management practices to ensure that there are policies that govern their issuance, storage, and distribution.
- Data loss—Losing access to all copies of a key such that data that were encrypted by it cannot be decrypted.
- High administrative costs—Certificates and private keys require four hours per year to manage on average. SSH keys must be continuously tracked and managed on all systems. Additionally, because of the lack of identifying information about SSH keys, keys that have not been used in some predetermined period must be removed. Conversely, a baseline must be maintained of all keys and certificates so that anomalies can be identified.
- CA compromise—Certificate authorities are attractive targets for hackers. The compromise of a CA enables attackers to conduct much broader attacks, a number of instances of which have been described in Section 3.6.

3.5.1 Trust models

CAs, as described earlier, can be public or private. The list of publicly trusted CAs is maintained and managed by the CA Browser (CAB) forum (https://cabforum.org/) and is dependent on the community to control which CAs are trusted by default on various computer systems. With private CAs, it is typically up to the enterprise that owns and operates the CA to govern where they are trusted. Regardless of public or private CAs, one of the biggest advantages of a PKI is the ability to *scale* trust within and outside the enterprise. There are essentially two trust models:

- *Direct trust model*—The public key certificate of the entity is directly trusted by the relying party. Any changes to the certificate (upon renewal, reissuance, etc.) will require that trust be reestablished, *manually*. SSH is an example of a direct trust model, where every SSH (public) needs to be explicitly trusted in order to gain access with the private key.

- *Derived or delegated trust model*—The issuing authority (CA) is what is trusted by the relying party—in essence, *any* (server or client) certificate that chains up under the trusted CA is considered trustworthy. This allows for certificate reissuance or new identities to be established without having to redefine the trust relationship (as long as the issuing CA continues to operate within its defined parameters). Trust is established at the CA level and *inherited* by any entity that chains up under that CA. This is why digital certificates are so essential to managing trust within and outside the enterprise. Assuming the list of trusted CAs is secured and controlled, new certificates or identities can be established seamlessly allowing the system to scale, effectively infinitely.

3.5.2 Attack vectors

Keeping the aforementioned trust models in mind, there are a few well-understood attack vectors that malicious actors have tried to exploit to circumvent the security afforded by the PKI trust models:

- CA compromise and fraudulent certificate scenarios
  - Impersonation: Trick RA into issuing a fraudulent certificate.
  - RA compromise: Infiltrate RA or steal credentials and authorize fraudulent certificates.
  - CA system compromise: Malware or other infiltration used to get fraudulent certificate signed by a trusted CA (without getting copy of the CA's private key).
  - CA key theft: Stolen or derived copy of the CA private key is used to issue fraudulent certificates.
  - Root CA compromise: Issue fraudulent certificates from the root CA (via system or signing key compromise).
- Impersonation

  Bob authenticates Bank.com using the server's certificate. Eve routes Bob to a fraudulent site by poisoning the domain name system (DNS) (typically by attracting Bob to connect to a rogue Wi-Fi network) or sending him a message with a specially crafted address (URL). By using an unauthorized certificate, Eve fools Bob into thinking that he is interacting with Bank.com. Most phishing campaigns attempt to exploit this vulnerability.
- Forged digital signatures

  Bob digitally signs documents authorizing fund transfers. Eve is able to forge Bob's signature using the fraudulent certificate.
- Encryption eavesdropping (man-in-the-middle)

  Bob normally connects to Gmail.com directly and verifies the authenticity of the server using its certificate. Bob is redirected through Eve's server and presented with a fraudulent certificate that claims to represent Gmail.com. Bob (more accurately, the application Bob uses) accepts the certificate that was presented. Eve can view all encrypted data.

  As instances of inbound malware continue to threaten corporate security, enterprises have resorted to implementing "controlled" man-in-the-middle (MITM) to inspect all inbound traffic. This is facilitated by the distribution of asymmetric key pairs to specialized SSL/TLS inspection devices, which then use the (shared) keys to intercept inbound traffic and scan for malware.

3.5.3 Code signing

Just as with web sites, a code also needs to be identified before applications or operating systems allow the code to be executed. The goal is to provide assurance about the publisher of the code *and* that the originator code has not been tampered with (this mitigates the threat of a malicious actor modifying otherwise legitimate code for nefarious purposes). As such, the signing of code is considered to be an important business function, but is rarely a centralized process today. Instead, most organizations enable developers or build systems to have direct access to code-signing keys, which leads to very little visibility about what is being signed and by whom. Even worse, the distributed nature of these credentials makes it hard to secure them, and thus, they are a popular target for malware. As such, code-signing keys are the most attacked type of cryptographic credentials, which is borne out by data—the total number of *signed* malicious binaries exceeded *20 million* in 2016 (McAfee Threat Report 2016). In response to this threat vector, the CA Security Council released recommendations on best practices (Code-signing requirements 2016), which include the need for the following:

- Minimizing direct access to code signing keys
- Protecting private keys by storing them on hardware security modules
- Time-stamping all signed codes
- Validating and virus scanning code before signing
- Rotating code-signing keys periodically, to prevent the overuse of key material

3.6 Attacks on trust

As we have established in previous chapters, cryptographic credentials are foundational to cybersecurity. It is no surprise, therefore, that they are high-value targets for malicious actors. In a recent survey conducted by the Ponemon Institute, *100%* of respondents from Global 2000 companies reported that they have had at least one attack on keys and certificates in the last 2 years.

Digital keys and certificates are being stolen—and increasingly often. Individuals and state-sponsored organizations are targeting them with the specific goal of misrepresenting themselves to steal sensitive information. Given the fact that keys and certificates are the most widely used means for establishing online identity, they are constantly under attack. This is borne out by the analysis of stolen data that is available on the dark web. While a stolen social security number goes for as little as $3, a stolen code-signing certificate goes for much as $1500—orders of magnitude more than what is considered PII.

As recent research from McAfee Labs (Q1, 2017) shows, SSL/TLS vulnerabilities account for as much as 33% of network attacks today, and this number is only growing as encryption is being rolled out across almost all web-facing infrastructure (McAfee Threat Report 2017).

3.6.1 Significant events

Here is a brief history of significant key- and certificate-related attacks and breaches—mostly as a result of lax security controls implemented by the issuers of these credentials (CAs):

- 2001—VeriSign issues Microsoft Corporation code-signing certificate to a non-Microsoft employee. VeriSign, Inc. advised Microsoft that on January 29 and 30, 2001, it issued two VeriSign class 3 code-signing digital certificates to an individual who fraudulently claimed to be a Microsoft employee. The common name assigned to both certificates is "Microsoft Corporation" (VeriSign mis-issuance 2001).

- 2008
  - Thawte issues certificate for Live.com to a non-Microsoft employee.
  - Attacker uses an unauthorized e-mail address to obtain a valid certificate for a Microsoft domain (PKI Failures 2017).
  - Comodo issues unauthorized Mozilla certificate.
  - Certstar, a Comodo reseller, failed to perform basic validation checks as part of its RA responsibilities prior to issuing a publicly trusted certificate to http://www.mozilla.com (Mozilla mis-issuance 2008).
- 2009—code-signing certificates used to sign malware ("StuxNet")
  In what was probably the first certificate-based attack that was attributed to a nation state as part of a cyberwarfare campaign, stolen certificates from RealTek Semiconductor, a hardware manufacturer in Taiwan, and JMicron Technology, a circuit maker, also located in the same business park in Taiwan, were used to mask and trust malware. The malicious code was then used to allegedly conduct attacks that caused significant disruption and damage to Iran's nuclear program. The level of sophistication used to craft these attacks was stunning and served as a blueprint for a number of cyberattacks for years to come (StuxNet 2011, StuxNet 2013).
- 2011—Comodo issues nine counterfeit certificates (Google, Yahoo, Live, Skype, and Mozilla).
  - Much like in 2008, Comodo depended on resellers to validate the ownership of domains prior to issuing certificates, instead of enforcing these requirements or doing it themselves. At least two of Comodo's resellers (GlobalTrust and InstantSSL) were among those.
  - StartSSL CA compromise.
  - While the details of the compromise were not made public, the StartSSL CA essentially suspended services for a week because of a breach. The breach did not impact the validity of previously issued certificates (StartSSL compromise 2017).
  - Dutch CA DigiNotar compromised.

    In what was probably the most disruptive of CA-related breaches, DigiNotar ended up issuing hundreds (531 to be exact) of rogue certificates for highly trafficked websites (to a total of 344 domains, including Google.com), which were subsequently abused in a large scale attack in August 2011 to conduct mass surveillance of Internet users in Iran (DigiNotar 2012). A hacker was able to exploit a vulnerability in DigiNotar's externally deployed application servers to get on to the company's network. Once there, the hacker took advantage of poor network security controls, including the Windows Remote Desktop and weak domain administrative passwords to get access to the CA infrastructure. DigiNotar was a relatively popular CA, with multiple CA instances, including one that issued certificates to the Dutch government. The impact was so severe that the Dutch government allegedly asked their citizens to return to using pen and paper in their interactions with the government (DigiNotar 2011b). As a result of the compromise, browser vendors Microsoft (Internet Explorer), Google (Chrome), Mozilla (Firefox), and Apple (Safari) blacklisted and removed the Diginotar CA hierarchy from their trust stores. In the wake of this devastating hack, DigiNotar filed for bankruptcy (DigiNotar 2011c).

    In what was a troubling footnote, the hacker also claimed to have hacked into other CAs, including GlobalSign and four others that he would not name (DigiNotar 2011a).

- Turktrust

  The Turkish CA inadvertently issued two unauthorized *intermediate* CA certificates to its customers (TurkTrust Fiasco 2013). While one was immediately found and revoked, the other continued to operate with the trusted intermediate cert. As part of its efforts to inspect encrypted outbound traffic, this customer then used the intermediate CA certificate to issue certificates to popular sites like Google.com (again!). Google Chrome's "public key pinning" feature surfaced this issue, and manifested itself as warnings to the user in their browsing experience.
- 2012—Microsoft CA certificates forged by exploiting MD5 (Flame)

  Malicious actors were able to forge a fake intermediate CA, taking advantage of a by then weak digital signature algorithm (MD5) (Flame Collision 2012a, MD5 Hash Clash 2008). They then used this forged CA (Microsoft unauthorized issuance 2012) to issue certificates that made it look like they were approved by Microsoft (Flame Collision 2012b). These certificates were used to sign a malicious code that was then distributed as Microsoft software updates, which was then used to perform an MITM attack against victims (Flame Collision 2013d). Much like with the StuxNet attack from a couple of years earlier, analysis of the attack vector by the Laboratory of Cryptography and System Security (CrySyS) at the Budapest University of Technology and Economics suggested a "a government or nation state with significant budget and effort, and may be related to cyber warfare activities" (Flame Collision 2012c).
- 2013
  - DigiCert issues bogus code signing certificate to Buster Paper Comercial, a Brazilian company.

    In yet another instance of malware that was found signed with a legitimate code-signing certificate, malicious actors used a signed PDF document to steal banking passwords. Unlike incidents like StuxNet or flame, the certificates did not appear to be stolen or forged. Instead, the attackers exploited lax security controls at the issuing CA (DigiCert) to obtain a certificate that was issued to Buster Paper, a Brazilian company—probably by spoofing the e-mail address that was used to validate ownership of the domain that the certificate was issued to (DigiCert-BusterPaper 2013).
  - ANSSI issues rogue certificates for Google domains.

    The Agence nationale de la sécurité des systèmes d'information (ANSSI), affiliated with the French Ministry of Finance, and chartered with the responsibility to protect government systems against cyberattacks, was found to have issued several certificates to Google domains without authorization (ANSSI mis-issuance 2013b). ANSSI operated an intermediate CA that chained under a root authority that is trusted by most browsers, and it signed certificates that were allegedly used to inspect encrypted traffic on a private network. ANSSI stated that the certificates were issued as a result of human error and preemptively revoked the issuing CA certificate (ANSSI mis-issuance 2013a).
- 2014—Indian National Informatics Center (INIC) issues unauthorized Google domain certificates.

  In July, security engineers from Google stated that they had identified several certificates for Google domains that had been issued without authorization by the National Informatics Centre, a branch of the Indian Ministry of Communications and Information Technology. In all, at least 45 SSL certificates were found to be improperly issued to Google and Yahoo web properties, through the issuance of an unauthorized intermediate CA. In the aftermath of this incident, Microsoft, Google, and other vendors updated their list of trusted issuers to revoke trust in this particular CA (INIC Compromise 2014).

- 2015—Chinese Root CA China Internet Network Information Center (CNNIC) issues subordinate CA cert to MCS Holdings, which then issues certificates for Google domains

  In an incident that had a lot of parallels to the INIC incident from 2014, a root authority operated by the CNNIC) issued a certificate to an intermediate entity, MCS Holdings, based out of Egypt. The intermediate CA issued a number of unauthorized certificates to Gmail and several other Google web properties in efforts to implement controlled MITM SSL inspection (CNNIC 2015a). Much like with the INIC incident, browsers such as Google Chrome and Mozilla Firefox proceeded to revoke trust in the entire CA chain, starting with the CNNIC root CA (CNNIC 2015b).

  Symantec (formerly VeriSign) issued certificates to unregistered and unauthorized domains.

  At the onset of an event that would have much bigger ramifications down the line, Symantec was found to have generated a number of internal test certificates for popular websites such as Google and Yahoo in a manner that was not consistent with its certificate issuance policies. At the time, 162 unauthorized certificates were found to be issued, but this number seemed to grow substantially later (Symantec mis-issuance 2016). In what was an indicator of things to come, Symantec fired the engineers responsible for the incident, but all signs pointed to a bigger, more systemic problem (Symantec mis-issuance 2015).

- 2016—more (allegedly as many as 30,000) misissued certificates by Symantec

  In what has thus far been the most disruptive certificate authority-related event, with global ramifications, an audit of Symantec's certificate issuance practices revealed a much bigger problem with how the CA ran its operations. Specifically, the CA, which has historically been one of the world's well-known and trusted issuer of certificates, employed poor validation practices at its registration authorities. A number of registration authorities around the world were found to be not vetting the identities of certificate requestors with potential impact on nearly *30,000* certificates that could have been mis-issued.

  In reaction to this event, and a brewing battle between the CA and browser community [both part of the CAB forum (https://cabforum.org/)], which serves as the custodians of what can be trusted by browsers, applications, and devices around the world) over the control of trust lists, Google took the drastic step of requiring Symantec to reissue *all* of its certificates. Additionally, the browser vendor revoked the CA's ability to issue extended validation (EV) certificates and imposed the restriction of issuing all new certificates with shorter validity times. The CA took exception to this course of action, disputed the alleged extent of the mis-issuance, and argued for a more collaborative approach that would not impact their ability to continue to run as a business while not being as disruptive to the ecosystem as a whole. Most importantly, organizations that had contractual agreements with Symantec to issue their digital certificates, and thus their reputation and brand on the Internet, were left in the uncomfortable position of having to either reissue all of certificates and then update all their websites and applications that depended on these certificates or have to move off Symantec and to another certificate issuer. Both options were severely disruptive and surfaced lots of talk in the industry about both the dependency on the prevalent hierarchical trust model and the ability of a few vendors to impact so many that depended on the Internet for critical functions.

  At the time of writing, this incident has yet to be played out to its conclusion, but it does speak to the foundational impact of how trust needs to be established and maintained on the Internet.

While not related to certificate mis-issuance, there are a couple of other noteworthy events that speak to the overall impact of encryption technology on our lives today that cross the boundaries between normal, day-to-day business or personal interactions and the trade-off between security and privacy in general.

3.6.2 *Edward Snowden*

Edward Snowden was a former Central Intelligence Agency employee and contractor for the US government who copied and leaked classified information to several news publications, without authorization, in 2013, from the National Security Agency (NSA) while employed by Booz Allen Hamilton (Snowden 2017). The contents of the classified documentation revealed several global surveillance programs that were run by the US government, with the cooperation of telecommunication companies and several European governments.

While the intent of his actions has been under a lot of debate, what was noteworthy was this comment:

> Encryption works. Properly implemented strong crypto systems are one of the few things that you can rely on. Unfortunately, endpoint security is so terrifically weak that NSA can frequently find ways around it. (Snowden 2013)

Put another way, Snowden disclosed aspects of encryption that were being exploited by nation states to monitor individuals and organizations around the world. Yet Snowden himself leveraged encrypted channels to exfiltrate data from the NSA, one of the most security-conscious organizations in the world. His preceding comment was to highlight the difference between the theoretical and practical aspects of implementing encryption.

A number of the tools and applications that were used by Snowden became popular in the aftermath of this incident—so much so that James Clapper, then director of National Intelligence, had this to say:

> As a result of the Snowden revelations, the onset of commercial encryption has accelerated by seven years. (Snowden 2016)

Asked to explain this comment, he then followed up with the following:

> The projected growth maturation and installation of commercially available encryption—what they (the NSA) had forecasted for seven years ahead, three years ago, was accelerated to now, because of the revelation of the leaks.

Indeed, the Internet hit a seminal milestone early in 2017—*more than 50% of all web traffic is now encrypted*, per the Electronic Frontier Foundation (EFF) (Encrypted Web 2017). Most popular messaging applications have implemented "end-to-end" encryption today, with the intent of reassuring their customers about the security and privacy of their data and communications. This extra layer of security afforded by encryption has in turn been leveraged by malicious actors and terrorists to escape detection in their attempts to cause damage. One such incident is outlined next.

3.6.3 *Apple vs FBI*

In response to the San Bernadino terror attacks of 2015 in the United States, the Federal Bureau of Investigation (FBI) needed to inspect the communication of the perpetrators to identify enablers of the attack and determine the possibility of other terrorist cells within the country. The data in question were encrypted on the iPhone of one of the attackers, and because of how encryption is implemented within the Apple ecosystem, getting access to the data proved to be especially hard for the FBI that was tasked with identifying the origins of the attack and potentially others in the future. As such, the FBI requested Apple's help in decrypting the encrypted content, but the vendor refused, citing privacy concerns about the implications of providing a "backdoor" to government agencies that could then exploit these data to spy on other citizens or foreign nationals. A six-week-long legal battle ensued, potentially setting the landscape for the battle for the balance between privacy and law enforcement in an age where communications are increasingly conducted online. On one hand, bad actors are exploiting the advantages of end-to-end encryption built in to popular apps such as WhatsApp, iMessage, and Signal to cover their tracks and advance their agenda. On the other, governments around the world including the United States, United Kingdom, and Russia are stipulating requirements that grant them access to sensitive data that can be used to work around privacy stipulations. In a worrisome twist to the proceedings, the Department of Justice ultimately was able to decrypt the communications via as yet undisclosed means, precluding Apple from having to implement and hand over a decryption mechanism to the government. While this particular incident ended without requiring cooperation with the government, more events will surely occur in the not too distant future, which will reinvigorate this debate.

The United States is not alone in their attempts to crack encryption—a number of governments around the world have *required* access to technology that can inspect encrypted traffic, leading to intense debate around the line that exists between protecting data . . . and exploiting it.

3.7 *What does the future look like?*

While keys and certificates are here to stay as evidenced by the use cases that they address, and the lack of viable alternatives, there are a number of initiatives, at varying degrees of maturity, that are poised to impact the growth of machine identities, and the resulting dependency on them, even further.

3.7.1 *Virtualization and cloud*

While the notion of virtualization has been around since the age of the mainframe (originally conceived in the 1960s), the use of cloud computing to improve the efficiency of resource utilization is relatively newer. As organizations increasingly look to the use of DevOps and cloud (Gartner expects that by 2020, a corporate "no cloud" policy will be as rare as a "no Internet" policy today) to accelerate innovation without incurring significant expenses, they have started to look at ephemeral, "immutable" infrastructures that are deployed only for as long as they need to exist. How this impacts machine identities is that there is now a need for rapid certificate issuance, in response to elastic workloads, and that machine identities are relatively short lived. Organizations are now building up and tearing down their IT infrastructures multiple times a day as they embrace the idea of "infrastructure as code" as they attempt to realize efficiencies of operation and cost savings.

This has surfaced a need for better management frameworks for keys and certificates—how do you effectively address the trust issues caused by continuously changing identities, wherein new credentials are being issued, and reissued, at very high frequencies? Additionally, since certificates will not need to "live" long, hitherto recommended best practices such as expiration monitoring, instance validation, etc. will cease to be as important as these credentials will be replaced well before they expire. Conversely, machine identities that are no longer current will need to be effectively destroyed to mitigate the impact of unauthorized access to a trusted credential.

Popular cloud providers such as Amazon (AWS), Microsoft (Azure), and Google (GCP) now offer some level of certificate issuance and management capabilities as they bolster their infrastructure-as-a-service and platform-as-a-service offerings to address customer needs.

3.7.2 *Certificate transparency*

As was referenced in previous sections, one of the most viable, and hence exploited, threat vectors relevant to keys and certificates is to attack the issuers of these machine identities—the CAs that are trusted by computing systems of all types and sizes. As malicious actors and nation states attempt to get access to sensitive data, e-mail communications are a popular target. E-mail providers such as Google and Yahoo are as such among the most targeted vendors, as evidenced by the attacks listed earlier in this section. In response to these attacks, most notably the DigiNotar CA compromise, Google proposed the Certificate Transparency (CT) initiative (https://www.certificate-transparency.org/) in 2011.

CT is essentially a framework for monitoring and auditing the issuance of certificates in near real time. Google has helped promote this by requiring that certificates have publicly accessible issuance records in order to be treated as the most trustworthy within the Chrome browser—currently indicated by a green lock icon in the browser's address bar, which can be interpreted by Internet users as Google-provided confidence in the validity of the site that they intend to visit. Conversely, websites that operate with certificates that do not have CT records (in essence, those without a publicly auditable record of issuance) will be treated as potentially insecure. Google currently only requires that EV certificates be logged in CT log servers, but has signaled an intent to extend CT requirements to cover all types of certificates, including domain validation and organization validation certificates by early 2018. This will force all public CAs to log all their certificates to Google-approved CT log servers to prevent a degraded user experience, which will in turn impact their customers.

One of the effects of recording the issuance of certificates and having this be publicly auditable is that anyone can monitor the generation of these credentials. This is particularly revealing in the case of machine identities as the certificate provides information about both who it was issued to and who issued it, in effect exposing the list of CAs' customers to potential competition in a market which is increasingly commoditized and hence fiercely competitive. This was understandably met with opposition by the CA community when first proposed to the CAB forum—which the vendor wants to publish their rolodex of customers?—but because of Google's significant market share and increasing frequency of incidents of certificate misissuance, CT has started to gain traction, with Firefox and Opera being popular browsers that support the initiative and a number of CAs starting to participate in it as well.

While CT is still in "experimental" status with the Internet Engineering Task Force (IETF), it has garnered widespread support, partly because of the popularity of the Google Chrome browser. CT is both an open standard and an open-source framework, and this has helped other vendors, most notably Mozilla, both contribute to and benefit from this initiative. There is now an ecosystem of CT log operators and CT log monitors that help sustain the viability of this initiative and have helped surface instances of unauthorized issuance around the world, including the aforementioned incidents with CNNIC, INIC, and Symantec.

In addition to certificate transparency, there are other initiatives such as pubic key pinning, DNS-based authentication of named entities and CA authorization records that have been implemented with varying degrees of success, all with intent to improve the digital certificate system that serves as the backbone (of trust) for the Internet.

3.7.3 Let's Encrypt

As part of their efforts to transition the Internet to run over secure, encrypted channels, the EFF, an international nonprofit digital rights group that is funded by industry heavyweights such as Cisco, Akamai and Mozilla, launched a *free* CA in 2016—Let's Encrypt (the web) (Let's Encrypt 2014). In a relatively short period, Let's Encrypt has risen to become, by far, the most popular CA for Internet-facing systems—in early 2017, they hit the milestone of *1 million* issued certificates *a day*, which is orders of magnitude more than the number issued by any other public CA. As a means of comparison, the next most popular CA issues 100,000 certificates a month.

In addition to the no-cost aspect to getting a certificate, what has heavily accelerated the adoption of this particular CA is a protocol that they introduced that essentially automates the certificate issuance (and, down the line, renewal) process—Automatic Certificate Management Environment (ACME spec 2017). ACME is currently an IETF draft and was designed to simplify the process by which certificates were issued, while attempting to validate the legitimacy of the requestor. As the need for ubiquitous encryption rises, the number of ACME implementations has exploded, as is evidenced by the wide variety of development frameworks that support this protocol. ACME/Let's Encrypt issues only domain-validated certificates today, as certificates with a higher level of assurance require a level of verification that is currently not possible via automated means.

This access to a free, publicly trusted CA with limited vetting capabilities has made Let's Encrypt attractive to exploits as well. Over 15,000 unauthorized certificates have been issued thus far to Paypal, a popular online payments platform, as part of phishing campaigns intended to trick Paypal customers into revealing their access credentials (Let's Encrypt 2017). Figure 3.4 captures the criticality of this problem very well.

To the average Internet user, there is no difference between the website on the left and the one on the right—both are treated as trusted by the browser, as they were issued by trusted CAs. Yet only one of these is a legitimate PayPal property, while the other has been issued to "paypal.com.summary-spport.com" that may look like a PayPal website, especially on a resource-constrained device such as a smartphone, but has nothing to do with the PayPal brand. What is more worrisome is the fact that the phishing campaign used perfectly legal techniques to register their domain and obtain a trusted certificate in an automated manner that allows the perpetrator to, in effect, repeat the attack (by registering a different domain) and incur very little cost. Even if this certificate were to be detected, there is very little PayPal or the CA can do to mitigate its use. Partly in response to these attack vectors, the US government enacted the Anti-cybersquatting Consumer Protection Act in 1999 to prevent using domain names that are confusingly similar to, or dilutive of a trademark belonging to, another entity. CT, as previously noted, can help surface these kinds of issues since the certificates that are used in these sort of attacks will be publicly logged, monitored, and audited.

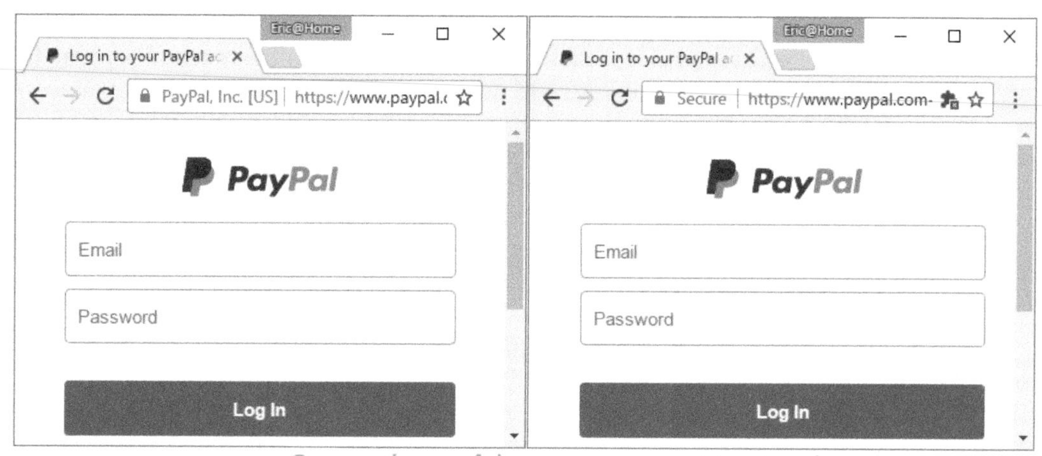

Figure 3.4 Top network attacks. (From Lawrence, Eric, *Certified Malice*, Text/Plain, https://textslashplain.com/2017/01/16/certified-malice.)

3.7.4 IoT

The IoT is a rapidly growing market, poised to exceed 20 billion devices by 2020, per estimates from Gartner. IoT is the quintessential machine-to-machine interaction use case, and while that is not a new concept, what is unique is the dependence on the Internet to serve as the backbone for these communications—and the possibilities and challenges it affords. On one hand, access to the Internet has significantly accelerated innovation in markets such as home automation, autonomous vehicles, and "smart" utilities. On the other hand, the rate at which these devices are being rolled out has made it necessary to depend on the Internet to scale to administer and manage these devices. This is where unrestricted access to these devices by anyone on the Internet has significantly increased their risk of compromise while exposing another threat surface—hacking these devices and using them to attack other Internet-facing systems, essentially simplifying the capability to launch "distributed denial of service" attacks at very little cost to malicious actors.

The Mirai attacks of 2016 serve as a glaring example of the risks posed by Internet-enabled systems that have been improperly secured. Mirai is malware that infects consumer IoT devices such as IP cameras and home routers and utilizes them as part of a "botnet" to conduct large-scale network attacks. The source code for Mirai is freely accessible and has been adapted and used to disrupt a number of highly networked websites including those belonging to Twitter, Netflix, Airbnb, Reddit, and GitHub (Mirai Malware 2017). Mirai attacks IoT devices by attempting to gain administrative access to these devices, exploiting poor access credentials (typically weak usernames/passwords that are shared across multiple devices) to take control of the device and then redirecting its traffic to attack other Internet-facing applications. This is where the enhanced security afforded by keys and certificates can significantly reduce the probability of compromise:

- Using strong, cryptographically generated access credentials mitigates the possibility of brute force attacks that would allow an attacker to attempt access using common passwords (admin/password is a frequently used access credential) or to guess a password—the cost of cracking an RSA or ECC key renders the cost of such attacks unviable.

- Using unique, automatically generated machine identities mitigates the possibility of large-scale device compromise—each device would have to be attacked separately, which would make it cost prohibitive to get access to the number of devices necessary to construct a viable botnet.
- Requiring all software downloads to be cryptographically signed and verified prior to installation on these devices further reduces the possibility of malicious actors taking control of these devices—the malware would have to be signed by a valid code-signing certificate that is issued by a CA that is trusted by the device manufacturer or consumer before it would be allowed to run on these devices.

As such, IoT vendors need to look at machine identities to enhance their security postures and reduce the possibility of compromise and subsequent attacks. As this area of additional security afforded by cryptography is not very well understood, there is a need for IoT security vendors to offer strong machine identities as part of their platform offerings.

3.7.5 Distributed ledgers (Blockchain)

While still relatively new, the concept of Blockchain has received a lot attention in recent years, primarily because of the success of Bitcoin as a cryptocurrency, and the disruption it has caused to the peer-to-peer payment industry. This has taken out the need for an intermediary, typically a financial institution such as a bank, to enable the transfer of money between the participants in the transaction. Additionally, details of the transaction are then signed for integrity and logged into a readable "ledger" using nodes ("blocks") that can be distributed, but *cryptographically* connected (and secured) in that one block builds on top of another, providing an auditable, tamper-resistant system that is decentralized as well. It is for this reason that Blockchain is sometimes referred to as a distributed ledger. In reality, Bitcoin, which is a very compelling alternative to current payment networks because of the security, transparency, and financial implications, is just the most well-known implementation of a distributed ledger network.

The use cases for distributed ledgers are still being developed—in addition to payment networks, land registries are another practical example where a distributed ledger brings significant advantages over currently used methods.

3.8 In summary

Hopefully, this chapter has established that cryptographic keys and certificates are the bedrock on which trust is established in the digital world and serve to enable many mission-critical use cases, including access to the Internet through wired/wireless networks, encryption of data (both at-rest and in-motion), messaging/communication platforms, and pretty much every application that needs to be secured for privileged access.

Keys and certificates are poised for explosive growth fueled in part by trends in virtualization, cloud, DevOps, and IoT. Yet they are also among the least understood concepts of cybersecurity, as evidenced by the low investment in good cryptography management practices. This is especially stark when compared to the billions of dollars we spend as an industry to protect older, less secure technology such as usernames and passwords.

It is because of this lack of awareness that malicious actors, whether it be private groups of individuals or nation states, target vulnerabilities in the implementation of cryptographic security—SSL/TLS attacks dwarf all other forms of network attacks today, and this difference will only continue to grow. The technology itself is mature and, when implemented correctly does what it is supposed to do, it secures sensitive data, protects its integrity, and authenticates the participants in any digital interaction. This robustness of the technology is now being utilized for malicious purposes, leading to an ongoing debate between the benefits afforded by security/confidentiality on one hand and the threats they pose to citizens from all walks of life on the other.

References

ACME spec, 2017: Automatic Certificate Management Environment (ACME) draft-ietf-acme-acme-06, by R. Barnes, J. Hoffman-Andrews, and J Kasten—Mar 13, 2017. Accessed on Jun 3, 2017. https://tools.ietf.org/html/draft-ietf-acme-acme-06.

ANSSI mis-issuance, 2013a: French Intermediate Certificate Authority Issues Rogue Certs for Google Domains, by Lucian Constantin—Dec 9, 2013. Accessed on May 28, 2017. http://www.computerworld.com/article/2486614/security0/french-intermediate-certificate-authority-issues-rogue-certs-for-google-domains.html.

ANSSI mis-issuance, 2013b: Further Improving Digital Certificate Security, by Adam Langley—Dec 7, 2013. Accessed on May 28, 2017. https://security.googleblog.com/2013/12/further-improving-digital-certificate.html.

App statistics, 2017: Number of Apps Available in Leading App Stores as of Mar 2017, May 23, 2017. https://www.statista.com/statistics/276623/number-of-apps-available-in-leading-app-stores/.

CNNIC, 2015a: Google to Drop China's CNNIC Root Certificate Authority after Trust Breach, by Owen Williams—Apr 1, 2015. Accessed on May 28, 2017. https://thenextweb.com/insider/2015/04/02/google-to-drop-chinas-cnnic-root-certificate-authority-after-trust-breach/#.tnw_hkMNtTBS.

CNNIC, 2015b: Google Chrome Will Banish Chinese Certificate Authority for Breach of Trust, by Dan Goodin—Apr 1, 2015. Accessed on May 28, 2017. https://arstechnica.com/security/2015/04/google-chrome-will-banish-chinese-certificate-authority-for-breach-of-trust/.

Code-signing requirements, 2016: Minimum Requirements for the Issuance and Management of Publicly-Trusted Code Signing Certificates—Version 1.1, Sep 22, 2016. Accessed on May 25, 2017. https://casecurity.org/wp-content/uploads/2016/09/Minimum-requirements-for-the-Issuance-and-Management-of-code-signing.pdf.

DigiCert-BusterPaperm 2013: Malware Strikes with Valid Digital Certificate, by Thor Olavsrud—Feb 4, 2013. Accessed on May 28, 2017. http://www.cio.com/article/2388640/security0/malware-strikes-with-valid-digital-certificate.html.

DigiNotar, 2011a: Comodo Hacker: I Hacked DigiNotar Too; Other CAs Breached, by Peter Bright—Sep 6, 2011. Accessed on May 25, 2017. https://arstechnica.com/security/2011/09/comodo-hacker-i-hacked-diginotar-too-other-cas-breached/.

DigiNotar, 2011b: DigiNotar Certificate Authority Breach Crashes e-Government in the Netherlands, by Robert Charette—Sep 9, 2011. Accessed on May 25, 2017. http://spectrum.ieee.org/riskfactor/telecom/security/diginotarcertificate-authority-breach-crashes-egovernment-in-the-netherlands.

DigiNotar, 2011c: Hacking Scandal Roils Dutch Public, by John W. Miller and Maarten Van Tartwijk—Sep 7, 2011. Accessed on May 25, 2017. https://www.wsj.com/articles/SB10001424053111903648204576554630259605332.

DigiNotar, 2012: Black Tulip—Report of the Investigation into the DigiNotar Certificate Authority Breach. Published by Fox IT—August 13, 2012. Accessed on May 25, 2017. https://www.rijksoverheid.nl/binaries/rijksoverheid/documenten/rapporten/2012/08/13/black-tulip-update/black-tulip-update.pdf.

Encrypted Web, 2017: EFF: Half of Web Traffic Is Now Encrypted, by Sarah Perez—Feb 22, 2017. Accessed on Jun 3, 2017. https://techcrunch.com/2017/02/22/eff-half-the-web-is-now-encrypted/.

Flame Collision, 2012a: Flame Malware Collision Attack Explained, by Microsoft SWI—Jun 6, 2012. Accessed on May 28, 2017. https://blogs.technet.microsoft.com/srd/2012/06/06/flame-malware-collision-attack-explained/.

Flame Collision, 2012b: Flame Attackers Used Collision Attack to Forge Microsoft Certificate, by Dennis Fisher—Jun 5, 2012. Accessed on May 28, 2017. https://threatpost.com/flame-attackers-used-collision-attack-forge-microsoftcertificate-060512/76648/.

Flame Collision, 2012c: sKyWIper (a.k.a. Flame a.k.a. Flamer): A Complex Malware for Targeted Attacks, by Laboratory of Cryptography and System Security (CrySyS Lab)—May 31, 2012. Accessed on May 28, 2017. http://www.crysys.hu/skywiper/skywiper.pdf.

Flame Collision, 2013d: Flame Windows Update Attack Could Have Been Repeated in 3 Days, Says Microsoft, by Kim Zetter—Mar 1, 2013. Accessed on May 28, 2017. https://www.wired.com/2013/03/flame-windows-update-copycat/.

INIC Compromise, 2014: Microsoft Revokes Trust in Certificate Authority Operated by the Indian Government, by Lucian Constantin—Jul 11, 2014. Accessed on May 28, 2017. http://www.pcworld.com/article/2453343/microsoft-revokes-trust-in-certificate-authority-operated-by-the-indian-government.html.

IOT stats, 2015: Gartner Says 6.4 Billion Connected Things Will Be in Use in 2016, Up 30 Percent from 2015, by Gartner—Nov 10, 2015. Accessed on Jun 3, 2017. http://www.gartner.com/newsroom/id/3165317.

Let's Encrypt, 2014: Launching in 2015: A Certificate Authority to Encrypt the Entire Web, by Peter Eckersley—Nov 18, 2014. Accessed on Jun 3, 2017. https://www.eff.org/deeplinks/2014/11/certificate-authority-encrypt-entireweb.

Let's Encrypt, 2017: 14,766 Let's Encrypt SSL Certificates Issued to PayPal Phishing Sites, by Catalin Cimpanu—Mar 24, 2017. Accessed on Jun 3, 2017. https://www.bleepingcomputer.com/news/security/14-766-lets-encrypt-ssl-certificates-issued-to-paypal-phishing-sites/.

McAfee Threat Report, 2016: McAfee Labs Threats Report, March 2016. Accessed on May 25, 2017. https://www.mcafee.com/us/resources/reports/rp-quarterly-threats-mar-2016.pdf.

McAfee Threat Report, 2017: McAfee Labs Threat Report, April 2017. Accessed on May 25, 2017. https://www.mcafee.com/enterprise/en-us/assets/reports/rp-quarterly-threats-mar-2017.pdf

MBUN, 2017: An Overview of Public Key Certificate Support for Canada's Government On-Line (GOL) Initiative, by Mike Just. Accessed on May 23, 2017. https://pdfs.semanticscholar.org/5571/65d3d33efc691a262221fdeb6fcd0604e921.pdf.

MD5 Hash Clash, 2008: MD5 Considered Harmful Today—Creating a Rogue CA Certificate, by Alexander Sotirov, Marc Stevens, Jacob Appelbaum, Arjen Lenstra, David Molnar, Dag Arne Osvik, and Benne de Weger—Dec 30, 2008. Accessed on May 28, 2017. http://www.win.tue.nl/hashclash/rogue-ca/.

Microsoft unauthorized issuance, 2012: Unauthorized Digital Certificates Could Allow Spoofing, by Microsoft—Jun 3, 2012. Updated: Jun 13, 2012. Accessed on May 28, 2017. https://technet.microsoft.com/library/security/2718704.

Mirai Malware, 2017: Mirai (Malware). Published by Wikipedia. Accessed on Jun 3, 2017. https://en.wikipedia.org/wiki/Mirai_(malware).

Mozilla mis-issuance, 2008: CA Issues No-Questions Asked Mozilla Cert, by John Leyden—Dec 29, 2008. Accessed on May 25, 2017. http://www.theregister.co.uk/2008/12/29/ca_mozzilla_cert_snaf/

PKI Failures, 2017: Timeline of PKI Security Failures. Accessed on May 25, 2017. https://sslmate.com/certspotter/failures.

Snowden, 2013: Edward Snowden: How to Make Sure the NSA Can't Read Your Email, by Dylan Love—Jun 17, 2013. Accessed on Jun 3, 2017. http://www.businessinsider.com/edward-snowden-e-mail-encryption-works-against-the-nsa-2013-6.

Snowden, 2017a: Edward Snowden. Published by Wikipedia. Accessed on Jun 3, 2017. https://en.wikipedia.org/wiki/Edward_Snowden.

Snowden, 2017b: Spy Chief Complains That Edward Snowden Sped Up Spread of Encryption by 7 Years, by Jenna McLaughlin—Apr 25 2016. Accessed on Jun 3, 2017. https://theintercept.com/2016/04/25/spy-chief-complains-that-edward-snowden-sped-up-spread-of-encryption-by-7-years/.

SP800-63, 2017: SP 800-63: Digital Identity Guidelines. https://pages.nist.gov/800-63-3/. Accessed on May 23, 2017.

StartSSL compromise, 2017: CA STARTSSL Compromised, but Says Certificates Not Affected, by Dennis Fisher—Jun 21, 2011. Accessed on May 25, 2017. https://threatpost.com/ca-startssl-compromised-says-certificates-not-affected-062111/75351/.

StuxNet, 2011: How Digital Detectives Deciphered Stuxnet, the Most Menacing Malware in History, by Kim Zetter—Jul 11, 2011. Accessed on May 25, 2017. https://www.wired.com/2011/07/how-digital-detectives-deciphered-stuxnet/.

StuxNet, 2013: The Real Story of Stuxnet—How Kaspersky Lab Tracked Down the Malware That Stymied Iran's Nuclear-Fuel Enrichment Program, by David Kushner—Feb 26, 2013. Accessed on May 25, 2017. http://spectrum.ieee.org/telecom/security/the-real-story-of-stuxnet.

Symantec mis-issuance, 2016: Update on Test Certificate Incident, by Symantec—May 25, 2016. Accessed on May 28, 2017. https://www.symantec.com/page.jsp?id=test-certs-update.

Symantec mis-issuance, 2015: Symantec Fires Staff Caught Up in Rogue Google SSL Cert Snafu, by John Leyden—Sep 21, 2015. Accessed on May 28, 2017. https://www.theregister.co.uk/2015/09/21/symantec_fires_workers_over_rogue_certs/.

Trust—changingminds, 2017: What Is Trust. Published by changingminds.org. http://changingminds.org/explanations/trust/what_is_trust.htm Accessed on May 23, 2017.

TurkTrust Fiasco, 2013: The TURKTRUST SSL Certificate Fiasco—What Really Happened, and What Happens Next? by Paul Ducklin—Jan 8, 2013. Accessed on May 25, 2017. https://nakedsecurity.sophos.com/2013/01/08/the-turktrust-ssl-certificate-fiasco-what-happened-and-what-happens-next/.

VeriSign mis-issuance, 2001: Erroneous VeriSign-Issued Digital Certificates Pose Spoofing Hazard, by Microsoft—Mar 22, 2001, Updated: Jun 23, 2003. Accessed on May 25, 2017. https://technet.microsoft.com/en-us/library/security/ms01-017.aspx.

section two

Trust and privacy

chapter four

New challenges for user privacy in cyberspace

Adam Wójtowicz and Wojciech Cellary

Contents

4.1 Introduction

Each new technology, always, is related to opportunities and threats. To take advantage of opportunities but eliminate or at least reduce threats, legal solutions are applied, which forbid some practices that are technically possible but socially inacceptable. A technology that has provided people with endless opportunities and deeply changed human lives is information technology (IT). It, however, is not free of threats, which are particularly hard to deal with. Among them, one of the most significant is breach of privacy. The concept of *privacy* is broad. It may concern individuals, organizations, businesses, public institutions, and states. It is also multifaceted. Among the facets are, for example, the following:

- *Statutory aspects of privacy*—e.g., the United Nations Universal Declaration of Human Rights, the Data Protection Directive of the European Commission, the US Health Insurance Portability and Accountability Act, and the US Family Educational Rights and Privacy Act
- *Technical aspects of privacy*—e.g., information gathering, information flow control, and information leakage
- *Societal aspects of privacy*—e.g., what information is private and how it is handled
- *Political aspects of privacy*—e.g., the surveillance state and population control

In this chapter, we focus on the privacy of individuals immersed in a world saturated with IT. In such a world, the problem of people's privacy has become more important than ever due to the accessibility of digital data describing not only a person's possessions, actions, and relationships with other people, but even their wishes, intentions, and emotions. The problem of privacy breach is critical since it may lead to restrictions of individual liberty and erosion of our society's foundations of trust.

The concept of user privacy has no precise definition that is commonly accepted. Certainly, user privacy is related to the concepts of "personal data" [1] and personally identifiable information (PII) [2]. Privacy concerns the right of a person to not disclose specific information about himself or herself, or more precisely, to disclose that information only to selected entities, but not to others. As such, privacy is inherently related to data confidentiality. Data confidentiality is usually related to strict secrets, e.g., a bank account password. The notion of privacy is broader; it may concern, for example, a person's medical history. A breach of confidentiality may lead to a breach of privacy. User privacy is also related to concepts of a user's anonymity, unobservability, and unlinkability. Anonymity is defined as the state of not being identifiable; unobservability is defined as a state of being undistinguishable; and unlinkability is defined as the impossibility of the correlation of two or more actions/items/pieces of information related to a user [3].

In general, privacy concerns the will of persons to control the disclosure of information about them. The awareness of threats to privacy performed by an agency or entity via intrusion or eavesdropping is nowadays high and constantly raised by many organizations collecting private data, e.g., financial institutions, telecommunication operators, and e-services providers. For example, a bank's business depends much on the trust of customers. A hack over the Internet into a bank's system may heavily impact the bank's business. Banks know that the weakest point in their security systems are naive customers, so they constantly raise the awareness of customers against hackers. Unfortunately, this is only one side of the privacy problem. The other side is the risk related to violating people's privacy by persons and organizations to which customers confidently entrust their private data. Continuing the example of banks, privacy may be breached by banks as organizations and by their employees as individuals. The problem of privacy breach by trust abuse is different from common security issues and—unfortunately—is not fairly highlighted by organizations collecting private data.

To throw light on the privacy problem, we present the points of views of individuals, businesses, and states in the following. We start with explaining the reasons why an individual's privacy should be protected.

The first reason why private data should be kept secret is to reduce the possible distress caused by the change in social relations: a person who has lost some aspect of his/her privacy can consequently be subject to judgment by other people, hardly ever favorable. The problem is amplified by the fact that it is difficult to stop the mass spread of disclosed private information.

The second reason for privacy protection is to reduce vulnerability to business-related attacks, such as (1) aggressive marketing, (2) refusing to enter certain contracts, or (3) aggravating contractual provisions. It is possible to imagine a scenario where a suffering patient calls a doctor for help and the doctor first analyzes the patient's financial situation and then sets the price of medical care based on patient's savings. In other words, if privacy is not protected, the price of a good or service paid by the customer may depend on the customer's wealth, instead of on the value of the good or service equal for every customer. The disclosure of private medical data of an important person, e.g., a chief executive officer of a company listed on the stock exchange, may influence the valuation of that company. Increased vulnerability may also be used to initiate attacks for political or social reasons.

The third reason to protect privacy is to minimize the probability of criminal attacks. Private data may be used by criminals to target potential victims and to minimize risk when planning a crime.

Finally, the last, but not least, reason to protect privacy is to minimize vulnerability to identity theft. Identity theft has serious consequences for a victim. It is very hard to prove that decisions, such as bank transfers, were made by an identity thief, instead of a true bank client, while the credentials used were true and correct. Banks are rightly afraid of a fraud—a dishonest client withdrawing money from his/her account and then claiming that it was not done by him/her but by an identity thief.

The reason why a business is interested in violations of the privacy of its clients is, unsurprisingly, to reduce the business risk and increase profits. A privacy breach is intended to detect person's needs and vulnerability to arguments and suggestions to purchase goods or services to meet those needs. A privacy breach is also used for price discrimination, i.e., charging varying prices when there are no cost justifications for the differences [3], as illustrated earlier by the example of a doctor having access to a patient's private financial data. Business often argues that permitting access to private data will enable it to better inform the client about the possibilities of meeting his/her needs. Also, a client is not forced to take advantage of the advertised offers. Although the latter may be formally true, business hides the risk of privacy abuse, because a reason of privacy breach by businesses is also the identification of vulnerabilities aiming at weakening a client's negotiating position—making them more susceptible to arguments for adopting a worse proposal or for refusal to conclude a contract [4]. Private data and other knowledge about a client may also be used by, or sold to, untrusted and unauthorized parties.

Most nations include the protection of privacy into law. Surveillance, and hence a reduction in privacy, is legally possible only with regard to particular citizens who are formally suspected of committing crimes, and only with the consent of the court guarding civil liberties and supervising law enforcement authorities. In practice, surveillance is used by different governments in a legal, semilegal, or illegal way to prevent activities deemed undesirable, from criminal acts to civil disobedience or political opposition. However, recently, an approach of governments to privacy is undergoing change driven by the phenomenon of terrorism, in particular suicide attacks. The current legal system is based on the assumption that punishment follows the committed crime. This assumption obviously does not work for suicide terrorists, because when they commit the crime of killing inadvertent innocent people, they inflict the highest punishment on themselves—death. Facing the danger of suicide terrorist attacks, the only way for the state to assure public safety is to preventively isolate suspects. This, however, implies the need to violate suspects' privacy to find out about their plans, in advance of the criminal act. Hence, there is a change in the attitude about the state's surveillance of its citizens. In the age of terrorism, the state tries to collect all possible data about all citizens—in other words, to keep the whole society under surveillance—and to analyze collected data when a suspect appears. With such approach, a person is treated as the sum of his/her social relationships, electronic interactions, and favorite content. A citizen becomes suspicious not because he/she has committed an illegal act, but just because his/her online activity patterns indicate that he/she is more prone to commit a crime than an average citizen [5].

The remainder of this chapter is organized as a set of sections devoted to privacy risks that are specific to various emerging IT and electronic business trends. Section 4.2 concerns privacy issues specific to the Internet of things; Section 4.3, to augmented reality (AR); Section 4.4, to biometrics; Section 4.5, to cloud computing; and Section 4.6, to big data. Section 4.7 summarizes these considerations by adopting a holistic view that considers the mutual dependencies of various technologies and trends.

4.2 *Privacy and the Internet of things*

The Internet of things (IoT) is defined as "a global infrastructure for the information society, enabling advanced services by interconnecting (physical and virtual) things based on existing and evolving interoperable information and communication technologies" [6]. As the IoT evolves from the early experiment phase to become a ubiquitous infrastructure processing sensitive data, the number of various privacy challenges to be met increases. One of the main obstacles, seen from the business point of view, is related to the fact that in the rush to provide the market with new IoT solutions before the competition does, the privacy issues are perceived only as a roadblock to productivity and cost-effectiveness. Many IoT investments initially focus on functional requirements only for solutions that can be rapidly marketed and are expected to produce the desired return on investment [7].

From the technical perspective, every sensor or device that collects, sends, stores, and/or processes sensitive data provides a potential privacy risk. As many as 70% of the most commonly used IoT devices contain security vulnerabilities [8]. Furthermore, the diversity of IoT devices makes achieving privacy protection challenging. This is related to the problems of the growing number of unstandardized devices and technological/business fragmentation [9]. Standardization allows developers to build on fewer software/hardware platforms and have more resources allocated for security protection. In turn, in standardized environments, privacy breaches affect a bigger number of devices and users [7]. Above all, the scale of the IoT networks alone is challenging: 11.2 billion connected devices will be in use in 2018, 20.4 billion in 2020 [10], usually containing several sensors and complex software-based logic.

IoT devices are often physically accessible to intruders. Since IoT interconnects physical "things," not only can the intruders perform usual digital privacy-targeted attacks, e.g., stealing data, but they can also take advantage of tampering with devices or attack networks [11] (e.g., healthcare devices, electrical grids, or traffic signals). For example, if a smart thermostat is not able to protect data from eavesdropping and unauthorized usage, specifically when transmitting energy usage data to the utility operator for dynamic billing or real-time power grid optimization, then the sensitive data leak could contain the information that the power usage level has decreased, which indicates that a person's home is left empty [12], which may provoke burglary. The network connections that the devices use may also give subsequent access to central applications and databases.

Another IoT privacy threat comes from the lack of adopting a privacy-focused approach to build systems. A strong focus on security from the beginning of the project is often missing, especially when dealing with emerging technologies and underdeveloped markets [7]. Trade-offs, e.g., a choice of solid security at the cost of compromising user experience, are very challenging. If a company plans to develop its own IoT infrastructure, or deploy an existing solution, it must do research and stay as informed as possible while putting much effort to the training for their personnel. For instance, software designers of IoT solutions, specifically "smart home" systems, who build connections between various devices, face new security engineering challenges specific to the new domain they are often not familiar with. In a recent work [13], four IoT smart home devices (a Sense sleep monitor, a Nest Cam Indoor security camera, a WeMo switch, and an Amazon Echo) have been analyzed. The results of this research prove that the network traffic rates of the devices can reveal a user's physical behavior even if the traffic is encrypted. Preserving user privacy would require special network traffic obfuscation to hide variations that reflect real-world interactions, which is not specified as a requirement by designers of such systems.

At the same time, users of connected devices practically do not realize that their security is in play, or at least, at risk. For an average user, a smart TV or smart watch is still just a TV set or watch. Users are not aware that it is a fully equipped network node, which can be used to collect data describing its owner and his/her environment, or that smart wrist wearables (particularly smartwatches and fitness trackers with embedded sensors such as accelerometer, gyroscope, or magnetometer that are paired with a networked smartphone) might be exploited to steal user's automated teller machine password [14]. Consequently, the user behavior creates another group of privacy risks.

Solid IT security controls that have been developed over the past three decades should be adapted to the specific constraints of the embedded devices popular in the IoT. Applying existing security practices to these devices requires significant reengineering to address device constraints [7]. This is caused by the fact that they are designed for low power consumption, typically have only as much processing power and storage as needed for their purpose, and often have limited connectivity. More powerful processors needed for encryption, and other data security functionalities can be used in some smart products, but it is impractical for disposable devices with no displays and with limited power consumption. More powerful processors have bigger size, and they need additional appliances and space for heat dissipation. Moreover, they need more power, which requires bigger, heavier, and more expensive batteries. In order to reduce weight and size, higher costs of research and materials are required as well as longer time-to-market [7]. With higher prices and more complex builds, such devices could not be considered disposable. Creating access control methods that can be implemented in cheap and compact IoT devices without compromising the user experience, or without adding additional hardware, represents an engineering trade-off challenge. Outsourcing computationally intensive encryption to the cloud is not a privacy-preserving solution either (cf. Section 4.5).

The other IoT privacy-related problem is a result of the fact that in M2M usage scenarios (e.g., telemetry or traffic control), there is no human operating the IoT devices who can input authentication credentials or decide whether an application should be trusted or not. The devices must make their own decisions about whether to disclose their data or trust in some process or other device. In turn, in IoT usage scenarios with a potential presence of human operator (e.g., telemedicine or wearables), connected devices have little or no interfaces that clearly present choices and explain their privacy-sensitive consequences; or even if the choices can be effectively presented in the initial setup of the devices, they can shortly become too hard to understand and remember for average user, because of the highly dynamic nature of IoT networks. Often, even if it is technically possible, service providers do not provide users with clear messages and choices for unexpected collection or uses of their data and a choice to opt out of data collection is not given [15]. For instance, not only do users not know that a smart meter is collecting data about their air-conditioning habits and that a smart watch is collecting data about their physical habits, but also they do not know to what extent this information is shared with data brokers or marketing companies.

The next privacy issue is an effect of ubiquitous data collection, which is possible with IoT devices. Service providers that collect PII do not follow the principle of data minimization. This principle states that only the data needed for a specific purpose should be collected and then safely disposed [15]. Data that have not been collected or have already been deleted cannot be used for unintended purposes. Conversely, of course, collecting and storing large amounts of data increases potential privacy risks that could result from a data leak. Unfortunately, there is an increasingly popular trend that promotes unlimited collection and storing of data because of the high value expected from its potential, but yet unknown future uses—cf. Section 4.6. This leads to putting user privacy at real risk on the off chance an organization might discover a valuable use for the private data at some future point in time.

The aforementioned business hopes related to future data uses decrease the willingness of data operators to deidentify consumer data where possible. Many IoT data uses could still be accomplished by using deidentified data [15]. However, even once anonymized, data can still be reidentified [3]. Therefore, a technical means for data anonymization should be coupled with administrative controls. Organizations should legally commit that they will not try to reidentify personal data. They should also require this commitment from those with whom they share data [15].

In the world of regular mobile devices, there are millions of unpatched and insecure devices in use. Even reputable vendors of costly devices, such us Apple and Google [16], do not update their software on devices that are only a few years old. In the case of inexpensive disposable IoT devices that can operate on the network for years, the lack of updates to past generations can be an even more significant issue, in terms of its scale as well as its consequences [7]. According to a recent report [17], 26% of IoT professionals, including developers, vendors, and enterprise users, find long-term support as the "biggest immediate challenge faced by IoT professionals." Further magnification of this challenge can be expected, when the IoT market forces the rapid release of new products based on emerging technologies and when the well-known phenomenon of "planned obsolescence" becomes common in this market.

The next challenge is related to the technical difficulty to apply a patch if a vulnerability is known or even just to provide users with a message about a new fix. The difficulty lies in receiving software updates or security patches in a timely manner without consuming the bandwidth, impairing functional security or causing significant recertification costs every time a patch is published [12]. Service providers need to publish patches, and devices need to authenticate them, in a seamless and secure way. This is the problem of thousands of devices processing sensitive data that are dependent on security patches to protect against attacks on the confidentiality of these data. Secure IoT devices must either be secure "by design" and protected from the beginning of operation or be able to receive updates throughout their life cycle. Neither option is realistic [7].

In the IoT, as in conventional networks, devices need a firewall or packet analysis to control traffic incoming to and outgoing from the device, to protect data confidentiality. A host-based firewall or intrusion prevention system is required even if network-based devices are installed. The problem results from the fact that often embedded devices use specific protocols, distinct from standard Internet protocols [12], and at the same time, there is lack of industry-specific protocol filtering tools, which could identify attack schemes or malicious payloads in nonstandard protocols at IoT devices. IoT devices do need to filter the incoming data in a way that makes optimal use of its limited computational resources.

As the described privacy risks result from the inherent specificity of IoT, no single solution can be ultimate for every deployment either currently or in the future. However, depending on the particular system, the protection strategies can be applied by combining solutions from different categories, to mitigate those risks. The solution categories include technical, organizational, and legal measures. Technical solutions include system design methods that minimize the attack surface and enforce data anonymization, such as the one presented by Wójtowicz and Wilusz [18], new access control models (e.g., supporting context awareness and addressing embedded device constrains), new encryption schemes, new methods for IoT patching, and reengineering network threat detection/prevention systems to be suitable for the IoT. Organizational solutions include IoT project management focused on user privacy from the requirement specification phase to the long-term support provisioning and end user training as well as continuous training for developers facing new threats. In turn, legal or administrative solutions include introducing obligatory information and opt-out options for users regarding the collection or usage of their data, forbidding violations of principle of data minimization, and data reidentification.

The privacy threats described in this section are common for many different applications of IoT. Moreover, each IoT application has its specific threats not covered here. However, one group of them, i.e., mobile AR systems, has particularly distinct characteristics and high impact potential for the personal data flow. Therefore, the next section is exclusively devoted to privacy threats resulting from the usage of AR systems.

4.3 Augmented reality applications

An AR system is defined as one that "combines real and virtual objects in a real environment; runs interactively, and in real time; and registers (aligns) real and virtual objects with each other" [19]. Modern AR systems operating on mobile devices require access to several data streams coming from a number of sensors. They include not only camera-captured images and audio streams, but also geolocalization data, accelerometer/gyroscope data, temperature, or data from peripheral devices. A significant risk related to AR applications results from the difficulty to set trade-off boundaries for the required levels of application access to those streams and the probability of data confidentiality violation, their unintended usage, and, therefore, violation of the users' privacy. Limited privacy controls in the AR domain are a good example of the classic usability vs. security dilemma. This risk can be illustrated with the following cases:

- AR application uploading the user video stream or geolocalization data to its server-side software components
- Shape detector reading credit card numbers or text on electronic displays or on a bottle of medicine that reveals a medical condition or identifying a person
- Object or gesture tracker recording user's activity; even anonymized skeleton stream allows the inference of potentially sensitive gestures, movements, proximity of faces, bodies, etc.
- Collecting geometrical three-dimensional (3D) data to create models of users' indoor spaces
- Face recognizer intended for device user identification, gathering data about other persons in the camera's field of view
- Quick response (QR) code scanner, apart from code scanning, records data about its environment

These examples imply that privacy risks are much higher in AR than in conventional systems because of the continuous mode in which AR systems operate. Complex AR applications require an always-recording feature, e.g., an AR application that automatically recognizes and decodes QR codes requires continuous access to a video stream. The always-on sensing of AR applications and wearables can disclose sensitive data such as personal images, health information, or enterprise intellectual property. This privacy risk is called data aggregation [20] and is related mostly to temporal and spatial accumulations of raw visual data. Apart from privacy issues enabled by data aggregation followed by applying reasoning and data mining techniques, the aggregation alone inherently introduces privacy breaches, since the human consciousness of the presence of always-on recording devices can alter one's "attitude, behavior, and physiological state" [21]. Also, the accumulation of spatial data in AR services raises the risks related to location disclosure (identity privacy, user's position privacy, and user's movement path privacy) in the context of user anonymity, unlinkability of the user's actions, and the strongest requirement of complete unobservability of user actions [3].

Today, AR applications perform data collection, rendering, and user input interpretation, aided by third-party software libraries or cloud-based recognition services. These applications provide some level of functional access control, but users do not have fine-grained control [22] over the confidentiality of particular pieces of the data against third-party applications. The main reason for that is related to the fact that today's operating systems (desktop and mobile) are built without AR applications in mind. Only coarse-grained controlling of access to data streams is offered, instead of AR-specific privileges [23]. For example, an application should only be provided with an access limited to specific objects that are recognized by the operating system with skeleton tracker, without access to the whole video stream.

Therefore, it is difficult to build an AR application that follows the "principle of least privileges," i.e., to ensure that every application and user is able to access only the data and resources that are necessary for their legitimate purpose. Policy-based mechanisms applied by the software distribution services tend to be ineffective against applications that collect users' PIIs at the back end [20]. It is not probable that these threats can be fully mitigated, except for specific classes of applications, e.g., requiring only numerical data aggregated from multimedia streams. Only in such cases could privacy-enhancing techniques that have been developed for years be utilized, such as differential privacy [24].

AR systems employ new input techniques such as voice, gaze-tracking technologies, or glove-based haptic sensors. The use of these input methods while running multiple applications simultaneously produces new privacy threats related to the inaccurate identification of the application that is seeking the input and should receive it [25]. Malicious applications can steal user input intended for another application, e.g., they could attempt to register a verbal command that sounds similar with that of another privacy-sensitive application. This threat is even more significant since multiple AR applications expose their application programming interfaces (APIs) to each other and users share multimedia content between these applications. Cross-application data sharing can also be implicit, e.g., in AR systems that automatically use video streams of nearby users to build a 3D model of the given user at the runtime [25].

An individuals' personal privacy (as well as information-gathering rights or device ownership rights) can be lessened by AR services that selectively disable sensing capabilities according to server-side rule-based logic, e.g., prohibiting the ability of AR devices to record during a music concert [26]. On the other hand, AR applications can provide users with correct information that cannot be legally used to make business decisions. This could be a case of an AR application collecting face images and performing face recognition during a job interview followed by mining in candidate social media profiles in jurisdictions where discrimination based on marital status, arrest history, etc. is illegal [27]. By providing a number of informational elements about a real person in real time, AR applications increase the risk of conscious or even unintended discrimination in various aspects of life. Data aggregation also creates privacy risks for bystanders, who are not able to opt out or be anonymized in AR streams. AR services are not able to consider anonymization requests from other users' devices or the environment.

Data processed with AR sensors can be used for human body detection and subsequently for user identification and authentication with biometric means. Biometrics, although potentially useful due to its convenience and usually robust security properties, brings additional privacy-related concerns, which are described in the following section.

4.4 Biometric access control

The use of biometric methods for user identification or authentication introduces various privacy concerns. In this section, they have been classified into six main groups of risks. The first group of risks is related to the fact that biometric attributes encode the biological properties of parts of the human body (physiological biometrics) or some human behavior (behavioral biometrics). It is relatively easy for access control systems designed for acquiring biometric samples and processing encoded templates to perform the additional analysis of templates or sample data and infer information describing users based on these data [28]. The inferred information can be deterministic or stochastic. Not only may the information refer to the body, or the medical condition of the user, but it can even be used to estimate cultural or social characteristics of the user. Examples of biometric attributes and their impact on privacy are listed in the following:

- Voice sequences—language spoken (nationality), accent (cultural/social characteristics), age, gender, and emotional state
- Face images or 3D head models—medical condition, age, gender, race, estimated cultural/social characteristics, and emotional state
- Fingerprints—medical condition (e.g., malformed fingers can be correlated with genetic disorders [28])
- Iris—medical condition
- Vein patterns and electrocardiogram patterns—medical condition
- Behavioral biometrics such as gait—medical condition
- Behavioral biometrics such as style of typing or style of touchscreen usage can reveal, directly or indirectly, privacy-sensitive input.

The second biometric-related risk of privacy breach comes from the side of service providers with regard to the unambiguity of biometric identifiers. The presence of biometric identifiers (even if they are not originally referring to any PII in the system) makes it possible to bind a person's virtual identity (anonymous or pseudonymous) used in cyberspace with his/her real-world identity, or to bind several persons' virtual identities with each other [29]. Moreover, biometric identifiers could be used not only to bind identities themselves, but also to bind data (and metadata) describing actions of a particular user biometrically authenticated in various distributed services if service providers collude. In emerging ubiquitous services which are naturally decentralized and untrusted on the one hand, and require new seamless and convenient access control methods (such as biometrics) on the other, this threat is of special significance. The derived information could become a basis for discrimination against a person if their characteristics are considered unwanted [28].

The third group of privacy risks is also related to the anonymity and pseudonymity of users of biometric systems. It follows from the fact that not all biometric attributes are as difficult to collect without one's knowledge or permission as vein patterns or electrocardiogram patterns. Generally, biometric systems cannot rely on the secrecy of biometric samples [30]. Samples such as face images, fingerprints (left frequently, e.g., on a glass), or various behavioral biometrics are relatively easy to collect without a person's knowledge. Subsequently, they can be used for two groups of purposes. The first purpose is to instantly infer additional information (e.g., emotional state from face images or voice samples) and take advantage of them (e.g., in dynamic marketing applications). The second group of purposes is related to identity theft (cf. Section 4.1): collecting one's biometric samples can be followed by preparing fake authenticators imitating corresponding parts of the human body (artificial finger, face mask, high-resolution iris image, etc.), in order to conduct unauthorized authentication and ultimately to violate the confidentiality of user's private data within the system.

To protect the biometric access control against attacks by authentication imitations, some researchers and systems engineers propose so-called liveness detection functionality [31] (e.g., based on the presence of the pulse and eye blinking detection). Although liveness detection can indeed reduce the likelihood of success of the attacks, it introduces new privacy-related risk, since it increases the amount of sensitive data that are collected in a continuous manner. Similarly, multimodal biometric systems combine multiple biometrical recognizers. They have been developed to reduce the false acceptance rate (FAR) and false reject rate (FRR) and to collect more streams of sensitive data. Multimodal biometric systems therefore increase the possibility of data cross analysis accompanied by the associated increased risks of privacy breach. They constitute the fourth group of privacy-related risks that also includes privacy risks following from the unexpected (from the users' point of view) cross analysis of voluntary biometric databases created for user verification purposes with mandatory screening databases [32].

The fifth, similar but distinct, privacy concern is related to the fact that as opposed to conventional authenticators such as passwords, once the biometric sample or template is eavesdropped or disclosed by an attacker, the countermeasures are not straightforward. Compromised password, digital certificate, or credit card data can be effectively revoked and reissued. In the case of a biometric pattern reflecting an immutable attribute of a person's body, the act of eavesdropping on the pattern has permanent consequences. Revocation or cancellation is possible only in specific cases with a priori use of special techniques of cancellable biometrics and/or biometric cryptosystems. However, these techniques cannot assure both provable security and practical FAR/FRR at the same time, and they introduce new issues [33]. Thus, a person's sensitive biometric templates are at constant risk while employed for practical access control. Despite obvious advantages, the fact that biometric patterns are immutable over time can also introduce privacy-related risks beyond just compromising the system. Potentially, there are many circumstances in which a user might want to change his/her identifier, but its biological uniqueness persists even though the sample as well the template are recoded to different digital representations.

The sixth risk arises from the practical limitation of a great majority of biometric access control systems that assumes the existence of nonzero FAR. Biometric systems usually allow their managers to adjust the sensitivity level and find an optimal trade-off between FAR, FRR, and other recognition parameters for a given application. However, the adjustment rarely allows these rates to be reduced to zero, especially in large-scale systems [30]. Disclosing private data because of false acceptances and allowing authentications followed by unauthorized penetration of the system (intentional or accidental) are inherent risks that cannot be omitted in the consideration of privacy.

Finally, a design solution that allows the reduction of some of the aforementioned privacy risks is a shift toward "distributed architecture." In this solution, biometric templates are stored in an encrypted form within devices (e.g., smartcard or smartphone) over which a user has full control [34]. Each device has a biometric sensor built in. User identification, authentication, or transaction authorization is performed locally by comparing the acquired sample with the stored template (according to a more robust verification instead of identification scheme). In some applications, such an approach is possible to implement and effective from the privacy perspective. However, unfortunately, the current dominant trend is just the opposite—to store and process as much data as possible in the cloud-based, centralized manner that is potentially privacy destroying. The cloud computing problem seen from the user privacy perspective is the subject of the following section.

4.5 Cloud computing

Cloud-based data processing requiring privacy assurance can be successfully deployed in private clouds [35]. However, it is the public cloud model that is the most popular architecture when cost reduction is concerned. Relying on a public cloud service provider to store and process user data raises serious privacy concerns since the user is forced to cede control to the cloud provider on many issues affecting data privacy.

The first group of privacy risks follows the cloud operator's difficulty in providing privacy controls required to protect the users' data. These risks result from technical, organizational, and legal limitations of the public cloud model. The loss of control over the physical as well as logical aspects of the system and data reduces the user's ability to keep actual knowledge about the processes and to make accurate and aware decisions regarding the privacy protection of his/her data or of the data of his/her organization. Also, verifying the functional requirements of the service and the effectiveness of privacy controls is not feasible to the same extent as with an internal organizational system [36]. The knowledge of a cloud provider's privacy protection measures and controls is needed if the user is to perform continuous privacy risk assessments. However, cloud providers are not eager to provide users with descriptions of their privacy measures and controls for several reasons. One of them is the fact that such descriptions are considered proprietary and could be used to develop an efficient attack scheme on cloud infrastructure. Providing detailed system-level monitoring by a cloud user is not part of most service-level agreements (SLAs), which limits the user's ability to conduct audits [36].

The result of migrating to a public cloud infrastructure is usually losing a direct point of contact with the entities responsible for data management and losing an influence over decisions made about the data environment. This makes the user dependent on the cooperation of the cloud provider to perform the responsibilities of both parties, such as the passive and active protection of data confidentiality. Also, compliance with data protection laws is an area of joint responsibility that requires cooperation and coordination with the cloud provider [36]. Consequently, there may be data security breaches of which the controller is not notified by the cloud provider and possibly unrecognized conflicts between cloud customer data security procedures and the cloud environment.

Redundant data storage in multiple physical locations is a common feature of cloud computing services. This can lead to the data proliferation phenomenon. Detailed information about the location of a user's data is unavailable or not disclosed to the user. Therefore, often, it is unclear which party is responsible for ensuring legal requirements and data handling standards for PII processing or whether it is possible to audit them for compliance with these requirements and standards. Moreover, it is not clear to what extent cloud subcontractors involved in processing can be identified and verified as trustworthy, particularly in a dynamic environment [35]. Trust is not transitive, which requires disclosing such third-party contracts in advance of reaching an agreement with the cloud provider, and maintaining the terms of these contracts throughout the agreement. In practice, it is rarely fulfilled, so privacy guarantees can become an issue with composite cloud services [36]. If cloud computing providers outsource certain tasks to third parties, the level of privacy protection of the cloud provider depends on the level of privacy protection of each "supply chain" link and the level of dependency of the cloud provider on the third party.

Any corruption in this chain or a lack of coordination of responsibilities between any parties involved can lead to loss of data privacy [37].

Moreover, business events such as an acquisition of the cloud provider could increase the probability of business strategy modification and introduce data privacy risks [37]. In turn, in the event of the confiscation of physical hardware because of a subpoena by law enforcement agencies or civil suits, the centralization of storage as well as shared tenancy of physical hardware results in a higher number of users at risk of the disclosure of their data to third parties [37]. If data centers are located in high-risk countries, e.g., lacking the rule of law and having an unpredictable legal framework and enforcement and states that do not respect international agreements, sites could be raided by local authorities, and private data could be subject to enforced disclosure [37]. Thus, a cloud computing service, which combines outsourcing and offshoring may raise very complex issues; hence, it can be difficult to ascertain privacy compliance requirements [35]. Moreover, a sealed search warrant served at the cloud provider may allow law enforcement to search the tenant's systems while forbidding the cloud provider from notifying the tenant that a search took place [38].

The threat of a "malicious insider" is considered especially important in the case of the cloud computing model, since cloud architectures employ user roles, which are particularly high risk. As cloud services use increases, employees of cloud providers increasingly become targets for criminal groups [37]. Insider threats also include business partners, contractors, and other parties that have any access to a cloud provider's systems. Incidents may involve various types of fraud, sabotage of information resources, and theft of sensitive information. Incidents may also be caused unintentionally. From a user's point of view, moving data and applications to the cloud computing environment operated by a cloud provider expands the circle of insiders not only to the cloud provider's staff and subcontractors, but also potentially to other customers using the service, thereby increasing risk [36].

Multitenancy and shared resources are two of the main attributes of the cloud computing model. Since computing power, storage capacity, and network resources are shared between multiple users, an attacker can exploit vulnerabilities from within the cloud environment, overcome the separation mechanisms, and gain unauthorized access to private data. This class of risks includes the failure of mechanisms separating storage, memory, routing, and reputation between tenants of the shared resources. An attacker can compromise the service engine by hacking it from inside a virtual machine, the runtime environment, the application pool, or through its APIs. The probability of this incident scenario depends on the cloud model considered; it is likely to be low for private clouds and higher in the case of public clouds [37].

An infrastructure of public cloud computing is complex compared with that of a conventional data center. Many software components (for both general computing purposes and management purposes) comprise a public cloud service, which results in a large attack surface. Components evolve in time as new features are deployed and existing ones are upgraded. Data security depends not only on the correctness and effectiveness of particular components, but also on interactions among them. Challenges exist in understanding and securing APIs that are often proprietary to a cloud provider. The complexity also results from the fact that the number of possible interactions between components increases proportionally to the square of the number of software components. The increasing complexity is followed by an increasing number and probability of privacy risks related to the loss or unauthorized access, deletion, use, modification, or disclosure of sensitive data [36].

The privacy risk related to ineffective data deletion can occur in several ways, e.g., when a provider is changed, resources are scaled down, or physical hardware is reallocated.

Also, fundamental cloud-related features impact this risk: data may be available beyond the lifetime specified in the security policy since in-depth data removal requires destroying its physical carrier, which frequently stores data from other users at the same time. When in-depth data removal from the cloud is requested, standard procedures that were developed before cloud emergence (e.g., certification requirements) are inefficient if only the software API is applied to data removal [37]. Also, the risk is impacted by the lack of knowledge of who controls retention of data or what the regulatory requirements are in that respect [35].

Probably the biggest privacy risk related to cloud services is related to the information that the cloud provider accumulates or calculates about user-related activity in the cloud. This would include data collected to measure and charge for resource consumption, logs and audit trails, and application-specific data. Such data, if sold or leaked, or in case of their release in the form of user-scoring service or organization-rating service, are a huge threat to user privacy. For example, in the case of organizations, the data could be used to infer the status and outlook of an organization's initiative [36]. At present, there are no technological barriers to such secondary uses [35]. Encrypting stored data is straightforward, but despite advances in homomorphic encryption, there is no prospect of commercial systems being able to maintain this encryption during real-time processing of large datasets [39]. This means that nowadays, and probably also in the foreseeable future, cloud customers doing anything other than storing encrypted data in the cloud must trust the cloud provider [37] or put their trust in ex post law enforcement. While the focus is mainly on protecting application data, cloud providers also store and process metadata. Regardless of whether the metadata is stored within or outside the cloud resources, metadata includes details about the accounts of cloud users that could be used by the cloud provider for unauthorized purposes or compromised by a third party and used in subsequent attacks [36].

Threats to the user's ownership rights over the data constitute the next group of privacy risks. Rarely does the service contract state clearly that the user or organization retains exclusive ownership over all its data and that the cloud provider acquires no rights or licenses through the agreement. Specifically, the service contract should exclude intellectual property rights and licenses to use the user's data for the cloud provider's purposes. It should also exclude any interests in the data even for security purposes and exclude any cloud provider's unilateral amendment to these data ownership rights [36]. Furthermore, SLAs are expressed in natural languages, as opposed to machine-readable formal languages, making automatic assessment whether data usage rules are respected by cloud service provider impossible. Also, it is hard to prevent data rights transferability to other third parties upon bankruptcy, acquisition, or merger, and it is hard to ensure that a data subject can get access to all his/her PII [35]. To summarize, the usage of public cloud infrastructures makes it difficult to assure effective controls of privacy compliance verification in an automated way, so the end user has no means to verify that his/her privacy requirements are fulfilled [35].

4.6 *Big data*

The big data phenomenon is a consequence of cheap data storage and transmission, and the explosion of digital data sources. There are two categories of digital private data, related to their source: data collected explicitly with the awareness of the person affected and data collected implicitly, without personal awareness that data are being collected. In the latter case, he/she just has the general knowledge that private data collection might happen, but has no specific knowledge if it really has happened, which data have been collected, how long they have been stored, who had or has access to them, and for which purpose they are processed.

Within the category of explicitly collected data, three cases are distinguished. In the first case, the user is fully aware of his/her disclosure of private data to a service provider in order to use the service's core functionality. In the second case, the service provider generates or collects the private data of a stakeholder/customer in a situation where the customer is, or should be, aware of and a participant to the data being used, e.g., a doctor generating the medical history of a patient or a bank keeping track of customer financial transactions. In those cases, providing private data is an unquestionable requirement for being served. The doctor cannot help the patient without being able to collect samples and analyze or process results. Without providing a courier with the private address, a parcel cannot be delivered, etc. The problem of privacy arises when data collected for the sake of a particular service are used for other purposes without the consent of the concerned person.

The third case of the disclosure of private data by a person includes the data collected by a wide range of digital services: social media, media sharing, games, education, training, coworking, etc. People voluntarily disclose their private data, in general, to be in contact with other people or to get a higher social position which is considered a benefit. The problem of privacy arises when such data go beyond people for whom they were intended. In practice, the spread and processing of such data are impossible to stop. It is also important to notice that the definition of benefits resulting from private data publishing evolves during one's lifetime. A video or collection of photos leveraging the popularity of a registered high school student may be an obstacle for his/her professional, social, or political career 20 or 30 years later.

In the category of implicit private data collection, automated data collection and big data analysis are distinguished. Data are collected automatically when digital services are provided. Examples are tracking credit card payments, mobile phones locations, websites visited, web user's queries inserted to search engines, recognizing objects in camera images, and data coming from sensors. Storing these historical private data is not a necessary condition for providing digital services. These data are stored for marketing or safety reasons, but they may be legally or illegally abused for violating people privacy. It is worth to stress that data retention and processing that is illegal in one country may be legal in another one.

Big data opens new possibilities for data analysis. Due to massive computational powers of modern computers, the availability of raw datasets from multiple sources, and the development of new data processing methods, instead of analyzing a random data sample, as it is done with classic statistical methods, it is possible to analyze all available data. Since all data are analyzed, it is possible to accept a higher level of their disorder and lower level of exactitude. It turns out that such approach often yields more objective and accurate knowledge [32]. Finally, one of the most important characteristics of big data analysis is the possibility of a paradigm shift. Instead of discovering knowledge by searching for causality, one can discover it by searching for correlation. Knowledge obtained in this way can form the ground for effective actions; however, it does not provide understanding. In other words, by analyzing big data, it is possible to learn with high probability what is happening, and even what will happen (predictive big data analysis), but not why it happens or why it will happen [5]. Correlation is a statistical relationship between two data values. If one of the data values changes, then the other data value (the correlate) is also likely to change. The correlate's change can be preceding, which permits predicting (with some probability, not with certainty). However, correlations may be meaningless and spurious.

A collection of such correlations is presented by Tyler [40]. It is also worth emphasizing that big data analysis can be personalized, and as such, it is different from profiling. Profiling, as this term is used in IT, is based on the classification and on assumptions that people belonging to the same class will behave in the same way so that they can be treated in the same way. For example, every man over 50 is at risk of heart attack, so every man over 50 should visit a cardiologist. Predictive big data analysis is based on calculating the probability of an event that will happen to a particular person. So, some men over 50 will be at high risk of heart attack, while others will not, depending on case-specific variables. Big data analysis is used to predict what decisions an individual will make in the future, e.g., what product he/she will buy as the next one, what holiday destination he/she will choose, or what next word he/she will type when texting. Big data analysis permits us to go beyond profiling due to the personalization of prediction. However, it must be stressed that both big data analysis and profiling are based on machine-learning techniques [41].

Big data analysis not only increases the risk of privacy violations, but it also changes the character of the risk. The value of big data analysis is in the data reuse for purposes different from the primary use. Some types of big data analysis may undermine the current, broadly used, legal principle of notice and consent of individuals for using their personal data for a specific purpose and a prohibition of using these data for any other purpose. One cannot consent in advance to processing his/her data in a way that does not exist yet. Due to the massive volume of data, the number of data owners is often counted in the millions. Due to their dispersion, an individual who is the owner of his/her private data cannot be asked again for consenting to processing those data when a secondary purpose arises. However, the lack of consent does not necessarily protect privacy. People protested showing their houses in Google Street View for fear of burglaries; however, blurring the image of a particular house could in itself provide a clue for the burglars [5].

One of the fundamental means of protecting user anonymity in the datasets that are to be published (e.g., medical or census data) is a process called data deidentification, which is composed of removing explicit or implicit identifiers, such as name, Social Security number, or driving license number. However, the efficiency of this process is brought into question by big data analysis which—to a large extent—permits to reidentify previously deidentified data. Reidentification tends to be persistent: once data are linked to an identified person, they become difficult to separate from his/her identity. Reidentification applied on a mass scale will gradually erode an individuals' privacy [42]. A significant challenge arises from the fact that it is difficult to develop formal constraints for deidentification that would prove to be robust enough to protect data from the threats of both present and future techniques for reidentification.

To bypass the current legal regulations forbidding personal data processing without explicit consent of a person concerned, service providers make such consent a condition for beginning service. This strategy for gaining consent is particularly significant in digital service markets operating according to the "winner takes all" rule, which leads to their monopolization (observed single social network, single search engine, single online auction service, etc.). Therefore, a person is faced with the hard dilemma: either surveillance acceptance or digital/social exclusion. What is even worse, the personal data collected are subject to trade. To mitigate laws limiting personal data interchange, these data are not traded directly, but shared in a form of recommendation services.

Big data analysis permits systems to go beyond the reidentification of raw data, namely, to create derivative datasets describing sensitive attributes of an individual [43]. This is done through the analysis of relationships of an individual with other persons, products, services, themes, opinions, etc. based on publicly available information and cross-referencing of different datasets. As such information is not directly collected from the individual, companies analyzing big data have no legal obligation to gather his/her consent or give notice in the way required by the laws regulating conventional PII collection [43].

Data derivation permits systems to generate a detailed picture of different aspects of an individual's life, including information that he/she has never explicitly disclosed [43], which is a real threat to his/her privacy. Taking advantage of the individual's sensitive data, a service provider gets to know the preferences of the transacting person. The service provider can therefore takeover the entire "added value" of the transaction by dynamic service pricing in an optimal (from the service provider perspective) distance from the user's reservation price [42]. These information asymmetries and "price discriminations" have been present in the online markets for years, but now, they are further escalated by big data techniques.

Narrowing the users' access and choice (called the "filter bubble" phenomenon) is also a consequence of the big data analysis of sensitive data. When searching the Internet, a person will be always limited to the same fragment of information and knowledge resources considered the most appropriate to him/her. So he/she will never get information from outside the "glass walls." In the long term, it may influence person's cultural capital and even impact free information interchange and the freedom of speech. Clustering Internet users leverages the division of the society into groups of similar-thinking clones [42].

An important problem of big data analysis, which belongs to probabilistic approaches, is that predictive algorithms are often themselves unpredictable. Techniques of machine learning including neural networks, which are the basis of predictive big data analysis, run in two phases. The first phase is devoted to training from examples; the second phase is devoted to prediction. The quality of prediction is highly dependent on examples used for training the network. If a real case does not conform to training examples, the prediction will be false; thus, decisions based on that prediction will be wrong. The consequences of such wrong predictions may be different, from negligible or severe. If a person is not properly prompted when texting, consequences are negligible, but if a person is wrongly qualified as a potential terrorist, the consequences may be very severe. In practice, it is impossible to prove that a prediction is right or wrong, because, as we mentioned earlier, big data analysis is not based on a cause-and-effect relationship but on correlations among different datasets and the analysis of a big number of training examples. Moreover, it is impossible to know in advance when a learning algorithm will predict a user's PII. Therefore, it cannot be planned where and when to assemble privacy protections related to these data [43].

The legal status of different datasets is ambiguous. Some of them are publicly or semi-publicly accessible, some of them are owned by communication and digital service providers. It is unclear whether an individual's data can be simply used (alone or as a part of a larger aggregate) without requesting permission whether the data can be taken out of the context and analyzed in a way likely opposite the subject's will, who can benefit from the access to big data; who is responsible for ensuring that individuals are not harmed by the big data analysis, how informed consent should be defined and how it can be executed, and what constitutes the set of best ethical practices for data analytics [44].

Even with the use of nonsensitive data, predictive big data analysis may have a discriminatory effect on individuals. Those who have a privileged digital history since their childhood, e.g., having well-situated family in their social network, having digital interests (likes and clicks) more attractive from the commercial point of view, or having more promising financial prospects will automatically receive even more measurable benefits in the digital society; for those with questionable or ambiguous digital records, even the social status that they have had thus far will be hard to maintain. "Predictive analysis becomes a self-fulfilling prophecy that accentuates social stratification" [42].

People attempting nonstandard behaviors or brave enough to take challenges, but who failed, will risk immediate decline in the digital society based largely on big data analysis. What is even worse, "as the ramifications of big data analytics sink in, people will likely become much more conscious of the ways they're being tracked, and the chilling effects on all sorts of behaviors could become considerable" [45]. What will result is a gradually emerging "surveillance society, a psychologically oppressive world in which individuals are cowed to conforming behavior by the state's potential panoptic gaze" [42]. The worst kind of censorship is autocensorship.

4.7 Conclusions

As follows from the preceding sections, emerging information technologies including the IoT, AR, biometrics, cloud computing, and big data analysis increase the risk of privacy breaches and, in many cases, make current approaches to protecting privacy inefficient and insufficient. Moreover, within the e-society, all these technologies are used simultaneously, so their cumulative and reinforcing effects apply to each person. Thus, it is required to adopt a more holistic view of privacy protection.

The IoT, AR, and biometrics may be seen as data providers. Those data are stored in the cloud. Data aggregated in the cloud, coming from different sources, are perfect objects for big data analysis.

The IoT provides new challenges for privacy because it follows the principle of ubiquitous computing. Sensors and actuators deployed in a particular environment (smart home, smart car, smart road, etc.) adapt that environment to individual or group needs automatically, without explicit human interaction. Therefore, there is no space for explicit consent. Adaptation requires knowledge of preferences, i.e., private data. If a smart home or a smart car is adapted to the needs of its owner, the owner's private data are not disclosed to third parties. If a road, an office, or a public building is adapted to the needs of an individual, his/her private data have to be entrusted to companies managing them, so the risk of trust abuse is much higher. It is also worth noting that IoT extends the risk of private data abuse from the digital world to the real world, i.e., real installations deployed in buildings, cars, roads, etc. The malfunctioning of these installations may cause physical damages.

AR provides private data coming from information-rich raw multimedia data streams that are temporally and spatially interrelated. These data concern not only the owner of an AR device, but all the people who surround him/her in a particular place and moment, who are captured in photos and videos.

Biometric access control provides unique personal identification for life. As such, it eliminates a possibility of privacy protection by several virtual identities of the same person devoted to different services. As mentioned in Section 4.4, data collected for biometric access control may be used to infer information describing persons, not only his/her body or medical condition, but even cultural or social characteristics.

Cloud computing is currently the most economical option of providing computing power and storage capacity. It is particularly useful in the case of small electronics devices of limited capabilities including power supply such as sensors, actuators, and mobile devices. The application of cloud computing requires entrusting private data to cloud computing providers, i.e., pass control over them. A client of cloud computing services can only trust that his/her data are not processed for purposes that he/she never agreed to. The risk of trust abuse is increased by the fact that data stored in the cloud are often replicated and spread among different locations in different countries governed by different law regulations. To reduce the risk associated with could computing, private or community clouds are used which restrict clients to one company or several companies of similar characteristics, e.g., only banks. In a contract with cloud a computing service provider, a client may consider restricting the storage of his/her data to one data center or centers located in one country governed by one system of privacy protection regulations.

As mentioned earlier, massive data aggregated in the cloud, coming from different sources, partially identified, are perfect objects for big data analysis based on correlations instead of on a cause-and-effect relationship. Big data technologies have the potential to bring together all the specific risks following from the technologies described earlier and amplify them. Big data analysis permits us to not only reidentify data deidentified prior to release, but also generate with high probability a detailed picture about different aspects of a person's life, including information the person has never disclosed to any service. Moreover, big data analysis permits—again with high probability, not certainty—the prediction of future behavior and future actions of a person. In cases when certainty is not required and a person retains the right to free choice, predictive big data analysis may provide a person with advantages, otherwise—not. If an individual is an object of a decision made by somebody else, e.g., he/she may get a loan or not, may get a job or not, or may be invited to an event or not, big data analysis may lead to discrimination.

As explained by Mayer-Schonberger and Cukier [5], predictive big data analysis challenges the current justice system. Thinking about committing a crime is not illegal, only progressing from the thought to the criminal act is. Individual responsibility is linked to the individual choice of an action. Finding someone guilty of an anticipated crime he/she has not yet committed is a mistake made by using predictive big data analysis based on the correlation with making decisions about one's individual responsibility that requires a proof of a cause-and-effect relationship. The abuse of big data analysis leads to a society in which there is no individual choice of action based on free will, but the individual moral compass is replaced by predictive algorithms, and individuals are exposed to the unlimited coercion of collective decisions made in the past used to calculate the probability of their actions to be made in the future. This poses the risk of enslaving of the society.

As follows from this chapter, in the era of emerging technologies, in particular predictive big data analysis, a new approach to protect individuals' privacy has to be developed. The risk of wrong predictions may be mitigated by providing access to data and algorithms for their verification and certification by trusted third parties. The same big data should be analyzed by a regulatory authority, to limit the monopolization of benefits coming from predictive analysis. Also, it is necessary to create legal procedures for rebutting a prediction about a specific person. Organizations using predictive big data analysis should be legally responsibility for its effects. Also, if provided open access to big data, governments or third-sector organizations could use the same big data techniques to discover who is being discriminated against and by whom the discrimination is being activated.

References

1. European Parliament and the Council of the European Union, Regulation (EU) 2016/679 of the European Parliament and of the Council of 27 April 2016 on the protection of natural persons with regard to the processing of personal data and on the free movement of such data, and repealing Directive 95/46/EC (General Data Protection Regulation), *Official Journal of the European Union*, L 119, 2016.
2. Stevens, Gina Marie. Data security breach notification laws. Congressional Research Service, Washington, DC (2012).
3. Cremonini Marco, Chiara Braghin, and Claudio Agostino Ardagna. *Privacy on the Internet Computer and Information Security Handbook* (Second Edition), Morgan Kaufmann, Burlington, MA (2013).

4. Cellary, Wojciech, and Jarogniew Rykowski. Challenges of smart industries—Privacy and payment in visible versus unseen Internet. *Government Information Quarterly* (2015).
5. Mayer-Schonberger, Viktor, and Kenneth Cukier. *Big data: A Revolution That Will Transform How We Live, Work, and Think*, Houghton Mifflin Harcourt, Boston, MA (2013).
6. International Telecommunication Union, Recommendation ITU-T Y.2060 (06/2012), 2012. http://handle.itu.int/11.1002/1000/11559.
7. Hajdarbegovic, Nermin. Are We Creating An Insecure Internet of Things (IoT)? Security Challenges and Concerns, 2015, Accessed April 5, 2017. https://www.toptal.com/it/are-we-creating-an-insecure-internet-of-things.
8. Hewlett-Packard Development Company. HP Study Reveals 70 Percent of Internet of Things Devices Vulnerable to Attack, 2014, Accessed April 5, 2017. http://www8.hp.com/us/en/hp-news/press-release.html?id=1744676.
9. Light Reading. Poll: Standardization Biggest Challenge in IoT, 2015, Accessed April 5, 2017. http://www.lightreading.com/iot/iot-strategies/poll-standardization-biggest-challenge-in-iot/a/d-id/714062.
10. Gartner. Gartner Says 8.4 Billion Connected Things Will Be in Use in 2017, up 31 Percent from 2016, Gartner Press Release, 2017, Accessed September 11, 2017. http://www.gartner.com/newsroom/id/3598917.
11. Verizon. State of the Market: The Internet of Things 2015, 2015, Accessed April 5, 2017. http://www.verizonenterprise.com/resources/reports/rp_state-of-market-the-market-the-internet-of-things-2015_en_xg.pdf.
12. Wind River System. Security in the Internet of Things. Lessons from the Past for the Connected Future, 2015, Accessed April 5, 2017. https://www.windriver.com/whitepapers/security-in-the-internet-of-things/wr_security-in-the-internet-of-things.pdf.
13. Apthorpe, Noah, Dillon Reisman, and Nick Feamster. A smart home is no castle: Privacy vulnerabilities of encrypted IoT traffic, *Workshop on Data and Algorithmic Transparency (DAT'16)*, 2016.
14. Wang, Chen, Xiaonan Guo, Yan Wang, Yingying Chen, and Bo Liu. Friend or foe? Your wearable devices reveal your personal pin. In *Proceedings of the 11th ACM on Asia Conference on Computer and Communications Security*, pp. 189–200. Association for Computing Machinery, New York (2016).
15. Ramirez, Edith. *Privacy and the IoT: Navigating Policy Issues*. US Federal Trade Commission, Washington, DC (2015).
16. Lobao, Martim. Software Updates: A Visual Comparison of Support Lifetimes for iOS vs. Nexus Devices., Android Police, 2015, Accessed September 11, 2017. http://www.androidpolice.com/2015/09/17/software-updates-a-visual-comparison-of-support-lifetimes-for-ios-vs-nexus-devices/.
17. Canonical, Defining IoT Business Models. Monetising IoT investments, maximising IoT skills and addressing IoT security, 2017, Accessed September 11, 2017. https://pages.ubuntu.com/IOT_IoTReport2017.html.
18. Wójtowicz, Adam, and Daniel Wilusz. Architecture for adaptable smart spaces oriented on user privacy, *Logic Journal of the IGPL*, vol. 25, issue 1, pp. 3–17 (2017).
19. Azuma, Ronald, Yohan Baillot, Reinhold Behringer, Steven Feiner, Simon Julier, and Blair MacIntyre. Recent advances in augmented reality. *IEEE Computer Graphics and Applications* vol. 21, issue 6, pp. 34–47 (2001).
20. Jana, Suman, Arvind Narayanan, and Vitaly Shmatikov. A Scanner Darkly: Protecting user privacy from perceptual applications. In *2013 IEEE Symposium on Security and Privacy (SP)*, pp. 349–363. Institute of Electrical and Electronics Engineers, Piscataway, NJ (2013).
21. Calo, Ryan. People can be so fake: A new dimension to privacy and technology scholarship. *Penn State Law Review* vol. 114, p. 809 (2009).
22. Raval, Nisarg, Animesh Srivastava, Kiron Lebeck, Landon Cox, and Ashwin Machanavajjhala. Markit: Privacy markers for protecting visual secrets. In *Proceedings of the 2014 ACM International Joint Conference on Pervasive and Ubiquitous Computing: Adjunct Publication*, pp. 1289–1295. Association for Computing Machinery, New York (2014).
23. Jana, Suman, David Molnar, Alexander Moshchuk, Alan M. Dunn, Benjamin Livshits, Helen J. Wang, and Eyal Ofek. Enabling fine-grained permissions for augmented reality applications with recognizers. In *USENIX Security*, pp. 415–430. USENIX Association, Washington, DC (2013).

24. Dwork, Cynthia. Differential privacy. In *Proceedings of the 33rd International Conference on Automata, Languages and Programming*, pp. 1–12 (2006).
25. Roesner, Franziska, Tadayoshi Kohno, and David Molnar. Security and privacy for augmented reality systems. *Communications of the ACM* vol. 57, issue 4, pp. 88–96 (2014).
26. Bell, Michael, and Vitali Lovich. Apparatus and methods for enforcement of policies upon a wireless device. US Patent 8,254,902, issued August 28, 2012.
27. Roesner, Franziska, Tamara Denning, Bryce Clayton Newell, Tadayoshi Kohno, and Ryan Calo. Augmented reality: Hard problems of law and policy. In *Proceedings of the 2014 ACM International Joint Conference on Pervasive and Ubiquitous Computing: Adjunct Publication*, pp. 1283–1288. Association for Computing Machinery, New York (2014).
28. Prabhakar, Salil, Sharath Pankanti, and Anil K. Jain. Biometric recognition: Security and privacy concerns. *IEEE Security & Privacy* vol. 99, issue 2, pp. 33–42 (2003).
29. Grijpink, Jan. Privacy law: Biometrics and privacy. *Computer Law & Security Review* vol. 17, issue 3, pp. 154–160 (2001).
30. Crompton, Malcolm. Biometrics and privacy the end of the world as we know it or the white knight of privacy? *Australian Journal of Forensic Sciences* vol. 36, issue 2, pp. 49–58 (2004).
31. Derakhshani, Reza, Stephanie AC Schuckers, Larry A. Hornak, and Lawrence O'Gorman. Determination of vitality from a non-invasive biomedical measurement for use in fingerprint scanners. *Pattern Recognition* vol. 36, issue 2, pp. 383–396 (2003).
32. Bolle, Ruud M., Jonathan Connell, Sharath Pankanti, Nalini K. Ratha, and Andrew W. Senior. *Guide to Biometrics*. Springer, Berlin (2013).
33. Rathgeb, Christian, and Andreas Uhl. A survey on biometric cryptosystems and cancelable biometrics. *EURASIP Journal on Information Security* vol. 2011, issue 1, p. 3 (2011).
34. Rejman-Greene, Marek. Privacy issues in the application of biometrics: A European perspective. In Wayman, James, Anil Jain, Davide Maltoni, and Dario Maio (eds), *Biometric Systems*, Springer, London, pp. 335–359 (2005).
35. Pearson, Siani. Privacy, security and trust in cloud computing. In *Privacy and Security for Cloud Computing*, pp. 3–42. Springer, London (2013).
36. Jansen, Wayne, and Timothy Grance. *Guidelines on Security and Privacy in Public Cloud Computing*. NIST Special Publication 800-144. National Institute of Standards and Technology, Gaithersburg, MD (2011).
37. European Union Agency for Network and Information Security. *Cloud Computing: Benefits, Risks and Recommendations for Information Security*. European Network and Information Security, Heraklion (2009).
38. Molnar, David, and Stuart E. Schechter. Self hosting vs. cloud hosting: Accounting for the security impact of hosting in the cloud. In *Proceedings of the Ninth Workshop on the Economics of Information Security (WEIS 2010)*. Microsoft Research, Cambridge, MA (2010).
39. Zang, Wanyu, Meng Yu, and Peng Liu. Privacy protection in cloud computing through architectural design. In Vacca, John R. (ed), *Security in the Private Cloud*, pp. 319–343. CRC Press, Boca Raton, FL (2016).
40. Vigen, Tyler. *Spurious Correlations*. Hachette Books, New York (2015).
41. Alpaydin, Ethem. *Introduction to Machine Learning*. MIT Press, Cambridge, MA (2014).
42. Tene, Omer, and Jules Polonetsky. Big data for all: Privacy and user control in the age of analytics. *Northwestern Journal of Technology and Intellectual Property* vol. 11, p. xxvii (2012).
43. Crawford, Kate, and Jason Schultz. Big data and due process: Toward a framework to redress predictive privacy harms. *Boston College Law Review* vol. 55, p. 93 (2014).
44. Boyd, Danah, and Kate Crawford. Critical questions for big data: Provocations for a cultural, technological, and scholarly phenomenon. *Information, Communication & Society* vol. 15, issue 5, pp. 662–679 (2012).
45. Stanley, Jay. *The Potential Chilling Effects of Big Data*. American Civil Liberties Union, New York (2012).

chapter five

Trust

David Schuster and Alexander Scott

Contents

5.1 Introduction

"Trust is good, but control is much better," attributed to Vladimir Lenin [1], suggests that being trusting means being vulnerable. Rather, trust is an adaptation to an uncertain, risky situation; humans apply trust to make decisions and minimize risk. At present, cybersecurity occurs in a context characterized by high risk, uncertainty, time pressure, and an almost inconceivable number of agents potentially affecting the security of a network. This environment challenges cybersecurity professionals, who defend organizations against threats, and consumer users of the Internet, who are willing or unwilling participants in their own personal and organizational security. These challenges will only increase as computer networks, and attacks on computer networks, grow in their size and sophistication.

Through their decision-making, users can mitigate or exacerbate threats. Human decision-making is critical to cybersecurity across roles and contexts. Poor decisions that affect security outcomes are the core reason why users are cited as the weakest link in security [2] or why social engineering is often an easier vector for attacking an organization than through electronic means [3]. Symantec's report on cyberthreat trends [3] found that malicious actors are increasingly utilizing social engineering tactics, citing effectiveness and ease of use. In this chapter, we focus on how trust affects decision-making, and ultimately, information security. By seeing trust as an adaptive process that can affect decision-making, we can better understand why social engineering works. The problem is not that humans trust, but that trust can be misplaced; we argue that designers should encourage the development of *appropriate* trust, which facilitates good decision-making. Further, we will describe how trust can be incorporated into a user-centered design process.

While human–computer interaction (HCI) in cybersecurity applies to an operator and a computer, interactions among members of a team, or between an individual and an attacker, also affect security outcomes. Complicating matters, many interactions between individuals are computer mediated. Sometimes, these interactions take the form of computer-mediated communication, as in sending an e-mail. Other times, an interaction could be one person's observation of another's behavior observable only via the computer network. For example, a cyberdefender may watch for changes in the dashboard of a security tool to determine whether an attacker was successful at infiltrating the network.

We offer these examples to distinguish cybersecurity, in its size and complexity, from other domains of interest to HCI. Methods traditionally used to understand less complex interactions, such as a heuristic evaluation of an application, are still used in cybersecurity domains. However, the increasing capability of automated tools in other domains has blurred the lines between people and computers in some ways. While continuing increases in computer processing power, network bandwidth, and storage space drive some of these improvements, cybersecurity is a challenging societal problem of unprecedented scale. The expected cost of data breaches by 2019 is $2.1 trillion [4]. The need for improved cyberdefenses by organizations will also drive the increasing sophistication of cybersecurity tools targeting cybersecurity professionals and other users at work and in the home. As the complexity of these tools grows, the interactions between users and the sociotechnical system may resemble interpersonal interactions in some ways.

5.1.1 *Defining trust*

Trust has been defined differently across disciplines. In this chapter, we take a user-centric definition of *trust* appropriate for HCI. From our perspective, *trust* is the "willingness to be vulnerable to the actions of another party" [5]. Our daily lives, professional and personal, are characterized by many decisions that are based on trust in others. We trust cashiers to accurately total our purchases, firewalls to remain active, and other vehicle drivers to also follow traffic laws. Trust is one way we adapt to an inherently risky and uncertain world that requires time-critical decision-making. Under these circumstances, we could not use our computers if we required certainty to proceed. Trust allows us to accept and limit risk in these situations despite uncertainty and dependence on others [5].

The application of trust is not universal across all people, computers, and situations. Trust in another person is called interpersonal trust while trust in a nonhuman agent is called trust in automation [6]. While the word *trust* dates to the year 1200 describing a general concept of reliance [7], trust has been applied to technology relatively recently. The application of trust to technology has led to a new construct in the literature. *Trust in automation* has been defined by Lee and See [8] as "the attitude that an agent will help achieve an individual's goals in a situation characterized by uncertainty and vulnerability." Trust in automation is particularly relevant to HCI in cybersecurity because untrusted technology might not be used; the recommendations of trusted technology are followed more than untrusted technology [8]. Empirical research has consistently shown that trust in automation predicts decisions to use tools [9–11] such that people are more likely to use and follow the guidance of technology that they trust [12].

To date, research has identified both differences and similarities between interpersonal trust and trust in automation. A theoretical explanation for differences in trust in automation versus people is that automation lacks intentionality, a quality of thinking being directed toward a known goal [8]. Machines execute instructions but do not consider goals. Our interactions with other people can benefit from social cues, such as symmetry of trust, which occurs when each partner understands how they will be perceived by the other member [13].

Interpersonal trust and trust in automation both suffer when the trustee is unreliable [13]. However, unreliability in machines is perceived differently than unreliability in humans [14]. Humans tend to expect machines to perform predictably, leading to more rapid declines in trust when unreliability is encountered [15]. According to Madhavan and Wiegmann [14], automation trust develops from a schema of perfect automation, making users more aware of errors and heavily basing trust on the perceived performance of the automation. Further, not all automation errors affect trust in the same way; the easy-errors hypothesis says that when people observe automation making mistakes on tasks they consider to be easy, trust falls even if the aid is otherwise quite reliable [16]. For example, if users cannot find a file known to exist using a desktop search feature and then find it manually, they may be disinclined to use the search feature in the future when searching for other files. In contrast, trust in humans develops from a schema of imperfection (i.e., to err is human), making users more forgiving of errors and basing trust on knowledge of the trustee.

An important consideration in HCI in cybersecurity, then, is whether the trustee is a human or a machine. In the cybersecurity context, interpersonal communication is often mediated by a computer instead of taking place face to face; that is, words written by a person are transmitted by computer. For example, crowd-sourced ratings of a product are written by individual people but are aggregated, filtered, and sorted by machines. Consequently, the distinction between trust in a human or machine is likely to be driven by the perception that an agent's behavior is dictated by a human or determined by an algorithm. One present author and colleagues explored this distinction in an experiment manipulating the origin of app security ratings that were presented as being algorithmically generated or human generated. They found that users were more likely to follow algorithmically generated ratings than human-generated ones when both provided similar levels of risk [17].

Despite differences in people's trust of humans versus machines, people tend to anthropomorphize, or apply human qualities to machines [18]. In other words, people sometimes treat technological agents the same way as they would treat human beings [19]. For example, people prefer technological agents that they feel are like themselves [18].

Designers may capitalize on anthropomorphism so that trust in automation more closely resembles interpersonal trust. Automation that supports more naturalistic interactions with people, and possibly attitudes resembling interpersonal trust, is called human–automation teaming.

Human–automation teaming [20] can be thought of as a style of interaction that describes how the nature of HCI will change as automation becomes more capable of participating in decision-making processes and interacting with humans in naturalistic ways. Human–automation teaming includes automation capable of participating in team process, the dynamic coordinating behaviors that humans engage in as part of human teamwork [21]. Human–automation teaming has been applied to the development of military robotics (called human–robot teaming; [22]) and in airspace operations research [23]. In both domains, the technological capability that would support human–automation teaming is nascent. This is similarly true in cybersecurity, but advances in machine learning and artificial intelligence applied to big data will lead to new methods to detect security threats. The continuous and rapid changes to the threat landscape pressures solution developers to leverage big data so that previously unseen or emerging threats can be detected. It is unlikely that such tools will be able to operate independent of human intervention. Cybersecurity professionals will still need to make decisions about identified threats and bridge between security outcomes and organizational impact to mitigate risk. For example, if a cybersecurity tool based on machine learning provided recommendations to a human operator for mitigation actions while describing the available evidence for the suggested decision with a chat interface, the operator would be able to hold a conversation in natural language to further explore the evidence for the threat along with possible mitigation strategies. Subsequently, with a more holistic view of the situation augmented by the security tool, a human operator would synthesize the machine-derived intelligence with their own knowledge of the context, such as whether the threat affects a system critical to business operations, to make an action decision. Interactions that support human–automation teaming further blur the line between interpersonal trust and trust in automation. If human interactions with automation begin to resemble human interactions with humans, the differences between trust in humans and trust in automation may decrease. However, there is a need for more research; research on trust in automation has so far focused on relatively simple interactions, decision selections, and information analysis in domains such as combat identification, decision aids, monitoring, visual inspection, route planning, and collision warning [24].

Trust is an intuitive concept and still emerging as a measurable construct of interest in HCI. Following a review of current theoretical perspectives on trust, we apply trust constructs to two user populations: cyberdefenders and consumer users of the Internet.

5.1.2 *Models of trust*

Trust reflects a relationship between an entity doing the trusting (the trustor) and the entity being trusted (the trustee). Trust implies that the trustor is invested in an outcome or goal and that there is a possibility of failure [25], which provides risk. In organizational contexts, goals are layered and may reflect individual, team, or higher-level goals [5]. Trust has been distinguished from similar constructs such as prediction, which also reduces risk; cooperation, which can occur with or without trust; and confidence, which can occur independently of decision-making [5].

Schaefer, Chen, Szalma, and Hancock [18] described trust as a relationship with three components. The first is propensity to trust, a relatively stable trait of the trustor. Propensity to trust is the initial likelihood to trust in an entity before having any experience with it [12]. Propensity to trust is a "generalized expectation about the trustworthiness of others" (p. 715) and reflects a willingness to trust [5]. Individuals with a higher propensity to trust may be more likely to maintain trust than individuals with less propensity to trust [26].

The second property is a situationally defined state of trusting. The third property reflects that trust changes over time, and the change over time is a property of the interaction between the trustor and the trustee. Both the trustor and trustee affect the bidirectional trust relationship. Trust is adaptive when it is justified, when there is a match between the trustor's trusting and the trustee's trustworthiness [6].

The bidirectional nature of the trust relationship presents challenges to our understanding of trust. Trust is both a predictor of successful human–technology interaction and an outcome of human–technology interaction. Trust is determined by properties of the trustee, suggesting terms such as *trustworthiness*, *dependability*, and *reliability*. But trust is also determined by properties of the trustor, such as propensity to trust and the trustor's understanding of the technology [24].

Another continued challenge frequent in the scientific study of HCI is the complex nature of cognition in human–technology interactions in operational environments. It is difficult to identify the difference between, and relationship between, dynamic processes and outcomes. In operational environments, trust is employed within complex and parallel decision-making processes with no clear start or end [6]. History-based trust changes over time as a user gains experience with automation and varies across entities being trusted [12]. Another research question is whether the loss of trust is distinct from the building of trust, with some evidence that these are separate constructs [27].

Although a research need remains, it is established that trust is a dynamic attitude held by the trustor, and the impacts of trust are observable in the behavior of the trustor over time. In HCI, we are primarily concerned with the trustor as a user or customer. Trust impacts the quality of an interaction across performance, safety, and user satisfaction outcomes. At the most surface level, trust may provide an explanation for user outcomes.

5.2 *How trust affects decision-making*

Users affect cybersecurity through their decision-making. *Decision-making* is defined as the selection of an option in a situation with some ambiguity or risk, provided the decision takes longer than 1 s to make [28]. This definition distinguishes decision-making from faster reactions to perceptual stimuli. Decision-making is critical to successful security outcomes, both for cybersecurity professionals and consumer users.

As a process, decision-making varies depending on the decision maker's strategies, experience, and resolution of conflicts, which reflect uncertainty [29]. In their integrative decision-making model, Lehto, Nah, and Yi [29] identified the conflicts resolved through decision-making (listed in Table 5.1).

Generally, decision-making can be understood as a process that seeks to resolve one or more of these conflicts. For example, conflicting objectives are resolved through judgments about the importance of objectives, leading to prioritization as an outcome. How human decision makers perform this process has been extensively studied, resulting in two broad approaches to modeling the cognition behind human decision-making: (1) normative models of decision-making and (2) descriptive models of decision-making. While both are useful and informative, neither provides a comprehensive model of how external information predicts decision-making.

Table 5.1 Conflicts resolved through decision-making

Lack of consensus
Uncertain consequences
Uncertain preferences
Conflicting objectives (bad consequences)
Uncertain aspirations
Need to compare alternatives
Unidentified conditions, alternatives, or consequences

Source: Lehto, Mark R., Fiona Fui-Hoon Nah, and Ji Soo Yi. *Handbook of Human Factors and Ergonomics*, John Wiley & Sons, Hoboken, NJ, 191–242, 2006.

Normative models, with origins in behavioral economics, reflect an information processing approach that maximizes utility. That is, people acquire information relevant to a decision; weigh the quality, importance, and reliability of this information; and select the option that maximizes their utility. Decades of research have demonstrated that normative models do not sufficiently predict how people make decisions in operational contexts. Decision-making in a naturalistic environment is not a linear process of weighting and synthesizing complex decision criteria [30].

Descriptive models of decision-making explain how decisions are made in context. A major feature of descriptive decision-making is satisficing, which is choosing the option that is good enough, even if it is not the optimal decision. One reason that normative models do not hold across contexts is that the amount of information required to make a normative decision grows exponentially as more factors are considered and as more options become available. Humans are adapted to be efficient in their cognition [31]; that is, we take shortcuts in our mental processing of information. Decision heuristics are an example of a descriptive approach to decision-making [32]. Decision heuristics are shortcuts that reduce the information-processing load by selectively disregarding some information that could inform a decision [33]. Heuristics can be adaptive in that they support faster decision-making with acceptable outcomes much of the time, especially when employed by experts, who are able to focus on the most relevant cues [34]. For example, a cyberdefender may use the representativeness heuristic to match critical cues in an investigated event to an attack observed earlier rather than considering the probability of the attack, which may be rare or difficult to identify.

Together, normative and descriptive models of decision-making show that humans are rational information processors willing and able to shortcut this process when some uncertainty is difficult or impossible to resolve. This process is adaptive to decision makers in a complex world, as is trust. Understanding the impacts of decision-making, and being able to predict user decisions, is critical to HCI in cybersecurity.

Cyberdefense, by both professionals and consumer users, is characterized by the use of multiple tools to defend against and respond to threats. Professionals and consumer users face a multitude of security-critical decisions about which tools should be used under what circumstances and what actions to take because of alerts from these tools. For example, a warning about an expired certificate suggests to a user that they should not continue to the intended site. From a normative decision-making perspective, the user must gather and appropriately weigh the relevant evidence to decide if this warning is spurious or consequential and then take appropriate action. However, consumer users often lack the expertise necessary to interpret security warnings [35,36], and therefore, a normative decision-making model is unlikely to predict the user's decision. Further, these decisions are made quickly and are not part of the user's primary task. As a quick decision characterized by uncertainty, trust may determine whether the user will follow the recommendations of the warning or not.

Trust is useful for describing or predicting other decision-making factors, especially those that are difficult to include in a normative decision-making model. For example, while the reliability of a tool predicts trust in it, trust may be inertial [37]. That is, occasional false alarms or misses by an otherwise reliable tool might not reduce the use of the tool. Further, trust in a security tool can affect use in two ways. Dixon and Wickens [38] demonstrated that user compliance and reliance can be affected independently. Compliance is the operator's use of automation when it has presented a signal, such as when security software indicates that a threat is present. Reliance is the use of automation when no threat signal has been presented, as occurs when security software indicates no threats. People separately consider the potential for misses, when automation fails to detect a threat, and false alarms, when automation falsely suggests that a threat is present. Trust can affect compliance, reliance, or both.

5.3 *How to incorporate trust into design for security in HCI*

Trust can be incorporated into the design of cybersecurity tools in both research and practice contexts. Trust predicts and explains user decisions, making it a useful variable in research studies. Designers may also increase user satisfaction and security when they ensure that tools are appropriately trusted. In this section, we propose strategies that designers can employ to address challenges related to trust in cybersecurity tools.

5.3.1 *Challenge 1: Either too much or too little trust can be problematic*

A challenge in designing for trust is that the designer must design for *appropriate* trust. We have described how trust is an adaptive process to facilitate decision-making in cybersecurity; trust is something to be attended to in design, not something to be avoided. It is adaptive, and when trust is placed in trustees who deserve to be trusted, better decisions can be made with less time and effort. Trust is based, in part, on observations of the behavior of targets of trust. When trustors observe potential trustees acting reliably and transparently, their trust increases. Similarly, trust decreases when potential trustees act in unpredictable ways or seem to make mistakes. This is not a perfect process, however; trust can become miscalibrated when trustors' perceptions of reliability are inaccurate. When trustors award less trust than deserved to those who would be trustees, called distrust, decisions may be suboptimal. Trustees, both human and technological, can be underutilized if they are undertrusted. This could be one reason that a tool that provides an important warning or alert goes ignored.

More trust does not always lead to better cybersecurity outcomes. Overtrust is an inappropriately applied trust that leads to overreliance on automation. When users overrely on automation, they give too much weight to automation in their decision-making and do not sufficiently anticipate automation failures. Overreliance predicts complacency, a state of insufficient monitoring of the output of automation [39]. Complacency may go undetected if automation only sporadically fails or misses a threat that occurs rarely. Thus, even if overtrust seems to satisfy users in the short term, performance and trust will eventually be harmed when user expectancies are violated, possibly in a security-critical situation. Because of the complex nature of threat defense, cybersecurity tools often provide an incomplete picture or provide results with a limited level of confidence. Thus, overtrust

Table 5.2 Summary of challenges in leveraging trust for security in HCI

Challenge	Proposed solution
1. Either too much or too little trust can be problematic.	Strive for appropriate trust: Identify targets of trust, describe trustworthiness criteria, and ensure that users understand when and why they should trust.
2. Trust is context and user dependent.	Incorporate trust measurement in each iteration to understand the effect of design decisions on trust.
3. Poor usability leads to mistrust.	Identify and match user expectancies to improve usability.
4. Trustworthiness depends on reliability, which may be unknown.	Provide transparency about the reasons for automation failure to support accurate inferences about reliability.

can negatively impact security outcomes when users do not investigate or respond to an alert from a tool.

5.3.1.1 *Proposed design strategy*

Designers must strive for appropriate trust. To assess appropriate trust, first identify the entities being trusted and identify them as human or technological. Second, describe the criteria under which each entity should be trusted. As a simple example, a firewall may be an entity that requires no monitoring, and thus a high level of trust, under regular operational conditions. If the organization is the victim of an attack, however, verifying the configuration and performance of the firewall may be important. Finally, designers can measure trust and employ strategies to increase or decrease trust that reflect the nature of the entity (human or technological) and the circumstances under which users should rely on (Table 5.2).

5.3.2 *Challenge 2: Trust is context and user dependent*

Trust, especially in automation, is contingent on the goals of the task or user. Just as performance is the degree to which user goals are achieved, trust in automation is contingent upon a goal [6]. A user might trust a virus scanner to find a known virus but not to monitor their home for a break-in. Consequently, HCI methods that measure or manipulate trust must consider to what degree the trust is task specific and whether findings of trust in automation to do one task may hold as the task parameters vary, especially as changes are made to the design or functionality of the automation. For example, the circumstances under which users trust virus scanners to detect viruses may differ from the circumstances under which users trust password managers to protect their login information.

Designers must also consider how user characteristics affect trust. Factors related to the trustor include emotive factors, cognitive factors, traits, and states [18]. Emotive factors include subjective outcomes such as user satisfaction, comfort, and attitudes toward the trustee. Emotive factors closely tie to user satisfaction outcomes, with the implication that user satisfaction may predict trust (see the following challenge).

Cognitive factors relate to the trustor's understanding of the automation, the ability to use the automation, and expectancies of the automation. These factors are relevant in security contexts where there may be variability in users' expertise.

As user characteristics, states and traits are distinguished by how stable they are within one individual over time. States are dynamic individual differences such as mood. States could affect the trust development, although there is a need for more research specific to trust. Traits are distinguished by their stability over time. Traits are relatively stable across task situations, including age, personality, and propensity to trust. Traits interact with task characteristics [40], preventing broad recommendations for design based on specific characteristics. However, designers should understand that user traits may interact with automation characteristics to impact trust. Most directly, propensity to trust is a characteristic that users bring to an interaction before trust is affected by the use of an automated tool.

5.3.2.1 Proposed design strategy

Because context matters, appropriate trust is not easily predicted a priori. As with other outcomes of interest to HCI practitioners, such as user satisfaction, an iterative, user-centered design process will lead to the most optimal solution as a product evolves. HCI practitioners can measure trust in each iteration using one or more of the methods described later in this chapter. Specifically, state measures of trust in a specific entity complement measures of propensity to trust.

5.3.3 Challenge 3: Poor usability leads to mistrust

The ability to use automation is closely related to usability, such that highly usable products may be trusted more [41,42]. Hoff and Bashir [24] presented several literature-based design recommendations to increase trust. They suggested that increasing anthropomorphism, creating usable interfaces, adopting a polite communication style, providing accurate feedback about and context for automation failure, and avoiding errors during early interactions or on easy tasks would increase trust. In a security context, this means that poor usability can lead to miscalibrated trust, especially distrust of automated systems.

5.3.3.1 Proposed design strategy

Automation that is compatible with users' expectancies is trusted more, especially when the consistency helps users to predict its behavior [43]. Users have expectancies of the automation, based on their understanding of the relationships among relevant system concepts [18,44]. When automation matches expectancies, it is trusted more. Trust is harmed when the behavior of automation is unexpected or incompatible with the internal representation held by the user. Assessments of user expectancies can be used to suggest modifications to automation behavior.

5.3.4 Challenge 4: Trustworthiness depends on reliability, which may be unknown

In security design, automation reliability and predictability are engineering challenges, especially in security technologies that use behavior-based methods to respond to novel threats. In contrast to signature-based methods, behavior-based methods identify anomalous user or software behavior that might indicate a threat or differ from expected behavior [45]. If security technology allows for the detection of previously unknown threats, then it may be impossible to quantify the probability of future threat detection. Consequently, HCI practitioners face a challenge when implementing an interface for an aid with limited reliability. Because trust is subjective, the users' subjective perceptions of automation reliability and predictability may be just as important as the true reliability of the tool. However, users make attributions of the reliability of technology in their interactions with it over time [12], so attempting to mislead users about reliability is unlikely to be effective. Not all errors have the same effect, however. High false alarm rates in cybersecurity tools may lead to reduced compliance with alerts. While research has suggested that operators will devote additional attentional resources to automation with high miss rates [38], this may not be possible for cybersecurity professionals who perform network defense; the high workload due to the volume of network traffic that must be investigated could limit the attentional resources available to compensate for an aid that misses threats. Further, security solution vendors may be motivated to adopt a liberal criterion because of the criticality of catching threats.

5.3.4.1 Proposed design strategy

In professional environments, providing transparency about the reasons and types of automation failure is likely to be useful [21,46] if cybersecurity professionals understand the feedback provided. That is, more detailed feedback is helpful if it is compatible with users' understanding and does not put excessive cognitive demands on the user. For interfaces designed for consumer users, the minimal understanding of how a tool works may challenge the building of appropriate trust and limit the usefulness of transparency if such feedback is not understandable to users. This is a challenge in designing tools for widespread use, but so too is the technical problem of automation reliability.

5.4 Three ways to measure trust

Several measures have been published that can be used to measure trust in HCI. Trust is commonly measured using self-report. Jian, Bisantz, and Drury [47] developed a subjective trust in automation scale known as the Checklist for Trust between People and Automation. The survey asks users to rate the intensity of their feeling of trust across 12 items scaled from 1 to 7. Five items are reverse coded. This survey focuses on trust in a specific entity, such as a software application, so it is suitable as a subjective trust measure in usability testing. It can also be applied to a variety of automated tools without needing to adapt the measure. However, professionals who use this instrument should take care to ensure that participants understand what entity is being assessed. This is especially important in the usability testing of complex systems in which the trust of only one agent, such as an intelligent assistant, is being evaluated. Practitioners may also consider the contingent nature of trust in automation. The capability of automation affects trust, but this capability is relative to the users' goals. For example, automotive cruise control would have high capability to maintain a set speed, but it would have low capability to autonomously drive a car across an intersection.

Rotter's Interpersonal Trust Scale [48] has been used as a measure of *propensity to trust*, but its items are not specific to automation. The scale has 25 items; 13 are positive statements about trust, leaving 12 items as reverse-coded negative statements about trust. For example, one item asks participants to rate their agreement with the statement, "Parents usually can be relied upon to keep their promises" [48, p. 654]. As we described earlier, empirical research has supported theoretical differences in the way people build interpersonal trust and trust in automation [14]. However, these differences are at least related to automation capability and, at most, will lessen as human–automation teaming becomes a viable interaction paradigm.

An automation-specific measure for the propensity to trust trait is the Automation-Induced Complacency Potential Rating Scale [49]. This 12-item survey asks users to rate their agreement on statements of trust in automated devices generally. For example, one item asks whether participants agree that medical automation saves time and money in the diagnosis of disease [49]. The measure was demonstrated to have high internal consistency

(a = 0.90) and test–retest reliability measured after three months (a = 0.87) [49]. A present limitation of this instrument is that the automated systems encountered in daily life have changed since the measure was published. Items about making purchases electronically rather than over the phone, or recording television shows using a videocassette recorder, may not be diagnostic today. Modifying this measure to include current technology could provide a better measure of propensity to trust at the cost of losing the ability to compare scores to researchers using the original measure.

5.5 Applications of trust in automation for cybersecurity professionals

5.5.1 What defines a cybersecurity professional?

Cybersecurity professionals are a diverse group of individuals who are responsible for ensuring the ongoing security of the computer networks in their organization. As a whole, information security professionals are in great demand; in 2014, Cisco estimated a shortage of more than a million information security professionals worldwide [50]. Critically, commonalities of knowledge, skills, and attitudes across job roles and organizations are only starting to be defined. One reason for this is the rapid evolution of cyberthreats and cyberthreat actors. Of primary focus are cybersecurity professionals who "protect, monitor, analyze, detect and respond to unauthorized activity," a task called computer network defense (CND) [51]. Because of the large and growing volume of network activity, the unaided performance of this task is impossible in large organizations. To reduce the human information processing requirements, automated tools are used. One example is an intrusion detection system (IDS), which examines server log files to find patterns associated with anomalies. When such a pattern is found, cybersecurity professionals can be alerted to investigate. However, IDSs are limited in their sophistication and reliability; this has been true for most forms of automation for CND. Because of this, CND is a joint human–machine collaborative task in which people depend on automated tools to perform their jobs but must remain in the loop as an information processor and decision maker. Consequently, the cybersecurity professional is a critical line of defense in CND. Effective human decision-making is a determinant of successful cybersecurity.

To address the high and growing demand for cybersecurity professionals, the US National Institute of Standards and Technology has led the development of the National Initiative for Cybersecurity Education Cybersecurity Workforce Framework (NICE NCWF) to describe the work of cybersecurity professionals. Five core functions underlie cyberoperations across an organization: identify, protect, detect, respond, and recover. Seven high-level categories are defined that span cyberdefense, intelligence, and forensics across an organization and include two or more of the core functions. These categories are listed in Table 5.3 [52].

Applications of trust are most relevant to the job roles that most require automated tools: protect and defend, analyze, collect and operate, and investigate. These job roles all have an intelligence activity that relies on the synthesis of large amounts of data. At present, these activities typically require the use of a suite of tools, each with a limited purpose [53]. In some cases, information from multiple tools may be integrated into a dashboard interface. Increasingly, cyberdefense tools leverage the vast amounts of data collected across the network of the organization to participate to a greater degree in decision-making.

Table 5.3 NCWF workforce categories and their definitions

Secure provision	Conceptualizes, designs, and builds secure information technology (IT) systems, with responsibility for aspects of systems and/or networks development
Operate and maintain	Provides the support, administration, and maintenance necessary to ensure effective and efficient IT system performance and security
Oversee and govern	Provides leadership, management, direction, or development and advocacy so the organization may effectively conduct cybersecurity work
Protect and defend	Identifies, analyzes, and mitigates threats to internal IT systems and/or networks
Analyze	Performs highly specialized review and evaluation of incoming cybersecurity information to determine its usefulness for intelligence
Collect and operate	Provides specialized denial and deception operations and collection of cybersecurity information that may be used to develop intelligence
Investigate	Investigates cybersecurity events or crimes related to IT systems, networks, and digital evidence

Source: National Institute of Standards and Technology. *NICE Cybersecurity Workforce Framework.* Last modified July 7, 2017. https://www.nist.gov/itl/applied-cybersecurity/nice/resources/nice-cybersecurity-workforce-framework.

5.5.2 *Cybersecurity professional characteristics*

Cybersecurity professionals develop their expertise over time as they engage in goal-directed practice. For example, this knowledge might include specific threats and vulnerabilities or maintain awareness of the organization's local area network/wide area network pathways [54].

Because network attacks can occur within milliseconds, and cybersecurity professionals' decisions are critical to the security of the organization, their work is categorized by uncertainty and time pressure. Miscalibrated trust is problematic at both ends. If a tool is capable of informing security decisions but the decision maker does not trust it, the tool is distrusted. Distrusted tools may be underutilized, increasing threat vulnerability when automation that could detect or mitigate a threat is ignored or disabled. Further, the workload of defenders increases when automation is not given the authority to manage detection tasks it is better suited for than a person. On the other end, overtrusted tools may lead to complacency, an insufficient checking of the results of the tool. This could manifest in a variety of outcomes. For those responding to low-level alerts, there may be insufficient research before alerts are escalated, causing overload at higher levels, or the misses of a tool may go undetected when it is overtrusted to just work without intervention. Fortunately, designers can design for appropriate trust by understanding how their design changes are more or less trusted by their cybersecurity professional users. Because trust predicts compliance and complacency, it should be measured as an outcome as part of the user-centered design process. Doing so can identify problematic levels of trust, or unanticipated changes in trust, as a design evolves.

Designers of tools for cybersecurity professionals must ensure that trust is appropriately calibrated. Cybersecurity professionals must understand the true reliability of the tool. In many cases, the reliability cannot be expressed as a simple percentage (e.g., "this IDS misses 10% of threats"). The circumstances surrounding automation failure in cybersecurity may be both complex and unpredictable. For example, an attacker may directly target a specific automated tool to knock it offline or have it provide inaccurate output. Designers should communicate these circumstances to cybersecurity professionals by way of the interface or by training.

Table 5.4 Recommendations to improve appropriate trust in design for cybersecurity professionals

1. Measure trust as a user-centered design outcome.
2. Provide evidence of the known reliability of automated tools whenever available. Provide transparency about the reasons for automation failure when reliability is unknown.
3. Ensure that the functionality of tools is understood through mental model elicitation techniques.
4. Incorporate human–automation teaming to facilitate trust calibration.

Cybersecurity professionals have a greater capacity for understanding complex reasons behind automation failures than consumers. However, the cybersecurity professional's knowledge of the function of a tool, and the circumstances for its failure, must match reality to build appropriate trust. Mental models, which describe structural knowledge, are particularly relevant. Mental models are "the mechanisms whereby humans are able to generate descriptions of system purpose and form, explanations of system functioning and observed system states, and predictions of future states" [44, p. 351]. Originally described as an internal representation of external reality by Craik [55], mental models support trust by providing a framework for understanding, interpreting, and integrating environmental cues [56]. Mental model elicitation techniques, such as concept mapping, can be used to evaluate how a cybersecurity professional understands the relationships between a new tool, existing tools, and the threat landscape (see Crandall, Klein, and Hoffman [57]).

Finally, the concept of human–automation teaming offers promise to facilitate trust calibration by incorporating naturalistic communication and team process in a way that allows humans to better leverage metaphors of interpersonal interaction. For example, a future automated tool may be able to explain why it made an error, minimizing the loss of trust that is likely in present automation (Table 5.4).

5.6 *Applications of trust among consumer users*

An understanding of how trust impacts use and performance has implications in consumer cybersecurity applications. Researchers in the cybersecurity domain have begun to investigate factors influencing trust; their research has shown that consumers who are not cybersecurity professionals lack security knowledge and frequently overtrust the capabilities of their security software, which lessens their computer security [58]. An illustration of the state of users' cybersecurity knowledge can be found in a 2017 McAfee survey. The study found that 41% of their users were not aware of how to check whether their devices are compromised [59].

Considerations of trust should be used by cybersecurity product designers to design products that foster appropriate user automation trust. As with many HCI guidelines, we offer solutions to leverage changes in the design as a first-line intervention. Failing that, leveraging the characteristics of users, or modifying user behavior through training, could be an option. Compared to cybersecurity professionals, however, consumer user tool use is highly discretionary, occurs with low frequency, and is characterized by mental models that may not reflect deep or accurate knowledge of security technology.

5.6.1 *Increase trust in features that promote cyberhygiene*

By fostering appropriate trust through interface design, researchers have sought to promote security-conscious user behavior and subsequently bolster the overall security of their systems. Through the manipulation of message framing, length, and the use of an anthropomorphic messenger, Rodríguez-Priego and van Bavel influenced user behavior on a consumer shopping website [60]. The researchers found that by manipulating the presentation of security warnings, they could manipulate the trust and the resulting behavior of site visitors. The researchers found that a lengthier message paired with a male anthropomorphic character led to safer online behavior. They also found that a loss-framed security message led users to engage in safer behavior. The loss-framed message emphasized what users stood to lose from cybersecurity threats rather than a gain-framed message emphasizing what users stood to gain by staying safe. This research demonstrates that design can lead to proactive behaviors, which can improve security for all through a reduction of attack surface. The concept of users proactively participating in their own security has been called cyberhygiene [61]. Cyberhygiene depends on the appropriate trust in tools that support cyberhygiene. Users may undermine their own security by underutilizing or circumventing security technologies that they do not trust to work toward their goals. For the security of users, it is imperative for the design to support trust in features that support cyberhygiene behaviors.

5.6.2 *Design for user diversity*

Defined by their stable nature, traits include dimensions such as age, gender, ethnicity, and personality [15,40,62,63]. As an illustration to the applicability of traits to the design of cybersecurity systems, we can consider a company that provides products to both Mexico and the United States. Hoff and Bashir [24] found that compared to their counterparts in the United States, Mexican users were more likely to trust automated systems. Thus, designers may need to design their product in a way that accounts for diversity in trust across various segments of the user population. More broadly, differences in traits and their resulting effects on trust should be represented in designs by increasing the individualization of system designs aimed toward different population segments. Indeed, designing around the needs and traits of individuals has been suggested by researchers as an effective way of improving system performance [64].

5.6.3 *Mental model compatibility*

Researchers have found that matching an interface with the mental model of their users increases the credibility and subsequent development of trust in the system [65,66]. Conversely, a mismatch between the mental model of the user and how technology operates can cause distrust in the system. Thus, practitioners should consider exploratory usability testing with the purpose of eliciting and describing the mental models of target users relevant to the product.

5.6.4 *Incorporating elements of human–automation teaming*

As with cybersecurity professionals, human–automation teaming may support appropriate trust in consumer users as interactions become more natural. Teaming may start to be incorporated in consumer-facing products in two ways: through naturalistic communication and adaptive coordination with people.

5.6.4.1 *Communication*

The reliability and consistency of a system over time have a significant effect on user trust. While interface designers might not be able to improve those metrics, implementing

Table 5.5 Recommendations to foster appropriate trust in consumer users

1. Identify features that support cyberhygiene and ensure they are usable and trustworthy.
2. Design for user diversity in trust.
3. Assess mental model compatibility to ensure that security technology is understandable.
4. Incorporate coordination and naturalistic communication features to facilitate trust calibration.

effective system–user communication through clear, appropriate, and accurate feedback to users can mitigate performance lost through poor reliability and inconsistent performance [67].

The appearance and sound of an interface have been correlated with likeability and subsequent trust. Specifically, more anthropomorphism has been shown to lead to greater trust [68]. Users tend to trust human speech over synthetic speech for the presentation of information [69]. Additionally, research by Parasuraman and Miller [70] found that a polite interface positively impacted trust development. These findings suggest that by utilizing more familiar and natural communication modes, and by increasing the politeness of communication, systems become more trustworthy. Thus, system designers should strive to design interfaces that communicate with their users in a polite manner that imitates human speech patterns and content.

5.6.4.2 *Coordination with users: Adaptive automation*

The level of the involvement of a tool in a task, and authority to act, is an antecedent to trust [26,71]. Automated tools can more effectively coordinate with users by providing the information users need at the time that they need it, called adaptive automation [72]. Users show more trust in systems that collaborate with them and account for their state [73]. In a study using a driving simulator, Cai and Lin [74] tested a form of driving automation that took control based on a function incorporating the criticality of the situation along with the user's present level of cognitive engagement as assessed using an eye tracker. They compared an adaptive condition to both strong involvement (i.e., automation taking action) and weak involvement (i.e., automation suggesting action) and found that the adaptive level of automation resulted in higher trust levels and was preferable to users than either fixed levels of involvement. Thus, interaction design for appropriate trust in security tools should facilitate collaboration; tools should adapt to the perceived state of the user. For example, security software should avoid alerts that are not time sensitive during periods of high user workload (Table 5.5).

5.7 *Conclusions*

Trust is an adaptation to uncertainty and risk. Cybersecurity is a domain with many participants who have diverse perspectives and interests, and participating in cybersecurity as a professional or consumer user means navigating a dynamic, risky, and uncertain environment. When trust is calibrated appropriately, it allows trustors to minimize risk and move forward with decision-making. When trust is not appropriately calibrated, however, security is likely to suffer. In describing trust and offering initial recommendations for its use in user-centered design, we hope that designers, researchers, and HCI practitioners will incorporate this construct as an outcome of interest. As cybersecurity tools, for professional and home uses, continue to evolve, additional research will be needed to clarify how trust is built and maintained in cybersecurity. We offer that trust is not something to be maximized nor minimized, but a feature of human information processing that interacts with other characteristics to explain how people can use tools most effectively and, ultimately, make decisions that maximize security outcomes.

Acknowledgments

This work is based upon work supported by the National Science Foundation under Grant No. 1553018. Any opinions, findings, and conclusions or recommendations expressed in this material are those of the authors and do not necessarily reflect the views of the National Science Foundation. The authors gratefully acknowledge Elizabeth Shallal and Steven Wu for their help in formatting citations.

References

1. Seligman, Adam B. Trust and sociability. *American Journal of Economics and Sociology* 57, no. 4 (1998): 391–404. doi:10.1111/j.1536-7150.1998.tb03372.x.
2. West, Ryan. The psychology of security. *Communications of the ACM* 51, no. 4 (2008): 34–40. doi:10.1145/1330311.1330320.
3. Symantec. *Symantec Internet Security Threat Report: Trends for 2016*. Mountain View, CA: Symantec Corporation, 2017. https://www.symantec.com/content/dam/symantec/docs/reports/istr-22-2017-en.pdf.
4. Juniper Research. *Cybercrime Will Cost Businesses Over $2 Trillion by 2019*. Press release, May 12, 2015, 2015. http://www.prnewswire.com/news-releases/cybercrime-will-cost-businesses-over-2-trillion-by-2019-finds-juniper-research-503449791.html.
5. Mayer, Roger C., James H. Davis, and F. David Schoorman. An integrative model of organizational trust. *Academy of Management Review* 20, no. 3 (1995): 709–734.
6. Hoffman, Robert R., Matthew Johnson, Jeffrey M. Bradshaw, and Al Underbrink. Trust in Automation. *IEEE Intelligent Systems* 28, no. 1 (2013): 84–88. doi:10.1109/MIS.2013.24.
7. Harper, Douglas. Etymology of trust. *Online Etymology Dictionary*. Retrieved on May 22, 2018. http://www.etymonline.com/index.php?term=trust&allowed_in_frame=0.
8. Lee, John D., and Katrina A. See. Trust in automation: Designing for appropriate reliance. *Human Factors* 46, no. 1 (2004): 50–80. doi:10.1518/hfes.46.1.50.3039.
9. Muir, Bonnie M. Trust between humans and machines, and the design of decision aids. *International Journal of Man-Machine Studies* 27, no. 5–6 (1987): 527–539. doi:10.1016/S0020-7373(87)80013-5.
10. Parasuraman, Raja, Thomas B. Sheridan, and Christopher D. Wickens. A model for types and levels of human interaction with automation. *IEEE Transactions on Systems, Man, and Cybernetics—Part A: Systems and Humans* 30, no. 3 (2000): 286–297. doi:10.1109/3468.844354.
11. Sheridan, Thomas B. HCI in supervisory control: Twelve dilemmas. In *Human Error and System Design and Management*, edited by Peter F. Elzer, Rainer H. Kluwe, and Badi Boussoffara, 1–12. Vol. 253. London: Springer London, 2000.
12. Merritt, Stephanie M., and Daniel R. Ilgen. Not all trust is created equal: Dispositional and history-based trust in human-automation interactions. *Human Factors: The Journal of the Human Factors and Ergonomics Society* 50, no. 2 (2008): 194–210. doi:10.1518/001872008X288574.
13. Lewandowsky, Stephan, Michael Mundy, and Gerard Tan. The dynamics of trust: Comparing humans to automation. *Journal of Experimental Psychology: Applied* 6, no. 2 (2000): 104–123. doi:10.1037//1076-898X.6.2.104.
14. Madhavan, Poornima, and Douglas A. Wiegmann. Similarities and differences between human-human and human-automation trust: An integrative review. *Theoretical Issues in Ergonomics Science* 8, no. 4 (2007): 277–301. doi:10.1080/14639220500337708.
15. Dzindolet, Mary T., Linda G. Pierce, Hall P. Beck, Lloyd A. Dawe, and B. Wayne Anderson. Predicting misuse and disuse of combat identification systems. *Military Psychology* 13, no. 3 (2001): 147–164. doi:10.1207/S15327876MP1303_2.

16. Madhavan, Poornima, Douglas A. Wiegmann, and Frank C. Lacson. Automation failures on tasks easily performed by operators undermine trust in automated aids. *Human Factors: The Journal of the Human factors and Ergonomics Society* 48, no. 2 (2006): 241–256. doi:10.1518/001872006777724408.
17. Schuster, David, Mary L. Still, Jeremiah D. Still, Ji Jung Lim, Cary S. Feria, and Christian P. Rohrer. Opinions or algorithms: An investigation of trust in people versus automation in app store security. In *International Conference on Human Aspects of Information Security, Privacy, and Trust*, edited by Theo Tryfonas, 415–425. Springer, Cham, 2015. doi:10.1007/978-3-319-20376-8_37.
18. Schaefer, Kristin E., Jessie Y. C. Chen, James L. Szalma, and Peter A. Hancock. A meta-analysis of factors influencing the development of trust in automation: Implications for understanding autonomy in future systems. *Human Factors* 58, no. 3 (2016): 377–400. doi:10.1177/0018720816634228.
19. Nass, Clifford, and Youngme Moon. Machines and mindlessness: Social responses to computers. *Journal of Social Issues* 56, no. 1 (2000): 81–103. doi:10.1111/0022-4537.00153.
20. Christoffersen, Klaus, and David D. Woods. How to make automated systems team players. In *Advances in Human Performance and Cognitive Engineering Research: Automation*, edited by Eduardo Salas, 1–12. Columbus: Elsevier Science/JAI Press, 2002. doi:10.1016/S1479-3601(02)02003-9.
21. Cuevas, Haydee M., Stephen M. Fiore, Barrett S. Caldwell, and Laura Strater. Augmenting team cognition in human-automation teams performing in complex operational environments. *Aviation, Space, And Environmental Medicine* 78, no. 5 (2007): B63–B70.
22. Ososky, Scott, David Schuster, Elizabeth Phillips, and Florian G. Jentsch. Building appropriate trust in human–robot teams. In *AAAI Spring Symposium: Trust and Autonomous Systems, Palo Alto, CA, March 25–27, 2013*, 60–65.
23. Prevot, Thomas, Nancy Smith, Everett Palmer, Todd Callantine, Paul Lee, Joey Mercer, Jeff Homola, Lynne Martin, Connie Brasil, and Christopher Cabrall. An overview of current capabilities and research activities in the Airspace Operations Laboratory at NASA Ames Research Center. In *14th AIAA Aviation Technology, Integration, and Operations Conference (ATIO), National Harbor, MD, January 13–17, 2014*, 1–20.
24. Hoff, Kevin Anthony, and Masooda Bashir. Trust in automation: Integrating empirical evidence on factors that influence trust. *Human Factors* 57, no. 3 (2015): 407–434. doi:10.1177/0018720814547570.
25. Ekman, Frederick, Mikael Johansson, and Jana Sochor. To see or not to see: The effect of object recognition on users' trust in automated vehicles. In *Proceedings of the 9th Nordic Conference on Human-Computer Interaction, Gothenburg, Sweden, October 23–27, 2016*, 1–4. New York: ACM.
26. Merritt, Stephanie M., Heather Heimbaugh, Jennifer LaChapell, and Deborah Lee. I trust it, but I don't know why: Effects of implicit attitudes toward automation on trust in an automated system. *Human Factors: The Journal of the Human Factors and Ergonomics Society* 55, no. 3 (2013): 520–534. doi:10.1177/0018720812465081.
27. Yang, X. Jessie, Christopher D. Wickens, and Katja Hölttä-Otto. How users adjust trust in automation: Contrast effect and hindsight bias. In *Proceedings of the Human Factors and Ergonomics Society Annual Meeting, Washington, D.C., September 19–23, 2016*, 196–200. Los Angeles, CA: SAGE.
28. Wickens, Christopher D., John D. Lee, Yili Liu, and Sallie Gordon-Becker. *Introduction to Human Factors Engineering (2nd Edition)*. Upper Saddle River, NJ: Pearson, 2003.
29. Lehto, Mark R., Fiona Fui-Hoon Nah, and Ji Soo Yi. Decision making models and decision support. In *Handbook of Human Factors and Ergonomics*, edited by Gavriel Salvendy, 191–242. Hoboken, NJ: John Wiley & Sons, 2006. doi:10.1002/0470048204.ch8.
30. Klein, Gary. Naturalistic decision making. *Human Factors* 50, no. 3 (2008): 456–460. doi:10.1518/001872008X288385.
31. Fiske, Susan T., and Shelley E. Taylor. *Social Cognition: From Brains to Culture*. Thousand Oaks, CA: SAGE, 2013.
32. Tversky, Amos, and Daniel Kahneman. Judgment under uncertainty: Heuristics and biases. *Oregon Research Institute Research Bulletin* 13, no. 1 (1975): 141–162.
33. Gigerenzer, Gerd, and Wolfgang Gaissmaier. Heuristic decision making. *Annual Review of Psychology* 62 (2011): 451–482.

34. Garcia-Retamero, Rocio, and Mandeep K. Dhami. Take-the-best in expert-novice decision strategies for residential burglary. *Psychonomic Bulletin & Review* 16 (2009): 163–169.
35. Bravo-Lillo, Cristian, Lorrie Faith Cranor, and Julie Downs. Bridging the gap in computer security warnings: A mental model approach. *IEEE Security & Privacy* 9, no. 2 (2011): 18–26. doi:10.1109/MSP.2010.198.
36. Furnell, Steve M., Peter Bryant, and Andy Phippen. Assessing the security perceptions of personal Internet users. *Computer Security* 26, no. 5 (2007): 410–417. doi:10.1016/j.cose.2007.03.001.
37. Parasuraman, Raja, and Victor Riley. Humans and automation: Use, misuse, disuse, abuse. *Human Factors* 39, no. 2 (1997): 230–253. doi:10.1518/001872097778543886.
38. Dixon, Stephen R., and Christopher D. Wickens. Automation reliability in unmanned aerial vehicle control: A reliance–compliance model of automation dependence in high workload. *Human Factors* 48, no. 3 (2006): 474–486. doi:10.1518/001872006778606822.
39. Manzey, Dietrich, J. Elin Bahner, and Anke-Dorothea Hueper. Misuse of automated aids in process control: Complacency, automation bias and possible training interventions. In *Proceedings of the Human Factors and Ergonomics Society Annual Meeting, San Francisco, CA, October 16–20, 2006*, 220–224. Los Angeles, CA: SAGE.
40. Szalma, James L., and Grant S. Taylor. Individual differences in response to automation: The five factor model of personality. *Journal of Experimental Psychology: Applied* 17, no. 2 (2011): 71–96. doi:10.1037/a0024170.
41. Atoyan, Hasmik, Jean-Rémi Duquet, and Jean-Marc Robert. Trust in new decision aid systems. In *Conference on l'Interaction Homme-Machine, Montreal, Canada, April 2006*, 115–122. New York: Association for Computing Machinery.
42. Li, Yung-Ming, and Yung-Shao Yeh. Increasing trust in mobile commerce through design aesthetics. *Computers in Human Behavior* 26, no. 4 (2010): 673–684. doi:10.1016/j.chb.2010.01.004.
43. Biros, David P., Mark Daly, and Gregg Gunsch. The influence of task load and automation trust on deception detection. *Group Decision and Negotiation* 13, no. 2 (2004): 173–189. doi:10.1023/B:GRUP.0000021840.85686.57.
44. Rouse, William. B., and Nancy M. Morris. On looking into the black box: Prospects and limits in the search for mental models. *Psychological Bulletin* 100, no. 3 (1986): 349–363.
45. Clark, David, Thomas Berson, and Marjory Blumenthal. *At the Nexus of Cybersecurity and Public Policy*. Washington, DC: National Academies Press, 2014.
46. Dzindolet, Mary T., Scott A. Peterson, Regina A. Pomranky, Linda G. Pierce, and Hall P. Beck. The role of trust in automation reliance. *International Journal of Human-Computer Studies* 58, no. 6 (2003): 697–718. doi:10.1016/S1071-5819(03)00038-7.
47. Jian, Jiun-Yin, Ann M. Bisantz, and Colin G. Drury. Foundations for an empirically determined scale of trust in automated systems. *International Journal of Cognitive Ergonomics* 4, no. 1 (2000): 53–71. doi:10.1207/S15327566IJCE0401_04.
48. Rotter, Julian. B. A new scale for the measurement of interpersonal trust. *Journal of Personality* 35 (1967): 651–665.
49. Singh, Indramani L., Robert Molloy, and Raja Parasuraman. Automation-induced "complacency": Development of the complacency-potential rating scale. *International Journal of Aviation Psychology* 3, no. 2 (1993): 111–122. doi:10.1207/s15327108ijap0302_2.
50. Beliveau-Dunn, Jeanne. Tackling the cybersecurity skills gap. *Cisco Blog*. Last modified January 21, 2014. https://blogs.cisco.com/education/tackling-the-cybersecurity-skills-gap.
51. Computer Network Defense. *Defense Technical Information Center*. Last modified June 10, 2015. http://www.dtic.mil/doctrine/dod_dictionary/data/c/10869.html.
52. National Institute of Standards and Technology. *NICE Cybersecurity Workforce Framework*. Last modified July 7, 2017. https://www.nist.gov/itl/applied-cybersecurity/nice/resources/nice-cybersecurity-workforce-framework.
53. Goodall, John R., and Mark Sowul. VIAssist: Visual analytics for cyber defense. In *IEEE Conference on Technologies for Homeland Security, Boston, MA, May, 11–12, 2009*, 143–150. Waltham, MA: Institute of Electrical and Electronics Engineers. doi:10.1109/THS.2009.5168026.
54. National Institute of Standards and Technology. *The National Cybersecurity Workforce Framework*. 2016. http://csrc.nist.gov/nice/framework/.
55. Craik, Kenneth. *The Nature of Explanation*. Cambridge, MA: Cambridge University Press, 1943.

56. Cooke, Nancy J., Rene'e Stout, and Eduardo Salas. (2001). A knowledge elicitation approach to the measurement of team situation awareness. In *New Trends in Cooperative Activities: System Dynamics in Complex Settings*, edited by Michael McNeese, Mica Endsley, and Eduardo Salas, 114–139. Santa Monica: Human Factors and Ergonomics Society, 2001.
57. Crandall, Beth, Gary Klein, Robert R. Hoffman. *Working Minds*. Cambridge, MA: MIT Press, 2006.
58. Hausawi, Yasser M. Current trend of end-users' behaviors towards security mechanisms. In *International Conference on Human Aspects of Information Security, Privacy, and Trust*, edited by Theo Tryfonas and Ioannis Askoxylakis, 140–151. Cham: Springer, 2016. doi:10.1007/978-3-319-39381-0_13.
59. Davis, Gary. Navigating today's connected world. *McAfee Securing Tomorrow Blog*. Last modified February 27, 2017. https://securingtomorrow.mcafee.com/consumer/consumer-threat-notices/connected-life/.
60. Rodríguez-Priego, Nuria, and René van Bavel. *The Effect of Warning Messages on Secure Behaviour Online*. Luxembourg: Publications Office of the European Union.
61. O'Connell, Mary Ellen. Cyber security without cyber war. *Journal of Conflict and Security Law* 17, no. 2 (2012): 187–209. doi:10.1093/jcsl/krs017.
62. Donmez, Birsen, Linda N. Boyle, John D. Lee, and Daniel V. McGehee. Drivers' attitudes toward imperfect distraction mitigation strategies. *Transportation Research Part F: Traffic Psychology and Behaviour* 9, no. 6 (2006): 387–398. doi:10.1016/j.trf.2006.02.001.
63. Schaefer, Kristin. *The Perception and Measurement of Human–Robot Trust*. PhD diss., University of Central Florida, Orlando, FL, 2013.
64. Hancock, Peter A., G. M. Hancock, and J. S. Warm. Individuation: The $N = 1$ revolution. *Theoretical Issues in Ergonomics Science* 10, no. 5 (2009): 481–488. doi:10.1080/14639220903106387.
65. Beggiato, Matthias, and Josef F. Krems. 2013. The evolution of mental model, trust and acceptance of adaptive cruise control in relation to initial information. *Transportation Research Part F: Traffic Psychology and Behaviour* 18 (2013): 47–57. doi:10.1016/j.trf.2012.12.006.
66. Fogg, B. J., and Hsiang Tseng. The elements of computer credibility. In *Proceedings of the SIGCHI Conference on Human Factors in Computing Systems, Pittsburgh, PN, May, 15–20, 1999*, 80–87. New York: Association for Computing Machinery. doi:10.1145/302979.303001.
67. Stanton, Neville A., Mark S. Young, and Guy H. Walker. The psychology of driving automation: A discussion with Professor Don Norman. *International Journal of Vehicle Design* 45, no. 3 (2007): 289–306. doi:10.1504/ijvd.2007.014906.
68. Pak, Richard, Nicole Fink, Margaux Price, Brock Bass, and Lindsay Sturre. Decision support aids with anthropomorphic characteristics influence trust and performance in younger and older adults. *Ergonomics* 55, no. 9 (2012): 1059–1072. doi:10.1080/00140139.2012.691554.
69. Stedmon, Alex W., Sarah Sharples, Robert Littlewood, Gemma Cox, Harshada Patel, and John R. Wilson. Datalink in air traffic management: Human factors issues in communications. *Applied Ergonomics* 38, no. 4 (2007): 473–480. doi:10.1016/j.apergo.2007.01.013.
70. Parasuraman, Raja, and Christopher A. Miller. Trust and etiquette in high-criticality automated systems. *Communications of the ACM* 47, no. 4 (2004): 51–55. doi:10.1145/975817.975844.
71. Looije, Rosemarijn, Mark A. Neerincx, and Fokie Cnossen. Persuasive robotic assistant for health self-management of older adults: Design and evaluation of social behaviors. *International Journal of Human-Computer Studies* 68, no. 6 (2010): 386–397. doi:10.1016/j.ijhcs.2009.08.007.
72. Byrne, Evan A., and Raja Parasuraman. Psychophysiology and adaptive automation. *Biological Psychology* 42, no. 3 (1996): 249–268. doi:10.1016/0301-0511(95)05161-9.
73. Moray, Neville, Toshiyuki Inagaki, and Makoto Itoh. Adaptive automation, trust, and self-confidence in fault management of time-critical tasks. *Journal of Experimental Psychology: Applied* 6, no. 1 (2000): 44–58. doi:10.1037/1076-898X.6.1.44.
74. Cai, Hua, and Yingzi Lin. Coordinating cognitive assistance with cognitive engagement control approaches in human–machine collaboration. *IEEE Transactions on Systems, Man, and Cybernetics—Part A: Systems and Humans* 42, no. 2 (2012): 286–294. doi:10.1109/TSMCA.2011.2169953.

section three

Threats

chapter six

Insider threat

The forgotten, yet formidable foe

Maria Papadaki and Stavros Shiaeles

Contents

6.1 Nature of the insider threat

As a way of introduction, it would be useful to consider what the security community considers the insider threat to be and how it manifests itself. Mukherjee et al. (1994) define the *insider threat* as that who has legitimate access to the system but is abusing their privileges. Schultz (2002) subsequently considers insider attacks as deliberate misuse by those who are authorized to use computers and networks and identifies insiders as employees, contractors, consultants, temporary helpers, or personnel from third-party business partners. He pointed out how little was understood on insider threats at the time and discussed the many misconceptions that surrounded the issue. Bishop and Gates (2008) go a step further by considering insider threats in the context of trust and security policies, where levels of trust are expressed in a set of access control rules, which are in turn represented in a security policy. Specifically, they provide the following definition for an *insider*:

> A trusted entity that is given the power to violate one or more rules in a given security policy ... the insider threat occurs when a trusted entity abuses that power.... An insider can thus be defined with regard to two primitive actions:
> 1. violation of a security policy using legitimate access, and;
> 2. violation of an access control policy by obtaining unauthorized access.

Recognizing the lack of understanding in the area and the need for further research, Carnegie Mellon Computer Emergency Response Team's (CERT) research program began in 2000 with a US Department of Defense sponsorship on insider threats in military services and defense agencies. Since then, their research has expanded to documenting more than 1000 insider threat case files, constituting the CERT Insider Threat Database, which provide technical, behavioral, and organizational details of each crime (Cappelli et al. 2012; Collins et al. 2016). Based on this knowledge, CERT's definition of a malicious insider is as follows (Silowash et al. 2012):

> A malicious *insider is* defined as a current or former employee, contractor, or business partner who meets the following criteria: i) has or had authorized access to an organization's network, system, or data; ii) have intentionally exceeded or intentionally used that access in a manner that negatively affected the confidentiality, integrity, or availability of the organization's information or information systems.

Apart from malicious deliberate insiders, Hunker and Probst (2011) also recognize the significance of accidental threats and define an *insider* as an entity of trust who misuses their privileges, deliberately or accidentally, in a way that constitutes a threat. The CERT Insider Threat Center also recognizes unintentional insider threats and further elaborates by providing the following definition (Collins et al. 2016; Federal Infrastructure Protection Bureau 2013):

> An unintentional insider threat is defined as a current or former employee, contractor, or business partner who meets the following criteria:
>
> who has or had authorized access to an organization's network, system, or data and who, through their action/inaction without malicious intent cause harm or substantially increase the probability of future serious harm to the confidentiality, integrity, or availability of the organization's information or information systems.

In considering these definitions, it is relevant to note the various aspects that they encompass, legitimate access that is misused, the level of trust that leads to a security policy violation, the roles within an organization where this level of trust could be abused, the actions and intentions that could constitute a threat, and the potential impact to an organization's security. One could argue that all are correct, but each reflects a different dimension. They could even indicate how our understanding of the problem has evolved over time.

How about the range of threats that insiders could pose? Of the 1000 insider threat cases in the CERT Insider Threat Database, 734 involved malicious insider attacks (these cases did not include espionage or unintentional damage) (Collins et al. 2016). According to the same source, malicious insider attacks can be categorized into the following four classes:

- *Information technology (IT) sabotage*: an insider's use of IT to direct specific harm at an organization or an individual
- *Theft of intellectual property (IP)*: an insider's use of IT to steal IP from the organization; this category includes industrial espionage involving outsiders

- *Fraud*: an insider's use of IT for the unauthorized modification, addition, or deletion of data (not programs or systems) of an organization for personal gain or theft of information that leads to an identity crime (e.g., identity theft or credit card fraud)
- *Miscellaneous*: cases in which the insider's activity was not for IP theft, fraud, or IT sabotage

6.2 *Significance of the problem*

Having defined insider threats, it is important to understand how significant they are and why we need to care. Let us initially look at the number and frequency of insider attacks. Intel Security (2015) revealed that insiders were responsible for 43% of data loss incidents, half of which were malicious, and half, accidental. Ponemon Institute's 2014 survey reported that 88% of respondents believed the risks from privileged user abuse would increase within the subsequent 12–18 months. This was confirmed at Ponemon Institute's (2015) more recent study on the cost of cybercrime, which showed a 6% increase in frequency for malicious insider threat. Specifically, 41% of companies experienced malicious insiders, up from 35% the year before. Nonetheless, malicious insiders were, in fact, the least frequent attack type. In contrast, malware infections were at the top of the list and had affected 98% of respondents.

However, this does not necessarily mean they can be easily ignored. When it came to reviewing the cost of attacks in relation to their frequency, the order was almost reversed. Malicious insiders proved to be the costliest with an average annual cost of $167,890, whereas malware was featured at the penultimate position with the cost of $5,110. As the longer it takes to resolve an attack, the costlier it becomes, perhaps it is not surprising that the estimated average time to resolve different attack types also featured malicious insiders at the top with an average of 51.5 days, as opposed to 5.6 days needed for malware infections (Ponemon Institute 2016).

Looking at how each class might have affected different industry sectors, it is worth looking at Figure 6.1, which shows the three main insider threat classes, namely, IT sabotage, fraud, and IP theft, against the top six infrastructure sectors, as reported by CERT (Collins et al. 2016). We can see that financial motivation has been more prolific, mainly affecting the banking, finance, and local government sectors. Fraud has consistently featured as the top motivation even in previous editions of the study (Silowash et al. 2012). According to the 2016 Global Fraud Survey by the Association of Certified Fraud Examiners, which covers more than 114 countries globally, it is estimated that typical organizations lose 5% of their revenue to fraud each year. One fifth of the reported cases caused losses of at least $1 million. Interestingly, it is also reported that most of the fraudsters are first time offenders with clean employment histories (ACFE 2016).

Apart from fraud, IP theft and sabotage seem to feature at equal weights (Collins et al. 2016). IP theft has affected IT and healthcare sectors to a larger degree than others. Additionally, sabotage has been more prominent in the IT sector. Lastly, the banking and finance industry seems to have had the largest share of reported malicious insider cases, standing with twice as many incidents as the second sector, IT. Does this mean that the banking and finance industry is a prime target? Before rushing to such conclusion, it is worth considering the different legislative requirements on mandatory reporting across industries. Perhaps we are more aware of insider incidents in some industries, rather than others, due to such different notification requirements. In the absence of mandatory reporting in some sectors, it would be fair to suspect that the number of reported incidents only reflects the tip of the iceberg.

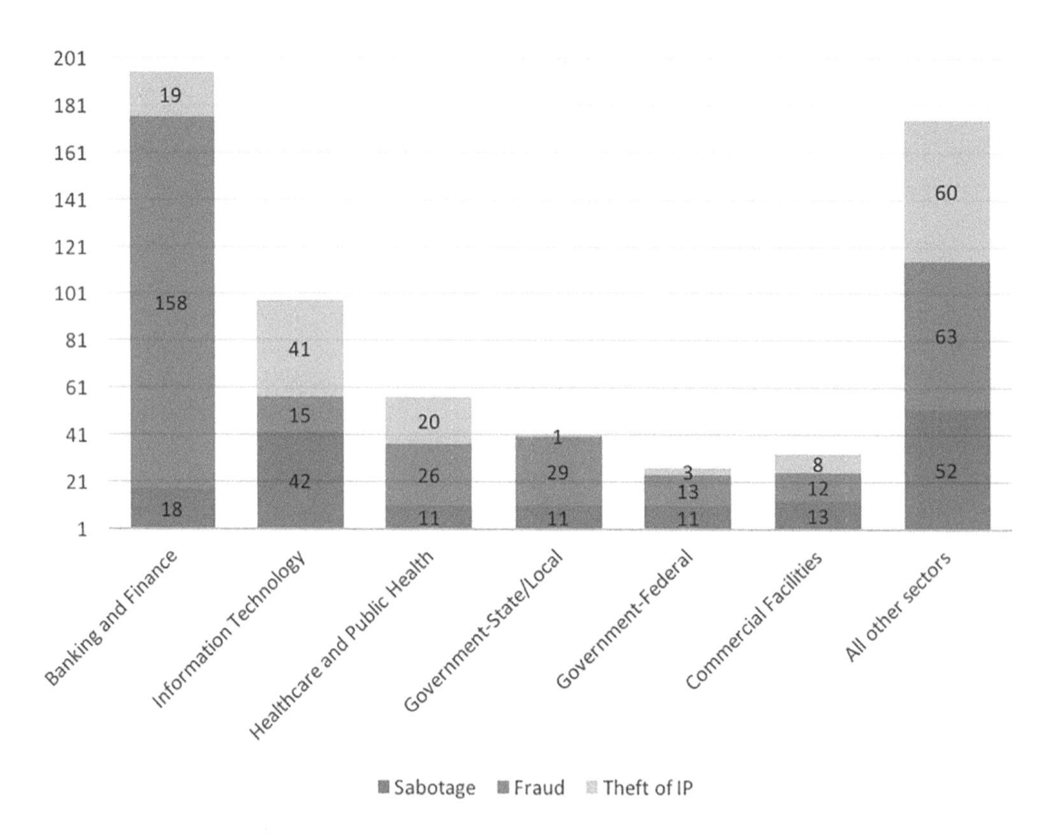

Figure 6.1 Top six infrastructure sectors for sabotage, fraud, and IP theft. (Data from Collins, M. et al., *Common Sense Guide to Mitigating Insider Threats*, Technical Report CMU/SEI-2016-TR-015, Software Engineering Institute, Carnegie Mellon University, Pittsburgh, PA, 2016.)

Further supporting this view, evidence from the 2014 US State of Cybercrime Survey reveals that 75% of respondents handled insider incidents internally, without involving legal action or law enforcement (PWC 2014). When asked about the reasons for not reporting, the answers raised concerns about their readiness to do so, as the lack of evidence and inability to identify the individual(s) responsible featured in the top two answers.

As such, if three out of four incidents are handled internally, without involving legal action or law enforcement, then perhaps it is worth reviewing the Verizon Data Breach Investigations Report, which is based on validated real data, reported and unreported, from 65 contributing organizations across the globe, including the CERT Insider Threat Center. Other participating organizations are Akamai, Checkpoint, Cisco, Dell, Kaspersky Lab, Mcafee, Qualys, etc. Their dataset provides approximate 95% confidence level for each statement (Verizon 2017). The 2016 Verizon report is based on more than 10,000 insider incidents and gives insight into the motivation behind insider threats. Although financial motives remain prominent with 34%, the gradual rise of espionage over time suggests that the threat landscape is changing (Verizon 2016).

Interestingly, when looking at organizations' discovery timeline for insider and privilege misuse, this is more likely to take months and years, rather than weeks or days (Verizon 2017). A highly reported insider case falls right into this category. It involved a decade of economic espionage within Canadian telecommunications company Nortel by Chinese hackers, leading to Nortel's ultimate collapse in 2009. It demonstrates the impact of insider threats and showcases the difficulties in detecting and responding to them. According to Brian Shields, Nortel's former senior systems security advisor, hackers stole at least seven passwords from top executives and subsequently downloaded sensitive information, including business plans, research and development reports, as well as employee e-mails. Suspicious activity was discovered in 2004, when a personal computer (PC) was found to be regularly sending sensitive data to Shanghai. Further investigation found evidence of insider attacks dating as far back as 2000. However, Nortel seized the internal investigation 6 months later, due to lack of progress and resorted to simply changing affected user passwords. Mike Zafirovski, Nortel's chief executive at the time, played down the importance of the breach, allowing it to carry on for years. Shields was certain that the extensive espionage attacks contributed to Nortel's downfall, and he is not alone in this view (CBC News 2012; Leyden 2012). Why did it take Nortel 4 years to detect that something was wrong? Is insider threat detection a technically difficult task or was Nortel lacking an appropriate insider threat program? Perhaps both are true. Ultimately, Nortel's consistent reluctance to acknowledge and address the significance of the problem led to their eventual demise and paved the path for the global dominance of Chinese telecoms competitors.

One could argue that high-profile cases, such as Nortel's attacks, have helped raise the profile of insider threats and have helped us understand their significance. More high-profile cases will be reviewed in Section 6.3. Perhaps, it is worth reviewing how organizations perceive the significance of the insider threat. The levels of concern among organizations seem to be rising, which might suggest that attitudes and priorities toward insider threats could soon be changing. Indeed, 89% of respondents in the 2015 Vormetric Insider Threat Report believed they were at risk, and one out of three felt very or extremely vulnerable. Reassuringly, 92% of respondents are planning to increase or maintain existing spending on IT security, whereas only 11% feel that their organization is safe (Vormetric 2015). The 2016 Insider Threat Spotlight report reveals similar findings (Schulze 2016). Specifically, three out of four organizations felt vulnerable to insider threats, and one in three felt very or extremely vulnerable.

6.3 *Insider threats in practice*

Having reviewed the significance of the insider threat and how it affects organizations, we also discussed the decade-long espionage attacks against Nortel. It is worth reviewing the literature for other high-profile cases, which attracted the media's attention and brought insider threats to the spotlight. It seems that IP theft cases are the most widely reported stories. The most notable example is that of Edward Snowden, a former US National Security Agency (NSA) contractor, who revealed the extent of the Internet and phone surveillance by US intelligence (BBC 2013). Snowden, a computer wizard without formal education, initially joined the US Central Intelligence Agency (CIA), working on IT security. He quickly rose through intelligence ranks, taking up a diplomatic post in Geneva in 2007. After leaving CIA in 2009, he subsequently worked as a US NSA contractor, through outside contractor consultancy, Booz Allen. As part of his role, he had access to sensitive information, which he downloaded and eventually handed to journalists. According to Szoldra (2016), he downloaded 1.5 million sensitive files from the NSA before fleeing the country, handing the documents to journalists in Hong Kong and eventually obtaining asylum in Russia. Since then, journalists have released 7000 of these documents, which represent only a small subset of the original set. Unsurprisingly, Snowden was fired from Booz Allen in June 2013 and has been charged by the US government and NSA for the unauthorized communication of national defense information and willful communication of classified communications intelligence. Snowden admitted in an interview that he only took the job at Booz Allen to obtain access to classified information and obtain evidence (BBC 2013). Nonetheless, looking beyond the exposed secrets, the organizations and individuals involved, which have certainly raised eyebrows and have attracted intense media attention, there are questions to be had on the insider threat. For example, how sensitive information was handled, how trust was managed with third-party contractors, and ultimately what detection and prevention mechanisms were put in place to mitigate information leakage. Wikileaks informant Manning (BBC 2017), as well as the leak of customer information at the Target incident, also received significant media attention and helped raise the profile of insider threats (Upton and Creese 2014).

Other cases motivated by IP theft involve Mike Yu (Reuters 2011a) and Kexue Huang (Reuters 2011b). Yu, a product engineer at Ford Motor Company during 1997–2007, was arrested in October 2009—and sentenced to 70 months in prison—on the basis of stealing a large number of secret documents from the company. According to the US Attorney's office, Yu accepted a job at the China branch of a US company in December 2006. However, before he notified Ford about his new job, Yu copied approximately 4000 sensitive documents onto an external hard drive, including sensitive design documents about engine transmission and electric power supply systems. Yu gave his termination notice to Ford via an e-mail from China and eventually worked for a Chinese automaker (Reuters 2011a). In a similar fashion, Kexue Huang led a scientific team developing organic insecticides at Dow Chemical Company subsidiary and subsequently for another multinational company, Cargill Inc. Huang was charged for stealing trade secret from Dow and Cargill, which he then transferred to China and Germany, giving them competitive advantage by saving them millions of dollars and years in research and development. The estimated losses from Huang's case ranged from $7 to $20 million (Reuters 2011b).

Sergey Aleynikov (WSJ 2015), a former Goldman Sachs programmer, stole the company's secret source code of high-frequency trading platform and was sentenced to 8 years in prison. During his last few days at the company, he transferred 32 MB of proprietary computer code, which could have cost his employer millions of dollars. Although he attempted to hide his activity, the company detected it through anomalies in its network monitoring system. In the last example of IP theft, Michael Mitchell (Steinmeyer 2010), a former engineer and salesman of DuPont, is a typical example of a disgruntled insider. After being terminated because of poor performance, Mitchell kept the numerous numbers of DuPont computer files and proprietary information, which he eventually used when he accepted a consultancy position at DuPont's Korean competitor. The case was discovered only when he attempted to contact current and retired DuPont employees to gather even more secrets, but ended up being reported by some employees instead. This example shows how insider cases could take a long time to be discovered by investigators.

One case of IT sabotage, as reported by CERT Insider Threat Center, involved a computer support technician, who had administrator-level password-controlled access to the organization's network. Three months after leaving the company, they logged in late at night using the administrator account to remotely access the organization's network. They changed the passwords of all other IT system administrators and shut down nearly all the organization's servers. They also deleted backup files, which could have enabled the organization to recover more smoothly. As a result, the organization and its customers were unable to access its services and data for days. As a result, the insider was arrested and convicted to a $30,000 penalty, almost 2 years in prison, and community service (Collins et al. 2016).

This case highlights the need for robust employee termination procedure, which could have removed or locked the employee's credentials.

Lastly, an example of preventable fraud case is provided by Collins et al. (2016). An employee embezzled $200,000 from their organization, a bookkeeper business, by writing checks from the organizational account to pay for personal expenses. They also modified the organizational records to hide their tracks and show different payees. The incident took place over 2 years and was only discovered through irregularities in the electronic check ledger. The insider was convicted for this crime, but the funds were never recovered. As it transpired, the same person had been convicted for similar fraud in the past, so background checks before hiring could have prevented this attack (Collins et al. 2016).

6.4 Detecting insider threats

Despite the fact that intrusion detection systems were initially conceived with the aim of detecting both internal and external attacks, the reality of detecting insider threats has proven to be more challenging. According to Schulze (2016), when asked about the difficulty of detecting and preventing insider attacks as opposed to external attacks, 66% of respondents found them to be more difficult, and only 7% thought they were easier. Specifically, respondents attributed the main challenges in effectively detecting and responding to the following parameters:

- Insiders already have credentialed access to the network and services (67%).
- There is increased use of applications with potential to leak data (i.e., web, e-mail, cloud data stores, and social media) (53%).
- Increased amount of data that leaves protected boundary/perimeter (46%).
- More end user devices capable of theft (33%).

This section will review the main approaches of detecting insider threats, drawing on their advantages and limitations.

6.4.1 Log and system analysis approaches

Magklaras and Furnell (2001) suggest a structured approach at predicting insider misuse by considering the insider's sophistication level, historical behavior, and their motivation. Its foundation is the insider threat prediction model (ITPM), which initially estimates the potential impact of the incident, as well as the suspected insider's role, the hardware and software tools they are capable of using, their historical behavior, etc. The resulting outcome will help predict the threat level from intentional and accidental insider incidents. Although largely theoretical, and not fully implementable at the time, ITPM presents interesting concepts and links the threat level (and hence possible outcome) to the attacker's level of sophistication.

In a similar fashion, Maybury et al. (2005) concentrate on highly sophisticated malicious insiders and propose an insider threat detection model, which distinguishes motives, actions, and associated observables. Observables for cyberactions include network, system, information reconnaissance, access to assets (e.g., media, hosts, and accounts), entrenchment (e.g., installing sensors or unauthorized software), exploitation (e.g., commanding and controlling entrenched assets such as software bots or zombie machines), extraction and exfiltration (e.g., of hardcopy, media, and information), communication (e.g., encrypted messaging, encoded messages, and covert channels), manipulation of cyberassets (e.g., changing file permissions and suppressing or altering information content), counterintelligence (e.g., wiping disks), and other cyberactivities associated with unethical or addictive behavior (e.g., online gambling). They test their approach against three types of malicious insiders, namely, an analyst, application administrator, and system administrator, involving a wide range of motives (e.g., financial, thrill, or ideological). The behavior of these three insiders is simulated and captured on MITRE's demilitarized zone network, which already has 75 online active users during the 3-month evaluation period. A heterogeneous and multilevel data collection approach is adopted, incorporating physical (e.g., card access records), network (e.g., Snort IDS, Stealthwatch, honeynet, and e-mail sensors), host (e.g., logins and password updates), and application [e.g., e-mail, secure shell (SSH), and web server logs] level data. The Common Data Repository (CDR) is the result of this evaluation, containing more than 11 million anonymized records. The proposed system is tested against the CDR and evaluated based on its accuracy and timeliness. The accuracy is determined in relation to false positives and false negatives, whereas timeliness is considered as the difference between the time malicious activity begins, the time it is put on a watch list, and the time an insider threat alarm is triggered. Stealthwatch alerts included an element of human analysis, whereas all other methods were autonomous. The goal of the proposed system is to reduce the time between defection to discovery from years and months to weeks and days. Their results are encouraging, as two insiders were detected within 1 week, and the third was detected within 2 weeks. Also, the heterogeneous, multilevel data collection and fusion approach is interesting. However, it is difficult to draw generic conclusions, based on three simulated case studies.

Agrafiotis et al. (2014) have focused on characterizing employee behavior based on their role within the organization. They use activity trees and sequentially based analysis for users in similar roles, where activity trees include the range of activities that an employee (potential insider) may conduct as part of their expected daily workload. A similarity measure indicates how branches compare to others, and detection is based on identifying unusual new activity branches with similarity between them and existing branches below a certain threshold. It is based on sequential tree-based profiling, through a proof-of-concept software tool that builds the tree profile, measuring the similarity of newly observed branches and seeing results. This concept is an extension to the general idea of attacking trees, which incorporates the sequence of events that will result in attack. Attack trees are the first tier of a tree system, which can provide insight into the objectives of internal attacks, receiving information about the attackers, and formulate a hypothesis about the occurrence of an attack. Behavior trees can reflect dynamic behaviors including patterns of conducts. The system is written in Python programming, which facilitates entering new set of data and achieving a dynamic approach.

The corporate insider threat detection (CITD) system is designed for large-scale data repositories and activity logs and incorporates user- and role-based profilings. It combines technical activities and behavioral actions to assess the threat posed by individuals (Legg et al. 2015). It is an anomaly-based detection system, which compares a user's observed behavior against their user and role profiles, flagging significant deviations. User and role profilings feature selection concentrates on unusual device, activity, and attribute access. For example, one observation could be that an e-mail was sent (activity) to john.davis@mycompany.com (attribute) from PC-012 (device). The CITD is designed to operate in supervised and unsupervised modes, where the feedback from a human analyst can significantly help reduce false alarms. The experimental results using various synthetic datasets in unsupervised mode have been highly encouraging, and its deployment and evaluation on a large multinational corporation is expected in the future.

Liu et al. (2005) also proposed an anomaly detection system for insider threats. It aims to apply anomaly detection methodologies, widely used for external threats, to the domain of insider threat detection. The proposed methodology consists of three steps: data collection, feature extraction, and internal threat detection using the *k*-nearest neighbor algorithm. Three types of feature collection are used to evaluate the performance of the proposed methodology: *n*-grams of the programs used, the frequency with which the programs are executed, and the parameters given in the executable programs. The *k*-nearest neighbor algorithm belongs to the supervised learning class. Supervised learning algorithms assume that it is possible to collect training data that is totally free from malicious actions or real data containing the label of normal or malicious energy. Such a case in the case of detecting internal threats can be fatal. Liu et al.'s (2005) paper outlines the insider threat detection method, which tracks system calls at the operative level to monitor inside level to monitor activity. The system calls have a higher degree of information reliability due to the capacity of monitoring *all* system activity, and the monitoring becomes more closely attached into the operating system, becoming more demanding in the user permissions to access, use, or delete monitoring information without traces of tampering.

A comparison between non-upervised and supervised learning techniques for detecting masquerading attacks is investigated by Wolff (2010). Two nonsupervised techniques were used, specifically the minimum distance technique and the compact cluster technique. A synthetic dataset of 15,000 system calls per user over a 6-month period, including synthesized masquerade data, was used for the comparison (Schonlau et al. 2001). The focus is on two classes of insiders, specifically masqueraders and traitors. The training dataset contains the first 5,000 calls for each user, whereas the remaining 10,000 calls were split into blocks of 100 commands each to create 100 blocks for each user. Initially, the unsupervised algorithms were compared. The unsupervised algorithms are used to take into consideration all the users' history when generating a user profile, while the supervised algorithm considers only a preestablished set of data. The basis for the nonsupervised learning techniques of the author is the minimum distance, which is a technique proposed for the creation of a user profile, based on the premise of the minimum distance algorithm aiming to find the block of data closer to other blocks of individual's user dataset. To attain it under a nonsupervised algorithm, the author makes adjustments to the uniqueness algorithm, removing the update technique because the user profiling algorithm is already controlled by all data when creating a profile, eliminating the need for dynamic updating of the profile while classifying the data. The results were promising for unsupervised approaches. For small to medium classification thresholds, supervised algorithms outperformed the others, but for the larger thresholds of 400 anomalous blocks, unsupervised methods matched those of supervised ones.

Salem et al. (2008) also conducted a comparison of different machine-learning techniques, for insider threat, using the same Schonlau dataset (2001). Specifically, they tested against the following algorithms: uniqueness of commands; Bayes one-step Markov approach based on one-step transitions from one command to the next; a hybrid multistep Markov method; compression method based on differences in compressing test data from the same user rather than a masquerader; a similarity measure for each sequence command; incremental probabilistic action modeling (IPAM); naive Bayes classifier; semiglobal alignment; Eigen cooccurrence matrix (ECM); and self-consistent naive Bayes classifier, which combines naive Bayes and the EM algorithm. Their experimental results are shown in Table 6.1. It is fair to say that no algorithm clearly outperforms all the others, whereas using a combination of approaches seems to yield marginally better results. The Schonlau dataset has limited scope and is therefore insufficient to reflect true insider

Table 6.1 Accuracy of machine-learning techniques in insider threat detection

Method	False alarms (%)	Missing alarms (%)
Uniqueness	1.4	60.6
Bayes one-step Markov	6.7	30.7
Hybrid multistep Markov	3.2	50.7
Compression	5.0	65.8
Sequence match	3.7	63.2
IPAM	2.7	58.9
Naive Bayes (updating)	1.3	38.5
Naive Bayes (no updating)	4.6	33.8
Semiglobal alignment	7.7	24.2
ECM	2.5	28.0
Naive Bayes + EM	1.3	25.0

Source: Salem, M.B. et al., A survey of insider attack detection research, In Stolfo, S.B. et al. (Eds), *Insider Attack and Cyber Security*, Springer, Boston, MA, 2008, pp. 69–90.

threat detection rates. However, it is still useful in comparing different machine-learning techniques.

As seen earlier, applying machine-learning techniques for insider threat detection can have various degrees of success. As insider threat behavior can vary widely, it is difficult to characterize it without considering the human factor. It is often approached as a generic anomaly detection task, where a combination of multiple techniques is likely to yield slightly better results. Chandola et al. (2012) provide a comprehensive review of various anomaly-detection approaches in generic intrusion-detection scenarios and note that most techniques deal with univariate discrete sequences. To understand online and multivariate sequences, more needs to be done. It would be interesting to investigate the applicability of deep learning approaches and whether they could produce better results. One such approach is that of Tuor et al. (2017), who use deep belief networks against the CERT Insider Threat Dataset v6.2 (Glasser and Lindauer 2013). Their approach considers normal behavior for a user, a role, or a project team and inspects the stream of system logs to infer user metadata and network activity. Their research is still in the early stages, but it does highlight the potential for further research in the area.

6.4.2 Psychological and technical approaches

Looking into the human factor, the opportunity in psychological models is a widely used feature for detecting insider threats (Gheyas and Abdallah 2016). In order to determine the level of this opportunity and identify or mitigate the attack, most studies focus on two broad categories of features: the role of the internal threat in the system and activity-based characteristics. A significant part of this research is conducted around the psychological profiling of company employees. The outline of the psychological profile enables the user's history to be recorded as well as the psychological status at specific times.

Under the psychological approach, the theoretical work of Kandias et al. (2010) states that an attack is highly dependent on three main factors: motivation, ability, and opportunity. The authors study the psychosocial perspective and the implications of the prediction of the insider threat through the use of social media, under open-source intelligence, and the content generated by the user, through an inductive methodology. The model is based on the assumption that psychological traits, such as negative beliefs toward authority, can be monitored through the use of social media tools, and those who hold such beliefs are more prone to be insider intruders given the opportunity. As such, it focuses on detecting users with a negative attitude toward authorities by profiling social media channel YouTube. They used a free flat representation data technique to target the user's attitude and improve the scalability of their method. Their research showed a correlation between psychological traits and malevolent insiders. The advantage of such model lies in the ease of gathering data, which are mostly public, and the fact that YouTube is a space where users tend to show their opinions. Apart from the privacy concerns of monitoring employees' social media data, another weakness resides on the fact that it cannot demonstrate with clarity the escalation of behavior, from malevolent attitude toward authority to the actions that cause an internal threat.

The work of Greitzer et al. (2010) combines traditional cybersecurity audit data and psychosocial indicators that can be used to predict internal threats. They also consider legal and ethical issues in the type of activity that one could collect in an organizational context. Traditional cybersecurity audit data, which consist of events that are normally used to determine policy violations or outlier behavior, are incorporated with demographic/organizational data about the employee. Information sources that informed various psychosocial factors included staff performance evaluations, competency tracking, disciplinary tracking, timecard records, proximity card records, and preemployment background checks. The proposed approach still raises privacy issues on employee monitoring, which could lead to more disgruntlement toward the organization and eventually more severe malicious insider events. On the other hand, it is possible for an organization to fall into the "trust trap," where the longer employees stay in an organization, the safer the organization feels about them, giving them a false sense of security.

Brdiczka et al. (2012) also use a combination of psychological profiles (PPs) with structural anomaly (SA) detection from social and information networks. SA uses graph analysis, dynamic tracking, and machine learning to detect anomalies, whereas PPs are dynamically constructed from behavioral patterns. Outcomes from both SA and PP are fused and ranked to detect insider threats, targeting the reduction of false positives and false negatives at the same time. The creation of dynamic psychological profiles uses observable indicators to reflect the emotional state and personality of a perpetrator. As such, they are based on psychometric assessments (e.g., extraversion) and temporal work patterns. The analysis includes online and PC usage behaviors, sentiment analysis of a user's communications, and social network features and analysis. The authors argue that creating a psychological profile has the effect of shrinking the volume of data as well as reducing false alarms. Quite interestingly, the proposed approach was tested on a real dataset retrieved from a multiplayer online game, World of Warcraft (WoW), and consisted of behavior traces of over 350,000 game characters over a 6-month period. Although the data are not derived from an organizational context, the authors claimed that their approach showed promising results and could be used to predict when characters would quit their guild (a form of gaming club) and cause possible damage to their group. One could argue that there are similarities between insiders in an organization and WoW players. Still, the proposed method has not been tested against organizational real data yet, and hence, it is difficult to evaluate its applicability in corporate scenarios.

Legg et al. (2013) also attempted to achieve the detection of internal threats through surveillance techniques and psychological tests, developing a conceptual model that incorporates an all-encompassing organizational view of the problem. The novelty is the priority rankings, known as lanes. There are four *lanes*, namely, enterprise, people, technology and information, and physical. Initially, users are ranked based on the level of access they hold, as well as using psychological tests. Psychological tests try to detect the user's level of knowledge, whether they have a predisposition for illegal energy and whether the user has high levels of stress. This is combined with real-time collected data, which feed into the decision-making system to extract the level of risk for each user. According to the authors, the algorithm that should be used in the decision-making system is different for each organization, depending on the data that each organization decide to use. The conceptual model is three dimensional and incorporates in descending order the following tiers: hypothesis, measurement, and real world. The analyst in charge of detecting potential insider threats is put above the hypothesis tier, capable of supervising all the tiers below it. The real-world tier contains sets of elements, such as activity logs, building access logs, and psychological mind-set among others. The measurement tier records measures of real-world elements, whereas the hypothesis tier relies on these measurements to test different hypotheses. The model considers high levels of confidence for directly observable elements, such as system access logs for the measurement of whether an insider could have downloaded sensitive IP. Psychological elements, such as the stress level of an employee, are only observable by an indirect mode based on a small set of indicators, therefore providing lower confidence level. The output of the model is the probability of the hypothesis being true. This reasoning-based model is important, as it allows the analyst to explore and test different hypotheses based on suspicions for specific employees, while it allows the hypotheses to be formed by the underlying reasoning component from the real-world tier.

While previous approaches have associated personality traits with social media behavior, Alahmadi et al. (2015) concentrated on web-browsing behavior. Their hypothesis is that the detection of insider threats could be based on the linguistic features of the websites users regularly visit. As such, they suggest that the detection of insider threats could concentrate on the user's Internet browsing behavior and the associated personality traits used for the detection of insider attacks. The associated personality traits include the OCEAN (openness, conscientiousness, extraversion, agreeableness, and neuroticism) and the dark triad (Machiavellianism, narcissism, and psychopathy) characteristics. The foundation of the proposed research is to consider how browsing behavior relates to personality traits and how browsing behavior deviates over time to proactively identify potential insider threats, before significant damage is caused. To this end, the proposed system collects the content of each website. Irrelevant content from the retrieved hypertext markup language (HTML) source code is initially removed (e.g., HTML and JavaScript tags), and then relevant keyword features are extracted, such as money, loan, debt, hire, tax, and love. The resulting dataset is passed through an algorithm calculating its dimensions and is then fitted to a *k*-means algorithm that tries to extract the personality traits of the user. The combination of user personality traits can indicate if a user is likely to be an insider threat, whereas the deviation of such behavior can signal an internal threat. The benefit of such model is the possibility of calculating the probability of deviant behavior based on widely available information.

Apart from psychological profiling, and the technical or behavioral indicators of potential attacks, Nurse et al. (2014) also consider the wider context of an attack and specifically the motivations of malicious attackers as well as human factors affecting unintentional incidents. They propose a framework for conceptualizing and characterizing insider attacks, based on four areas, namely, *catalyst, actor characteristics, attack characteristics*, and *organization characteristics*. The proposed framework identifies a catalyst as a key event, which could tip the insider over the edge into committing the offense. Catalyst events include, among others, the following: demotion, dismissal, dispute with employees, family problems, blackmail, a new job offer, or even the lack of training in unintentional attacks. Actor characteristics encompass the following: psychological state, personality traits, attitude toward work, motivation to attack, skill set, opportunity to attack, as well as previous history of rule violations. Dark triad traits Machiavellianism, excitement seeking, and narcissism were more closely related to malicious actors, whereas OCEAN traits (especially agreeableness and openness) can indicate susceptibility to scams. Overall, significant attention is given toward understanding the propensity to attack and how it could be influenced by different elements. The proposed framework recognizes that none of these elements in isolation is sufficient to detect insider attacks. However, in combination, they could provide invaluable insight into characterizing and understanding insider threats.

To understand the attack itself, one could consider elements, such as the overall objective, as well as the specific steps and goals needed to achieve the attack. For example, an attack that plants a logic bomb could be driven by revenge and have the objective to sabotage a company's mission-critical function. In the case of unintentional threats, the business task, which was the reason for the activity, could be considered the attack objective. For example, a time-critical task, which has to be completed within strict time constraints, could lead an employee to copy sensitive data on a universal serial bus (USB) key and subsequently lose them in public transport. As for attack steps and goals, in a similar fashion to attack trees, they represent the individual steps that are needed to achieve the attack. For example, in order to steal sensitive IP, an insider might follow several stages: initially gather intelligence on who has access credentials; figure out a way of extracting them through blackmail, charm, financial means, etc.; extract credentials to access sensitive information; and finally cover their tracks. The last area within the framework includes organization characteristics, such as assets under attack and their vulnerabilities. Several known scenarios from the CERT Insider Threat Dataset and the United Kingdom's Centre for the Protection of National Infrastructure were successfully characterized. The authors also used an additional set of cases within their broader research, as directly collected by Whitty and Wright (2013). Overall, although not a detection system in itself, the proposed framework is very useful in understanding different elements and contextual factors influencing insider attacks.

The potential of informing insider threat detection with psychological and psychosocial factors can be beneficial and a significant body of research has informed insider threat detection with personal characteristics of potential insiders. It is particularly useful as it has the potential to proactively identify potential insiders and escalate monitoring and response before any significant damage is caused. Psychological and psychosocial characteristics alone are not enough to identify insider attacks, but if combined with suspicious behavior, opportunities, and catalysts, they could play an important role in insider threat detection. However, a significant challenge of these approaches is that their accuracy is heavily influenced by the availability of such data. Quite often, psychological profiles might not be readily available within organizations or even shared across different departments. Although personality traits and psychological features can be dynamically inferred from online user behavior, there are still privacy issues with such approaches at various degrees (Greitzer et al. 2010). Although one could argue, for example, that monitoring online browsing data would pose fewer privacy concerns than social media activity, the potential for employees' privacy to be invaded cannot be ignored. As such, organizations and solutions would need to consider the ethical and legal issues surrounding the issue. Although far-fetched, another consideration is how malicious insiders, who are aware of the importance of psychological profiling in insider threat detection within an organization, could potentially modify their online behavior accordingly to evade monitoring controls. Despite these concerns, it is important to recognize that insider threat detection can be greatly enhanced by considering the human factor and its strong relationship with the propensity to attack.

6.5 Best practices for insider threat mitigation

Preventing and detecting insider threats is a complex issue, especially as malicious insider online activity is often similar to what insiders do as part of their normal role. Hence, it is important to recognize that the mitigation of insider threats cannot involve technical controls alone and cannot be the sole responsibility of the IT and security personnel (Cappelli et al. 2012). Instead, best practice guidelines require the cooperation of management, human resources, legal, physical security, data owners, information technology, and software engineering (Collins et al. 2016). Organizations are advised to adopt the following 20 best practice guidelines, which will enable them to prevent, detect, and respond to such threats effectively, incurring as little costs and disruption as possible. Specifically, these guidelines and practices are summarized in the following (Collins et al. 2016):

1. *Know and protect your critical assets*: This includes identifying and protecting all assets that provide the organization with a competitive advantage. As a quick win solution, organizations are advised to conduct physical assets inventory, understand what data the organization process, identify the software configurations of all assets, and prioritize assets and data to identify high-value targets.
2. *Develop a formalized insider threat program*: A formalized insider threat program provides designated resources to deal with the problem and allows the effective prevention, detection, and response to insider incidents. An insider threat program needs to consider issues, such as whom to involve, who has authority, whom to coordinate with, whom to report to, what actions to take, and what improvements to make.
3. *Clearly document and consistently enforce policies and controls*: All organizational policies should include a clear and consistent message that aims to reduce the chance of employees damaging or lashing out at the organization for a perceived injustice. Policies and punishments ought to be fair and consistent and not disproportional to the violation that occurred. Quick win solutions for enforcing and advocating this clear message include the following: adoption of policies and practices by everyone, including senior management; regular briefing of all employees on policies and procedures, accompanied by signing of acceptable use and/or nondisclosure agreements; making policies and procedures easily accessible to all employees; making annual refresher training mandatory for all employees; and facilitation of clear and concise enforcement of policies, free from favoritism and injustice.
4. *Beginning with the hiring process, monitor and respond to suspicious or disruptive behavior*: Organizations should proactively deal with suspicious or disruptive employees, as this could reduce the risk of insider threats. This could include a thorough background investigation on criminal or credit records, encouragement of employees to report suspicious behavior, and investigation and documentation of incidents involving suspicious or disruptive behavior.
5. *Anticipate and manage negative issues in the work environment*: Additional monitoring of employees with an impending or ongoing personnel issue, based on consistent organizational policies and procedures, will allow easier detection and response to a potential insider incident. User privacy would need to be considered, and access to such logs would need to be on a need-to-know basis. Additionally, organizational

changes would need to be communicated clearly and transparently to allow them to better plan for their future.

6. *Consider threats from insiders and business partners in enterprise-wide risk assessments*: This practice includes a comprehensive risk-based security strategy for the protection of critical assets against internal and external threats, including trusted business partners. High-impact solutions would include nondisclosure agreements upon hiring and termination of employment or contracts, enforcing background investigation of business partner employees, performing background checking against all employees to be acquired during company merging or acquisitions, preventing unnecessary printing of sensitive documents, avoiding direct connections of business partners to information systems, and restricting access to backup systems only to relevant employees.
7. *Be especially vigilant regarding social media*: Policies, training, and procedures ought to define how employees, contractors, and business partners should use social media to avoid intentionally or unintentionally threatening critical assets. Apart from providing social media policy and training program, users can be encouraged to report suspicious e-mails or phone calls to the information security team.
8. *Structure management and tasks to minimize insider stress and mistakes*: An organization is encouraged to provide an environment conducive to positive behavior, by understanding the psychology of employees and the demands placed upon them. For example, establish success metrics that are relevant and appropriate to the work environment; encourage focusing on one thing at a time rather than multitasking; offer opportunities for employees to destress; routinely monitor employee workloads to ensure that they are appropriate; and encourage employees to think through projects, actions, and statements before committing to them.
9. *Incorporate malicious and unintentional insider threat awareness into periodic security training for all employees*: Periodic security training for employees and contractors will support a successful security culture within the organization. An anonymous or confidential reporting mechanism of security incidents will also help toward this goal.
10. *Implement strict password and account management policies and practices*: This aims to prevent malicious insiders from circumventing manual and automated control mechanisms by compromising user accounts. Account management policies, strong password selection requirements, training on secure password practices, access to shared accounts on a need-to-know basis, and regular auditing of account creation and password changes would help toward this practice.
11. *Institute stringent access controls and monitoring policies on privileged users*: Privileged users have the technical ability to commit and conceal malicious activity, so organizations should conduct periodic reviews to avoid privilege creep and ensure that they follow the principle of least privilege as employees change roles.
12. *Deploy solutions for monitoring employee actions and correlating information from multiple data sources*: Relying on network activity alone is not enough, as the number of data sources could significantly enhance insider threat analysis and response. Relevant technical and nontechnical data sources could include among others authentication logs, firewall logs, telephone records, file access logs, physical access records, performance evaluations, physical violation records, or travel reporting.
13. *Monitor and control remote access from end points, including mobile devices*: Remote access is often perceived by insiders as a less risky option. Also, mobile workforce and employees' own devices often connect via the same route. Monitoring remote access, disabling it once an employee or contractor leaves the organization, considering the

use of personal devices in the risk planning, and limiting the use of cameras in sensitive areas are relevant good practices.

14. *Establish a baseline of normal behavior for both networks and employees*: Building on from practice 12, it is good practice to analyze the collected data to establish normal behavior per role and deviations from these profiles. Network behavior could include bandwidth utilization, usage patterns, and protocols, whereas employee behavior could encompass typical working hours, resource usage patterns, and resource access patterns.
15. *Enforce the separation of duties and least privilege*: In a similar fashion to privileged users, account management ought to consider the separation of duties and follow the principle of least privilege, especially as they change roles.
16. *Define explicit security agreements for any cloud services, especially access restrictions and monitoring capabilities*: Data access control and monitoring ought to be included in any agreement with cloud service providers. In addition, good practices include risk assessment for any data or services that are subject to cloud outsourcing, verification of the cloud service provider's hiring process, and remote access restrictions to hosts providing cloud services.
17. *Institutionalize system change controls*: Change management controls across systems and applications could help prevent the introduction of backdoors, keystroke loggers, or logic bombs. In addition, the periodic review of configuration baselines could help discover any undocumented discrepancies.
18. *Implement secure backup and recovery processes*: Regular backup processes will enhance the ability of an organization to recover from incidents and enhance its resilience. Maintaining off-site backups and including network infrastructure devices in the backup and recovery plans are both good practices.
19. *Close the doors to unauthorized data exfiltration*: Understanding data exfiltration vulnerabilities is important to establish relevant mitigation strategies. USB flash drives, printers, cloud, and e-mail are relevant vectors, each with its unique challenges. Relevant practices include establishment cloud computing policy; monitoring of printer, fax, copier, scanner usage; defining of data transfer policy; defining of and enforce removable media policy; and restricting of data transfer protocols, such as file transfer protocol and SSH file transfer protocol.
20. *Develop a comprehensive employee termination procedure*: Termination procedures should ensure that all accounts are closed, access tokens and equipment are collected, remaining personnel are notified, and nondisclosure agreements are reaffirmed.

6.6 Conclusions

The problem of insider attacks cannot be ignored, as the Nortel case highlights. The insider threat might not be as frequent as malicious software, but its impact can be costly. Our understanding of the insider threat problem, the relevant indicators, and how various factors can influence one another are increasing. Still, the detection of insider threats is more likely to take months and years, rather than hours and days. Although attitudes are starting to change, and organizations have started to recognize the significance of the issue, they are still a long way from adopting successful insider threat programs. In terms of detection, the research community has concentrated on both technical and hybrid solutions. Detecting insider threats is not a purely technical solution, and the human factor can play an important role. Recent research has recognized its importance and has incorporated personality traits, psychological and psychosocial data, as well as motivations and possible catalysts of insider events. Beyond prevention and detection, though, best practices and guidelines recognize that insider threat is a multifaceted problem, and the success of insider threat mitigation strategies depends on the cooperation of various groups within an organization. Specifically, specific emphasis is given on management, human resources, legal, physical security, data owners, IT, and software engineering.

References

ACFE (Association of Certified Fraud Examiners). *Global Fraud Study: Report to the Nations on Occupational Fraud and Abuse*. ACFE, Austin, TX (2016). Available at: https://s3-us-west-2.amazonaws.com/acfepublic/2016-report-to-the-nations.pdf.

Agrafiotis, Ioannis, Philip Legg, Michael Goldsmith, and Sadie Creese. Towards a user and role-based sequential behavioural analysis tool for insider threat detection. *Journal of Internet Services and Information Security* 4, no. 4 (2014): 127–137.

Alahmadi, Bushra A., Philip A. Legg, and Jason R. C. Nurse. Using Internet activity profiling for insider-threat detection. In *International Conference on Enterprise Information Systems* (2), (2015), pp. 709–720.

BBC. Profile: Edward Snowden. (2013). Available at: http://www.bbc.co.uk/news/world-us-canada-22837100.

BBC. Chelsea Manning: Wikileaks source and her turbulent life. (2017). Available at: http://www.bbc.co.uk/news/world-us-canada-11874276.

Bishop, Matt, and Carrie Gates. Defining the insider threat. In *Proceedings of the 4th Annual Workshop on Cyber Security and Information Intelligence Research: Developing Strategies to Meet the Cyber Security and Information Intelligence Challenges Ahead*. Association for Computing Machinery, New York (2008), p. 15. Available at: http://nob.cs.ucdavis.edu/bishop/papers/2008-csiiw/definsider.pdf.

Brdiczka, Oliver, Juan Liu, Bob Price, Jianqiang Shen, Akshay Patil, Richard Chow, Eugene Bart, and Nicolas Ducheneaut. Proactive insider threat detection through graph learning and psychological context. In *2012 IEEE Symposium on Security and Privacy Workshops (SPW)*. Institute of Electrical and Electronics Engineers, Piscataway, NJ (2012), pp. 142–149.

Cappelli, Dawn M., Andrew P. Moore, and Randall F. Trzeciak. *The CERT Guide to Insider Threats: How to Prevent, Detect, and Respond to Information Technology Crimes (Theft, Sabotage, Fraud)*. Addison-Wesley, Boston, MA (2012).

CBC News. Nortel collapse linked to Chinese hackers. (2012). Available at: http://www.cbc.ca/news/business/nortel-collapse-linked-to-chinese-hackers-1.1260591.

Chandola, Varun, Arindam Banerjee, and Vipin Kumar. Anomaly detection for discrete sequences: A survey. *IEEE Transactions on Knowledge and Data Engineering* 24, no. 5 (2012): 823–839.

Collins, Matt., Michael Theis, Randall Trzeciak, Jeremy Strozer, Jeremy Clark, Daniel Costa, Tracey Cassidy, Michael Albrethsen, and Andrew Moore. *Common Sense Guide to Mitigating Insider Threats*, 5th Edition (Technical Report CMU/SEI-2016-TR-015). Software Engineering Institute, Carnegie Mellon University, Pittsburgh, PA (2016). Available at: http://resources.sei.cmu.edu/library/asset-view.cfm?AssetID=484738.

Federal Infrastructure Protection Bureau. *Unintentional Insider Threats: A Foundational Study*. (Technical Report CMU/SEI-2013-TN-022). Software Engineering Institute, Carnegie Mellon University, Pittsburgh, PA (2013). Available at: http://resources.sei.cmu.edu/asset_files/TechnicalNote/2013_004_001_58748.pdf.

Gheyas, Iffat A., and Ali E. Abdallah. Detection and prediction of insider threats to cyber security: A systematic literature review and meta-analysis. *Big Data Analytics* 1, no. 1 (2016): 6.

Glasser, Joshua, and Brian Lindauer. Bridging the gap: A pragmatic approach to generating insider threat data. In *Security and Privacy Workshops (SPW) 2013*. Institute of Electrical and Electronics Engineers, Piscataway, NJ (2013), pp. 98–104.

Greitzer, Frank L., and Deborah A. Frincke. Combining traditional cyber security audit data with psychosocial data: Towards predictive modeling for insider threat mitigation. In Probst, Christian W., Jeffrey Hunker, Dieter Gollmann, and Matt Bishop (Eds), *Insider Threats in Cyber Security*. Springer, New York (2010), pp. 85–113.

Hunker, Jeffrey, and Christian W. Probst. Insiders and insider threats—An overview of definitions and mitigation techniques. *Journal of Wireless Mobile Networks, Ubiquitous Computing, and Dependable Applications* 2, no. 1 (2011): 4–27. Available at: http://isyou.info/jowua/papers/jowua-v2n1-1.pdf.

Intel Security. *Grand Theft Data: Data Exfiltration Study*. McAfee, Santa Clara, CA (2015). Available at: https://www.mcafee.com/us/resources/reports/rp-data-exfiltration.pdf.

Kandias, Miltiadis, Alexios Mylonas, Nikos Virvilis, Marianthi Theoharidou, and Dimitris Gritzalis. An insider threat prediction model. In *International Conference on Trust, Privacy and Security in Digital Business*. Springer, Berlin (2010), pp. 26–37.

Legg, Philip A., Nick Moffat, Jason R. C. Nurse, Jassim Happa, Ioannis Agrafiotis, Michael Goldsmith, and Sadie Creese. Towards a conceptual model and reasoning structure for insider threat detection. *Journal of Wireless Mobile Networks, Ubiquitous Computing, and Dependable Applications* 4, no. 4 (2013): 20–37.

Legg, Philip A., Oliver Buckley, Michael Goldsmith, and Sadie Creese. Caught in the act of an insider attack: Detection and assessment of insider threat. In *2015 IEEE International Symposium on Technologies for Homeland Security (HST)*. Institute of Electrical and Electronics Engineers, Piscataway, NJ (2015), pp. 1–6.

Leyden, J. Whistleblower: Decade-long Nortel hack "traced to China." *The Register*. (2012). Available at: https://www.theregister.co.uk/2012/02/15/nortel_breach/.

Liu, Alexander, Cheryl Martin, Tom Hetherington, and Sara Matzner. A comparison of system call feature representations for insider threat detection. In *IAW'05. Proceedings from the Sixth Annual IEEE SMC: Information Assurance Workshop*. Institute of Electrical and Electronics Engineers, Piscataway, NJ (2005), pp. 340–347.

Magklaras, George B., and Steven Furnell. Insider threat prediction tool: Evaluating the probability of IT misuse. *Computers & Security* 21, no. 1 (2001): 62–73.

Maybury, Mark, Penny Chase, Brant Cheikes, Dick Brackney, Sara Matzner, Tom Hetherington, Brad Wood, Conner Sibley, Jack Marin, and Tom Longstaff. *Analysis and Detection of Malicious Insiders*. Mitre Corporation, Bedford, MA (2005).

Mukherjee, Biswanath, L. Todd Heberlein, and Karl N. Levitt. Network intrusion detection. *IEEE Network* 8, no. 3 (1994): 26–41.

Nurse, Jason R. C., Oliver Buckley, Philip A. Legg, Michael Goldsmith, Sadie Creese, Gordon RT Wright, and Monica Whitty. Understanding insider threat: A framework for characterising attacks. In *2014 IEEE Security and Privacy Workshops (SPW)*. Institute of Electrical and Electronics Engineers, Piscataway, NJ (2014), pp. 214–228.

Ponemon Institute. *Privileged User Abuse and the Insider Threat*. Commissioned by Raytheon Company, Independently conducted by Ponemon Institute LLC. Ponemon Institute, Traverse City, MI (2014).

Ponemon Institute. *2016 Cost of Cyber Crime Study and the Risk of Business Innovation*. Sponsored by Hewlett Packard Enterprise. Ponemon Institute, Traverse City, MI (2016). Available at: http://www.ponemon.org/local/upload/file/2016 HPE CCC GLOBAL REPORT FINAL 3.pdf.

PWC (PricewaterhouseCoopers). *US Cybercrime: Rising Risks, Reduced Readiness*. PWC, London (2014).

Reuters. Ex-Ford engineer sentenced for trade secrets theft. (2011a). Available at: http://www.reuters.com/article/us-djc-ford-tradesecrets-idUSTRE73C3FG20110413.

Reuters. Chinese man pleads guilty for US trade secret theft. (2011b). Available at: http://www.reuters.com/article/us-crime-china-theft-idUSTRE79H78R20111018.

Salem, Malek Ben, Shlomo Hershkop, and Salvatore J. Stolfo. A survey of insider attack detection research. In Stolfo, Salvatore J., Steven M. Bellovin, Angelos D. Keromytis, Shlomo Hershkop, Sean W. Smith, and Sara Sinclair (Eds), *Insider Attack and Cyber Security*. Springer, Boston, MA (2008), pp. 69–90.

Schonlau, Matthias, William DuMouchel, Wen-Hua Ju, Alan F. Karr, Martin Theus, and Yehuda Vardi. Computer intrusion: Detecting masquerades. *Statistical Science* 16, no. 1 (2001): 58–74.

Schultz, Eugene E. A framework for understanding and predicting insider attacks. *Computers & Security* 21, no. 6 (2002): 526–531.

Schulze, H. *Insider Threat: Spotlight Report, Crowd Research Partners Report.* (2016). Crowd Research Partners. Available at: http://crowdresearchpartners.com/wp-content/uploads/2016/09/Insider-Threat-Report-2016.pdf.

Silowash, George, Dawn Cappelli, Andrew Moore, Randall Trzeciak, Timothy J. Shimeall, and Lori Flynn. *Common Sense Guide to Mitigating Insider Threats,* 4th Edition (No. CMU/SEI-2012-TR-012). Software Engineering Institute, Carnegie Mellon University, Pittsburgh, PA (2012). Available at: http://resources.sei.cmu.edu/asset_files/technicalreport/2012_005_001_34033.pdf.

Steinmeyer, Peter A. *Former DuPont Employee Sentenced to 18 Months For Trade Secret Misappropriation.* (2010). Epstein Becker & Green, P.C. Available at: http://www.tradesecretsnoncompetelaw.com/2010/03/articles/trade-secrets-and-confidential-information/former-dupont-employee-sentenced-to-18-months-for-trade-secret-misappropriation/.

Szoldra, Paul. This is everything Edward Snowden revealed in one year of unprecedented top-secret leaks. *Business Insider UK* (2016). Available at: http://uk.businessinsider.com/snowden-leaks-timeline-2016-9.

Tuor, Aaron, Samuel Kaplan, Brian Hutchinson, Nicole Nichols, and Sean Robinson. *Deep Learning for Unsupervised Insider Threat Detection in Structured Cybersecurity Data Streams.* Artificial Intelligence for Cybersecurity Workshop at AAAI, North America (2017). Available at: https://aaai.org/ocs/index.php/WS/AAAIW17/paper/view/15126.

Upton, David M., and Sadie Creese. The danger from within. *Harvard Business Review,* September 2014 issue. (2014) Available at: https://hbr.org/2014/09/the-danger-from-within.

Verizon. *2016 Data Breach Investigations Report,* 9th Edition. Verizon, Basking Ridge, NJ (2016). Available at: http://www.verizonenterprise.com/resources/reports/rp_DBIR_2016_Report_en_xg.pdf.

Verizon. *2017 Data Breach Investigations Report,* 10th Edition. Verizon, Basking Ridge, NJ (2017). Available at: http://www.verizonenterprise.com/resources/reports/rp_DBIR_2017_Report_en_xg.pdf.

Vormetric. *2015 Vormetric Insider Threat Report: Trends and Future Directions in Cyber Security.* Vormetric, San Jose, CA (2015). Available at: https://www.vormetric.com/campaigns/insiderthreat/2015/pdf/2015-vormetric-insider-threat-press-deck-v3.pdf.

Wall Street Journal. Ex-Goldman programmer guilty of stealing code. (2015). Available at: http://www.wsj.com/articles/jury-gives-split-verdict-in-trial-of-former-goldman-programmer-sergey-aleynikov-1430496291.

Whitty, Monica, and Gordon R. Wright. *Corporate Insider Threat Detection Internal Deliverable 3.1: Report of Findings from Case Studies.* University of Oxford, UK (2013).

Wolff, Matt. Unsupervised methods for detecting a malicious insider. *Proceedings of the 7th International ISCRAM Conference–Seattle* 2 (2010).

chapter seven

Social engineering

Abbas Moallem

Contents

7.1 Introduction

Social engineering is "any act that influences a person to take an action that may or may not be in their best interest" (social-engineer.org 2017). "Social engineering, in the context of information security, refers to psychological manipulation of people into performing actions or divulging confidential information" (wikipedia.org 2017). These deception techniques have been used throughout human history. They were used for financial gain, access to power, and spying on enemies and especially as war techniques for victory on the battleground. From olden times, there is the tale of Greeks using a Trojan horse to enter the city of Troy (*Encyclopedia Britannica* 2017) and win the war. Also, we can refer back to Victor Lustig, the man who sold the Eiffel Tower in 1925 (wikipedia.org 2017). Certainly, all of history is full of examples of a human deceiving his/her fellow human. The most memorable films are where viewers find there to be a question of whether the character using deception is good or bad; there are certain moral claims that can serve to justify these otherwise illegal or illicit actions. One might remember the movie *The Sting*, directed by George Roy Hill (1973) telling the story of a young con man, in September 1936, seeking revenge for his murdered partner, who teams up with a master of the big con to win a fortune from a criminal banker or a more recent movie based on the true story of Frank Abagnale. He was one of the most famous impostors claiming to have assumed multiple identities. *Catch Me If You Can*, directed by Steven Spielberg (2002), tells the story of how Frank successfully conned millions of dollars' worth of checks as a Pan Am pilot, a doctor, and a legal prosecutor. Social engineering, and deception techniques now possible with the digital age, have started new lives. There is now the story of Kevin Mitnick (Mitnick and Simon 2002), who used his sophisticated skills to worm his way into many telephone and cell phone networks and vandalize government, corporate, and university computer systems. Arrested in 1995 (BBC 2002), after five years in prison for various computer and communications-related crimes, he wrote about his experience and illustrated the massive scale of social engineering and the effect on the computer security system as a whole.

Using social engineering techniques to access and break into any targeted computer system becomes almost a routine phenomenon with the expansion of the Internet, globalization, and computerization. The scale of usage goes way beyond a few famous cases and rather becomes an entire industry (legal—such as penetration testers—and illegal—such as phishing e-mails). The goal of social engineering techniques is to gain unauthorized access to systems or information to commit acts such as fraud, network intrusion, industrial espionage, or identity theft. Who are these social engineers? The groups or individuals who are or may be "social engineers" include but are not limited to the following:

- Hackers
- Penetration testers
- Spies
- Identify thieves
- Disgruntled employees
- Scammers

In addition to the preceding groups who are using social engineering with criminal intents, some reputable professionals might use social engineering techniques for a legitimate purpose by collecting information for their cases or customers. These professionals might include the following:

- Executive recruiters
- Salespersons
- Governments
- Doctor or psychologists
- Lawyers

7.2 Social engineering techniques

Social engineers use different techniques to acquire sensitive information from the legitimate users of a system. Since targeted individuals or organizations who are the victims of social engineering tend to not admit that they were attacked, it has been hard to find real social engineering scenarios, tools, or statistics about cases, outside of the few reported cases by social engineers who wrote and published books or article about their practice (Mitnick and Simon 2002). Another complexity is, of course, that social engineers do not report the tools or technology that they use or the statistics about their successes or failures. The information published is most often the scenarios used in penetration testing or security audits performed by security experts. Social engineers usually take advantage of the flaws in the security design of a technology product to then manipulate people. In a study conducted in New Zealand, Janczewski and Fu (2010) report that 40% of interviewed participants emphasized that in general, security strategies of the organization had poor security policies because they overlooked people errors. Social engineers often use the following techniques to collect needed information for their attacks:

- *Physical location:* Access to physical location such as workplace or home.
- *Phone:* One of the ways that social engineers attack is through phone calls. Social engineers may call using various scripts and pretexts to obtain information.
- *Trashing:* The collection of information through the targeted entity's trash (residential or place of work), which includes old computers, papers, reports, credit card bills, utility bills, medical insurance, bank statements, and similar items.
- *Mail theft:* This occurs when someone targets other people's mailbox and removes mail that has pertinent confidential information on it. As in trashing, a social engineer can obtain credit card bills and bank statements, anything that can be used to obtain detailed information about the targeted individual.
- *Social networking:* Social networking these days is easily accessible, and social engineers use social networking sites to collect information about any target (individual or organization). Technologies such as Google applications, online social networking websites (Facebook, Twitter, LinkedIn, etc.), discussion forums, and blog sites where people and organizations self-disclose all sufficiently feed the social engineer's information needs. According to one study, 93% of respondents in a global survey of 853 information technology (IT) professionals agreed that new technology products and social networking sites are used by social engineers as information-gathering tools (Chitrey et al. 2012). In the same survey, 72% of respondents believe that social engineers use the support of Google applications, 47% for social networking sites (79% for Facebook, 29% for Twitter, and 32% for LinkedIn), 60% for discussion forums, and 38% for blog sites. Huber et al. (2009) substantiated these public impressions in an experiment that confirmed that the information-gathering stage of social engineering can be automated to collect data from Facebook users without being blocked by the system or in another author-automated retrieval of the profile's attributes and list of top friends from MySpace by examining and extracting the relevant tokens in the parsed hypertext markup language code (Alim et al. 2009). Then, the information collected can be used for targeted sending of attached or phishing e-mails.

Sites such as Craigslist (http://craiglist.com) are also favorites for social engineers. The following are two typical cases of how Craigslist is used to hack people.

- Internet fraud case using Craigslist [case captured by A. Moallem (2016)]

> I needed to find a large home in San Diego for a large family gathering. All the places on Airbnb were either booked or too small. I searched all vacation rental sites with no luck, and I was running out of time. Then I remembered that we had once found a very nice vacation rental off Craigslist and had liked the place. Someone had rented out their timeshare. So I checked Craigslist and lo and behold there was a perfect large home at the right price available!
>
> I checked the address on Google Earth against the pictures of the home. All matched and looked great. I emailed, and he gave me a US phone number. I called, and he said since it's close to the rent date I'd have to deposit the check in the account of the co-owner to secure my reservation. Since it was a local Bank of America, I felt secure and did it. Then I sent the good news to the whole family!
>
> As I emailed about the time, when we would get there, and how to get the keys, etc., his answers were a little strange. Suddenly I was alarmed! I Googled the name of the co-owner and found a bunch of fraud cases showing up. I called my bank, but the check had already cleared. I called B of A, and they said they could not do anything. I called the police and they came to my home. I shared the names, emails, phone numbers, Bank of America account number, etc. But they said they could not do anything. The phone number was an Internet disguised one. The account was probably owned by a foreign account they could not prosecute. I could not believe that nothing could be done! At least we found out before all of us showed up at the place.

- Internet fraud case using Craigslist [captured by A. Moallem (2016)]

> I wanted to sell my motorcycle, so I put an ad on Craigslist. I had some inquiries, mostly text messages, and a few phone calls. Among those inquiries, came a text message asking about the condition of my motorcycle and the reason for selling it. After responding to his concerns, the "buyer" said he was satisfied with the condition and the price and willing to purchase it.
>
> Following my policy for doing any internet transactions, I asked him to call me on the phone; he texted back "I'm at work, and calling is restricted." I gave him the benefit of the doubt. Then he texted me: "Okay I'll take it. I'll have to pay you through PayPal because I am currently at the Hanscom Air Force Base in Bedford, Massachusetts. I have a mover that will come for pick up once the payment has cleared in your PayPal account." Then he texted "I added the agent fee of $475 and extra $50 for the MoneyGram charges to make it easier and faster for pick up" so that you have to pay out of your pocket now but would later be compensated. He asked me to send the money through MoneyGram.

> In the meantime, he bombarded me with text messages asking me if I had sent the money. That made me concerned about the whole thing. Therefore, I checked my PayPal account to see if the deposit was made, and I found out there was no fund deposited into my account. I became suspicious and told him that I had not received the money yet. He texted me to check my spam box, I did so, and I saw an email with PayPal logo look, but I was not convinced, therefore I called the PayPal to verify the email and to confirm the deposit, the agent told me there was no email sent from them and no funds were deposited into my account. So I became sure that it was a scam. I was so happy for not continuing the transaction and ended my communication with him by texting him: f*** you.

- *E-mail:* Phishing e-mails are another common way social engineers use targeted attacks on a larger scale to get access to information or the victims. The phishing e-mail continues to be a favorite technique and an easy way to collect credentials, access a user computer, and commit fraudulent e-commerce activities.
- *Shoulder surfing:* Using direct observation techniques, such as looking over someone's shoulder, to get information.
- *Texting:* With two-factor authentication, many people use their mobile phone and texting to set passwords or receive notification alerts from a bank account. Since it is easy to find the cell phone number of people, this has become a favored way to trick users.

No matter which techniques social engineers use to target their attacks, they rely on social psychology and different methods of persuasion to convince their victims. The bottom line is that they must choose the scenarios, pretext, and method of gaining trust and persuasion to be successful. The famous case of Shane MacDougall, who in front of a live audience called a Walmart store manager in a small military town in Canada and obtained a tremendous amount of information about the Walmart store through a convincing pretext and persuasion technique, is a good illustration (Cowley 2012).

In the following section, we will review theoretical foundations used in persuasion.

7.3 Theoretical foundations used in social engineering attacks

The central question one might ask is why and how are social engineers successful? To answer this question, we need to better understand the principles of behavioral and cognitive psychology that social engineers use for their success. In this section, we will review a few fundamental frameworks explaining the principles of human behavior that social engineers successfully use to acquire users' credentials.

7.3.1 Schema theory

The schema theory, first introduced by Frederic Bartlett (1932), works from the perspective that concepts have meaning if they relate to knowledge that an individual already possesses. A schema guides both information acceptance and information retrieval: it affects how humans process new information and how they retrieve old information from long-term memory. Piaget and Cook (1952) called schema the core building blocks of intelligent "units" of knowledge, each relating to one aspect of the world, including objects, actions, and abstract (i.e., theoretical) concepts. Over our lives, as we learn and discover the world around us, our schemas expand and get complex. The more we know, the bigger and more complex our schemas become. However, the more we are aware, the easier it is to remember new information related to the schema. Thus, since the information exists in our heads, we can relate to it and organize and predict our actions. Activity schemas are called scripts. All of us have many scripts in our long-term memory for a variety of activities. The script will put us in a context, prepare our brain to respond, and prepare our consequent actions based on what we expect to happen. Let us look at the following script as an example:

> Person 1: Johnny of computer support, how may I help you?
> Person 2: Yes, I seem to have trouble with my laptop computer.
> Person 1: What sort of trouble?
> Person 2: Well, it wouldn't turn on.
> Person 1: Are you in front of your computer right now?
> Person 2: Yes.
> Person 1: Kindly check if the computer is properly plugged in.
> Person 2: Yes, it is.
> Person 1: Now try to push the power button on your laptop computer.
> Person 2: Ok, nothing happens.
> Person 1: Can you please turn on the light in your room?
> Person 2: Damn, I think we have no electricity.
> Person 1: Sir, I guess that's why you can't turn on your computer.
> Person 2: I guess so, thanks a lot.
> Person 1: Thank you for calling and we are glad to be of service.

Most people using a computer are very familiar with this type of script and immediately understand that this is a call between a support agent and a customer or user and prepare themselves to answer. Now look at this script (Mitnick and Simon 2002):

> Person 1: Good afternoon, this is Mary. How can I help you?
> Person 2: Can you connect me to the transportation department?
> Person 1: I am not sure if we have one. I'll look in my directory. Who is calling?
> Person 2: It's Didi.
> Person 1: Are you in the building? Or?
> Person 2: No, I am outside the building.
> Person 1: Didi who?
> Person 2: Didi Sands. I had the extension for transportation, but I forgot what it was.
> Person 1: One moment.
> Person 2: What building are you in—Lakeview or Main Place?
> Person 1: Main Place (pause). It's 805-555-6469 x123.
> Person 2: I also want to talk to real estate.
> Person 1: 805-555-6469 x456.
> Person 2: How about accounts receivable at corporate in Austin, Texas?
> Person 1: 805-555-6469 x789.

The social engineer created a script within a context that is very familiar and feasible to the victim, so it is then easy to convince the victim to provide the information needed. Our brains sort out a schema of a person who needs help with information rather than a social engineer to trick for getting information. Social engineers research and prepare a script that fits the norm for their victim's context, so that they will act accordingly. The central goal is ensuring that all the elements are compatible, logical, and natural.

7.3.2 Cialdini's six principles of influence

In social engineering, like many other areas such as marketing, sales, politics, or negotiations, Cialdini's principles for influence are knowingly or unknowingly used to acquire information (Cialdini 1994). A thorough understanding of these principles can not only help prepare people to thwart social engineers but should also result in awareness and behavior modification in cybersecurity employee training.

7.3.2.1 Reciprocity or obligation to repay

The first principle of Cialdini (2009) is "reciprocity or obligation to repay." According to anthropologists, the rule of reciprocity is apparent in all human societies. Cialdini (2009, p. 23) believes that "one of the reasons reciprocation can be used so effectively as a device for gaining another's compliance is its power. The rule possesses awesome strength, often producing a 'yes' response to a request that, except for an existing feeling of indebtedness, would have surely been refused."

The reciprocity principle is widely used in many professional relationships within politics, marketing, and sales. It is also a common design behavior among all people, shown through actions such as gift giving and receiving or doing a favor and expecting to receive one in return. People feel obligated to reciprocate a gift even if the gift was unwanted. According to Cialdini (2009, p. 33), "a small initial favor can produce a sense of obligation to agree to a substantially larger return favor." For example, a small free token (dinner or drink) at an exhibition might facilitate a much bigger deal. "The rule allows one person to choose the nature of the indebting first favor and the nature of the debt-canceling return favor; we could easily be manipulated into an unfair exchange by those who might wish to exploit the rule."

The reciprocity rules are widely used by social engineers to gain trust and get a favor returned. Imagine that you are driving and got lost somewhere. After asking and receiving help from a passerby, you would be pleased to answer a question about your car or even offer up a bottle of water. Now imagine that you are sitting in a coffee shop struggling with your computer. Somebody seated next to you helps you solve your computer problems and in return asks you to let them use your computer to view an address. Probably, you would not refuse that request. Here is a scenario reported by Mitnick and Simon (2002, p. 248):

> An employee receives a call from a person who identifies himself as being from the IT department. The caller explains that some company computers have been infected with a new virus not recognized by the antivirus software that can destroy all files on a computer, and offers to talk the person through some steps to prevent problems.
>
> Following this, the caller asks the person to test a software utility that has just been recently upgraded to allow users to change passwords. The employee is reluctant to refuse because the caller has just provided help that will supposedly protect the user from a virus. He reciprocates by complying with the caller's request.

In this example, the social engineer gives some help and then puts the victim in a situation that obligates them to "repay."

7.3.2.2 *Consistency and commitment*

Cialdini's (2009, p. 52) second principle states that "once we have made a choice or taken a stand, we will encounter personal and interpersonal pressures to behave consistently with that commitment." This applies to a variety of human behaviors in politics and the pattern of voting for a specific party, commitments to accept certain deals and policies, and so on. Social engineers use this type of behavior to push victims into revealing information or behaving in certain ways. For example, assume that you and a coworker are coming back from lunch at your office, and one of you scans the smart card to open the door. When you enter, a couple of other people walk in at the same time, or when you enter the office, out of courtesy, you keep the door open for the person after you. Social engineers might use these tailgating techniques to enter a protected zone. Tailgating is a simple scenario for a social engineer to enter a restricted area. If someone has learned the behavior of keeping the door open to let people come in behind them due to politeness, then he/she will consistently do it even when the security policy requires the door to be closed and for the next person to use his/her smart card to open it.

Here is an example of an attack according to Mitnick and Simon (2002, p. 248):

> The attacker contacts a relatively new employee and advises her of the agreement to abide by certain security policies and procedures as a condition of being allowed to use company information systems. After discussing a few security practices, the caller asks the user her password "to verify compliance" with policy on choosing a difficult-to-guess password. Once the user reveals her password, the caller makes the recommendation to construct future passwords in such a way that the attacker will be able to guess it. The victim complies because of her prior agreement to abide by company policies and her assumption that the caller is merely verifying her compliance.

This often happens because people might have experienced the same type of situation in the past with no bad results or consequences. After all, when your computer is not working, and you take it to a repair shop, they ask for your password and you give it voluntarily.

7.3.2.3 *Social proof*

Cialdini's third principle is social proof, or the power over what others do. It states that "one means we use to determine what is correct is to find out what other people think is correct" (Cialdini 2009, p. 116). A variety of examples is provided in this very widely used principle, but we will use a simple one regarding the ways in which security policies are applied in an organization.

People follow personas that are accepted and considered to be correct. In the application of security policies and when people should follow the rules, social proof is fundamental. Imagine a situation where all coworkers instinctively lock their computer when leaving it behind. If someone does not follow the set behavior, everybody who follows that person will be observing and possibly copying an incorrect behavior. That is the reason why security policy execution is more effective when there is a security culture in an organization. In Western societies, everybody who leaves their home locks their door behind them; this is considered to be a normal behavior. Now consider whether we do the same for computer or mobile devices and lock them when not using then. After all, we might have even more assets that we want to keep safe than home furniture.

7.3.2.4 Liking

Liking, or the obligations of friendship, is the fourth principle. According to Cialdini, we like to say "yes" to people whom we like and know on a personal level. The liking principle is primarily used by the salesperson who tries to create a sense of friendship with potential customers. That is why they start by asking some personal questions, such as how many children you have. Then they may acknowledge certain commonalities with potential buyers. During the day, we use this liking approach to get certain activities done more effectively, in customer support calls or whenever we have a request, or even when we want to get a better price with the mechanic's shop that is repairing our car.

Social engineering has become a champion in applying the liking approach. It is meticulously used in scripts and conversations down to the way the social engineers might be dressed or exhibit particular nonverbal behaviors. An effort is made to create a relationship with a victim through physical attractiveness, commonalities (showing shared interests or circumstances), compliments to the victim, cooperating (showing the victim that they have similar beliefs), and finally conditioning and association (showing the victim they hold the same beliefs).

7.3.2.5 Authority

The fifth principle is that of authority. We obey those in charge. "A multilayered and widely accepted system of authority confers an immense advantage upon society. It allows the development of sophisticated structures to produce resources, trade, defense, expansion, and social control that would otherwise be impossible" (Cialdini 2009, p. 180). Authority is the principle probably used most frequently among all social engineers in phishing attacks or voice calling. The effect of authority is one of the areas in social psychology that is largely observed and backed by research from the Milgram (2009) experiment (1974–2009) on the effect of authority on obedience, up to Zimbardo's Stanford Prison experiment (Wikipedia, 2017), where research concluded that people obey either out of fear or out of a desire to appear cooperative—even when acting against their better judgment and desires. Social engineers use these tactics in voice calling and extensively in phishing e-mails. In voice calling, the caller may pretend to be calling from the Internal Revenue Service (IRS), prosecutor's office, police station, or a law office. Here is a script of a voice call that the author has received on his home phone number:

> Voicemail from (877) 719-4201
>
> Hello, we have been trying to reach you. This call is officially a final notice from IRS internal revenue services. The reason of this call is to inform you that IRS is filing the lawsuit against you. So please call immediately on our department number 8777194201.
>
> I repeat 8777194201.
>
> Thank you.

The following is another example of a phishing e-mail message:

> After the last annual calculation of your financial activity, we have concluded that you are eligible to get the tax refund of $645.

> You may submit the tax refund application and give us 3-9 days to process it.
> A refund can be hindered by many different reasons.
> E.G. submitting invalid records or not meeting a deadline.
> To get information about your tax refund, please Open this link.
> Sincerely,
> Tax Refund Department
> Internal Revenue Service

Again in this e-mail, the IRS, that is, the Internal Revenue Service in the United States, is used to create fear and seek obedience. In fact, the IRS retains the legal authority to enforce liens and seize assets without obtaining a judgment in court.

7.3.2.6 Scarcity

The sixth principle is scarcity. It relies on the observation that we want what may not be available. The scarcity tactic operates on the value that people attach to things. Scarcity suggests that things are more valuable when they are less available. The scarcity tactic involves the "limited-number" or "deadline technique." The deadline technique works because it puts an official time limit on the product availability. The "limited-number" tactic works because it creates added value to a product by reducing the availability of the product. This tactic is widely used in marketing and particularly in e-commerce very effectively by underlying perhaps a discounted flight or hotel reservation with a limited number of seats or rooms at the discounted prices. Since people are very familiar with this pattern, it is primarily used by social engineers when sending a phishing e-mail about winning a prize, limited time offers of reduced subscription rates, and so on, using very famous e-commerce sites such as fake Amazon or Netflix offers. This is also an effective tactic for fake e-commerce sites. The fake sites may offer a huge discount on a specific product so people impressed by the discount place orders, not knowing that the site was not secure and their data are compromised by identity theft and fraud.

7.3.3 Stajano and Wilson principles

To understand the general principles of human behavior that explain how scams worked, Stajano and Wilson (2009) studied a variety of scams and "short cons" that were investigated, documented, and recreated for the BBC TV program "The Real Hustle" and then extracted from them some general principles about the recurring behavioral patterns of victims that hustlers have learned to exploit. This study extrapolates several principles helping us understand where people are vulnerable to specific attacks. The followings are the suggested principles:

7.3.3.1 Distraction principle

While you are distracted by what retains your interest, social engineers can obtain what they want without you noticing. Here is an example:

> The young lady who falls prey to the recruitment scam is so engrossed in her task of accurately compiling her personal details into a form to maximize her chances of finding a job that she utterly fails even to suspect that the whole hiring agency might be fake.

In this case, the user is eager to get a job that is set to be really attractive; thus, she would provide all sorts of information to get a job without questioning the validity of the agency and why she should be answering the questions that are not relevant to the job. Distraction is at the center of many fraud scenarios; it is also a fundamental ingredient of most magic performances. The authors underline the use of distraction as successful tools used to divert attention the same way as magicians manipulate attention and awareness to show their magic. (Macknik et al. 2008)

Stajano and Wilson believe that "it's not that the users are too lazy to follow the prescribed practice on how to operate the security mechanisms, but rather that their interest is principally focused on the task, much more important to them, of accessing the resource that the security mechanisms protect. The users just see what they're interested in (whether they can conveniently access the resource) and are blind to the fact that those annoying security mechanisms were put there." Thus, social engineers, like magicians, use these human vulnerabilities to achieve their goals. They divert the attention of the victim and manipulate them so that they provide answers to the attacker's questions.

7.3.3.2 *Social compliance principle*

Social compliance works like Cialdini's principle of consistency and commitment in that it is based on the way society trains people not to question authority. In phishing, this may present itself as a website that replicates the appearance of a bank's site and directs customers to it to steal their online banking credentials. The lesson for the security architect is that people are trained as a citizens to obey commands from authorities, who can be government agents (i.e. police officers), doctors or, in this case, a system administrator managing the network. This behavior can be a double-edged sword. Although people are pretty good at recognizing people they already know (by face, by voice, by shared memories, etc.), they are not very good at all at authenticating strangers, whether over a network, over the phone, or even in person. Incentives and liabilities must be coherently aligned with the overall system goals. If users of a product are expected to perform extra checks rather than subserviently submitting to orders, then social protocols must change to make this acceptable. Conversely, if the product's users are expected to obey authority from the company unquestioningly, those who exercise the authority must offer safeguards to relieve users of liability and compensate them if they fall prey to attacks that exploit the social compliance principle. The fight against phishing and all other forms of social engineering can never be won unless this principle is understood and taken on board.

7.3.3.3 *Herd principle*

"Even suspicious marks will let their guard down when everyone next to them appears to share the same risks" (Stajano and Wilson 2009, p. 13). This principle is based on the common recognition that people look at people around them to gain confidence in their actions. If you, as a customer, are seeing many people shop around you or purchase from a specific e-commerce site or if you are seeing that large numbers have reviewed a product or service, then these actions of others become a source of increased confidence motivating toward a purchase decision. However, all of this information may have been falsely created. A fraudulent site can easily simulate many reviews, shoppers, and almost all aspects of a legitimate site review or activity credential to gain consumer confidence. It is not hard for a hacker to create multiple aliases and set up a fake Facebook site with hundreds of friends or likes to gain the trust of his/her victims.

7.3.3.4 *Dishonesty principle*

Anything illegal you do will be used against you by the fraudster, making it harder for you to seek help. A prime example of this principle is the gadget scam. Imagine individuals who bought a machine that can create prepaid credit cards. If it worked, it would clearly be illegal. Therefore, once they discover that it does not work, they cannot go to the police and complain about the seller, because the police might ask questions about what they intended to do with the device.

7.3.3.5 Need and greed principle

Your needs and desires make you vulnerable. When social engineers know what you want, they can easily manipulate you.

7.3.3.6 Time principle

When you are under time pressure to make an important choice, you use different decision-making strategies. Time, like "scarcity," pushes people to use less reasoning and rush to make unreasoned judgments.

7.3.4 Cialdini principles and phishing mails

The phishing e-mail, despite the advent of some technological tools to capture them, is still a very useful tool for social engineers. The main reason is that phishing e-mails use deception techniques or, in the case of targeted attacks, very particular persuasion techniques. Social engineers, through a collection of data (manually or automatically) from social networking, can create a very specific context that seems very realistic to a recipient. Let us use a very simple example: Imagine you get an e-mail to your personal or professional e-mail address. The e-mail seems to be coming from your manager, informing you that in the management meeting the previous day, they talked about your efforts and performance. The manager says that they insisted on giving you a special bonus. They also provide a link to self-report some of your achievements arguing that the human resource (HR) department needs this information to process your bonus. When you click the link to enter your performance, the form requires you to log in with your username and password before reporting your achievements. You might not pay attention that this is not your usual HR application, but rather a Google form owned by social engineers outside your company who only wish to gain access to your authentication and password. Social engineers will effectively use some of the preceding persuading principles to acquire your trust and collaboration so that everything will work well for you. The decision-making in these cases is not governed by how smart the person may be, but by the contexts created through the psychological environment and how information is presented.

One study (Ferreira et al. 2015) investigated how the principles of Cialdini, Gragg, and Stajano (2003) relate to one another and tried to create more general principles of persuasion in social engineering. To do this, they organized a collection of phishing e-mails according to their goals: data theft, malware, and fraud. Then they tried to find patterns based on the principles of persuasion in social engineering. The results demonstrated that the most commonly used principles of social engineering are liking and similarity and deception followed by distraction. The next most common principles are authority for data theft; e-mail and malware; and commitment, reciprocation and consistency for malware and fraud e-mails.

7.4 Defense against social engineering attacks

In the expansion and evolution of social engineering attacks against people and organizations, it is essential to be preventive and protected. Contrary to technological attacks such as malware, viruses, or other types of hacking, protecting people and organizations against social engineering attacks is not that simple. The complexity comes from the fact that people have a hard time distinguishing lies from truths and perform very poorly at detecting deception. This poor performance happens despite the existence of ample cues that people can consider in determining deception. There is substantial psychological literature relating why people lie and how lies can be detected. Vrij (2000) has reviewed the relevant research on lying and detection in detail and shown that people are particularly poor at detecting lies (44% accuracy rate). DePaulo et al. (1985) and Bond and DePaulo (2006) report that the percentage of lie detection ranges from 45 to 60% when 50% accuracy is expected by chance alone. The psychological studies suggest that the poor performance in distinguishing lies from truth is related to factors such as cognitive biases (Burgon and Levine 2009). People tend to reason in a way that confirms their assumptions and ways of thinking even when the conclusion leads to systematic deviations from rationality or good judgment. In very recent history, we can observe how fake news (during the 2016 election in the United States) was used to influence and shape public opinion even when sufficient cues were available to differentiate facts from lies. (Allcott and Gentzkow 2017) There is the use of "heuristics" or "mental shortcuts," processes where one focuses on one aspect of a complex problem while ignoring other pertinent facts. Since honest behavior, in general, is most frequent, people tend to expect to hear the truth and are reluctant to think that the "truth" they are hearing is actually a "lie."

In differentiating attributes of a lie from the truth, research suggests that 62.2% of people think that individuals who lie tell longer stories than usual, pause in the middle of speaking, and use terms with less emotion or feeling (Bond 2006).

Since people have a hard time detecting lies or identifying if a social interaction comes from social engineers or reliable sources, we become motivated to find a technology or an automatized tool that can be used to filter fake, fraudulent interactions. Qin and Burgoon (2007) designed an experiment on deception detection where the performances of the human and an automated system were compared. The attributes in each deceptive case included the following:

- Vocal cues (talking time and speech disturbance such as vocalized pause and nonvocalized pause) and verbal cues
- Quantity [number of words, verbs, and complexity (syntactic complexity or average sentence length)]
- Diversity (lexical diversity, content word diversity, and redundancy)
- Specificity (temporal immediacy and temporal nonimmediacy)
- Affect (activation, pleasantness, and imagery)
- Uncertainty (modal verbs) and verbal (nonimmediacy: passive voice)

The results showed that the automated system using discriminant analysis to classify deception performed significantly better than humans in detecting deception. Humans in this study tended to judge all messages from the perspective of "senders as truthful," even when they knew about the possibility of error in human judgment. The result of this study confirms the previous study about the application of "cognitive heuristics" or "preconceived expectations for truthfulness" by people in detecting the deceptions (Chaiken 1980).

However, we still lack the general availability of reliable automated systems to augment human judgment in detecting deceptive calls, e-mails, or even face-to-face communication. The main issue continues to be the vulnerability of individuals who can be manipulated in different ways. There is not and cannot be one type of protection that can be applied to all people because every human may be vulnerable in different areas. Thus, security needs to first identify the vulnerabilities for each group of people and then establish policy, training, and awareness measures. According to Gragg (2003), the defensive measure against social engineers should be "a multi-layered defense" that would include the following:

- Functional level: security policy
- Parameter level: security awareness training for all
- Fortress level: resistance training for key employees
- Gotcha level: social engineering land mines
- Offensive level: incident response

7.4.1 *Information security policy*

One of the foundations of protection against social engineering is having an information security policy (ISP). An ISP addresses data, programs, systems, facilities, tech infrastructure, users of technology, and third-party organizations. It provides employees with guidelines concerning how to ensure information security when they utilize information systems to perform their jobs. For such a policy to be effective, however, employees must comply with the ISP. Thus, the effectiveness of the ISP is not measured by how well the policy is defined and written, but rather by the degree of employee compliance with the policy. The employee's motivation to comply with their organization security policy is the key foundation for the success of the ISP. It is important that motivating factors and why it is important to comply with security policy are explained to employees, convincing them to comply with the ISP of the organization.

In addition, the organization must ensure that the ISP being implemented does not solely rely on technology-based solutions (Ernst and Young 2008). Both technical employees and human resource staff are authorized to use information systems. Several researchers have found that employees' abusive behavior and misuse of information systems, together with noncompliance with security policy, are information security risks. (Stanton et al. 2005) Some research applied the theory of planned behavior (TPB) (Ajzen 1991), to study the employees' behavior and intention in complying with the ISP. According to TPB, the attitude toward behavior, subjective norms, and perceived behavioral control together shape an individual's behavioral intentions and behaviors.

People have a positive attitude toward the actions when they perceive that they can control behaviors. This attitude also applies to the employee's intention to comply with the ISP of the organization. The results provide some evidence of the significant impact of motivational factors other than rewards and sanctions that reinforce an employee's compliance behavior. Bulgurcu et al. (2010) suggest that information security awareness (ISA) programs should be designed to focus employees' beliefs on cost and benefit, safety, and vulnerability and to create a security-aware culture within the organization to improve information security. Therefore, organizations should create security awareness training programs to ensure sufficient awareness and understanding on risk issues.

7.4.2 *Security awareness training*

In general, most people do not have comprehensive training on computer security. Their knowledge is limited to some sporadic sources of information. For example, a user might buy a router to equip his/her home or office with a wireless network and follow some general steps to set its password and connect to the Internet. The user does not necessarily learn about home networking and security issues, how hackers attack, and so on. Similarly, companies invest a tremendous amount on technology solutions. However, the budget to train and educate employees on security issues is extremely limited, often only covering online trainings that employees are required to take.

Several studies have investigated the effectiveness of company awareness programs and information security training. The study conducted by Bulgurcu et al. (2010) suggests that ISA influences employee attitudes, directly and indirectly, to directly comply with the ISP. A meta-analysis conducted by Hauch et al. (2016) showed that training improved the overall ability to detect deception. Burgoon et al. (2008) conducted two studies with IT specialists in the US Armed Forces. They found that that their developed e-training system was successful for both teaching managers and employees to improve their participants' ability to detect deception. These training approaches focused on web-based trainings that taught people how to recognize the "cues" that deceivers unwittingly give out. They also showed that people who had been trained were able to "recognize the tactics that deceivers might use to hide the truth. Deception tactics are domain-specific, but deception cues are not context-specific. This means that cue-based training is applicable in a wide variety of organizations" (Burgoon et al. 2008).

7.4.3 *Resistance training for key employees*

Training, awareness, and knowledge by all individuals or employees are essential to prevent social engineering attacks. However, the training and awareness of some key employees is even more important since they might be more targeted than others. According to Gragg (2003), key employees include help desk personnel, customer service, business assistants, secretaries, receptionists, and system administrators/engineers. It is fundamental that people in these roles be trained and adequately prepared not to be persuaded in giving information away to social engineers. Gragg identified these techniques as follows:

- Inoculation: Employees would be exposed to the weakened arguments that will be used by the social engineer and anticipate the arguments of the social engineer (Sagarin 2002, p. 527).
- Forewarning: by warning employees how social engineers might use their vulnerabilities to persuade them to give them information that they want to acquire.
- Reality check: by making employees aware of their unrealistic optimism. Thus, they should not ignore legitimate risks.

7.4.4 *Persistence level*

One training and awareness program is not sufficient to protect people against social engineers who are creative and constantly designing new techniques. The key employees should particularly be regularly reminded and retrained about the new techniques and social engineering approaches.

7.4.5 *Gotcha level*

The Gotcha level defense is performed by alerting targeted employees about an attack that is in progress and how they should be prepared to address it and suggesting techniques that they need to use. The techniques include but are not limited to the following:

- Be on the lookout for a security risk in the form of the physical presence of a social engineer—for example, an unapproved or expired badge or unescorted visitor.
- Monitor a centralized security log of events.
- Not calling back an individual who requests a password reset or requests information.
- Verifying the identity of anyone who is calling and trying to get information about the company.

7.4.6 *Offensive level*

This level involves creating a well-defined process that an employee can begin as soon as they suspect that something is wrong and go aggressively after the hacker and proactively inform potential victims, security professionals, and IT in the company.

7.5 Conclusion

Social engineering is an evolving practice with many sources of new perpetrators. Social engineers use well-known techniques and continually explore to find new ways to use human behavior to exploit the weakness in people unable to distinguish lies from truth to acquire information. Until technology offers automatic solutions to help users in detecting the lies and protecting people from being victims of social engineers, users will be required to gain awareness and knowledge to protect themselves from deception techniques. Cybersecurity experts should be constantly evaluating and detecting tactics of social engineering and providing efficient warning and training to protect personal and organizational assets.

References

Ajzen, I. (1991): The theory of planned behavior, *Organizational Behavior and Human Decision Processes*, vol. 50, no. 2, 179–211.

Alim S., Abdul-Rahman R., Neagu D., and Ridley M. (2009): Data retrieval from online social network profiles for social engineering applications. *2009 International Conference for Internet Technology and Secured Transactions*, 1–5.

Allcott H., and Gentzkow M. (2017): *Social Media and Fake News in the 2016 Election*, January 2017. https://web.stanford.edu/~gentzkow/research/fakenews.pdf.

BBC (2002, October 14): *Mitnick, Kevin. How to Hack People*. http://news.bbc.co.uk/2/hi/technology/2320121.stm.

Bond, C. F. (2006): A world of lies: The global deception research team. *Journal of Cross-Culture Psychology*, vol. 37, no. 1, 60–74.

Bond, C. F., and DePaulo, B. M. (2006): Accuracy of deception judgments. *Personality and Social Psychology Review*, vol. 10, no. 3, 214–234.

Bulgurcu, B., Cavusoglu, H., and Benbasat, I. (2010, September): Information security policy compliance: An empirical study of rationality-based beliefs and information security awareness, *MIS Quarterly*, vol. 34, no. 3, 523–548.

Burgoon, J. K., and Levine, T. R. (2009): Advances in deception detection. In S. W. Smith and S. R. Wilson (Eds), *New Directions in Interpersonal Communication Research* (pp. 201–220).

Burgoon, J. K., George, J. F., Biros, D. P., Nunamaker Jr., J. F., Crews, J. M., Cao, J., Marett, K., Adkins, M., Fruse, J., and Lin, M. (2008): The role of e-training in protecting information assets against deception attacks. *MIS Quarterly Executive* 7(2) June 2008. https://www.researchgate.net/publication/220500702_The_Role_of_E-Training_in_Protecting_Information_Assets_Against_Deception_Attacks.

Chaiken, S. (1980): Heuristic versus systematic information processing and the use of source versus message cues in persuasion. *Journal of Personality and Social Psychology*, vol. 39, no. 5, 752–766.

Chitrey, A., Singh, D., Bag, M., and Singh, V. (2012, June): A comprehensive study of social engineering based attacks in India to develop a conceptual model. *International Journal of Information & Network Security*, vol. 1, 45–53.

Cialdini, R. B. (1994): *Influence: The Psychology of Persuasion*. Collines Business. New York, NY: Harper Collins Publishers.

Cialdini, R. B. (2009): *Influence: Science and Practice* (5th ed.) (pp. 19, 52, 116, 180, 248). Boston, MA: Pearson Education.

Cowley, S. (2012, August 8): How a lying "social engineer" hacked Wal-Mart. *CNNMoney*. http://money.cnn.com/2012/08/07/technology/walmart-hack-defco.

DePaulo, B. M., Stone, J. I., and Lassiter, G. D. (1985): Deceiving and detecting deceit. In B. R. Schlenker (Ed.), *The Self and Social Life* (pp. 323–370). New York: McGraw-Hill.

Encyclopedia Britanica (2017): Trojan horse, Greek mythology (update 2015). https://www.britannica.com/topic/Trojan-horse.

Ernst & Young (2008): *Moving Beyond Compliance: Ernst & Young's 2008 Global Information Security Survey*. Ernst & Young, London. http://130.18.86.27/faculty/warkentin/SecurityPapers/Merrill/2008_E&YWhitePaper_GlobalInfoSecuritySurvey.pdf.

Ferreira A., Coventry L., and Lenzini G. (2015): Principles of persuasion in social engineering and their use in phishing. In *Human Aspects of Information Security, Privacy, and Trust*, vol. 9190 of the series Lecture Notes in Computer Science (pp 36–47).

Gragg, D. (2003): *A Multi-Level Defense Against Social Engineering*. SANS Institute InfoSec Reading Room, Singapore. https://www.sans.org/reading-room/whitepapers/engineering/multi-level-defense-social-engineering-920.

Hauch, V., Siegfried, L., Sporer, S. L., and Meissner, C. A. (2016): Does training improve the detection of deception? A meta-analysis. *Communication Research*, vol. 43, no. 3, 283–343.

Hill, G. R. (1973): *The Sting*, Written by David S. Ward, starring Stars: Paul Newman, Robert Redford, Robert Shaw. http://www.imdb.com/title/tt0070735/.

Huber, M., Kowalski, S., Nohlberg, M., and Tjoa, S. (2009): Towards automating social engineering using social networking sites. *IEEE International Conference on Computational Science and Engineering. CSE '09 Proceedings of the 2009 International Conference on Computational Science and Engineering*—Volume 03, 117–124.

https://www.social-engineer.org/.

https://www.wikipedia.org/.

Janczewski, J. L., and Fu, L. (2010): Social engineering-based attacks—Model and New Zealand perspective. *Proceedings of the International Multiconference on Computer Science and Information Technology*, 847–853.

Leahey, Th. H., and Harris, R. J. (2001): *Learning and Cognition* (p. 234). Upper Saddle River, NJ: Prentice Hall.

Macknik, S. L., King, M., Randi, J., Robbins, A., Teller, Thompson, J., and Martinez-Conde, S. (2008): Attention and awareness in stage magic: Turning tricks into research. *Nature Reviews Neuroscience*, vol. 9, 871–879, http://dx.doi.org/10.1038/nrn2473.

Milgram, S. (2009): *Obedience to Authority: An Experimental View* (9th ed.). New York: Harper and Row.*

Mitnick, K. D., and Simon W. L. (2002): *The Art of Deception: Controlling the Human Element of Security* (pp. 22, 246, 248). Hoboken, NJ: Wiley.

Piaget, J., and Cook, M. T. (1952): *The Origins of Intelligence in Children*. New York: International University Press.

Qin, T., and Burgoon, J. (2007): An investigation of heuristics of human judgment in detecting deception and potential implications in countering social engineering. In *Intelligence and Security Informatics*. Piscataway, NJ: Institute of Electrical and Electronics Engineers.

* An excellent presentation of Milgram's work is also found in Brown (*Social Forces in Obedience and Rebellion. Social Psychology: The Second Edition*. New York: Free Press, 1986).

Sagarin, B. J., Cialdini, R. B., Rice, W. E., and Serna, S. B. (2002): Dispelling the illusion of invulnerability: The motivations and mechanisms of resistance to persuasion. *Journal of Personality & Social Psychology*, vol. 83, no. 3, 526–541.

Spielberg, S. (2002): *Catch Me If You Can*, written by Jeff Nathanson and Frank Abagnale Jr, Starring Leonardo DiCaprio, Tom Hanks, Christopher Walken. http://www.imdb.com/title/tt0264464/.

Stajano, F., and Wilson, P. (2009, August): Understanding scam victims: Seven principles for systems security. University of Cambridge Computer, United Kingdom UCAM-CL-TR-754, ISSN 1476-2986 (Page 36). https://www.cl.cam.ac.uk/techreports/UCAM-CL-TR-754.pdf.

Stanford prison experiment (2017): Stanford prison experiment, Wikipedia, Accessed March 2017, https://en.wikipedia.org/wiki/Stanford_prison_experiment

Stanton, J. M., Stam, K. R., Mastrangelo, P., and Jolton, J. (2005, March): Analysis of end user security behaviors. *Computers & Security*, vol. 24, no. 2, 124–133.

Victor Lustig (2017): Wikipedia, Accessed March 2017, https://en.wikipedia.org/wiki/Victor_Lustig

Vrij, A. (2000): *Detecting Lies and Deceit: The Psychology of Lying and the Implications for Professional Practice*. Chichester: Wiley.

What is Social Engineering (2017): social-engineer.org, Accessed March 2017, https://www .social-engineer.org/about/.

chapter eight

Money laundering and black markets

Brita Sands Bayatmakou

Contents

8.1 Introduction

This chapter will provide an overview of money laundering and the interplay with cybercriminal activity. The subject of today's black markets that increasingly facilitate and enable these crimes will be covered in the second half of the chapter.

As described in previous chapters, cyber-related crimes have increased exponentially in recent years. After gaining an understanding of who and what is behind the data breaches and cyberattacks as well as the tactics used to commit cyber-related crime, it is important to understand how the stolen information, assets, or even identities are bought and sold and how the proceeds are laundered. Choking off a criminal organization's ability turn a profit through underground markets is arguably one of the most important deterrents. If the venue and methods to obscure and move illegally obtained funds are removed, the benefit to pursue such criminal activity no longer outweighs the cost. Case examples will be provided in the second half of this chapter.

Almost all profit-generating crimes involve money laundering as a means to disguise illicit proceeds and integrate them into the legitimate economy. Organized criminal networks that engage in a range of activities including, but not limited to, drug trafficking, human smuggling, embezzlement, insider trading, bribery, terrorism, illegal gambling, and cybercrime employ money laundering techniques to maintain their highly lucrative operations. The laundering of proceeds is arguably the most important aspect of any criminal enterprise as it ensures that profits make their way back to the criminal's pocket without being identified, reported, or ultimately seized by law enforcement. Criminals do this by disguising the sources, changing the form, or transferring the funds where it is unlikely to attract attention. According to the United Nations Office on Drugs and Crime (UNODC), criminals launder an estimated $1.6 trillion, or 2.7% of the global gross domestic product in 1 year. Less than 1% of global illicit financial flows are currently seized by authorities (UNODC 2011). The magnitude of illicit funds generated and the extent to which they are laundered through today's globalized systems has certainly risen with the increasing technological advancements, connectivity, and integration of the financial services industry. Thus, the statistics quoted earlier should be considered conservative figures.

8.1.1 Money laundering defined

Money laundering is the act of concealing the illegal origins of money derived from criminal activities and making those illegally gained profits appear legal or "clean." Criminals attempt to make the proceeds appear legal by misusing financial institutions and by employing complex methods to mask the origin, movement, and destination of such ill-gotten gains (Sharman 2011). Money is introduced or "placed" into the legitimate financial system, often transferred from any institution that facilitates financial transactions to another to obscure the source of funds, and is ultimately integrated into the formal economy to appear clean or legitimate.

In addition to undermining the legitimacy of financial systems and governments, money laundering imposes adverse macroeconomic impacts on society, threatening the safety and soundness of the global financial system. These significant impacts include distorting markets, imposing odious debt, contaminating the financial sector, destabilizing and creating counterintuitive capital flows, asset price bubbles, undermining the reputation of local institutions, and unfair competition. Much of this has a corrosive effect on the economy, government, and social well-being of a country and can impede investment and economic growth. The adverse socioeconomic impacts of money laundering include the perpetuation and promotion of criminal activities; corruption; drug abuse (in the case of laundering drug proceeds); and transfer of power from citizens, the market, and government to criminals.

8.1.2 Money laundering in the twenty-first century

It is important to distinguish today's money laundering methods and popular crimes from those of the past. Today, an increasing portion of illicit funds are generated through cyber-enabled crimes, defined as an illegal activity that is carried out or facilitated through electronic systems and devices, such as networks and computers. Thus, professional money laundering services are in high demand. Perhaps the most disturbing trend is that cyberattackers are innovating much faster than those who defend against such attacks. Criminals today are reusing malware and adapting their products to stay ahead of the antimalware, antifraud, and anti-money laundering (AML) industries. This means that the role that governments and the private sector play in enhancing both AML and cybersecurity risk management policies will be critical to combating today's new and evolving threat landscape.

In many ways, however, cybercriminals are no different from other criminals involved in more "traditional" illegal activities such as drug or weapons trafficking. Like other organized crime groups, cybercrime enterprises function as well-funded businesses with an ultimate goal of making as large a profit as possible. Like a drug trafficker who must conceal the source and destination of drug sales, a cybercriminal must maintain strong partnerships and well-run supply chains. Unlike their legitimate counterparts, illegitimate enterprises must hire specialized employees to assist in laundering the business proceeds. Whether launching malicious software, harvesting an individual's credentials through a phishing e-mail, or demanding a ransom in Bitcoin in return for restored access to one's computer, these operations are only profitable if they are able to collect, obscure, and move the earnings. Also, like other traditional criminals, cybercriminals will always pursue the path of least resistance, migrating toward those activities that yield high rewards for relatively low risk and low cost. The lower risk offered in the cybercrime space has yielded a high demand for goods and services such as user credentials (usernames and passwords), personally identifiable information (PII), nonpublic information (NPI), exploit kits, and malicious software (malware), among many others.

The scale and size of cyber-related crime have grown dramatically and, according to a global economic crime survey, have become the second most reported crime (PwC 2018). Cyberattacks take a heavy toll on the economy, with corporate data breaches costing companies millions of dollars. The revenue from cyberrelated crime dwarfs that of the illegal drug trade, which clearly demonstrates how attractive this field is for criminals operating in the twenty-first century.

Without the dangers that accompany street-level criminal enterprises and a low likelihood that illicit profit will be seized, crimes perpetrated over electronic channels present a new frontier for professional money launderers. The reliance on electronic channels to move money and information has created opportunities for criminals. Indeed, many have pivoted their enterprises to leverage this shift by incorporating cyber tactics, techniques, and procedures (TTPs) into their illicit activities and by either committing cyber-enabled crimes themselves or using cybertools to facilitate unlawful acts such as money laundering. Technological advancements provide a greater reach for criminals to use the Internet and offshore servers to perpetrate crimes such as phishing, Internet auction fraud, romance scams, advanced fee fraud schemes, and fraudulent access to electronic digital media [computers, mobile devices, or Internet protocol (IP)-based phone systems] with a much broader impact. Thus, it should come as no surprise that business is good for cyberhackers and their support networks today. As society becomes more dependent on technology and innovation and as electronic channels are the primary medium through which crimes are committed, the scope for professional money laundering services will expand.

The most common methods to execute illicit transfers and attempt to launder proceeds today through financial institutions include the following (Figure 8.1):

- Wire and automated clearing house (ACH) fraud
- Credit card fraud
- Paper instrument fraud

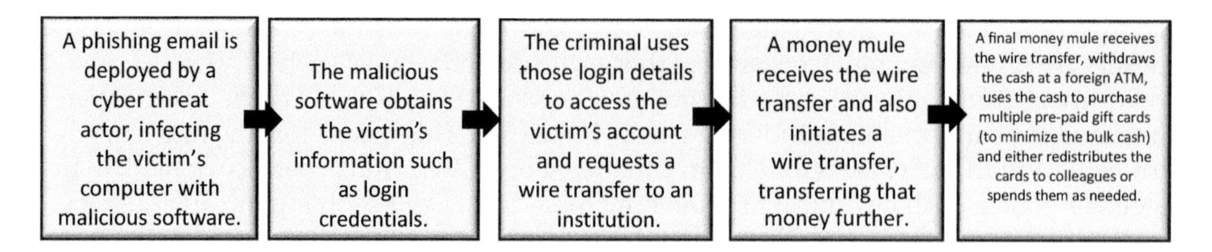

Figure 8.1 The scenario demonstrates how the financial proceeds of cybercrimes and fraud crimes may be laundered.

- Securities transactions
- Emerging payment networks such as prepaid access cards, peer-to-peer payments, crowdfunding, and microlending
- Cryptocurrencies and anonymous or encrypted transactions

One very popular way that cybercriminals can move their money is through micro-laundering, which makes it possible to launder a large amount of money in small amounts through thousands of electronic transactions. Financial institutions have systems that detect anomalous activity but may choose to focus on the relatively larger money movements. Emerging payment technology, such as PayPal and Venmo, or job advertising sites are popular for transferring small sums that may evade detective thresholds. Moreover, since online and mobile micropayments are interconnected with traditional payment services, funds can be moved through a variety of payment channels, increasing the difficulty to apprehend money launderers. One scenario might include the use of virtual credit cards as an alternative to prepaid mobile cards. The "card" is loaded using funds from a stolen bank account and an instant transfer can be sent using PayPal's network (Richet 2015).

8.2 *Anti-money laundering policy responses*

Given the devastating consequences of money laundering on society as a whole, it is increasingly important that organizations and governments work together to develop a response. The policies that have been developed to combat money laundering include legislative, regulatory, and investigative ones, which entail the collaboration of both the public and private sectors.

Many such policies have been similarly undertaken worldwide, and there are several bodies worth noting that are involved in anti-money laundering compliance and enforcement (Sharman 2011, p. 15). The Wolfsberg Group, an association of global banking institutions, aims to develop financial services industry standards relating to "know your customer (KYC)" anti-money laundering and counterterrorist financing policies. The Financial Action Task Force (FATF) is a policy-making body that promotes policies and measures to combat money laundering, the financing of terrorism, and proliferation of weapons of mass destruction. The FATF frequently publishes recommendations and papers on the topic. The Egmont Group is an international association of financial intelligence units that work to combat money laundering and terrorist financing through the exchange of expertise and information (Egmont Group 2018). The US Department of Treasury's Office of Foreign Asset Control administers laws that impose economic sanctions against entities that threaten US security. The UNODC, Basel Committee on Banking

Supervision, International Monetary Fund, and World Bank are other major institutions working to fight money laundering-related crimes.

Not surprisingly, financial institutions play an important role in helping investigative and regulatory agencies identify money laundering entities and take appropriate action. Because financial institutions are top targets, they are on the "front line" of defending the security and soundness of the economy and financial system by identifying and fighting fraud, money laundering, and the finance of terrorism. Compliance with US anti-money laundering legislation, particularly the Uniting and Strengthening America by Providing Appropriate Tools Required to Intercept and Obstruct Terrorism Act (USA PATRIOT Act), is a foundational component of risk management programs of banks, law firms, broker-dealers, asset management firms, auditors, money service businesses (MSBs), and other nonbank financial institutions. The USA PATRIOT Act has required banks and MSBs to comply with specific legislation. It has enabled the government to monitor information that helps counter criminal activity, thereby minimizing destabilizing impacts on the economy and security. While it provides a new layer of security, it also demands that banks implement strong internal compliance programs. Money laundering violations can bring serious penalties, may result in reputational damage, and, at worse, put a bank out of business.

The prevalence of cyber-related crimes in recent years has driven an industry-wide response with cybersecurity governance rising to the top of regulatory priorities alongside anti-money laundering compliance. Strengthening cybersecurity and anti-money laundering programs are key areas of focus for the Treasury Department (namely, the Office of the Comptroller of the Currency), the Financial Industry Regulatory Authority, the Securities and Exchange Commission, the Federal Reserve System, and the Federal Deposit Insurance Corporation (FDIC). The regulatory framework has shifted and expanded in the past 15 years to accommodate this ever-changing threat. To understand today's environment, it is worth outlining where the anti-money laundering regime started.

8.2.1 Money laundering regulation

The first legislation targeting money laundering crimes was the Bank Records and Foreign Transaction Reporting Act, also known as the Bank Secrecy Act (BSA). It was passed by the US Congress in 1970, requiring US financial institutions to collaborate with the US government in cases of suspected money laundering and fraud. According to the Treasury Department, the BSA established requirements for recordkeeping and reporting by private individuals, banks, and other financial institutions. It was designed to help identify the source, volume, and movement of currency and other monetary instruments transported or transmitted into or out of the United States or deposited in financial institutions. It imposed requirements for banks to report cash transactions over $10,000 using the currency transaction report, properly identify persons conducting transactions, and maintain a paper trail by keeping appropriate records of financial transactions. Under the BSA, financial institutions are obligated to assist the US government in the detection and prevention of money laundering, including by submitting suspicious activity reports (SARs) to report suspicious transactions or any series of transactions conducted or attempted that involve $5,000 or more in funds or other assets (Gup 2007). The BSA was implemented partly in response to income tax evasion and the concealment of assets that became commonplace through the use of bank accounts maintained in foreign jurisdictions. This was the first law of its kind and has served as one of the most important tools in combating money laundering by providing the Internal Revenue Service access to bank records, which has helped facilitate criminal and tax investigations that involve money laundering.

BSA data have also been used by a variety of US agencies such as the Federal Bureau of Investigation (FBI), the Department of Homeland Security, and the Drug Enforcement Administration. The Financial Crimes Enforcement Network (FinCEN), a bureau of the Treasury Department, acts as the designated administrator of the BSA and has contributed to its expanded use for investigations.

Many subsequent laws have enhanced and amended the BSA to provide law enforcement and regulatory agencies with the most effective tools to combat money laundering, and it remains the foundational framework with which institutions must comply. The following list of anti-money laundering laws have been implemented since 1970:

- BSA (1970)
- Money Laundering Control Act (1986)
- Anti-Drug Abuse Act of 1988
- Annunzio–Wylie Anti-Money Laundering Act (1992)
- Money Laundering Suppression Act (1994)
- Money Laundering and Financial Crimes Strategy Act (1998)
- USA PATRIOT Act
- Intelligence Reform and Terrorism Prevention Act of 2004

8.2.2 USA PATRIOT Act

The legislation introduced after the 9/11 terrorist attacks strives to further prevent pervasive financial criminal activity associated with terrorism. The USA PATRIOT Act was enacted to prevent and deter terrorist acts and to enhance law enforcement investigatory tools. The act strengthened special measures for certain jurisdictions, financial institutions, or international transactions of "primary money laundering concern" to detect and prosecute international money laundering and the financing of terrorism.

As noted, the BSA requires financial institutions to establish effective ways to detect, monitor, and report suspicious activity and to keep records and file reports that are determined to have a high degree of usefulness in preventing money laundering that may be a part of a criminal enterprise, terrorism, tax evasion, or other unlawful activity. Domestic and international law enforcement agencies leverage the documents filed under the BSA requirements to identify, detect, and deter money laundering (IRS 2017). Originally, this helped establish the "paper trail" that would build upon and enhance an investigation. With the advent of the Internet and the widespread use of technology to carry out electronic crimes, traditional reporting components have become less relevant. The paper trail has given way to the "digital footprint," a more useful forensic tool in today's environment.

8.2.3 *Recent public policy developments: Cybersecurity and the financial services industry*

Today's reliance on electronic channels to move money, communicate, and engage in financial transactions has created opportunities for criminals that are exacerbated by nascent security and cyber risk management programs. While the successful fraud or money laundering scheme still often relies on human error or the exploitation of a human component, the number of all financial crimes committed through electronic channels has outpaced any other method. The use of technology in financial services has expanded well beyond online banking and back-end computer systems. It now encompasses innovations in financial services such as mobile payment applications, roboadvisers, peer-to-peer lending, and distributed ledger technology. The good news is that today's illicit financial activity is accompanied by data associated with a computer user's digital footprint, which may help investigators build networks of illicit actors and their associates, their activity, and related suspicious transactions. Examples of such data include device identification, IP addresses, time stamps, and indicators of compromise (IOCs).

Cybercrime is considered as one of the most significant threats targeting the financial services industry, and as cyberattacks become progressively more sophisticated and frequent, there is increasing pressure from financial services regulators to tighten cybersecurity governance and risk management. Recent high-profile cyberattacks on financial institutions have caused significant damage and highlight the concerns and challenges around the management of external threats and vulnerabilities. In response to the changing criminal landscape, several regulatory authorities issued guidance and regulations relating to cyber-related crime and attacks and took action, reflecting a focus on improving cybersecurity in the financial services industry.

8.2.4 Federal Financial Institutions Examination Council cybersecurity assessment tool

The Federal Financial Institutions Examination Council (FFIEC), an interagency group that coordinates the federal supervision of depository institutions, released a cybersecurity assessment tool in June 2015. In 2014 and 2015, the FFIEC also issued several joint statements on various types of cyberattacks, citing the increased risks around distributed denial-of-service (DDoS) attacks, malware, cyberextortion, automated teller machine (ATM), and card authorization systems. According to the FFIEC (2017), the aim of the tool is to "help institutions identify their risks and determine their cybersecurity preparedness. The Assessment provides a repeatable and measurable process for financial institutions to measure their cybersecurity preparedness over time."

8.2.5 FRB, OCC, FDIC Advanced notice of proposed rulemaking

On October 26, 2016, the board of governors of the Federal Reserve Bank (FRB), the Office of the Comptroller of the Currency, and the FDIC jointly published an advance notice of proposed rulemaking of the Enhanced Cyber Risk Management Standards (FRB 2016a). The stated goal of the three federal banking regulatory agencies in considering these enhanced standards is to strengthen the operational resilience of large and interconnected financial entities and, by doing so, reduce the likely impact of a cyberevent on the financial system as a whole. Under consideration by the agencies are five categories of cyber standards: cyber risk governance, cyber risk management, internal dependency management, external dependency management; and incident response, cyber resilience, and situational awareness. At the time of writing, the rule had not yet been published or finalized.

8.2.6 New York Department of Financial Services

At the state level, on September 28, 2016, the New York Department of Financial Services (NYDFS) published its proposed cybersecurity regulations as "first-in-the-nation" rules that would require all supervised entities including banks, insurers, and other financial services institutions to establish and maintain cybersecurity programs to protect consumers from cyberattacks "to the fullest extent possible" (NYDFS 2017). A year prior, the

NYDFS also issued regulations with specific standards for suspicious activity monitoring and watchlist screening by certain entities under its jurisdiction.

8.2.7 FinCEN guidances

Recent high-profile cyberattacks on financial institutions caused significant damage and highlighted the concerns and challenges around the management of external threats and vulnerabilities. FinCEN issued a targeted advisory notice on e-mail compromise fraud schemes in September 2016. Focusing on the impact to financial institutions, this advisory discussed schemes commonly referred to as business e-mail compromise (BEC), which target commercial customers and individuals' e-mail accounts (referred to as EACs). FinCEN cited FBI statistics that 22,000 cases were reported of BEC and EAC since 2013 involving $3.1 billion. These schemes focus on impersonating victims in order to submit seemingly legitimate transaction instructions for a financial institution to execute, rather than taking over the victim's actual account. FinCEN provided 11 examples of suspicious activity red flags that could help employees of institutions monitor for suspicious activity involving e-mail correspondence and fraudulent transaction instructions (FinCEN 2016a).

In October 2016, FinCEN released a more far-reaching advisory regarding cyberevents, cyber-enabled crime, and obligations of financial institutions for reporting under the BSA (FinCEN 2016b). The advisory provides interpretive guidance to financial institutions on the expectations of FinCEN with regard to the reporting of cyber-enabled crime and cyberevents and how to properly do so through SARs. Cyber-enabled crime encompasses illegal activities such as fraud, money laundering, or identity theft carried out or facilitated by electronic systems and devices, such as networks and computers. Cyberevents are defined as an attempt to compromise or gain unauthorized electronic access to electronic systems, services, resources, or information. Financial institutions targeted by the advisory include banks, casinos, money services businesses, broker-dealers, mutual funds, insurance companies offering particular types of insurance, futures commission merchants, introducing brokers in commodities, nonbank residential mortgage lenders or originators, and housing-related government-sponsored enterprises. Expanding the reporting requirement to include cyberevents and cyber-related information now provides critical pieces of information to form a "digital trail" that serves as a valuable source of leads to initiate investigations, identify criminals, and disrupt and dismantle criminal networks. The advisory indicates that relevant and available cyber-related information should be included in SARs. This includes technical details of electronic activity and behavior such as IP addresses with timestamps, IOCs, device identifiers, virtual wallet information, and methodologies used.

The advisory also requests that SARs relating to cyberevents should include a description of the magnitude of the event and the following information:

- Source and destination information, including the following:
  - IP address and port information with respective date time stamps in coordinated universal time
  - Attack vectors
  - Command-and-control nodes
- File information, including the following:
  - Suspected malware filenames
  - MD5, SHA-1, or SHA-256 hash
  - E-mail content

- Subject usernames, including the following:
  - E-mail addresses
  - Social media account/screen names
- System modifications, including the following:
  - Registry modifications
  - IOCs
  - Common vulnerabilities and exposures
- Involved account information, including the following:
  - Affected account information
  - Involved virtual currency accounts
- Known or suspected time, location, and characteristics or signatures of the event
- Other relevant IP addresses and their time stamps
- Device identifiers
- Methodologies used
- Other information the institution believes is relevant

It also encourages the collaboration within regulated institutions between BSA/AML, in-house cybersecurity departments, and fraud prevention teams to identify suspicious activity (Greene et al. 2016). Likewise, financial institutions should work with their peers, as outlined in Section 314(b), to identify threats, vulnerabilities, and criminals.

The following are examples of situations that would require the suspicious activity reporting to the Treasury Department:

- Malware intrusion that would put customer funds at risk
- Cyberevent that exposes sensitive customer information such as passwords, credit card numbers, or account numbers
- DDoS attack

Note that the standard may require the filing of SARs even in circumstances where no actual financial transactions ultimately occur or are attempted in connection with the cyberevent, indicating how crucial the government considers this information.

8.2.8 FRB, FDIC, and National Credit Union Administration guidances

Guidance was also issued by the board of governors of the FRB, the FDIC, and the National Credit Union Administration concerning the filing of SARs to report certain computer-related crimes (FRB 2016b; FDIC 1997; NCUA 1997).

The expanding requirements to enhance cybersecurity risk management efforts and leveraging of the anti-money laundering regulatory framework indicates an emphasis and priority by government institutions on these areas that is sure to remain for years to come.

8.3 Underground or black markets

A black market, underground economy, or shadow economy can be defined as an economic activity involving the buying and selling of merchandise or services illegally. The goods themselves may be illegal to sell, may be stolen, or may be otherwise legal goods sold illicitly to avoid taxes or licensing requirements (such as cigarettes or unregistered firearms). In this section, the terms *darknet markets, black markets,* and *underground markets* refer to nonlegitimate marketplaces, and *dark web* refers to the location online where those underground markets exist.

Underground markets employ nearly half of the world's working population and have been a part of modern governance for centuries so it is important to take the past into account when describing today's environment. As has long been the case, black markets flourish, in part, by circumventing the rules and regulations put in place to create order. In early American history, much of the economy was based on smuggling and illicitly importing technological equipment and workers from England. Both legitimate businesses and criminal enterprises alike benefitted from the industrial revolution and later innovations, which brought conveniences such as steam power; machine tools; and eventually the railroad, the automobile, the telegraph, the radio, and cellular phone technology. The prohibition period in the early twentieth century in the United States serves as a great example of the creation of a black market and its return to legal trade. Many organized crime groups benefitted from banned alcohol production and sales, taking advantage of the fact that much of the populace did not view drinking alcohol as a particularly harmful activity. Illegal speakeasies prospered, and organizations such as the Mafia grew tremendously powerful through their black market alcohol distribution activities. Thus, when trying to picture a black market actor of the twentieth century, easily retrieved images include Italian mafiosos; Colombian drug lord and narco-terrorist Pablo Escobar; or a figure such as Viktor Bout, a Russian arms dealer known widely as the "merchant of death" for his delivery of weapons that aided and abetted many wars across the globe. Conversely, if asked to describe today's typical underground criminal, most would be hard pressed to provide a physical description of the "average hacker" or counterfeiting fraudster operating in the shadows of the dark web. Nevertheless, despite the technical, computerized, and digitized aspects of online black markets, the human factor remains the central component to business operations.

As described earlier in this chapter, globalization, technological advancements, and the Internet have created a more integrated global economic order, reducing barriers between financial networks and systems, facilitating growth, and increasing the speed at which transactions take place. As legitimate businesses benefit from the broadening opportunities resulting from such changes, so too are those who trade in illicit goods and services. Narcotics trafficking, sex tourism, computer hacking, money laundering, stolen artworks and antiquities, illegally harvested timber, oil, diamonds, counterfeit medicine, software, luxury brands (think very visible knock-off purse vendors on the streets of New York), and a globally networked market for human organs represent a multitrillion dollar global complex. These criminal activities still occur on the street but are increasingly moving online where the growing presence of underground markets in "cyberspace" facilitate the exchange of products between buyers and sellers who never have to interact. This has made it easier than ever for anyone to access illicit goods and services in the shadow markets that operate in anonymized and informal spaces.

This unregulated space presents a growing threat to businesses, governments, and consumers operating online because the activity has evolved from being a disparate and ad hoc group of networks to highly complex and well-organized operations with an increasingly sophisticated offering of products, tools, and services that help individuals carry out crimes, hide evidence of those crimes, and facilitate the laundering of the illicit proceeds. Where geography, physical presence, and face-to-face interactions built the trusted networks that defined black markets of the past, today's online underground economy is governed by anonymity and technology-based solutions. Although physical contact is largely absent, and despite the anonymization, secrecy, and the "dark" nature of today's online black markets, as with traditional black market activity, the human component remains a pivotal component of the environment. Identity, alliances, and trust remain crucial to the markets' functioning.

Users establish relationships and gain comfort with a transaction through eBay-style or Yelp-style rating systems where an unreliable seller that delivers a "poor" product (or does not deliver at all) will be advertised as such. Market participants place orders and sell products, comfortably communicating thanks to the use of encryption, privacy, and cryptocurrencies (discussed more in the following sections). This allows users to build trust and feel more secure as they are protecting their communications and transactions. Web forums, e-mail, locked down social media, online stores, and instant messaging platforms exist as both private and open chat rooms. Transactions between suppliers, vendors, potential buyers, and intermediaries for goods or services are often fast and efficient with customer service that mirrors that of any legitimate business. In fact, buyers willingly pay a premium for good customer service and a guarantee that they are purchasing verified data such as credit card details or user logins. It is precisely these secure communications and the anonymization facilitating direct links between end users or customers all over the world that are attracting people to these markets and redefining the human dimension of today's "average" black market participant.

8.3.1 Anonymization, cryptography, and privacy

Defining the profile of today's underground economy actor presents a challenge for additional reasons. Given the rapid rise and fall of these markets in recent years allows little time for media, literature, or entertainment to study and document what the "typical" online criminal looks like. More importantly, the very feature making the business so attractive to a criminal also prevents the public from understanding his or her profile. The dark web keeps cybercriminals and money launderers afloat because it offers (often requires) anonymizing networks and untraceability. Such features impede the efforts of law enforcement, Internet service providers and financial intelligence units from detecting criminal activity, making this space a haven within which hackers and their criminal colleagues operate, almost with impunity. Obtaining a virtual private networks with strong encryption, using a proxy server, IP spoofing, and using the Onion Router (Tor) or the Invisible Internet Project (I2P) will hinder efforts to trace users' and administrators' online activity on unindexed sections of the web. TOR and I2P are free software and open networks that enable anonymous communication, file transfers, and Internet browsing that conceal a user's physical location preventing network surveillance or traffic analysis.

8.3.2 Crytpocurrencies

Digital cryptocurrencies and the use of distributed ledger technology (also known as the blockchain) are rapidly becoming the preferred method to value transfer for illicit goods and services, thus posing a number of money laundering risks. Developed in 2008, Bitcoin has become the most famous decentralized payment network. This cryptocurrency enables real-time, peer-to-peer payments to anyone in the world without the intermediation of a central authority or regulating country. This also means that individuals who hold cryptocurrencies have little to no recourse if fraud or theft occurs. Although Bitcoin is the most recognized cryptocurrency, there are hundreds of others, the most popular of which include Ethereum, Ripple, Litecoin, and Monero. Coinmarketcap.com (2017) provides the market capitalization and current price of the top cryptocurrencies in circulation today. Cryptocurrencies are held and encrypted in various types of "wallets," both online and offline. Although not completely anonymous (all transactions are permanently stored on a traceable, public ledger), there is no requirement to store PII or for administrators to perform customer identification on these transactions unless formally registered with the US Treasury Department. Likewise, the pseudonymous nature of the "currency" makes it difficult for financial intelligence units or US-based virtual currency exchangers that are obligated under the BSA to identify and verify clients, to determine the counterparties involved in such transactions. Like other anonymizing elements of the dark web, cryptocurrencies make it difficult for law enforcement to trace, seize, and freeze illicit profits. This makes it attractive for actors operating on the black market.

There are many measures to further obscure funds transfers. Tumbling or mixing services offer a way to comingle a cryptocurrency transaction with many other transactions with the intention of confusing the source or origin of the funds. Another convenient tool for money launderers that emerged a few years after Bitcoin gained an increasing user base was a Bitcoin application to protect users' identities launched by a company called Dark Wallet. The application was described by its founder as "money-laundering software" (Greenberg 2014).

8.3.3 *What is for sale?*

The market forces that prevail on the dark or deep web also do so based on the same microeconomic concepts that have always driven legitimate markets. Prices go up with rising demands for products and services or when there is scarcity and go down when demand is low or supply is high. There is a healthy appetite today on underground markets for both stolen data and tools used to carry out malicious attacks, steal private data, or launder money. There is a range of products, information, and services available on the black market that enable cybercrimes and money laundering and that can be customized based on a buyer's needs. There is a substantial interest in goods such as hacking tools, digital assets, or "as-a-service" products such as money handling, obfuscation, and evading detection (Ablon et al. 2014).

The following are examples of products available on the black market. This nonexhaustive list is limited to cyber-related products and services. It does not include the sale of illegal substances, for example.

Stolen data

- Credit card information and personal identification numbers
- NPI
- TTPs
- E-mail logins and passwords
- Protected health information
- Personal information (social security number, date of birth, and identifying information used for authentication)
- Stolen identities

Services

It is important to keep in mind that criminals engaged on the dark web are well funded with professional business models that outsource a wide variety of services. In a noncentralized online system, characterized by greater flexibilities and agility, anyone can rent, lease, or purchase an as-a-service," which is a flourishing business model run on black markets found on the darknet. Examples of an ever-expanding list are as follows:

- Data-as-a-service (data are stored in the cloud and is accessible by a range of systems and devices)
- Hacking-as-a-service
- DDoS-as-a-service
- Platform-as-a-services (to develop, run, and manage applications without the complexity of building and maintaining expensive infrastructure or launch applications)
- Ransomware-as-a-service
- Money laundering-as-a-service
- "Money handling services"
- Mule services (witting and unwitting money "mules")
- Fake website design (front companies)
- Shell company formation

Example products

- Malware
- Exploit kits
- Botnets for rent
- Hacking tools
- Tutorials on "how to blackmail for Bitcoin (ransomware)"

8.3.4 *Black market/dark web participants*

The ease with which market actors can get involved has increased over the years due to the array of available websites, forums, and chat channels that educate the average lay person on "how to hide your money trail," where to buy or sell credit card information, and other guides that lowers previous technical barriers to entry (Ablon et al. 2014, p. 4). Higher tiers of access to black markets require more vetting. Lower tiers might be publicly available open and require less vetting.

Indeed, hierarchies are established within these markets as well as specialized roles. Administrators occupy the most desirable position at the top and are followed by subject-matter experts who have sophisticated knowledge of particular areas (e.g., root kit creators, data traffickers, and cryptanalysts). Intermediaries, brokers, and vendors are next on the black market totem pole followed by the general membership. Each member may have a subsidiary cell of associated members (Ablon et al. 2014, p. 5). Other participants may include threat actors such as hacktivists or nonstate actors and freelancers.

As the incidence of data breaches and credit card theft online continues to rise, the general public has become virtually numb and apathetic to the high probability that their personal information such as that stored on a credit card has been stolen. When compromise occurs, a bank will promptly replace the card and reset the password, the consumer feeling little to no immediate impact, despite the fact that his or her credentials or personal information is now readily available for a relatively small price. To illustrate the demand, and thus the need, for awareness, Table 8.1 lists the estimated costs of credential and other data. Anyone who allows their data to be stored by a third party such as a healthcare provider or retailer (in other words, nearly the entire consumer population) should understand the high likelihood that their information could be compromised if a data breach occurs and will likely be sold on the black market. In 2016, one in three Americans had their healthcare data exposed. This is just one industry that happens to be more vulnerable

Table 8.1 Average underground prices for various hacker services

Hacker service	Average price
Visa or Mastercard credentials	$7.00
Credit card with magnetic stripe or chip data	$15.00
Premium American Express, Discover Card, Mastercard, or Visa with strip or chip data	$30.00
Bank account credentials (balance of $15,000)	$500
Bank account credentials (balance of $70,000–$150,000)	6% of account balance
Large US airline points accounts	1.5 million points cost $450
Large international hotel chain points account	1 million points cost $200[a]

Source: Cetera, M. *Prices Rise for Your Data on the Black Market,* http://www.bankrate.com/financing/credit-cards/prices-rise-for-your-data-on-the-black-market/, 2016.

than others, but the figure illustrates the scale and impact that these crimes can have on our society and the type of avenue for the compromise of highly personal information. Either an employee of the company will make a mistake resulting in data exposure or a determined cybercriminal will break through security defenses and steal sensitive information. The stolen information will likely be sold on a darknet market such as Alphabay, Silk Road 3, Dream Market, Crypto Market, or Ramp, among others.

8.3.5 *Pursuing today's criminal*

An advanced and evolving threat has also meant that law enforcement has had to shift its approach to combating these elements. Crime busting before the twenty-first century involved investigating individuals' patterns and associates as they moved through society. Today, an investigation is built around a digital footprint and the use of cyberforensics. Court-ordered search warrants to seize and attempt to search cellular phones, tablets, and other electronic devices, are helpful but often pose challenges for yet untrained investigators who may meet technical barriers. The following are a few examples for reference of how law enforcement has apprehended black market participants.

8.4 *Silk Road and Operation Onymous: 2014*

In 2014, an internationally coordinated law enforcement action, involving the Department of Justice, Department of Homeland Security, and law enforcement agencies of approximately 16 foreign nations working under the umbrella of Europol's European Cybercrime Centre and Eurojust, put an end to a half dozen top darknet sites, including Silk Road 2.0, targeting high-value actors of the dark web drug trade (FBI 2014). The bust occurred just a year after the takedown of the original Silk Road online marketplace that similarly offered a range of illegal goods and services including drugs, firearms trafficking, counterfeit goods, and fake passports. The website addresses, known as ".onion addresses," and computer servers hosting these websites, were seized as part of takedown, and 17 people were arrested. Although seen as a major success, other darknet sites, such as Agora, Evolution, and Andromeda remained online and intact, and warmly welcomed the new business that resulted from their competitor's demise. This "whack-a-mole" effect demonstrates the ongoing challenges faced by law enforcement in halting all underground market activity.

8.5 Bangladesh Bank: 2016

The largest bank robbery in history occurred in 2016 when $81 million was stolen from the Bangladesh central bank. Funds were sent from the account of the Bangladesh Bank at the Fed, then to several accounts that had been opened with false identification in May 2015 at Rizal Commercial Banking Corporation (RCBC) in the Philippines. Once received, RCBC wired them to a remittance company, Philrem Service Corporation, which then sent the money on to a number of Philippine casinos. From there, the authorities were unable to track the funds further. Experts believe the fraudsters installed malware at the Bangladesh Bank in January 2016. The company that investigated the attack discovered a malicious program in Bangladesh Bank's systems, which allowed the fraud actors to enter and systemically follow SWIFT transactions, disconnect a printer at the bank so that warnings regarding the transfers did not print, and hide or delete transactions in the system of the bank. The investigation linked the malicious code used in the heist to code used in both the Sony Pictures attack in 2014 and other bank attacks, including against several in South Korea (Fox-Brewster 2016). Many suspected that these attacks originated in North Korea.

Several other stories involving SWIFT-related wire frauds emerged in headlines shortly after the Bangladesh Bank events including an unsuccessful attempt through the Tien Phong Bank in Vietnam in December 2015 for $1.1 million and the theft of $9 million from a bank in Ecuador. Hackers also stole $10 million from an unnamed bank in Ukraine. These thefts all required the subsequent laundering of stolen funds, reflecting how anti-money laundering practices, cybersecurity, technology, and the international financial system converge.

8.6 Alpha Bay, Hansa, and Operation Bayonet: 2017

In the summer of 2017, Alpha Bay and Hansa, two of the world's largest dark web bazaars estimated to have generated more than a billion dollars in sales of drugs, stolen data, and other illegal goods in just a 3-year period, were taken down in a multinational sting called Operation Bayonet. Although customers on dark web sites are encouraged to encrypt their addresses so that only the seller of product can read it, many do not. These, or other sloppy online hygiene practices, are what create a digital footprint that ultimately helps law enforcement trace activity. In this case, the Alpha Bay administrator's personal e-mail was used in welcome message to new users, which led investigators to his PayPal account, front company, and ultimately his home (Greenberg 2017). Although the takedown threw the darknet participants into chaos momentarily, as in the 2014 Operation Onymous, many other sites remained online including Dream Market, Crypto Market, Ramp, Tochka, Trade Route, and Silk Road, all happily taking on the additional business.

8.6.1 Research challenges

Unfortunately, criminals and their associates do not share information about their methods or successes. Some cybersecurity firms focus on the collection of information from dark web sites and forums to better understand cybersecurity threats, but the maintenance of a low profile is critical to their success, thus making this a difficult area to research. One such firm followed dark web conversations on popular underground forums where specific tools and ransomware attacks against hospitals were discussed, mentioning the many systems that could be targeted, many in the medical industry. It is impossible to know who posted it, and it is not evidence that people who participated in the thread were responsible. However, this type of information provides insight into hackers' expertise and what future attacks might look like, which may ultimately help companies, law enforcement, and users defend against hacks.

8.7 Challenges and opportunities

While the body of research around money laundering and successful prosecutions is robust, the methods used to obscure and move illicit proceeds through black markets as well as the legitimate financial system are constantly shifting. Given the inherently anonymous nature of the online activities and limited number of firms engaged in this space, the data associated with underground or black markets today are limited. With that said, we know that the nexus between cybercrimes, black markets, and money laundering will continue to grow. The major challenge that anti-money laundering, cybersecurity professionals, and law enforcement face today is around whether the ability to hack, attack, or launder will outpace the ability to defend. Combating threats will require a collaborative approach from both the public and private sectors.

As more consumers shop and pay with connected devices, and commerce increasingly migrates to digital channels, industries must invest in new standards, technologies, and products. One of the best defenses is removing sensitive account data from the payment environment, putting it into a form that cannot be used by criminals for fraud. Products, services, and online platforms should develop built-in security and privacy features, thereby protecting both the product and the customer information from being hacked.

Although transnational organized criminal networks now operate in a new technological paradigm where old rules for combating the criminal element no longer apply, the human factor remains an important component in both executing and detecting crime. In other words, computers, robots, or malicious software is not wholly responsible for all illicit transfers. Conversely, automated systems that monitor transactions can only go so far. There is a limit to how much a machine can catch, and human intelligence remains crucial for developing methods to detect illicit activity. Nascent security and cyber risk management practices within organizations present challenges in fighting criminal actors. A simple step to create stronger defenses is the establishment of robust cybersecurity risk management programs within the private sector that closely intersect (but not necessarily merge) with financial crimes departments. No single person, department, or company has all the skills and resources needed to address these issues, so collaborative practices will be paramount to combating the threat. The anti-money laundering regulatory regime requires that institutions subject to the BSA implement systems and methods to gather certain client and transactional information and continually develop ways to detect and analyze unusual online behaviors. Criminal actors today are adapting to new technologies at rapid pace and seek out the path of least resistance. The lines between fraud, cybercrime, and the laundering of illicit proceeds will continue to blur. Thus, the division between anti-money laundering compliance and cybersecurity risk management professionals is narrowing. To effectively counter cyberattacks and cybercrimes, it is increasingly important that both anti-money laundering and cyber risk professionals in the public and private sectors proactively identify ways to integrate their functions by leveraging and sharing investigative information and monitoring and reporting tools. Regular collaboration and effective communication between financial crimes and cyberdepartments may equip financial institutions with an enhanced ability to refine detective methods, appropriately report suspicious activity to the government, and uncover data-driven solutions.

A well-defined culture of cyber risk awareness and compliance among all members of the public and private sector through annual trainings, information security programs, and governance frameworks will help protect the financial ecosystem, a critical component of our security infrastructure.

References

Ablon, L., Golay, A. A., Libicki, M. C. *Markets for Cybercrime Tools and Stolen Data: Hacker's Bazaar.* Santa Monica, CA: RAND Corporation. pp. 4, 5, 8. (2014). Available at: https://www.rand.org/content/dam/rand/pubs/research_reports/RR600/RR610/RAND_RR610.pdf.

Cetera, M. *Prices Rise for Your Data on the Black Market.* (2016, May 3). Accessed June 17, 2017. Available at: http://www.bankrate.com/financing/credit-cards/prices-rise-for-your-data-on-the-black-market/.

Coinmarketcap.com. CryptoCurrency market capitalizations. Available at: https://coinmarketcap.com/.

Egmont Group. *The Egmont Group of Financial Intelligence Units.* Toronto: Egmont Group. (2018). Available at: http://www.egmontgroup.org.

FBI (Federal Bureau of Investigation). *Dozens of Online "Dark Markets" Seized Pursuant to Forfeiture Complaint Filed in Manhattan Federal Court in Conjunction with the Arrest of the Operator of Silk Road 2.0.* Washington, DC: FBI. (2014, November 7). Available at: https://www.fbi.gov/contact-us/field-offices/newyork/news/press-releases/dozens-of-online-dark-markets-seized-pursuant-to-forfeiture-complaint-filed-in-manhattan-federal-court-in-conjunction-with-the-arrest-of-the-operator-of-silk-road-2.0.

FDIC (Federal Deposit Insurance Corporation). *Guidance for Financial Institutions on Reporting Computer-Related Crimes.* FDIC Financial Institution Letter FIL-124-97. Washington, DC: FDIC. (1997, December 5). Available at: https://www.fdic.gov/news/news/financial/1997/fil97124.html.

FFIEC (Federal Financial Institutions Examination Council). *Cybersecurity Assessment Tool.* Arlington, VA: FFIEC. Page last modified: September 1, 2017. Available at: https://www.ffiec.gov/cyber assessmenttool.htm.

FinCEN (Financial Crimes Enforcement Network) *Advisory to Financial Institutions on E-Mail Compromise Fraud Schemes.* Vienna, VA: FinCEN. (2016a, September 6). Available at: https://www.fincen.gov/sites/default/files/advisory/2016-09-09/FIN-2016-A003.pdf.

FinCEN. *Advisory to Financial Institutions on Cyber-Events and Cyber-Enabled Crime.* Vienna, VA: FinCEN. (2016b, October 25). Available at: https://www.fincen.gov/sites/default/files/advisory/2016-10-25/Cyber%20Threats%20Advisory%20-%20FINAL%20508_2.pdf.

Fox-Brewster, T. Crooks behind $81M Bangladesh Bank heist linked to Sony Pictures hackers. *Forbes.* (2016, May 13). Available at: https://www.forbes.com/sites/thomasbrewster/2016/05/13/81m-bangladesh-bank-hackers-sony-pictures-breach/#6a0de2142ee6.

FRB (Federal Reserve Bank). *Agencies Issue Advanced Notice of Proposed Rulemaking on Enhanced Cyber Risk Management Standards.* Washington, DC: FRB. (2016a, October 19). Available at: https://www.federalreserve.gov/newsevents/pressreleases/bcreg20161019a.htm.

FRB. *Guidance Concerning Reporting of Computer Related Crimes by Financial Institutions.* FRB Supervisory Letter SR 97-28. Washington, DC: FRB. (2016b, April 21). Available at: https://www.federalreserve.gov/supervisionreg/srletters/sr1609.pdf.

Greenberg, A. Dark wallet is about to make Bitcoin money laundering easier than ever. *Wired.* (2014, April 29). Available at: https://www.wired.com/2014/04/dark-wallet/.

Greenberg, A. Global police spring a trap on thousands of dark web users. *Wired.* (2017, July 20). Available at: https://www.wired.com/story/alphabay-hansa-takedown-dark-web-trap/.

Greene, C., Stinebower, C. N., and Wolff, E. D. FinCEN cybercrime advisory expands SAR requirements. *Law360.* (2016, November 29). Available at: https://www.law360.com/articles/864549/fincen-cybercrime-advisory-expands-sar-requirements.

Gup, B. E. *Money Laundering, Financing Terrorism and Suspicious Activities.* New York: Nova Science Publishers. p. 8. (2007).

IRS (Internal Revenue Service). *Bank Secrecy Act*. Washington, DC: IRS. Page last modified: August 6, 2017. Available at: https://www.irs.gov/businesses/small-businesses-self-employed/bank-secrecy-act.

NCUA (National Credit Union Administration). *Guidance for Reporting Computer-Related Crimes*. NCUA Regulatory Alert 97-RA-12. Alexandria, VA: NCUA. (1997, December 5). Available at: https://www.ncua.gov/Resources/Documents/97-RA-12.pdf.

NYDFS (New York Department of Financial Services). *Regulation Emphasizes Compliance Culture at Top Levels of the Institution*. Albany, NY: NYDFS. (2017, February 16). Available at: http://www.dfs.ny.gov/about/press/pr1702161.htm.

Office of the Comptroller of the Currency. *Bank Secrecy Act: Combating Money Laundering and Terrorist Financing*. Washington, DC: Office of the Comptroller of the Currency, US Treasury Department. Available at: https://www.occ.treas.gov/topics/compliance-bsa/bsa/index-bsa.html.

PwC. Global Economic Crime Survey. PwC. (2018). Available at: https://www.pwc.com/gx/en/services/advisory/forensics/economic-crime-survey/cybercrime.html.

Richet, J.-L. Laundering money online: An overview. *Harvard Law Blog* (2015, February 7). Available at: http://blogs.harvard.edu/jeanlouprichet/files/2015/02/Laundering-Money-Online_an-Overview.pdf.

Sharman, J. C. The Money Laundry: Regulating Criminal Finance in the Global Economy. Ithica, NY: Cornell University Press. pp. 14–16. (2011).

UNODC (United National Office on Drugs and Crime). *Estimating Illicit Financial Flows Resulting from Drug Trafficking and Other Transnational Organized Crime*. Vienna: UNODC. pp. 5, 39. (2011, October). Available at: https://www.unodc.org/documents/data-and-analysis/Studies/Illicit_financial_flows_2011_web.pdf.

US Treasury Department. *History of Anti-Money Laundering Laws*. Washington, DC: US Treasury Department. Available at: https://www.fincen.gov/history-anti-money-laundering-laws.

section four

Smart networks and devices

chapter nine

Smart home network and devices

Abbas Moallem

Contents

9.1 Introduction

With the speedy expansion of the Internet and Internet-based technologies in homes, the connection of a household to an Internet service provider has become as common as connecting to the basic utility providers for electricity and gas. According to Internet Live Stats (2016), 88.5% of the United States, and 40% of the world population, are Internet users. Users, moreover, are not typically satisfied with having just their computer connected to the Internet. They need wireless connections at home for their computer devices, and cross access connections that allow them to view, modify, and control their computer devices. Users also need to control and monitor security and problem-solve to maintain an uninterrupted Internet connection and a secure network.

Not long ago, home Internet connections were limited to computers. Now, gradually, within a "smart home" (Cheng and Kunz 2009), a wide range of computers and computerized devices can be connected and interact globally and with each other. However, after initial configuration, users are advised to conduct frequent monitoring of the interactions of each device and to frequently check security settings of each device (updating firmware, viewing the connected devices, changing passwords, and so on). Frequent monitoring provides a method of examining the router to ensure that it has not been hacked. For instance, monitoring can display all the devices connected to the user's network, so any unauthorized devices can be easily identified.

Today, roughly 9 in 10 American adults use the Internet (Pew Research Center 2017). It is estimated that households have 10 connected devices now, projected to rise to 50 connected devices per home by 2020 (Phadnis 2016). More devices mean a wider number of systems and accounts that require user management. For example, managing a switch box that turns one Ethernet connection into several, allowing multiple wired devices to connect to the Internet without overloading the router, or eliminating bottlenecks to make sure that the modem/router can handle the broadband speeds.

Despite the increasing value of the information stored on devices connected to a home wireless network, users are also presented with the threat of network privacy and security breaches. User habits, and knowledge of network features and safety, are thought to have a great impact on risk rates.

9.1.1 Passwords

Most users are under the impression that a complex password will keep them safe from breaches. However, reports indicate that even a complex password is not necessarily a secure password (McMilllan 2014). In addition, in October 2107, new research from security researcher Mathy Vanhoef of KU Leuven, in Belgium, found that a flaw in the cryptographic protocols of Wi-Fi protected access II (WPA2) could be exploited to read and steal data that would otherwise be protected. In some situations, this vulnerability even leaves room for an attacker to manipulate data on a Wi-Fi network, or inject new data into it. In practice, that means hackers could steal users' passwords, intercept their financial data, or even manipulate commands to, say, send their money to hackers (Newman 2017).

9.1.2 Beyond passwords

Users who follow the basic device setup guides, but who want to further configure the security of their devices, are struck with two deficiencies: the lack of the knowledge of what proper preventative measures are and the lack of the skill to implement preventive actions [how to do it using device user interface (UI)]. Users with limited networking experience generally claim that managing a home network beyond the basic setup protocols of the devices is a tedious, difficult task.

People who have a home networking device have more than likely experienced a variety of home networking UI issues. They might have even asked someone with technical knowledge to help with installing, connecting, and configuring the device that they acquired or had problems with. No matter the type of device—a wireless router or an interactive TV can very quickly become complicated.

9.1.3 Network usability problems

Edwards et al. (2011) concludes that "network usability problems run deep because the technology was initially developed for research labs and enterprise networks and does not account for three unique characteristics of the home: 1) lack of professional administrators, 2) deep heterogeneity, and 3) expectations of privacy."

9.1.3.1 No professional administrators

The issues that seem to be related to user difficulties with network security technology are that UIs are built for advanced users with information technology (IT) backgrounds, heterogeneity in users and home networking configurations, and the complexity of problem-solving home network issues. For example, Grinter et al. (2005) noted that home users are often unable to verbally articulate accurate information about their networks or even a mental model of their network, which has been shown to be related to their level of expertise. In addition, most people have neither the time nor the inclination to be continually vigilant for new threats on their home network; they are focused on getting their work done (Dourish et al. 2004).

There is a need to understand some of the network vulnerabilities in an average household and how to thwart a hacker's future attempts to attack. Households will have to improve their ability to detect and correlate attack activity to respond to increasingly sophisticated threats that accompany these high-growth technologies. With that said, security experts state that no single product or vendor can cover every possible threat angle. Home network users must understand our worldview of cybersecurity should be a systematic framework. With this perspective, homeowners can understand how to prevent access to digital spaces such as bank accounts as well as physical devices such as smart TVs. However, with today's technologies, this level of security is harder to obtain when the user does not have an advanced level of cybersecurity knowledge and skill.

One of the main demographic groups of Internet users in most developed countries is older adults. Older adults benefit from the expansion of the smart home. They seem to have an overall positive attitude toward this technology (Demiris et al. 2004). The usage of health monitoring sensors or security devices in the home to enhance their lives is a good example of how these users benefit from this technology. However, the ease of use for these applications is crucial for secure usage by older adults especially if they are using medical monitoring devices. Consequently, this population can be even more vulnerable to cyberattacks, identity theft, and social engineering.

Home network devices should make it easy to add computers, or any other smart appliances to the home network, and establish interactions by offering users easy-to-use secure settings. In the following section, we will provide a summary of common issues and discuss potential solutions.

9.2 Home networking routes and feature management

The modern home generally uses a router connected to an external network (Internet service provider or ISP) as the centerpiece of home networking. The router provides wired and wireless Internet connection to all devices in the home. Despite the extent of home networking, the large number of users, and the potential impact on home users' security by nefarious actors and schemes, the number of investigations into home security vulnerabilities and breeches is relatively very small.

In general, home security needs to be managed at each of three layers: by the ISP, by the home networking device (router), and by each connected device.

According to Cisco Systems, Inc. (2017), security breaches at the ISP provider level include the following:

- Denial-of-service and distributed denial-of-service (DDoS) attacks, which are aimed at disabling access to various Internet services for legitimate users
- Excessive traffic and resource depletion caused by infected machines, which can generate problems for service providers
- Attacking the border gateway protocol (BGP) routing and injecting faulty BGP routes for traffic redirection, one of several techniques attackers use to obtain "interesting" traffic

- Stealing of domain name system (DNS) information and using this to redirect Internet traffic to serve the needs of people with criminal intent
- Device compromise, including breaking into vital components of the infrastructure and modifying their configuration

These security issues are handled by ISP companies and will thus not be reviewed in this chapter.

The second level of security is the responsibility of end users in each home, or small businesses, through the UIs of the router.

9.2.1 *Acquiring a router*

An ISP provides a router or users acquire one from the multitude of brands available on the market. The selection of a router is always a tedious task since most of the time, people might not know exactly what the meaning of the information provided on the packaging of the device is. Figure 9.1 shows the information on three major brands of routers. Phrases such as "dual-band gigabit," "up to 340 bps," "300 + 300 Mbps," and "tri-band 2.2 Gbps combined speeds" used on the packaging are not likely to be understandable to an average user. A survey of home networking users conducted in 2011 by *TMC News* might still be relevant. This survey showed that only 1% of the respondents bought their particular router because of its reported ease of use, while 22% of respondents bought based on the speed of the router and 17% bought based on its price. When participants were asked about purchasing a new wireless router, speed remained the top factor and increased in importance to 37%, while ease of use moved up to the second priority at 17% (TMC News 2011). One might think that this still would be the case.

9.2.2 *Connecting router to network*

Over the years, most brands have improved the installation process, through simple wizard and smartphone and tablet apps that users can use for first-time and subsequent installations. Smartphone and tablets did not necessarily simplify the task. Having said that, the user following the instructions can be quickly connected to the Internet and with as password that is offered during the installation or a self-specified password. Once users are connected to the Internet, they consider that their task is completed and write the password for further use. However, from the security point of view, the issues

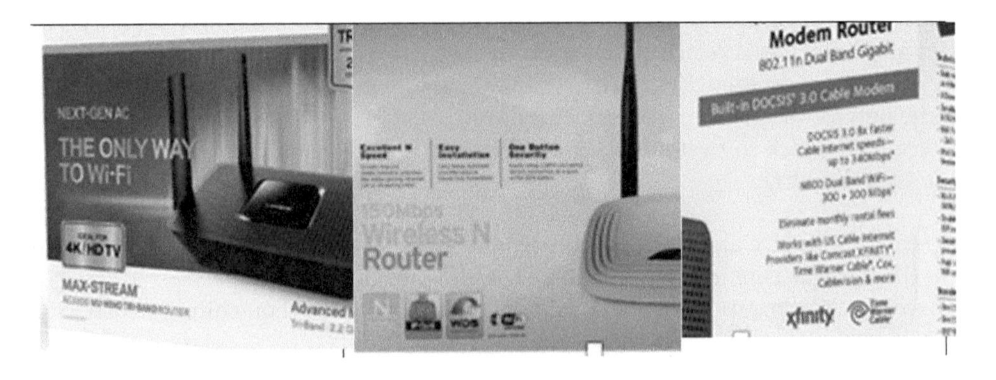

Figure 9.1 Information on packaging of three major brands of routers.

start from the moment that the user is connected to Internet. Despite continuous progress in hardware, improvements to managing the security settings of these devices are still complex.

9.2.3 Managing home network

Today, most people can easily connect their newly purchased home routers to the Internet using the Quick Guides and helpful Tech Support line that vendors provide. However, users commonly agree that managing home networks is still a difficult task that is out of reach for most users with limited networking experience.

It is probably unrealistic for router manufacturers to require such knowledge on the part of their customers. Some of the issues in the way of securing users' networks are (1) an extremely personal device and attack vector-filled environment where there is (2) no professional administrator to maintain a (3) heterogeneous collection of consumer technologies that (4) are increasingly cyberphysical and sensor rich. The combination of these factors leads to an array of attacks and complicates the design of defenses for home devices (Denning 2013).

Some research suggests that instead of having individuals manage their networks, they should "outsource the management and operation of these networks to a third party that has both operations expertise and a broader view of network activity" (Feamster 2010). Yang et al. (2010) suggest that it could be better to provide the home user with a conceptual model that can help them understand key aspects of networking and with visual tools to do a range of common tasks. Another approach has been to try and improve router UI usability, enabling users to perform complex system management tasks independently for their home network (Moallem 2014).

Let us look at a number issues when managing a home networking router. A user must access the UI of a router in order to set and manage a network. The UI always acts as the point of access, regardless of the brand, range, remote access, and other provided functionalities of the router such as universal serial bus (USB) ports or guest networks. To do so, in addition to username and password settings, users must deal with a variety of other router security settings for device management.

9.2.3.1 Setting user name and password

After initial password setting at installation, using the default router password or user-selected password, the user should be able to change their password. In a study conducted by the author with 104 undergraduate students (63 males and 41 females with 72% between the ages of 18 and 25), it was found that 91 (88%) owned a router. This group was asked if they knew how to change their network password. If they answered "yes" to knowing how to change their network password (50 participants or 48%), they were asked how they would go about making such a change. The explanations of how to change their password by those 51 who responded "yes" showed that 23 (43%) would do it by entering the IP address of the router and 8 (16%) would go to the provider site. The remaining 20 (41%) students did not provide a clear or accurate response, with answers such as "asking my brother" or "Google search." Considering that the people who responded with "IP address" and "service provider" as those who actually knew how to change the router password (23 participants—22% of all participants), even if we consider that the survey was a self-reporting one and participants' responses may have been completely wrong either way, we can conclude that among the group, approximately one-third might be able to change their router password. However, we can also extrapolate the level of understanding of this group of people about how to manage a router. If young college students do not have the understanding that they can log in their router user interface to change the password, then we can assume that the understanding will be even less among the general population that owns a router for home networking. According to PEW Research Center (2017), "undergraduate and graduate students differentiate themselves more clearly when it comes to home broadband access, as more than nine in ten undergraduates (95%) and graduate students (93%) are home broadband users—well above the national adult average of 66%."

Today's user is asked to set a more personalized password, keep the default settings, or switch to a desired name and password (surprisingly many users might not even remove the password or choose a very simple one). It is reported by Barker (2014) that more than half of all home routers are poorly protected using default or easily hacked password combinations such as admin/admin or admin/password.

The password setting has two layers. One sets a password for the Wi-Fi requiring, all devices to enter the password. The second layer sets a password to manage the router, access the admin UI of the device, and manage network security including the network password.

All the router brands have a preset username and password for the device and an IP address to access the device. For example, the Comcast Xfinity router generally uses the IP address http://10.0.0.1 and "admin" as the username, with the default password "password." According to a study (Moallem 2012), 9% of participants did not know what an IP address is, while 39% claimed that they thought they knew what it meant but were not sure (Figure 9.2).

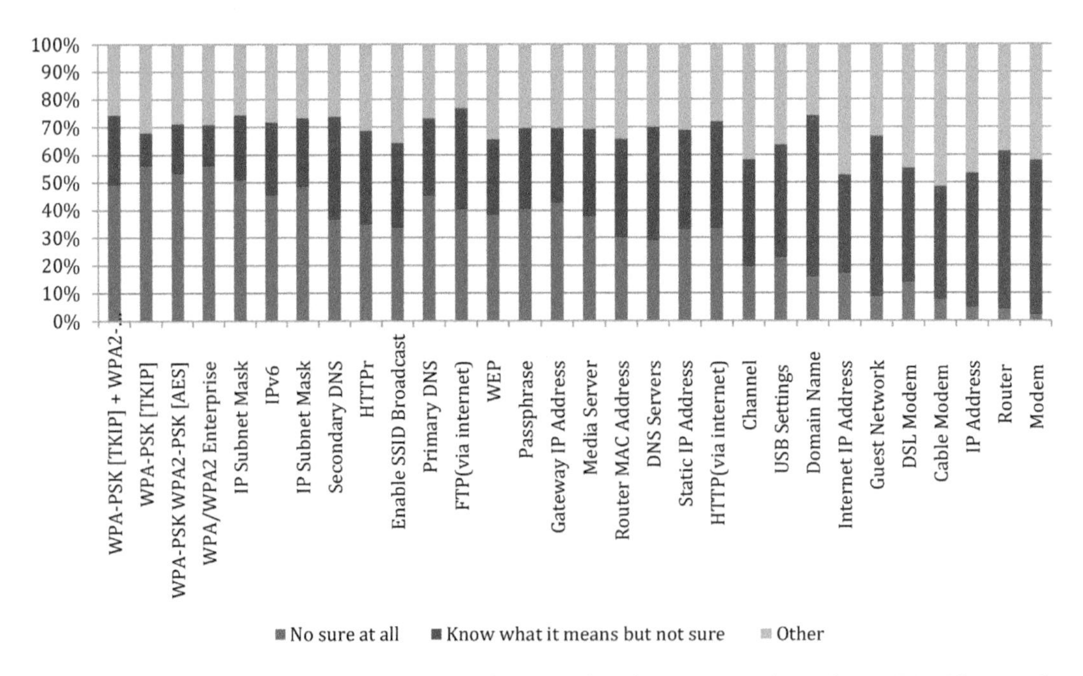

Figure 9.2 Understandability of the terminology used on home networking device UIs. The graph shows the percentage of participants who were not sure at all what the meaning of each term was (84 participants: 54% male, 46% female, 54% under 25 years).

9.2.3.2 Home networking configuration

There is a variety of parameters that make it hard for users to configure and manage their networking devices efficiently. These parameters include channel selection, remote access to routers and devices, and IP selection. The growing number of devices that will connect to home networking systems increases the difficulties presented by each parameter. Apart from regulatory concerns, users must also be aware of standard practices undertaken when managing a wide variety of systems and accounts (Phadnis 2016). Even for a relatively small home network, managing the security of network each time that a new device is added can be a substantial challenge, and each device can introduce a new, substantial challenge.

As we have previously mentioned, to manage a home networking device, user must log in to the router UI to change or set configuration.

Let us take a look at a router UI that is provided by a major Internet provider to their clients in the United States (Figure 9.3). Looking at the main page of this device requires the user to understand the concept of the password for Wi-Fi and password to manage the device. Even if they are logged in to administer the device, they still need to have a good understanding of the terms. If users do not understand the meaning of the words or concepts used in the UI, how would they be able to manage security aspect of their devices?

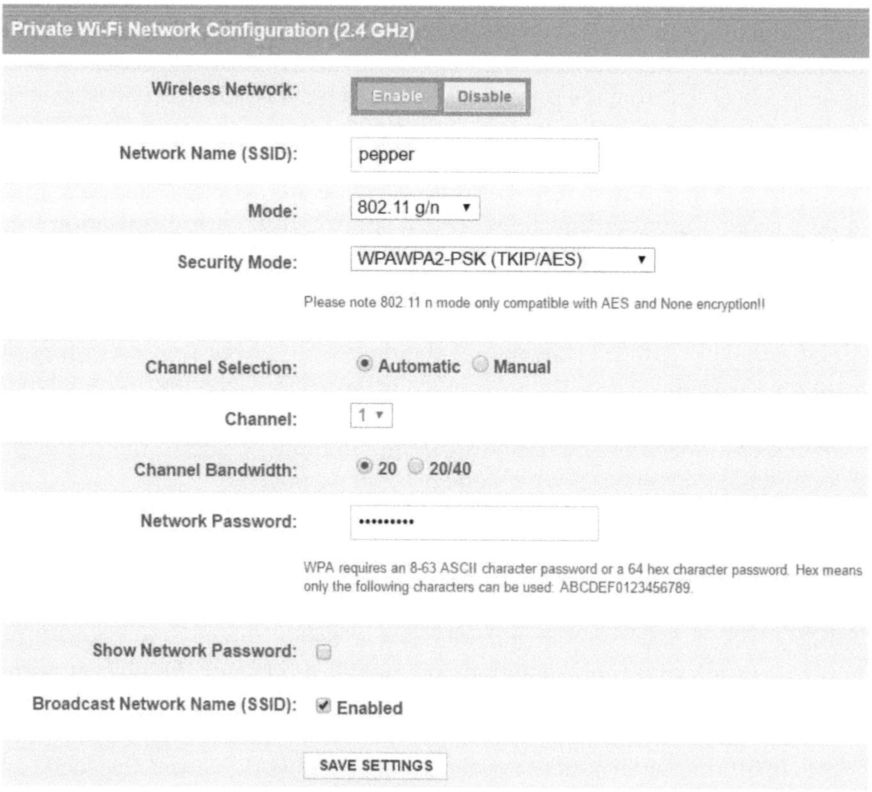

Private Wi-Fi Network Configuration (2.4 GHz)
Wireless Network: Enable Disable
Network Name (SSID): pepper
Mode: 802.11 g/n
Security Mode: WPAWPA2-PSK (TKIP/AES)
Please note 802.11 n mode only compatible with AES and None encryption!!
Channel Selection: Automatic Manual
Channel: 1
Channel Bandwidth: 20 20/40
Network Password: •••••••••
WPA requires an 8-63 ASCII character password or a 64 hex character password. Hex means only the following characters can be used: ABCDEF0123456789.
Show Network Password:
Broadcast Network Name (SSID): Enabled
SAVE SETTINGS

Figure 9.3 UI of the router provided by the main ISP in the United States. Most people do not have a clear idea about the security mode and relationship with password and channel selection.

Figure 9.3 shows the UI of the router provided by Comcast, an ISP in the United States. Most people do not have a clear idea about the security mode and relationship between password, mode, and channel selection. All routers provide WPA2-phase-shift keying (PSK) [temporal key integrity protocol (TKIP)], WPA2-PSK [advanced encryption standard (AES)], and WPA2-PSK (TKIP/AES) as options. These modes or options encrypt a network with an encryption key, and each provides a weaker or stronger encryption. In fact, we have observed that many average users did not even understand the concept of encryption. Even if you search the Internet or read the device documentation, you can hardly see an easy explanation of these options, just an indication of which one is better to use. Therefore, people might randomly use one versus another or even go with an easy option without knowing the consequences.

Here is how Wikipedia defines WPA and WPA2:

> Wi-Fi Protected Access (WPA) and Wi-Fi Protected Access II (WPA2) are two security protocols and security certification programs developed by the Wi-Fi Alliance to secure wireless computer networks. The Alliance defined these in response to serious weaknesses researchers had found in the previous system, Wired Equivalent Privacy (WEP).

WPA (sometimes referred to as the draft IEEE 802.11i standard) became available in 2003. The Wi-Fi Alliance intended it as an intermediate measure in anticipation of the availability of the more secure and complex WPA2, which became available in 2004 and is common shorthand for the full IEEE 802.11i (or IEEE 802.11i-2004) standard.

As it can be observed from the preceding terminology, the definition or explanations are very complex while it could just have said "strong password" like a simple lock or a more secure lock for better secure system. This would probably be more understandable for people.

All Wi-Fi broadband routers communicate over specific wireless channels. Like channels on a television, a number that represents a specific radio communication frequency designates each Wi-Fi channel. This type of UI is more appropriate for the IT professional than an average home networking user. Although routers allow user to choose the "channel," the typical default channel setting is generally is set to "Auto," presuming that most users will be satisfied with the selection of the device. To check this, one might just ask people who have a router about the difference between "5 GHz" and "2.4 GHz," and it will quickly be noticed that people hardly know their meanings. Many of us have looked through the list of a dozen or so channels that the manual option of the router provides and wondered what they are and which one is faster, to no avail.

One of the fundamental issues in understanding the UI of the router by the users is the fact that they do not have a correct mental image of how the device operates.

9.2.3.3 Home networking and user mental image

Mental models reflect how people understand a specific knowledge domain (Norman 1983). People form internal representations or mental models of the objects with which they interact with to create predictive and explanatory powers for understanding the interaction (Norman 1983). The mental models play a central and unifying role in representing objects, states of affairs, sequences of events, the way the world is, and the social and psychological actions of daily life (Johnson-Laird 1983). Norman (1983) supports the concept that mental models are based on the way people understand a specific knowledge domain. Staggers and Norcio (1993) perform an action, having an accurate mental image of how systems or devices are working is essential. However, when the mental image of a system or device is incorrect, it is referred to as a "folk model." Folk models are mental models that are not sufficiently accurate in the real world, leading people to incorrect decisions when their decisions are based on this type of conceptualized image. It has been observed that in technological contexts, users often operate with folk models rather than with a correct mental model of the system or device. If technology is designed on the assumption of a different mental model from that of the user, the desired behavior will also be different from what is expected when making decisions (Athhavanka 1997).

Wash (2010) conducted a qualitative study to understand users' mental models of attackers and security technologies. In this study, the researcher investigated the existence of folk models held by home computer users.

The fact that most user do not know that their router has a UI that they need to log in to manage the configurations indicate this mismatch from user mental image and how the device operates.

9.2.3.4 Home networking devices and firmware

The increasing number of devices connected to a home network system creates additional endpoints, which adds more complexity to protecting security vulnerabilities. With some embedded systems on the rise, cyberattacks that focus on firmware rather than application or operating system levels are also on the rise. Choi et al. (2016) summarizes several types of firmware attacks as follows and suggests solutions such as firmware validation and update schemes.

- *USB firmware*: A USB memory stick is plugged into the system. The compromised firmware presents itself as a normal mass storage device but shortly installs a Trojan from the web to the compromised firmware, which is able to send system commands to manipulate the system.
- *Network interface card (NIC) firmware*: The firmware of an NIC is modified to control packets.
- *Hard drive firmware*: A hacking tool designed to reflash the hard disk firmware of a system with malicious code. It obtains the malicious code from a command server and then flashes the obtained code to the existing firmware, i.e., replacing the current firmware with a malicious one. The boot process of the system is also able to capture the disk encryption password or other passwords at the operating system level.
- *Battery firmware*: A battery contains an embedded microprocessor just as USB devices, NICs, and hard drives do. By compromising the battery firmware, it is possible to overheat the battery and even cause the battery to catch on fire.
- *Printer firmware*: A firmware modification which attacks a printer. This vulnerability allows the arbitrary injection of a developed malware into the printer firmware. There are multiple ways that the can be used to access the computer connected to home networking or alter and reroute print jobs, open saved copies of documents, or reset the printer to its factory defaults, or wiping out all users' settings. This can happen by hacking into a network printer without security features. However, if the printer is accessible through the Internet, there is a variety of ways to hack into home networks. If the printer is not password protected, the task of the hacker is even much easier, although even a password-protected one will not stop hackers. One of the most efficient ways to avoid being hacked would be to acquire printers that support encrypted connections to and from personal computers (PCs) on the network and get rid of older printers that do not have security features (Geier 2012).

9.2.3.5 Internet of things

Gartner, Inc. (2015) forecasts that 20.4 billion connected things will be in use worldwide in 2020, and the total spending on endpoints and services (an endpoint device is an Internet-capable computer hardware device on a transmission control protocol/IP network) will reach almost $2 trillion in 2017. This development makes managing home networking even more fragile from a security perspective.

Botnets are a collection of Internet-connected devices, which may include PCs, servers, mobile devices, and Internet of things (IoT) devices that are infected and controlled by a common type of malware. Users are often unaware of a botnet infecting their system. A botnet might run one or more bots. Botnets can be used to perform a DDoS attack, steal data, send spam, and allow the attacker access to the device and its connection.

The Internet bot [a software application that runs automated tasks (scripts) over the Internet.], Kelihos, and Asprox botnet, created on connected devices (even appliances such as refrigerators) can, for example, start a spam e-mail attack. TVs with built-in cameras and microphones pose another attractive category of targets, as do other previously innocuous household devices. The possibilities for IoT attacks are truly endless, but ultimately, such attacks are likely to be about money/profit.

There is a need to look into understanding how some of the vulnerabilities present them for the average household should a hacker attempt to attack.

Households will have to improve their ability to detect and correlate attack activity to respond to increasingly sophisticated threats that accompany these high-growth technologies. Security experts state that no single product or vendor can cover every possible threat angle (Huffington Post 2013). We must look into understanding how our worldview of looking into cybersecurity must be in a systematic framework that helps homeowners control access from physical device points such as smart TVs, smart cars, Wi-Fi-connected washing machines to digital spaces such as social media and bank accounts. This requires the user to have a level of technical know-how in networking and security, which would be unrealistic to enforce and maintain, given the rapid changes and advancement of technologies. The combination of this variety of factors can lead to the vulnerability of home networking to an array of attacks such as ransomware (Kaspersky Lab 2016).

9.3 Solutions and challenges

9.3.1 Home networking and privacy

Home and home-based small business networks are used by many users to connect their computers; cell phones; tablets; music storage devices; photo storage devices; and appliances such as games, smart TVs, and IP cameras to the Internet. Obviously, these devices contain the digital heart and life of the household. If a small home business is also being operated within this framework, the digital life includes, but is not limited, to all legal, tax, and property documents, pictures, and videos. Consequently, if a hacker could access these digital assets, it would be a big loss for every single member of that household. With the expansion of ransomware and identity theft, every single unprotected house is a favorable target for criminals. Making the target safer is the job of the home network manager. Sometimes professional services working for different legitimate businesses, such as law offices, private investigators, or even sometimes law enforcement agencies are also intruders (Goldstein 2015).

The only rational solution to make home networking more secure is to provide easy-to-use software that makes the task of managing the network for all level of home users accessible. In this way, user will be able to secure their networks easily with an easy-to-use UI. Until this happens, which may be a while away, home users need to learn how to and make themselves responsible for securing their network. There are several main steps that users of the household need to learn and understand. These include viewing and changing the password for wireless connection, turning on and off the guest network or guest network password, and setting parental controls to view a router admin UI; there is no need to have an Internet connection since users log in to the router software. After logging in to the router, the user can make all configurations or needed changes.

The first step is upon arriving at the admin UI login page, one should find a default username and password to log in. The default username and password are presented on the label of the router box or other documentation. Sometimes, even a Google search of the router model and brand can turn up the username of each brand. These days the admin password may not be an easy set password such as "password" but another configuration of letters and numerical characters. Upon successful login, three parameters need to be changed to secure home or small business networking:

- Change IP address to log in to the admin UI from default to user-selected IP address.
- Change admin UI password to a difficult long password to prevent somebody else using the default IP and password to log in.
- If possible, change username.

Following these changes, the user should set a user-friendly wireless network name and configure a reliable security password for connection.

One of the security measures for a small home business networking should be to separate the business user's network and pertinent computers from the Internet connection utilized by guests. Creating a guest network separates guest connection to the Internet from the user's user home computers.

Educating all family members about possible home network vulnerabilities and cyberattacks is another good measure to take. This is especially important for families with children who may have friends over. Children these days may be more skilled with computers than their parents or simply have the curiosity to explore other computers in a network, which can consequently lead to the use of the protection of a guest network to access private files on a computer on the network.

Another important measure to protect home networking is regularly changing the Wi-Fi password about every 3–6 months.

It is important that all home networking users shutdown their laptop and desktop computers when not in use. Otherwise, ransomware attacks could encrypt all connected device data including the backup on external hard drives.

Many routers have a USB port that allows users to utilize its internal storage device or simply share pictures or music through the router. It is recommended not to have the storage device be connected at all times to the Internet. The user should disconnect the storage device after back up. This would prevent users from becoming a victim of ransomware.

A virtual private network should always be used when working from home, even if using a "secure" work laptop. Also, a USB drive or any media storage device found from the street should never be used, especially on work computers.

The router should be checked for having a file transfer protocol (FTP) server. If one exists, it should be secured with a password or shut down.

The firmware of the router, the operating system on cellular devices, and the operating system on computers should all be updated from time to time or as soon as these updates are available. The firmware updates often include the security patches that would protect the device with the new known issues.

Besides these minimum security settings, the rest of security might be more complicated, particularly in the network with several different devices such as smart TV, gaming devices, smart plugs, and IP cameras.

The other fundamental way of protecting the home network is setting a firewall. A firewall is a barrier that controls data types in and out of your network. It can prevent spam sites from installing unwanted programs on your computer, protect your personal information from theft, and much more. Consequently, setting firewalls to maximum security would be another good security measure.

Managing the access of a specific site can be a particularly beneficial security measure in a home with children. This may include blocking certain specific URLs or sites by setting keywords. This can be achieved by creating a black list and white list.

Lastly, using DNS services such as OpenDNS security services can filter unwanted content and unsafe phishing fraudulence websites.

9.4 Conclusion

The home networking router works as the front door to your digital data. An unlocked door makes everyone quite vulnerable to all levels of intruders. Putting on a reliable lock may not completely protect a family's digital assets, but at least, it will lock the system to most hackers. All research shows that unfortunately, due to the complexity of protecting systems, most home networks are very vulnerable. Security can be achieved first by making the settings and configurations of the home networking device more user friendly and intuitive to user needs. Secondly, there needs to be an increase in user awareness on security, along with basic trainings to how to properly manage a home network, at the very least to help users protect their system by enabling basic security settings.

References

Athhavankar U.A. (1997): Mental imagery as design *tool, An International Journal Cybernetics and Systems*, Volume 28, Issue 1, Pages 25–42.

Barker I. (2014, May 11): Badly secured routers leave 79 percent of US home networks at risk of attack. *Betanews*. http://betanews.com/2014/11/05/badly-secured-routers-leave-79-percent-of-us-home-networks-at-risk-of-attack/#comment-1677890437.

Cheng J. and Kunz Th. (2009, September): *A Survey on Smart Home Networking Inside the Smart Home*. Springer, Berlin. Carleton University, Systems, and Computer Engineering, Technical Report SCE-09-10.

Choi B.-C., Lee S.-H, Na J.-C., and Lee J.-H. (2016, February): Secure firmware validation and update for consumer devices in home networking, *IEEE Transactions on Consumer Electronics*, Volume: 62, Issue 1, Pages 39–44.

Cisco (Accessed 2017): *Service Provider Security*. Cisco, San Jose, CA. http://www.cisco.com/c/en/us/about/security-center/service-provider-infrastructure-security.html.

Demiris G., Rantz M., Aud M.D., Marek K.D., Tyrer H., Skubic M., and Hussam A. (2004, June): Older adults' attitudes towards and perceptions of "smart home" technologies: A pilot study, *Medical Informatics and the Internet in Medicine*, Volume 29, Issue 2, Pages 87–94.

Denning T., Kohno T., and Levy H.M. (2013): Computer security and the modern home. *Communications of the ACM*, Volume 56.1, Pages 94–103.

Dourish P., Grinter R.E., Delgado dela Flor, J., and Joseph M. (2004, November): Security in the wild: User strategies for managing security as an everyday, practical problem, *Personal and Ubiquitous Computing*, Volume 8, Issue 6, Pages 391–401.

Edwards W.K. (2011, June): Before building the network or its components, first understand the home and the behavior of its human inhabitants, *Communications of the ACM*, Volume 54, Issue 6, pp. 62–71.

Feamster N. (2010, September 3): Outsourcing home network security, *HomeNets 2010, New Delhi.*

Gartner (2015): Gartner Says 6.4 Billion Connected "Things" Will Be in Use in 2016, Up 30 Percent from 2015. Gartner, Stamford, CT. http://www.gartner.com/newsroom/id/3165317.

Grinter R.E., Ducheneaut N., Edwards W.K., and Newman M. (2005): *The Work to Make the Home Network Work*. Computer Supported Cooperative Work (CSCW), Springer/Kluwer, Paris, 469–488.

Geier E. (2012, April 25): *Your Printer Could Be a Security Sore Spot*. PCWorld, London. https://www.pcworld.com/article/254518/your_printer_could_be_a_security_sore_spot.html.

Goldstein M. (2015, March 6): Investigator admits guilt in hiring of a hacker. *New York Times*. https://www.nytimes.com/2015/03/07/business/dealbook/a-guilty-plea-in-a-hacker-for-hire-case.html?mcubz=3.

Huffington Post (2013, June 22): The Internet of things: By 2020, you'll own 50 Internet-connected devices. http://www.huffingtonpost.com/2013/04/22/internet-of-things_n_3130340.html.

Internet Live Stats (2016): Accessed on November 21, 2016, http://www.internetlivestats.com/.

Johnson-Laird P.N. (1983): *Mental Models*. Harvard University Press, Cambridge, MA.

Kaspersky Lab (2016, June 22): The Evolution of the Threat and Its Future. Kaspersky Lab, Moscow. Accessed November 21, 2016. https://securelist.com/analysis/publications/75145/pc-ransomware-in-2014-2016/.

McMilllan R. (2014): Turns out your complex passwords aren't that much safer. *Wired*. https://www.wired.com/2014/08/passwords_microsoft/.

Moallem A. (2012): *Why Should Home Networking be Complicated?* Taylor & Francis Ltd, Abingdon.

Moallem A. (2014): Home networking: Smart but complicated. In *Human–Computer Interaction*, Kurosu M. (ed.), Part III, HCII 2014, LNCS 8512, pp. 731–741. Springer International Publishing, Cham.

Newman L. (2017, October 15): The "secure" Wi-Fi standard has a huge, dangerous flaw. *Wired*. https://www.wired.com/story/krack-wi-fi-wpa2-vulnerability/.

Norman D. (1983): Some observations on mental models. In *Mental Models*, Gentner D. and Stevens A.L. (eds). Psychology Press, London UK.

Phadnis S. (2016, August 19): Households Have Ten Connected Devices Now, Will Rise to 50 by 2020. ETCIO.Com, Noida. https://cio.economictimes.indiatimes.com/news/internet-of-things/households-have-10-connected-devices-now-will-rise-to-50-by-2020/53765773.

Pew Research Center (2017): *Internet/Broadband Fact Sheet*. Pew Researc Center, Washington, DC. http://www.pewinternet.org/fact-sheet/internet-broadband/.

Staggers A., and Norcio A.F. (1993, April): Mental models: Concepts for human-computer interaction research, *International Journal of Man-Machine Studies*, Volume 38, Issue 4, Pages 587–605. http://userpages.umbc.edu/~norcio/papers/1993/Staggers-MM-IJMMS.pdf.

TMC News (2011, August 31): Cisco home networking issues consumer survey [professional services close-up]. http://www.tmcnet.com/usubmit/2011/08/31/5741669.htm.

Wash R. (2010): Folk models of home computer security, *Proceedings of Symposium on Usable Privacy and Security (SOUPS) 2010, July 14–16, 2010, Redmond, WA*. https://cups.cs.cmu.edu/soups/2010/proceedings/a11_Walsh.pdf.

Yang J., Edwards W.K., and Haslem D. (2010): Eden: Supporting home network management through interactive visual tools, *UIST'10, October 3–6, 2010, New York.*

chapter ten

Trusted IoT in ambient assisted living scenarios

Elias Z. Tragos, Alexandros Fragkiadakis, Aqeel Kazmi, and Martin Serrano

Contents

10.1 Introduction

Throughout the world and particularly in Europe, the average life expectancy of the population has dramatically increased in the last few decades [1]. European citizens are now living longer than ever before. According to the European Commission, it is expected that the percentage of older adults (65+ years old) will increase from 18% to 28% of the European Union (EU) population by the year 2060 [2]. This will indeed have a strong impact on the health and social care systems, as nations must find efficient ways to provide services to an increasing population of older people. The latter can be realized with the exploitation of new information and communication technologies (ICTs) such as the Internet of things (IoT).

The IoT has emerged over the last decade as a promising set of enabling technologies that will support the concept of the hyperconnected society. In our future, we now expect that everything around humans will be connected to the Internet, forming an enormous global network of digital, virtual, physical, and cyberobjects, with the goal of improving the peoples' quality of life [3,4]. IoT is providing unprecedented opportunities for businesses to improve their products and develop new "smarter" products and services in domains such as manufacturing, environment, agriculture, maritime, food, city services, home automation, and health [5]. The foundations of an IoT system are the physical devices that are either sensors, spread around an area gathering information about physical entities (such as the environment, the air, the users, the doors, and any other real-world objects), or actuators that are controlling physical entities with actions (such as closing a window, switching on the radiator, raising alarms, and moving objects). One main objective of IoT is to interconnect billions of devices in an efficient, scalable, and secure way so that they are able to provide the requested services with the highest quality of service [3,4].

Until recently, IoT systems were only providing sensing services, while the actuation was mostly performed with manual interaction from the users. However, lately, there are many attempts toward automating the actuation procedures, giving the devices the ability to learn and act themselves in an autonomous way. Despite the many benefits that automation provides, it also raises significant concerns with respect to the security and safety of those systems, especially in applications that directly involve human users [6].

The term *ambient assisted living* (AAL) has been extensively used lately with various similar (but many times ambiguous) meanings. Here, we adopt the definition given by IGI Global with regard to independent living: "It's a relatively new ICT trend to embed intelligent objects in the environment to support people (mostly elderly) in living independently and monitored" [7]. AAL applications can be considered as an integration of e-health and home automation applications, with the goal of supporting the independent everyday life of disabled people, patients, and the elderly. AAL applications support the execution of advanced user services for monitoring the health and vital signs of a user. They incorporate various sensors for temperature and air quality, and possibly cameras. Wearable on-body sensors can be integrated with various sensors in the surroundings of the user to monitor the ambience, the air quality, the room, and the appliances that are in the same area as the user. Moreover, actuators can be used for assisting users to open doors, set the temperature of the room, open windows for better air ventilation, inform the user to take their medication, move objects to avoid obstructing the movement of the user, and other similar activities [8,9].

Consequently, it is evident that AAL applications can greatly benefit from IoT technologies since they are the enabling technologies that support the gathering, analysis, and exploitation of measurements from the devices. They also allow the execution of smart applications for acting upon the physical environment using the actuators. The fact that IoT merges the digital with the physical worlds, allowing remote control of physical entities, raises concerns with respect to the safety of such AAL applications. Additionally, the fact that AAL applications may gather sensitive user information in centralized servers raises concerns for the privacy of the users of such systems, especially now with the effect of the EU General Data Protection Regulation (GDPR) [16].

We will now present a basic example as the basis of the discussions of the rest of the chapter. We will refer to this as the Basic_Example. Following the concept described by Neisse et al. [10], AAL systems need to have strong security and trust systems, which should also be at the same time dynamic and adaptive to the context of the situations. Imagine a situation when an AAL system is installed in a household, restricting access to the front door only to inhabitants. When there is an emergency at home, with an inhabitant having a heart attack not being able to react, the AAL system will be able to correctly identify the situation (using the trust mechanisms to ensure the reliability of the data of various sensors that indicate the emergency). Then, since the context of the home situation will be changed to "health emergency," the access policies for the front door can change so that the police and the doctors or the nurses (with the necessary identification that the trust manager can use to identify that they are indeed members of the emergency response teams) will be allowed to enter, while otherwise, they would not be able to get in the house and help the user. This example shows the importance of trust in AAL systems, because all actions have to be certified that they are "trusted." In this scenario, an unknown user will not be allowed access since he/she is not considered as trusted. Similarly, when the situation is not an emergency, the police and the doctors will not be allowed to enter the house, even if they are trusted users. Nevertheless, for an AAL system to maintain its high levels of security and trustworthiness, it has to be first protected against a number of risks and attacks [11].

In this chapter, we analyze the risks of IoT-based AAL applications, focusing on the elements that need to be protected, the types of attacks that can be launched by malicious attackers, and how such attacks may be mitigated. Additionally, a proposal for a distributed framework for improving the trustworthiness of IoT-based AAL systems using blockchain technology is presented.

10.2 Security, privacy, and trust issues in IoT-based AAL scenarios

10.2.1 Overview and methodology

AAL applications using IoT technologies are becoming mainstream lately due to the inherent advantages they can provide for remotely monitoring the health of patients in prehospital or posthospital scenarios. However, as with all systems that are based on information technologies, the AAL applications do not come without security, privacy and trust issues, and threats. AAL applications are vulnerable to many threats and attacks that can be potentially very harmful for the end users. For example, malfunctioning or hacked patient-monitoring devices may not be able to signal alarms for the health of the patient or may disclose false information to the doctors. Since AAL applications are critical applications that have immediate impact on the health of humans, they must be carefully designed, tested, and evaluated through a rigorous process. Additionally, IoT-based AAL applications must be protected against a number of IoT-originating attacks, and they have to follow the recommendations of well-acknowledged projects and initiatives (i.e., IoT European Research Cluster [12], IoT architecture (IoT-A) [13], RERUM [14]).

In this section, we present an overview of the security, privacy, and trust issues in IoT-based AAL scenarios, building on a thorough vulnerability analysis of AAL applications. It is not the goal of the chapter to analyze the vulnerabilities of specific commercial products; thus, the analysis here will be more generic, aiming to provide an overview of the threats and assets of AAL applications.

The first step toward a vulnerability and threat analysis is to identify which approach will be followed. The most commonly used approaches are analyses for confidentiality, integrity, and availability and authentication, authorization, and accounting (CIA/AAA) or the STRIDE/DREAD analysis that originated from Microsoft. In this chapter, we will follow the STRIDE/DREAD methodology [15] as it was adapted for IoT by the IoT-A project [13] in combination with the methodology used in RERUM [16].

The STRIDE methodology splits the threats into six major categories: (i) spoofing identity, (ii) tampering with data, (iii) repudiation, (iv) information disclosure, (v) denial of service, and (vi) elevation of privilege. After the analysis of the threats, the assessment of the risks is usually done, following the DREAD methodology, which helps evaluate the criticality of an identified threat. Each risk/threat is evaluated against (i) damage potential, (ii) reproducibility, (iii) exploitability, (iv) affected users, and (v) discoverability. For the DREAD methodology, usually, each risk/threat is being rated against the preceding five metrics, and the ratings can be either numbers (1–5) or low, medium, and high [15].

10.2.2 Risk sources

To conduct a threat analysis, the identification of the risk sources and the assets or elements that should be protected is the mandatory initial step. In IoT-based AAL scenarios, the risk sources for potential attacks or threats in the system can be originating from either humans or other phenomena. Human-based threats can be either malicious or due to faults, but from a security point of view, what matters is the result of the attacks on the AAL applications and users. Human-originated risk sources can be related to stealing information, hacking devices, or identities device loss, accidents, and errors. Nonhuman risk sources can be related to natural phenomena such as lightning, fire, heat, or device failures [15,16].

In AAL applications, human-based threats can be related to attackers that want to take control of AAL platforms, sensing devices, and communication channels. Malicious users may be stealing information regarding the health status of the patients that are being monitored, intercepting the measurements and identifying when the user is at home or is absent. This information can be stolen in multiple ways, such as monitoring the wireless channel and intercepting packets that are being transmitted, hacking applications, and accessing databases or hacking devices themselves.

A critical issue in AAL scenarios is also the potential of attackers to access actuator devices that are used for acting on the physical environment, simplifying the everyday activities of the elderly. This means that using some specific rules, doors and windows can open automatically (when the user is nearby), the air-conditioning will set the correct temperature by combining information for the outdoor temperature and the user preferences, and the alarms will notify the user when he/she has some health condition and should take a pill (for example, in low blood sugar or in high blood pressure). It is evident that malicious users that take control of such devices can create harmful effects on the human health.

Nonhuman threats can also have harmful effects on AAL applications, since they can affect the measurements and decisions of the overall system. For example, heat or fire can affect the measurements of health sensors and water and humidity can cause device malfunction, which in turn can affect the integrity of the measurements. Malfunctioning or hacked devices sending false measurements with respect to the health status of the users can decrease the reliability of the overall AAL platform, which in turn lowers the trust that humans put on the system.

10.2.3 Elements to be protected

In AAL scenarios, the elements to be protected via ICTs can be split into several categories and are a mixture between the physical and the digital world. Of course, the most important element is the *human user,* who is the end user of the AAL system that either monitors the user's health status or acts using actuators to provide everyday assistance. In the Basic_Example, the human user is the inhabitant of the house who is in either a critical or a normal state. Tampering with sensors or actuators can have devastating effects for the human, because sending false information and commands on an insecure system results in bad system decisions. In the Basic_Example, when the user is in a critical state, an attacker may stop the system from sending alarm to the ambulance or reject access to the house to the emergency teams. These attack scenarios are mainly considered as "safety" and not security, and there is a research domain that tackles functional safety and tries to identify solutions so that IoT or cyberphysical systems ensure the safety of the users. With the upcoming implementation of the EU GDPR [17], AAL systems and applications will have to comply with very strict privacy requirements, with respect to data gathering and processing. GDPR requires that the user's privacy has to be protected at the highest level. Users have to be informed about the type of data that are gathered and how these are processed and stored. All private user information, such as health status, phone number, name, social security number, and credit card number, have to be protected from unauthorized access, and the user should have full control over their usage.

The second physical element that has to be protected is the *IoT devices* that are sensors, actuators, or even gateways. In the Basic_Example, these devices can be the actuator controlling the door, a motion sensor to identify the user movement, or a wearable measuring the blood pressure or the pulse of the user. These devices are mainly the generators of AAL data and can include any type of sensors such as on-body sensors, ambient conditions sensors (temperature, humidity, and light levels), and air quality sensors that may affect the user (gases and dust). Gateways have to be protected too, since they are critical parts of the IoT network, gathering all data from the end sensors and sending them to the backbone servers and applications. Getting access to those devices will open up a Pandora's box for an AAL system. These devices are ICT devices, and their software, firmware, and applications and their communication with other devices or gateways have to be protected.

Another element to be protected are various types of *data* that can be application (sensing) data, actuation commands, or signaling data (for example, networking measurements for routing or channel assignment). Examples of data that have to be protected are user's heartbeat, blood pressure, temperature, blood sugar levels, status (walking, standing, and laying down), location, room conditions, air pollution, door/window status (open/closed), actuator's state, and actuator request and response.

Device and server *software* also has to be protected against tampering and hacking. This includes the operating system, the firmware, the drivers for the sensors/actuators, the implementation of the network stack, and any services that are running on devices or gateways. The software also includes AAL services and applications, such as end user applications (rule based for executing control loops as well as home automation), data collection, service discovery and lookup, identity management, as well as trust and reputation.

Authentication *credentials and user policies* are also critical for the secure operation of an AAL system. Credentials and policies are used by security and privacy mechanisms to identify users and grant them access to the AAL system. False or altered credentials and policies may provide access to malicious users, who can steal personal user information or send malicious commands to actuators. In the Basic_Example, the home user has the role of the owner and manager of the system and has access to everything. The AAL system should also have a role for the emergency response teams, which should be able to grant access to the front door "only in emergency situations." To be able to distinguish emergency response members, the AAL system, when sending an alarm to the emergency response teams, could also send an access code for the front door. In this respect, different response teams could also have different codes, which change dynamically.

The *wireless channel* in the IoT network has to be protected against a number of attacks, because common threats can affect the integrity and the confidentiality of the data. Eavesdroppers can monitor the wireless channel and intercept messages; masquerading attacks can allow the attacker to get access to sensitive information, and authentication; credentials and intermediate malicious nodes can alter measurements and actuation commands, i.e., opening doors and setting off alarms.

Finally, since IoT systems can have an impact on the physical environment, the actual physical entities, such as the house and the furniture have to be protected from the actions of the actuators. Toward this, examples of malicious commands to actuators for, e.g., flooding the apartment (in case of a false fire event) or overheating appliances and causing explosions, have to be avoided.

10.2.4 *Attacks against AAL elements*

The STRIDE analysis can be used to identify potential attacks against elements that need to be protected in an AAL system (see also Table 10.1). For the *human user* in the Basic_Example (discussed in the introduction), attackers *tampering* with the monitored user's data can alter them, sending false measurements to the AAL system, so that emergencies are not identified or false health status is noticed and wrong medical treatment is provided. *Falsifying* the user's health data can have devastating effects on the safety of a user, which is of utmost importance for AAL systems. Similar results can take place when there is a *denial of service attack* on an AAL system, so that services that are critical for the user's safety are disabled.

Attacking IoT devices is relatively easy because most of the current commercial products have very limited on-board security features. Especially due to the fact that these devices are lightweight, they can be an easy target for hacking, blocking, or altering their communication, changing their configuration, or sending them false commands. Apart from that, *physical attacks* such as destruction, theft, or reprogramming/controlling the devices through their universal serial bus interfaces are usual types of attacks [18]. *Masquerading* attacks on devices is also quite common in IoT or wireless sensor networks. In this attack, an attacker plays the role of a "gateway" so that all measurements go to

Table 10.1 Overview of attacks in IoT-based AAL systems

Element to be protected	Attacks	Impact	Mitigation procedure
Human user	Data tampering, repudiation, and user safety	User's health and private data	Access control, safety procedures, and user trustworthiness
IoT devices (sensors, actuators, and gateways)	Denial of service, data manipulation, device hacking, and spoofing identity	Access to sensitive information, access to the IoT system, and sending false information	Encryption, access control, physical protection, and device trustworthiness
Data	Data tampering and eavesdropping	Unauthorized access to data and breach of privacy	Encryption, integrity protection, and service trustworthiness
Software	Data tampering, device hacking, and denial of service	Loss of data and no access to services	Encryption, access control, and security management
Credentials	Elevation of privilege and identity spoofing	Breach of privacy and unauthorized access to the IoT services	Strong access control and identity management
Wireless channel	Eavesdropping, masquerade attacks, and man-in-the-middle attack	Loss of data, loss of communications, and breach of privacy	Communication encryption, integrity protection, and antijamming

his/her device instead of the proper AAL gateway. These attacks can be quite severe, resulting in the loss of communication so that AAL devices cannot send crucial monitoring data (or send false data) to the AAL application. Moreover, sending false commands to actuators can also have an effect on the physical environment. For example, opening doors/windows to intruders or even closing doors when the user is passing creating physical harm to the user.

Attacks against various types of data are also very common. Especially in an AAL system where measurements from various sensors are mostly sensitive carrying critical health information, attacks, such as tampering with data, false measurements reporting, or denial of service, can raise issues with regard to the trust of the overall system. Loss of the *integrity* of the data in an AAL system can occur both when the data are at rest (stored on a device) or in transit (when they are exchanged). When at rest, an attacker can modify the data by launching a malicious code on the device or by gaining remote control. When in transit, an attacker can modify the data by attacking the networking infrastructure using a man-in-the-middle attack. Loss of data *availability* can also take place either at rest or in transit. In the former case, an attacker might delete the data stored on a device, while in the latter case, an attacker can perform radio jamming, denial of service attacks, or sinkhole attacks. Similar attacks can also target command and control data, such as routing or wireless channel assignment data, which can have an effect on the overall performance of the system and the communication between the devices and the applications.

The *software* of both devices and the backbone servers is also a point of attack in AAL systems. Services and applications for monitoring the health of the users and acting for either assisting them in their everyday activities or notifying their doctors in cases of emergency can be attacked by malicious users so that they gain access to unauthorized data, perform denial of service, change measurements and data in databases, or impersonate services and applications. This will result in hacked user accounts, devices becoming unable to communicate with applications, or users denied access to their own applications.

Loss of *confidentiality of credentials and policies* can take place when an attacker specifically gets access to the security servers of an AAL system or if he/she spoofs the identity of devices and gains access to the applications. The access to credentials or policies can be achieved by executing malicious code on a gateway/server or at a device or by eavesdropping and performing a man-in-the-middle attack. In the Basic_Example, when an attacker gets credentials and access policies, then he/she can launch more attacks by getting unauthorized access to the servers, applications, and user data, not allowing the alarms to be sent, sending false alarms or even taking control of devices and harming the user with injection of medicines at wrong times.

Most of the preceding attacks take place when an attacker tampers with the communication channel. In most AAL scenarios, the communication channel is wireless, and it is easy for an attacker to perform man-in-the-middle, eavesdropping, or jamming attacks [19]. That way, the attacker can listen to transmissions gaining unauthorized access to data, can perform identity spoofing and pretend to be an authorized device/user or can disrupt the communication by degrading the link quality of the wireless connection between the devices and the gateways. Considering an AAL system in a home, if the communication link is not secured, a neighbor who is listening to the wireless transmissions of the devices will be able to identify the measurements and even pretend to be an authorized user and send commands to the actuators. Moreover, even if the transmissions are encrypted and the attacker is not able to extract the content of the transmissions, just the simple monitoring of the transmissions allows the extraction of information about which devices are operating, when the user is at home, or when there are emergencies.

10.2.5 Assessment of the identified risks

After analyzing what the risks against critical elements of an AAL system are, the important step is to see what the impact of these risks on such a system is and what can be done to mitigate this impact. The attacks against a human user have a high damage potential, cannot be easily reproduced, and can be very exploitable (even with financial gains for the attacker). Moreover, these AAL attacks are mostly targeted per person, so they cannot affect many users and they cannot be discovered very easily. These attacks can be mitigated with a security, privacy, and trust framework that employs strong security with cryptographic protocol, access control, and privacy-enhancing techniques. Such a system is also critical for emergency situations, as the one described in the Basic_Example. The main target will be to minimize the possibility for an attacker to gain access to the private information of the user or to send any type of malicious commands to the actuators that can harm the human or his/her surroundings.

Attacks against an IoT device may also cause significant damage, especially if the device has stored private user information or credentials that would provide access to the system to an attacker, if the device is used to gather real-time health-related information or raise alarms for emergency situations. These attacks cannot be easily reproduced or discovered, but can be exploitable by the attacker. They can be sometimes avoided with the physical protection of devices, such as installing the devices in unreachable locations, hiding the devices, or covering them. Additionally, these attacks can be mitigated with secure storage functionalities to prevent an attacker to gain to the filesystem of the device to read the files. Additionally, proper authentication can contribute to avoid device identity spoofing and masquerading. Moreover, software security updates when any vulnerabilities are identified are very critical to ensure the proper protection of the device.

Attacks against data and services normally have high damage potential due to the fact that in AAL systems, the data gathered by the devices are sensitive user health data that can be exploited by malicious attackers for knowing the medical history and possible diseases of the user. These attacks can be easily reproduced when launched at both the devices at the local level and at backbone servers, targeting services and applications. They can be easily discovered when they target services that actively use and transmit user data, but cannot be easily discovered when they are passive and only capture user data. To avoid these types of attacks, proper security management with a strong authentication and authorization framework should exist, together with a trust management framework to identify the trustworthiness of users and the data that are gathered by the devices. Moreover, the encryption of data and secure communications are of high importance to minimize the possibility for an attacker to eavesdrop or to decrypt the data that it gathers from the devices. Additionally, accountability mechanisms should exist so that when such an attack is launched, the system should be able to identify and isolate the attacker. A proper identity management scheme will also provide secure identities to the devices and the components of the system, so that they cannot be replicated or stolen by an attacker to be used for impersonation.

Attacks against communication channels, especially in AAL scenarios where most of the communications are based on wireless links, can have severe effects on the performance of the system, on the privacy of the user data, and on the safety of the system. The attacks can be easily reproduced, regardless of the number of the devices or of the type of wireless links, and they cannot be easily discovered, especially considering the specificities of the eavesdropping attack. An attacker who listens to a wireless channel can only intercept personal data, but cannot affect the system performance or the trustworthiness of the system. However, wireless jamming can be quite harmful since it affects data availability, which can cause issues such as missed emergencies or actuating commands not received by the devices. These attacks can be mitigated by end-to-end integrity protection on the measurements from the devices, to increase their trustworthiness. Additionally, secure communications with data encryption to avoid eavesdropping and anonymous communications to protect user data can also be used, in combination with privacy-enhanced techniques for improving the privacy of the user data. Antijamming techniques with automatic channel or spectrum reassignment using cognitive radio devices can be used, so that denial of service attacks have minimum to zero effect on the system performance. Finally, authorization mechanisms on the devices can be used to ensure that only authorized users send actuation commands.

Most AAL systems, as also described earlier, are mostly centralized (i.e., Sánchez-Pi and Molina [20]) with all the decision-making processes handled by a central server. However, this creates new threats such as a single point of failure or a single point of attack, which means that if an attacker wants to exploit such a system, he/she can only target to hack or jam this central point. To avoid this, lately, there is a shift toward more decentralized or distributed systems, to allow the devices and gateways to take cooperative decisions and enable the distributed storage of data. This approach increases the scalability, the security, and the trustworthiness of the system, but only when strong and efficient trust management mechanisms are employed. In the following section, such a framework for the decentralized management of AAL systems using the new technology of blockchains is introduced, aiming to set the foundations for improving the trustworthiness of AAL systems.

10.3 Decentralized trust management for robust and secure AAL scenarios using blockchains

Trust management and trust computation (computation of nodes' trust based on metrics such as average packet drop ratio and forwarding delay) are not trivial issues in an IoT ecosystem for several reasons: (i) the presence of resource-constrained devices, (ii) the lack of standardization, (iii) the presence of heterogeneous devices, (iv) protocol inefficiencies, (v) the lack of interoperability, and (vi) an unattended operating environment. In the special case of the AAL, IoT devices perform several specialized operations such as monitoring, alerting, on-demand data provisioning, and actuating. Moreover, wireless sensors are mainly used in related scenarios as more flexibility is provided [13]. For these reasons, the IoT networks for such scenarios are susceptible to several attacks launched by adversaries with various motives, as also presented in detail in Section 10.2.4. Countermeasures against such attacks can include cryptographic means in several layers (e.g., symmetric-based encryption on advanced encryption standard). However, given the broadcast nature of the wireless medium, countermeasures such as these cannot protect against several types of attacks such as routing attacks (black hole, gray hole, selfish behavior, etc. [21]). For tackling these issues, several trust management and computation schemes have been proposed by the research community. The main idea is that all nodes observe their neighbors by collecting various pieces of information such as the packet drop rate and the packet modification rate [13]. Related research contributions (e.g., Fragkiadakis and Tragos [22]) combine physical-layer measurements such as the signal-to-interference-plus-noise ratio to adjust the reliability of an observation based on the amount of interference when the specific observation takes place. Based on nodes' observations, a level of trust computation is performed aiming to assign a trust value for each node. In general, trust-based models are classified into three categories [22]: (i) *centralized*, where all nodes send their evaluation reports to a single node that has more advanced capabilities (in terms of processing), and performs the fusion of the reports, inferring about a potential attacker; (ii) *distributed*, where each node fuses the individual reports and estimates the reputation of its neighbours; and (iii) *hybrid*, where large portions of the wireless network are split into multiple clusters such that the elected cluster heads are responsible for the fusion. Each model has its pros and cons, and the design choice depends on several factors such as: (i) the network size, (ii) the hardware capabilities of the nodes, and (iii) security mechanisms to be deployed. With centralized schemes, processing can be performed by advanced devices (servers and gateways), but these single points of failure exist that can make the network collapse in case of failures or attacks against these advanced devices. Distributed models assume that all processing regarding the trust management and computation are performed by the nodes exclusively, something that cannot be always feasible considering the resource-constrained nature of IoT nodes (sensors). The hybrid schemes try to compare the benefits of the centralized and distributed ones; however, the consensus algorithms used may not be robust in case of failures or deliberate attacks (e.g., Sybil and Byzantine attacks).

In this section, we investigate the feasibility of trustless distributed schemes in the IoT domain for defending against three types of attacks: (i) black hole routing attack, (ii) gray hole routing attack, and (iii) integrity attack. We will consider the use of the so-called *blockchain*, a distributed data structure replicated and shared among all nodes in a network. The blockchain is used in Bitcoin, the famous cryptocurrency [23], and consists of a series of interrelated blocks as shown in Figure 10.1.

Each block can have several fields, depending on the implementation:

- The previous hash that contains the hash value of the previous block.
- The list of transactions (Transactions []) that will be executed within this block. Usually, this list is organized as a Merkle tree that is a binary tree using hash pointers. A transaction is used to describe a specific type of operation based on a specific asset.
- The nonce that is a one-time random value is used as one of the hash function arguments.
- The hash function H() that receives data of arbitrary length and produces a fixed-size output. For the transactions' list case, the hash function is used multiple times, depending on the level of the Merkle tree. Initially (Level-0), and for each transaction, a hash value is computed taking as input the body of the transaction. Next, in Level-1, pairs of the hash values are formed, and their concatenated values are used as input to the hash function. This continues to all upper levels, until a single hash value is computed.
- The hash value of the block is computed using the data of the block as input (depending on the implementation). A typical input can include the concatenated values of the nonce, the mrkl_root hash value, and the hash value of the previous block.

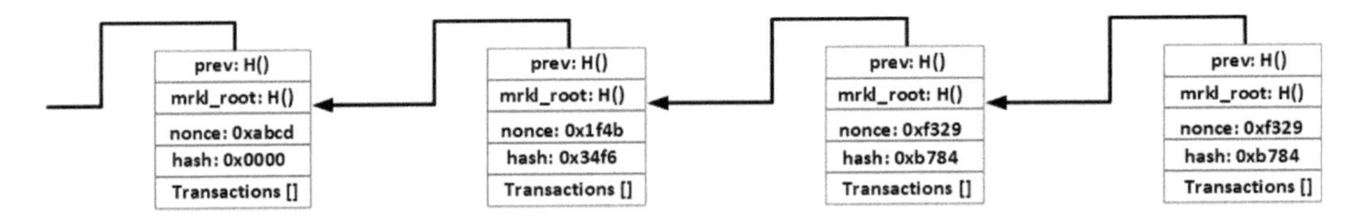

Figure 10.1 Blockchain structure.

The blockchain data structure is maintained by all nodes in the network. Its size is proportional to the number of blocks and the number of the transactions for each of the blocks. As this is a distributed system, the nodes are free to generate new transactions and blocks. However, certain rules have be and respected by all nodes; otherwise, chaos will rise:

- Strong cryptographic means are used, and each node that generates a transaction signs it with its private key; therefore, a public key cryptography system is required.
- Nodes forward only valid transactions that belong to other nodes. As each transaction is signed by its owner, it is easy to verify that a specific transaction belongs to a specific node by using its public key.
- Block creation can be potentially performed by any node; however, strict rules can apply on the requirements for building such blocks. When a new block is successfully added into the long-term blockchain, this is broadcasted to all nodes so as to refresh their local copies, and the miner who created this block is rewarded.

The question now is how robust this mechanism is against various types of attackers considering that no central authority exists. Suppose that an attacker (Bob) modifies a transaction created by Alice, while this transaction is on transit. Alice, who is a legitimate user, has properly signed this transaction with her private key. If Bob modifies this transaction, the modified transaction is not valid anymore, as he does not know Alice's private key, and hence, all other legitimate nodes will discard it. Therefore, the modified transaction will never reach the miners, so there are no chances for it to be included in a successful block. Moreover, by altering a single bit of a transaction, Bob must re-compute all hash values of the Merkle tree that is computationally difficult, with the difficulty increasing with the number of transactions.

In a different attack, Bob tries to drop or discard all packets created by Alice that carry valid transactions. This could happen if Bob acts as an intermediary in a wireless network, and instead of routing Alice's packets, he drops or discards them. This will not create any problems, given that the number of the legitimate nodes is high enough, so Alice's transactions can reach the miners as transactions are broadcasted to all users. Similarly, if Bob drops packets that notify about new valid blocks, the broadcast nature of the network guarantees that a fraction of the legitimate nodes will still be able to receive these packets.

In the special case of Bob being a miner, but still acting maliciously, he could manage to create a block in the long-term blockchain that contains an invalid transaction. Again, this attack cannot be finally successful because legitimate users will detect that the specific block contains an invalid transaction, and legitimate miners will continue the mining process based on the last valid block.

Next, we will discuss how the blockchain concept can protect an IoT network against black hole and gray hole routing attacks. Very often, IoT ecosystems are based on wireless sensor networks (WSNs) with many nodes (sensors) to provide data to backbone servers. The network topology can employ multihop links where data are forwarded from the source nodes to the ultimate destination (e.g., server) over multiple links. In this case, other nodes operate as routers by following an appropriate routing protocol (e.g., the IPv6 Routing Protocol for Low-Power and Lossy Networks (RPL) [24]). A malicious user can take advantage of this mechanism and do the following:

- Discard all packets he/she has to forward (black hole attack [21])
- Selectively drop packets based on several criteria such as type of information and data owner (gray hole attack [25])

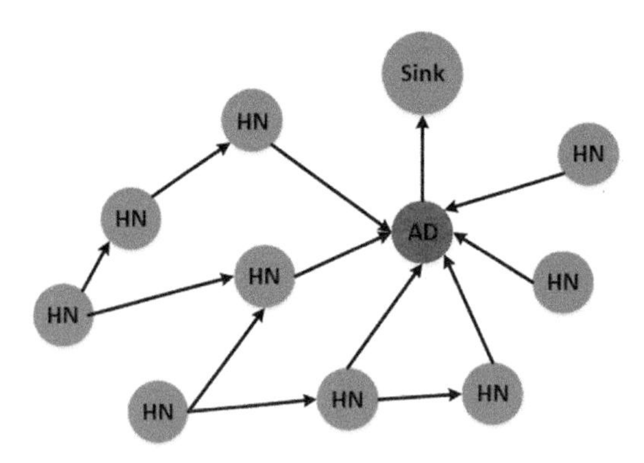

Figure 10.2 Successful sinkhole routing attack.

For detecting these types of attacks, several schemes (e.g., Zahariadis et al. [26] and Liu et al. [27]) have been proposed that attempt to compute a node's trust based on observations from other nodes. Therefore, a trust management and computation mechanism is required. Referring to the WSN shown in Figure 10.2, a schematic representation of a number of honest nodes (HNs), one adversary (AD), and a sink is displayed. The aim of the adversary is to drop either all or selective packets. To maximize the effect of his/her malicious action, and prior to packet dropping, he/she takes advantage of the routing protocol used and broadcasts an attractive routing metric (e.g., minimum number of hops from the sink and minimum delay) aiming to appear as the ideal neighbor to the rest of the nodes for forwarding their packets. This is usually referred as sinkhole attack [28], and if successful, the adversary can start dropping packets.

We now consider the use of blockchain in such a network. All nodes, based on the data they have to transmit, create transactions and sign them with their public key. All transactions are broadcasted into the network, and the miners (all nodes or a fraction of them) try to create a successful block. Despite the presence of the adversary who is dropping packets, at least one miner will manage to create a block and add it in the long-term blockchain.

If the adversary instead of dropping packets tries to modify their content, this will lead to invalid transactions as he/she is not aware of the public keys of the HNs. The invalid transactions will be rejected by all users, so they will not be included in their blockchain (each user maintains a copy of the blockchain).

10.4 Blockchain and smart contracts for AAL scenarios in smart building management

As described in the previous section, a blockchain, which is an immutable ledger, can be efficiently used to provide a decentralized trustworthy system with nontrusting peers. Along with the blockchain ledger, research groups around the globe work on *smart contracts* (SCs), which are scripts that are stored in the blockchain and used for multistep process automation [29]. An SC is a "computerized transaction protocol that executes the terms of a contract" and usually describes a list of conditions and the actions to be performed when one or more of such conditions occur. For example, if a user's device battery drops below a threshold, then do not use it in future blockchain transactions. Specific languages are used for the creation of the SCs, as for, example, the Solidity language used in Ethereum [30].

Next, we describe how SCs and blockchains can be used in smart building management (SBM) for AAL scenarios. SBM systems can provide efficient frameworks for data collection, monitoring, and actuation based on IoT architectures. SOrBet is an IoT architecture designed to securely interconnect smart devices that are equipped with intelligence, able to support AAL scenarios [31]. SOrBet functional architecture (Figure 2 in Tragos et al. [31]) consists of modules such as the service manager, communication manager, event manager, context manager, quality of service (QoS) manager, and security and trust manager. Given this architecture, SCs can be created by the communicating peers in several layers.

SCs are digitally signed and stored in the blockchain, and they execute automatically if certain conditions are met. For example, an SC can bind to the QoS and communication managers, stating that if a packet delay exceeds a threshold, then give priority to the specific flow. For this action, some reward can be defined through the exchange of digital assets (e.g., Ether or Bitcoin).

In another example, if strong privacy is required, then data must be encrypted before storing them to the blockchain. An SC can assign encryption duties to more advanced peers that will encrypt the data of less advanced ones (in terms of processing memory and storage), and they will be rewarded after encryption is completed.

As many AAL applications are heavily based on SBM systems, the use of SCs can in general automate several of their processes, enable the exchange of digital assets, and create a marketplace for the AAL ecosystem.

10.5 *Trust challenges in AAL systems: Testimonials from EU projects*

AAL applications have also been the focus of many EU research and innovation projects, aiming to improve the functionalities of the systems and the services they can provide to the end users. Since AAL systems also have inherent challenges with respect to security and privacy, most relevant EU projects are also considering components for improving the trustworthiness of AAL in their functional architectures. Since every project has its own objectives, we provide here example testimonials of trust challenges that two EU-funded research projects have identified with respect to their AAL systems and how they addressed these challenges.

10.5.1 *ACTIVAGE*

One part of the AAL community is the new active and healthy aging (AHA) community, which is wide and heterogeneous in terms of needs, demands, and living environments, which will use IoT-based services to address many of the challenges of everyday living of the elderly. ACTivating InnoVative IoT smart living environments for AGEing well (ACTIVAGE) is a European multicentric large-scale pilot on smart living environments [32], which aims to develop methodologies, while responding to the real needs of caregiver, services providers, and public authorities, to prolong and support the independent living of older adults in their preferred living environments. This will be achieved and validated through the real-world deployment of innovative and user-led large-scale pilots across nine IoT-enabled deployment sites, in seven European countries, involving up to 7000 users.

ACTIVAGE aims to build the first European AHA-IoT ecosystem, which is modeled as a technological infrastructure of hardware and software services and standard protocols and a constellation of stakeholders interacting with each other within a governance framework toward the achievement of common goals. ACTIVAGE will utilize IoT solutions through nine different use cases that address specific end user needs to improve their quality of life and autonomy. These use cases include daily activity monitoring, integrated care, monitoring assisted persons outside home, emergency triggering, exercise promotion, cognitive stimulation, prevention of social isolation, safety and comfort at home, and support for transportation and mobility.

In the conceptual model for AHA-IoT ecosystem, the components for trust, security, and privacy are core components. Data streams are the core asset of the ecosystem, which belongs to either private or public sources. As in all AAL applications, private data are produced by wearable and medical devices as well as smart sensors and devices in older adults' living environments. Public data, not necessarily linked to user interactions, are harnessed from public sources, including weather data, public transport timetables, and traffic situations. Both private and public data are processed at the edge and/or at cloud level. These data streams are then passed through different processes, such as anonymization, aggregation, and analysis that aim to increase the security and privacy of the overall system and thus its trustworthiness.

The sources of the ACTIVAGE AAL system introduce many issues related to the security of medical data and the protection of user privacy, which can be separated into several fundamental data management concepts, namely, trustworthiness of data sources, integrity of aggregated data, data privacy, anonymization of the data provider, location privacy, as well as the confidentiality of the network packets. As in many other areas, privacy and security are critical aspects of the general IoT environment, where multiple concerns are constantly being raised and compared to the privacy issues of traditional ICT systems.

ACTIVAGE proposes that AAL systems require inherent security, trustworthiness, and privacy by design. To this end, a modular framework will provide placeholders for incorporating security and privacy preserving algorithms, along with protocols ensuring that only trusted entities (i.e., "things") can become part of the deployment. The project suggests to investigate the economics of privacy and security in a cloud environment, with a view of associating them with the researched utility metrics of the cloud infrastructure. So privacy and security by design methodology and tools should be used for system design and risk management. The main novelty of ACTIVAGE in this area is the incorporation of utility-based schemes for negotiating and enforcing privacy and security.

10.5.2 SOrBet: Smart objects for intelligent building management

The SOrBet project [33] develops an IoT-enhanced intelligent building management system (BMS) based on the concept of reliability by design, and one of the main application scenarios considered is the provision of AAL applications. SOrBet built its system architecture considering various requirements that should be met to improve the reliability of AAL applications, especially in terms of security, privacy, and trust. SOrBet addresses the issues of security and privacy in AAL environments by considering heterogeneous wireless devices that are communicating in a trustworthy manner to ensure that only trusted devices are involved in the decision-making processes of the system and sensitive user data from the devices are sent only to authorized and trusted end users.

The requirements set by the project for the security, privacy, and trust of AAL systems are considered as key factors to ensure the overall reliability of the system. SOrBet considers that data encryption, strong authentication and access control framework, reputation management, and privacy-enhancing techniques are key requirements to protect AAL applications from attacks. In this respect, the SOrBet project has defined a dedicated functional

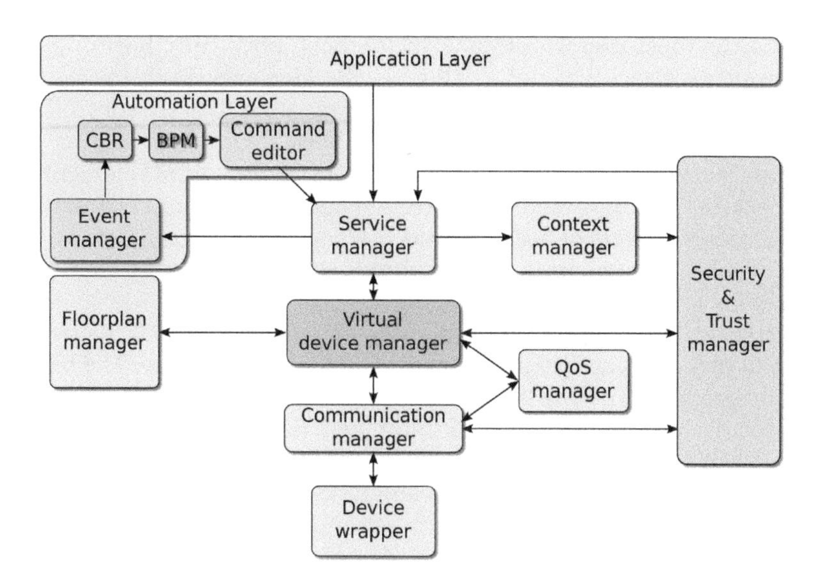

Figure 10.3 SOrBet functional architecture. (Reprinted from Tragos, Elias et al., D1.1: System requirements and architecture definition, SOrBet Deliverable D1.1, Dec. 2014.)

entity called "security and trust manager" in its functional architecture (Figure 10.3), which is responsible for (i) secure communications between devices and the system, (ii) the secure configuration of devices, (iii) the access control, and (iv) the trust management of the system.

It can be seen from Figure 10.3 that security and trust components are critical for AAL systems since they should be protecting all components: communications between the various sensors and devices, services that are provided by the system, the virtualization of the components, as well as the automation layer that handles the automated actions of the actuators, according to rules defined using case-based reasoning and business management processes.

It is important to note that as SOrBet proposes, the security and trust actions can be affected by the *context* of the situation. This can be of great importance to provide important assistance to people in need. In cases of nontrustworthy or insecure systems, a malicious user might be able to send false data to change the context of the system and then gain access to the house by tampering with the door access policies. Another scenario would be when a malicious user blocks the communication of the devices with the AAL application so that they cannot send the data about the emergency scenario and the emergency response teams are not notified about the user's health, which will have devastating results.

10.6 Conclusions and recommendations

AAL applications for supporting the aging and well-being of human users at their homes have attracted a high interest lately due to the explosion in the usage of IoT technologies. Although technology can greatly help monitor users' health status at home, it also introduces various risks for the security, privacy, and safety of users. In this chapter, building on the requirements and the testimonials of two EU-funded projects, we analyzed in detail the trustworthiness of AAL systems, focusing on the threats and risks of IoT-based AAL systems and discussing what has to be protected in these systems and how. As a summary of this discussion, we can briefly mention the fact that protecting users' safety is the most important goal. In the past, physically harming users using standard computer attacks was not at all possible. However, with the introduction of IoT technologies that allow the remote control of actuators that are controlling physical objects, the convergence of the cyberworld and physical world is creating new threats for the humans. Such safety risks can arise from attacks on various elements of an AAL system, which raises the importance of a strong trust management framework that will be combining strong aspects of security and privacy. Since the centralized control and management of such systems imposes significant security risks, we also discussed the use of blockchains for distributed management.

The use of blockchain can provide solutions to many security- and trust-related aspects of the IoT due to the inherent distributed nature and trust assurance. The blockchain can provide user and device trustworthiness ratings, so that only trusted users/devices are part of the system. Additionally, the decentralized nature of a blockhain-based AAL system will mitigate issues of a single point of failure and overloading. However, this new technology also introduces new issues and risks and has to be carefully adopted. A blockchain-based scheme requires nodes with advanced memory capabilities as the blockchain is replicated and maintained from all nodes. Blockchain is maintained by all nodes, so broadcast messages are transmitted by the nodes. If not carefully designed, packet collisions and network delay will decrease the performance of a network. Furthermore, extensive packet broadcasting can exhaust nodes with limited energy.

Overall, AAL systems are critical systems that are directly affecting the everyday lives of people in need. These systems are handling extremely private user information and should be designed to be reliable, secure, and trusted so that people can be reassured that they are safe to be used.

Acknowledgment

The authors acknowledge support from the ACTIVAGE project, cofunded by the European Union's Horizon 2020 research and innovation program under grant agreement number 732679. The research leading to these results has also received funding from the People Programme (Marie Curie Actions) of the European Union's Seventh Framework Programme FP7/2007-2013/under Research Executive Agency (REA) grant agreement number 612361.

References

1. He, Wan, Daniel Goodkind, and Paul R. Kowal. *An Aging World: 2015.* US Census Bureau, Suitland, MD, 2016.
2. European Commission. *The 2015 Ageing Report, Underlying Assumptions and Projection, Methodologies.* European Commission, Brussels, 2014.
3. Vermesan, Ovidiu, and Peter Friess, eds. *Building the Hyperconnected Society: Internet of Things Research and Innovation Value Chains, Ecosystems and Markets.* Vol. 43. River Publishers, Delft, 2015.
4. Vermesan, Ovidiu, and Peter Friess. *Digitising the Industry-Internet of Things Connecting the Physical, Digital and Virtual Worlds.* River Publishers, Delft, 2016.
5. Angelakis, Vangelis et al. eds. *Designing, Developing, and Facilitating Smart Cities: Urban Design to IoT Solutions.* Springer, Berlin, 2017.
6. Tragos, Elias Z. et al. Securing the Internet of things—Security and privacy in a hyperconnected world. In *Building the Hyperconnected Society: Internet of Things Research and Innovation Value Chains, Ecosystems and Markets.* River Publishers Series of Communications, Delft, 2015: 189–219.

7. Nalin, Marco, Ilaria Baroni, and Manuel Mazzara. A holistic infrastructure to support elderlies' independent living. *Encyclopedia of E-Health and Telemedicine*. IGI Global, 2016. 591–605.
8. Memon, Mukhtiar, Stefan Rahr Wagner, Christian Fischer Pedersen, Femina Hassan Aysha Beevi, and Finn Overgaard Hansen. Ambient assisted living healthcare frameworks, platforms, standards, and quality attributes, *Sensors*, 2014, 14(3), 4312–4341, doi: 10.3390/s140304312.
9. Mitseva, Anelia et al. "ISISEMD: Intelligent System for Independent living and self-care of SEniors with mild cognitive impairment or Mild Dementia." *The Journal on Information Technology in Healthcare* 7.6 (2009): 383-399.
10. Neisse, Ricardo et al. Dynamic context-aware scalable and trust-based IoT security, privacy framework. In *Internet of Things Applications—From Research and Innovation to Market Deployment*, IERC Cluster Book, River Publishers, 2014 216–241.
11. Tragos, Elias et al. Improving the trustworthiness of ambient assisted living applications. *Wireless Personal Multimedia Communications*, 2015.
12. Internet of Things Research Cluster. http://www.internet-of-things-research.eu/.
13. Bassi, Alessandro et al. Enabling Things to Talk. Springer-Verlag, Berlin, 2016.
14. Pöhls, Henrich C. et al. RERUM: Building a reliable IoT upon privacy-and security-enabled smart objects. *Wireless Communications and Networking Conference Workshops (WCNCW)*, Institute of Electrical and Electronics Engineers, Piscataway, NJ, 2014.
15. Swiderski, Frank, and Window Snyder (2004). *Threat Modeling*, Microsoft Press, Redmond, WA, July 14, 2004.
16. Mouroutis, Theodore et al. (eds). Use-cases definitions and threat analysis. RERUM Deliverable D2.1, 2014, May.
17. Regulation, General Data Protection. Regulation (EU) 2016/679 of the European Parliament and of the Council of 27 April 2016 on the protection of natural persons with regard to the processing of personal data and on the free movement of such data, and repealing Directive 95/46. *Official Journal of the European Union (OJ)*, 2016, 59, 1–88.
18. Becher, Alexander, Zinaida Benenson, and Maximillian Dornseif. Tampering with motes: Real-world physical attacks on wireless sensor networks. *International Conference on Security in Pervasive Computing*. Springer, Berlin, 2006.
19. Wu, Bing et al. A survey of attacks and countermeasures in mobile ad hoc networks. In *Wireless Network Security*, Xiao, Yang et al. (eds). Springer, Boston, MA, 2007: 103–135.
20. Sánchez-Pi, Nayat, and José Manuel Molina. A centralized approach to an ambient assisted living application: An intelligent home. *International Work-Conference on Artificial Neural Networks*. Springer, Berlin, 2009.
21. Mathur, Avijit, Thomas Newe, and Muzaffar Rao, Defence against black hole and selective forwarding attacks for medical WSNs in the IoT. *Sensors*, 2016, 16(1), 118.
22. Fragkiadakis, Alexandros, and Elias Tragos. A trust-based scheme employing evidence reasoning in IoT architectures. *World Forum on Internet of Things*, 2016.
23. Nakamoto, Satoshi. Bitcoin: A peer-to-peer electronic cash system. *Bitcoin.org*. https://bitcoin.org/bitcoin.pdf.
24. Oikonomou, George, Iain Phillips, and Theo Tryfonas. IPv6 multicast forwarding in RPL-based wireless sensor networks. *Wireless Personal Communications*, 2013, 73(3), 1089–1116.
25. Gaur, Meenakshi Tripathi MS, and V. Laxmi. Comparing the impact of black hole and gray hole attack on leach in wsn. *The 8th International Symposium on Intelligent Systems Techniques for AdHoc and Wireless Sensor Networks (Procedia Computer Science 19 (2013) 1101--1107. DOI=10.1016/j.procs.2013.06.155.*2013.
26. Zahariadis, Theodore et al. Efficient detection of routing attacks in wireless sensor networks. *International Conference on Systems, Signals and Image Processing*, 2009.
27. Liu, Yuxin, Mianxiong Dong, Kaoru Ota, and A. Liu. ActiveTrust: Secure and trustable routing in wireless sensor networks. *IEEE Transactions on Information Forensics and Security*, 2016, 11(9), 2013–2027.
28. Salehi, S. Ahmad, Mohammad Abdur Razzaque, Parisa Naraei, and Ali Farrokhtala. Detection of sinkhole attack in wireless sensor networks. *IconSpace*, 2013.
29. Christidis, Konstantinos and Michael Devetsikiotis. Blockchains and smart contracts for the Internet of things. *IEEE Access*, 2016, 4, 2292–2303.

30. Szabo, Nick. The idea of smart contracts. *Nick Szabo's Papers and Concise Tutorials*, 1997, 6.
31. Tragos, Elias et al. An IoT-based intelligent building management system for ambient assisted living. *IEEE International Conference on Communications*, 2015.
32. *ACTIVAGE Project*. http://www.activageproject.eu.
33. *EU FP7 SOrBet Project*. http://www.fp7-sorbet.eu.

chapter eleven

Smart cities under attack

Cybercrime and technology response

Ralf C. Staudemeyer, Artemios G. Voyiatzis, George Moldovan, Santiago Reinhard Suppan, Athanasios Lioumpas, and Daniel Calvo

Contents

11.1 Introduction

11.1.1 Cybercrime and its cost

Cybercrime is crime committed using computing and communication systems such as computers and networks. When cybercrime occurs, computer systems are either the direct target of the crime or the technological means that enabled it. Many reports have been commissioned to estimate the size of the cybersecurity market and the cost of cybercrime. Hemanshu "Hemu" Nigam, an expert in online safety and privacy, estimated* that the market will reach 170 billion USD by 2020.

Allianz Global Corporate & Specialty suggested† that cybercrime costs the global economy 445 billion USD every year. According to Juniper Research, cybercrime costs are projected‡ to reach 2.1 trillion USD by 2019, while Cybersecurity Ventures predicts§ the annual costs to grow from 3 trillion USD in 2015 to 6 trillion USD by 2021. The 2016 Norton Cyber Security Insights Report found¶ that more than 689 million people in 21 countries experienced cybercrime in 2015. In other words, one out of five Internet citizens** experienced some form of cybercrime already that year.

11.1.2 Smart city evolution

Typically, the desktop computer was the device exposing common users to potential cybercrime activities. With the advent of pervasive computing, however, users do not need to be explicitly or directly interacting with computing devices, as before. Many built-in and mobile sensors, smart appliances, and existing public monitoring infrastructure can track, report, and trigger events according to the needs of users in specific locations (e.g., at home, in transit, those using public transportation services, or utilities), and these are only the current incarnations which are supposed to further develop into smart cities, an even more networked infrastructure providing even broader attack vectors.

What makes a city "smart"? While there are numerous definitions, here we adopt the one provided by Cesar Cerrudo [1]: "A city that uses technology to automate and improve city services, making citizens' lives better."

A systematic literature review on smart city research for the period of 2008–2016 is provided in by Raaijen and Daneva [2]. As cities evolve, they are assumed to use more and more technology to improve the quality of life for their inhabitants. To quote†† Eberhard van der Laan, Mayor of Amsterdam:

> Smart Cities are a bit like football: Every city has a team working on a "Smart-city" and wants to be the Smartest City in the world, and at the start of every season every supporter thinks his or her team will be the global champion. Various ranking systems exist, comparing cities on indicators ranging from energy consumption per capita to life expectancy; from WiFi coverage to crime-rates. In other words: Smart

* http://rt.com/usa/315147-cybersecurity-market-growth-boom, accessed Oct. 5, 2017.
† http://agcs.allianz.com/assets/PDFs/risk%20bulletins/CyberRiskGuide.pdf, accessed Oct. 5, 2017.
‡ http://www.juniperresearch.com/press/press-releases/cybercrime-cost-businesses-over-2trillion, accessed Oct. 5, 2017.
§ http://cybersecurityventures.com/cybersecurity-market-report, accessed Oct. 5, 2017.
¶ http://www.symantec.com/content/dam/symantec/docs/reports/2016-norton-cyber-security-insights-report-en.pdf, accessed Oct. 5, 2017.
** http://www.internetlivestats.com/internet-users, accessed Oct. 5, 2017.
†† http://ercim-news.ercim.eu/en98/keynote-smart-cities, accessed Oct. 5, 2017.

> Cities are about everything and therefore about nothing. To be frank: I do not really believe all this Smart-city marketing. I do believe that innovation and technology gives us the opportunity to improve the quality of life of the citizens and make our cities more competitive.

Frost & Sullivan Research estimates* a combined market potential of 1.5 trillion USD globally for the smart city market in segments of energy, transportation, healthcare, building, infrastructure, and governance. Cities around the world are investing big budgets to become "smarter," such as a 7.4 billion USD smart city project recently announced in South Africa. Other cities, such as New York; San Francisco; Los Angeles; Washington, DC; Seattle; and Miami in the United States, are already there. Similar trends are observed in Europe: London, Barcelona (ranked as the world's smartest city), Amsterdam, Paris, Stockholm, and Berlin; in the Asia-Pacific: Singapore, Seoul, Tokyo, Sydney, Melbourne, and Hong Kong; in the Middle East: Abu Dhabi, Saudi Arabia (70 billion USD investment), Dubai, and Qatar; and in South America: Rio de Janeiro and Santiago [1].

11.1.3 Smart cities and cybercrime

Smart cities rapidly deploy infrastructure for information and communication technologies (ICTs) and digitize the available information, creating models reflecting specific interactions and domains (e.g., transportation models, relevant rules, and expectations). The "Internet of Things" (IoT) is a prominent key technology that enables connections and communications with a vast number of smart city "things." These things are used to collect, process, share, and distribute critical information for the sustainable and livable operation of a city. This is to allow the whole city to be managed more efficiently and effectively. At the same time, the ICT-enhanced infrastructure and all of these smart things introduce a huge attack surface for potential cyber-attacks. What does it take for a smart city to be safe and secure for its inhabitants and respect privacy?

This chapter provides a primer on cybercrime in smart cities and possible technological countermeasures to it. Section 11.2 introduces the threat landscape, both for citizen-facing systems and smart infrastructures, and a survey of real-world incidents of attacks against smart cities. Section 11.3 discusses the security and privacy requirements for IoT-based smart city components and the integration of citizen concerns and considerations in the design. Finally, Section 11.4 describes a list of recommendations for designing and deploying citizen-centric IoT-based systems for future smart cities.

11.2 Smart city threat landscape

11.2.1 Use cases for IoT systems

Smart cities employ ICT mechanisms to manage and improve on livability and cost concerns such as pollution, traffic congestion, safety, and the functioning of strained utilities. IoT-based systems are a key technology to improve the efficiency of handling available resources, improving the services provided to its citizens, and improving the living standards of their residents. The role of IoT-based systems is not only to monitor various changes and to report on them, but also to react to certain conditions and adjust accordingly. This enables a more efficient management and utilization of the resources within the cities.

* http://www.frost.com/sublib/display-report.do?id=M920-01-00-00-00

A smart city can monitor, for example, pollution levels and traffic changes. It can also detect certain threshold conditions and prioritize traffic to specific sections to avoid traffic jams, excessive air and noise pollution, or vibration development in certain infrastructure points. It can also equip buildings with structural and metering sensors measuring physical building dynamics, humidity, and energy consumption. Such capabilities will enable proactive maintenance and minimize energy costs for heating and/or cooling. Some further IoT-based examples of a smart city are the following:

- Smart structural monitoring: Sensors mounted to specific structures prone to deterioration, such as old buildings or buildings exposed to vibrations due to nearby construction, heavy traffic, or seismic movement, can report and alert on possible structural changes such as mechanical stress [3,4].
- Smart waste management: Intelligent waste bins and containers are able to detect their load level so that waste collection services can plan and optimize the routes used in order to collect the waste, minimizing unnecessary trips [5,6].
- Smart water management: Built-in sensors within water distribution systems (WDS) can detect and report minor leaks, adjust water pressure to protect the pipes, and prioritize distribution emergencies (e.g., draught). Fine-grained historical information for consumption can also result in significant cost savings for infrastructure operation [7,8].
- Smart environmental monitoring: Air quality and noise sensors can provide near-real-time measurements detailing the environmental conditions within the city, automatically adjust traffic restrictions, and suggest healthier routes for pedestrians, people performing physical activities, and citizens with chronic illnesses (e.g., asthma) [9,10].
- Smart traffic management: Traffic congestion can be detected by using road- and camera-based sensors. Traffic lights, specific road restrictions, and traffic police can all adjust and react in a timely and semiautonomous manner [11–13].
- Smart energy metering: Both service providers and consumers (citizens and the city itself) can adjust and optimize their energy behavior, toward reducing their energy costs and increasing their ecological profile [14–17].
- Smart lighting: Combined sensor measurements (e.g., weather, acoustics, and movement detection) can be utilized to adjust the light intensity in public spaces and roads based on weather conditions, physical presence, and historical data (e.g., no pedestrians or cars present) [18,19].
- Smart parking: Different kinds of sensors (e.g., cameras, radio-frequency identification, and magnetic) can collect parking information, especially for open spaces (e.g., street sideways) and provide near-real-time updates for free spots or even estimations for near-future availability [20–22]. This results in less idling, less drive time, less emissions, and improved quality of life for citizens [23]. Such systems are expected to become more prevalent with the advent of electric cars and the need to charge them for long hours.
- Smart public transport: Citizens schedule their trips better when up-to-date information is available regarding trip duration and recommendations for alternative means and routes in case of service disruption or personalized needs (e.g., carrying a bicycle or opting for more environment-friendly transport means) [24,25].

The list of services provided by a smart city to its citizens and visitors provides a set of use cases for building threat models [26]. A common theme underlying all these services is the blend of existing, physical entities of the private and public spaces with digital technology. There is no longer a clear boundary between the "physical" and the "digital" part. Rather, these systems evolve into "cyberphysical systems" (CPS) that are deeply integrated and interconnected. As such, the threat landscape becomes complex. We can no longer rely on the existence of "ground truth," as it was the case of solely virtual worlds and services (e.g., e-commerce platforms). A possible consequence is the emergence of malicious interventions in the physical realm itself. Consequently, novel protection mechanisms must be developed that protect the physical space of a smart city from digital threats.

11.2.2 *Infrastructure risks*

The US Department of Homeland Security identifies three crosscutting themes for security considerations of integrating CPS into existing city infrastructures [27]. The first relates to the changing seams. The physical and virtual seams are becoming increasingly permeable as cybersystems and physical systems become networked and remotely accessible, changing and stretching the borders that smart cities must secure. The second relates to inconsistent adoption. Different cities or city parts adopt and migrate to new technologies at different paces. Such inconsistencies pose challenges to developing consistent security policies for all of them.

The third relates to the increased automation. We are experiencing an era where cyberphysical infrastructure migrates control from human beings to algorithm-based systems. This introduces a level of security and resilience into a system by mitigating potential human errors. However, at the same time, algorithmic responses can be hard to predict and comprehend and can result in cascading failures that were not considered. The trillion-dollar stock market "Flash Crash" on May 6, 2010, in the United States, serves as a very good example of how fully automated, high-frequency decision systems can cause significant damage beyond our control [28,29].

11.2.3 *Real-world incidents*

The so-called Northeast blackout of 2003 was one of the first software-related incidents that affected the operation of whole cities and a total population of 55 million people [30]. A software bug in the alarm system at a control room of the FirstEnergy Corporation, located in Ohio, United States, was the primary cause. A sequence of events in a different order from that expected by the system was produced by exploiting a race condition in the control software, and this resulted in a series of cascading effects way beyond a local energy blackout. Infrastructures such as power generation, water supply, transportation, communication, and factories were severely affected in the Northeastern and Midwestern United States and the Canadian province of Ontario. During another event in 2012, a software bug summoned some 1200 people to jury duty on the same morning, resulting in a tie-up on Interstate 80, California, United States, as people drove to court.*

"Major computer problems" caused the San Francisco Bay Area Transit system to shutdown, affecting hundreds of thousands of daily riders while some 500–1000 passengers were trapped on trains in the late evening and early morning hours†. A software update in December 2016 caused a four-hour service outage affecting travelers entering the United Sates in the middle of the holiday week‡. A 14-year-old teenager turned the train system of Lodz, Poland, into a personal train set, "triggering chaos and derailing four vehicles in the process," as covered by Schneier on Security§. It took the teenager no more than a modified

* http://www.npr.org/2012/05/03/151919620/computer-glitch-summons-too-many-jurors

† https://www.bizjournals.com/sanfrancisco/blog/2013/11/bart-system-shut-down-by-software.html

‡ https://www.usatoday.com/story/tech/news/2017/01/03/border-outage-not-caused-hackers-customs-and-border-patrol-airlines-lines-airport/96107764/

§ https://www.schneier.com/blog/archives/2008/01/hacking_the_pol.html

television remote control unit to succeed in this, demonstrating how easy accessing the transport infrastructures of some city might be. Although the aforementioned events were merely proofs of concept and not caused by an intentionally malicious action, they still serve as a good reminder on the complexity of all these interconnected smart city infrastructures and how vulnerable systems can be. Turning to malicious incidents, it was officially acknowledged* that Iranian hackers remotely breached a water dam outside of New York, United States, in 2013. The attack did not cause any damage but still demonstrates that attackers can control physical infrastructure even when not physically close to their target.

In coordination with the involved authorities, researchers demonstrated that smart traffic control systems often used in intersections in the United States are vulnerable [31]. Three major weaknesses contribute to this: (1) lack of communication encryption at the network level, (2) lack of secure authentication at the system level (default usernames and passwords), and (3) controller software unpatched for known exploits. The attack scenarios showed that as many as 100,000 intersections in the United States and Canada could potentially be taken over maliciously. The researchers recommended defenses at the organizational and technical levels. The state of adoption is unreported to date [31].

On December 23, 2015, Ukrainian power companies experienced unscheduled power outages impacting 230,000 residents in Ukraine.† This well-prepared attack was the first confirmed hack to take down a power grid. The availability of rich system and firewall logs allowed the investigators to reconstruct the attack timeline and link it with the BlackEnergy3 malware [32]. A few months later, a nuclear power plant in Germany was infected with malware that could give remote access to attackers‡. The threat was considered low, as the systems were isolated from the Internet. Still, some 18 removable data drives were detected to contain malware, including portable universal serial bus (USB) storage drives (e.g., thumbdrives and sticks). Such devices are a known attack conduit; there are already attack examples that use USB drives to bypass even air-gapped systems [33,34].

On April 7, 2017, all 156 Dallas storm warning systems started blaring across the city at around midnight. It took more than 90 minutes to silence them by shutting down their radio system.§ While initially considered a malfunction, it was acknowledged later that a malicious actor had penetrated the radio system and initiated the alarm. Soon after, on May 12, 2017, the WannaCry ransomware [35] attack was launched worldwide, affecting more than 230,000 computers in 150 countries.¶ The attack exploited a Microsoft Windows operating system vulnerability, encrypted data, and demanded ransom payments in the form of Bitcoins [36]. The effects of WannaCry were felt globally**:

- In the United Kingdom, hospitals and doctors were unable to access patient data and medical appointments, and operations had to be cancelled.
- In Germany, electronic boards displaying train route information were disrupted.
- In France, the carmaker Renault was forced to stop production at a number of sites.
- In the United States, FedEx, a delivery company, was affected.
- In South Korea and Indonesia, hospitals suffered.
- In South America, the Brazilian telecom operator Viva and the LATAM Airlines Group reported effects.

* http://edition.cnn.com/2015/12/21/politics/iranian-hackers-new-york-dam
† https://ics-cert.us-cert.gov/alerts/IR-ALERT-H-16-056-01
‡ http://www.telegraph.co.uk/news/2016/04/27/cyber-attackers-hack-german-nuclear-plant
§ http://www.reuters.com/article/us-texas-sirens-idUSKBN17B001
¶ http://www.bbc.com/news/world-europe-39907965
** https://www.wired.com/2017/05/ransomware-meltdown-experts-warned

The wide range of targets affected by this attack demonstrates the strong dependence on ICT of modern and evolving smart cities. Even more worrying, a postincident scan two months later revealed that more than 60,000 hosts that spread in 130 countries were still vulnerable to this attack.*

A city-wide attack can happen without penetrating air-gapped systems or infecting servers and large computers. Constrained devices that form the IoT and the smart city of the future, such as vulnerable smart light bulbs, could also be used to spread malware across a whole smart city. It would take only 15,000 randomly located light bulbs to control an area as big as Paris, France [37].

Finally, in this set of examples, The Devil's Ivy† (officially: CVE-2017-9765) vulnerability was discovered in July 2017. Devil's Ivy again vividly reminded us that many IoT systems share the same software codebase for implementing communication and network stacks. This makes them equally vulnerable, despite their apparent diversity in function and shape.

11.3 Security and privacy requirements for smart cities

The impact of vulnerabilities on smart city operations can be significant. The numerical and geographical scale of computing makes postincident reaction and fixing time-consuming and costly or sometimes—infeasible. Hence, it is necessary, to integrate appropriate security and privacy mechanisms of a smart system, preferably beginning in the design and predeployment phases [38–40]. In this context, IoT-based systems are crucial to defend, as they are most likely to represent the first and most accessible link in the smart city security and privacy chain: IoT sensors are scattered through a vast area, left unsupervised, easily accessible, and usually without complex software and hardware mechanisms meant to limit their exposal to tampering. Although the adoption of smart technologies is done in a fast pace, we consider that there is the need for integrating proper security and privacy. In the following, we review the security and privacy requirements for smart cities.

Gaining access to an IoT computing system can take place by either active or passive attacks. Active measures might involve manipulating devices physically (e.g., by analyzing specific components and reverse engineering them) or simply infiltrating them by exploiting software vulnerabilities, the end goal being to gain logical control over the system. In addition, active attacks can also be formed on communication channels by the insertion of messages or by the downgrading of the parameters of the communication channel (e.g., bandwidth). In contrast, the eavesdropping of communications is an example for passive attacks [39–41].

It is of great importance to acknowledge that information security is not limited to the protection of the content that is stored and transmitted. Other sensitive (meta-) information are revealed when simply observing information flows and patterns. The protection of meta-information is essential in smart city scenarios for the security and privacy of citizens [39,42].

11.3.1 Requirements for confidentiality, integrity, and availability

Confidentiality is the property that protects information from being disclosed to unauthorized parties. It denies access to those not entitled—no matter whether the data are stored or transmitted. On a communication channel, an attacker is assumed to be able to eavesdrop on messages that are being exchanged. For data-at-rest, the attacker is assumed to have physical access to the device.

* http://omerez.com/eternal-blues-worldwide-statistics/
† http://blog.senr.io/blog/devils-ivy-flaw-in-widely-used-third-party-code-impacts-millions

Citizens may assume that data flowing between devices at their private home or in their immediate vicinity are confidential, that it is inaccessible to any other parties without consent or warrant, and that at least not everyone is able to collect and process this kind of data at will. In a smart city environment, this rather clear separation between private and public environments blurs.

To give an example, the city of Amsterdam, in the Netherlands, supports more than 40 smart city projects ranging from smart parking to the development of home energy storage for integration into the smart grid.* One of these projects concerns the installation of smart energy meters with incentives provided to households who plan to actively save energy. Smart meters record energy consumption in households and report these in short intervals (e.g., 2 s) to the provider. The benefit for the energy provider is that frequent reports allow demand management. Consumers could then benefit from possible lower rates during off-peak times. The downside is that these frequent measurements reveal detailed information on household activities, including presence, electrical devices in use, and even what content consumers watch on television.

Integrity is violated whenever data or a system is modified in an unauthorized way. Modification can occur due to transmission errors or due to an active attack. To ensure a correct and expected behavior of an information technology (IT) system, it is necessary that any modified data must become reliably distinguishable from unchanged data.

The protection mechanisms against malicious modifications differ from those that detect random transmission errors. Integrity protection is basic security functionality. Integrity protection mechanisms can be used to authenticate software and support, securing the distribution channels. Integrity protection detects erroneously or maliciously modified information and should support using it as input for further analysis. Thus, integrity violations can be used as early detectors for a fail-safe behavior.

For example, in smart cities, sensed data are gathered and used by algorithms to enable decision-making. Thus, the decisions are based on data gathered, and bad quality input data can lead to bad decisions. Imagine a smart city without integrity-protected messages: every sensor value can be potentially tampered with. A faulty or misbehaving air pollution sensor in one city area might cause that area to be declared a "zero emission zone" (ZEZ).

To visualize the consequences, assume that a high risk of pollution is detected in a smart city. As a response, the access to the inner city area is restricted. The inner city is declared a ZEZ, and only electric cars are allowed. This leads to the cars in the neighborhood being denied access to the inner city area. Citizens in the area are expected to experience far less noise due to the decreased traffic. The surrounding areas can however experience congested roads due to the increased traffic—caused by noncompliant cars having to avoid the ZEZ.

Likewise, all control messages could be manipulated, e.g., by changing an "access denied" message from the barrier control system into an "open-barrier" one.

A defective sensor could also send erroneous readings. In such a case cryptographic mechanisms would still recognize the data as being correct—which, looking purely at the security perspective, would be true. Hence, smart cities need to deploy additional processing logic to detect erroneous sensor readings and possibly send maintenance crews to investigate the actual sensor.

Availability ensures timely and reliable access to devices and services. The availability property is violated if an attack succeeds by degrading a computer resource or rendering it unavailable, i.e., a denial-of-service attack.

* http://amsterdamsmartcity.com/projects

When monitoring critical environments, sensor values and alert messages need to arrive in a timely manner; otherwise, detection fails and no alarm is triggered (recall the Northeast blackout of 2003 incident in Section 11.2). Critical values might be related to industrial contamination with potentially hazardous consequences; for example, air pollution, high radiation levels, or a decrease of water quality. For example, the city of Tarragona, in Spain, is next to chemical factories. Here, constant and reliable monitoring of air pollution for critical substances is vital to protect citizens. It is essential that potential air contamination can be detected before it reaches the closest households. To achieve this, the deployed detection sensor systems need to send their data to a monitoring server. The city can then detect and potentially react timely in any manner whatsoever. Should parts of the infrastructure, such as the sensors, the communication networks, or the monitoring servers, become unavailable, the detection will fail and the population may not receive the timely warning the system was designed to provide.

11.3.2 *Authentication, authorization, and accountability requirements*

Successful authentication and authorization enables accountability to be achieved for a certain action by a certain entity. The three goals are often grouped together and referred to as "AAA" or "triple A": authentication, authorization, and accountability as defined by Shirey [43].

Authentication is the "process of verifying a claim that a system entity or system resource has a certain attribute value." Authorization is the "process for granting approval to a system entity to access a system resource." Authorization controls who can do what to which objects, while authentication involves identifying who is seeking the access, often being a specific part of the authorization process. Accountability enables "the detection of actions to be traced to the potentially responsible entity." To achieve accountability, an entity first needs to be authenticated, and then the request for access is subject to authorization. To control access, systems must check if an entity is authorized to carry out a certain action.

There are authentication challenges at different layers of a system as complex as a smart city. Returning to a previous example, assume that today, only electric cars are allowed because the inner city is declared a ZEZ. In this setting, the first question is, "who is the entity of the system that you want to authenticate?," which can be hard to answer in technical detail, as discussed by Gollmann [44]. Do we want to authenticate a single car, the on-board device of the car, or the car passengers? Here it becomes obvious that peer entity authentication can happen on various layers.

As reported in 2004 by the World Health Organization, the city of Milan, in Italy, is one of Europe's most air-polluted urban centers. The problem is caused by very high downtown traffic volumes [45]. In 2008, the city introduced electronic road pricing to address traffic congestion, to promote sustainable mobility and public transport, and to decrease the smog levels. The so-called inner city Area C is a restricted zone.* The toll revenues from cars entering the area are reinvested into sustainable energy projects. The area is accessible via gates monitored by video cameras equipped with automatic number plate recognition technology.

In this use case, the attribute value is the car license plate number. If the road toll system recognizes the license plate of a car as registered, the system can charge the owner's account for its use of the toll road. Admittedly, one weakness of the system design is that it is not accounting for privacy requirements. The plate number data are stored and correlated, potentially enabling unauthorized citizen surveillance.†

* https://www.comune.milano.it/wps/portal/ist/en/area_c

† The interested reader can consult an example case reported in New York in 2013: https://www.forbes.com/sites/kashmirhill/2013/09/12/e-zpasses-get-read-all-over-new-york-not-just-at-toll-booths

There is no need for an application to realize authentication based on such unique identifier such as license plate numbers. For example, in the aforementioned case, it suffices to authenticate and distinguish between electric (allow) and fossil-fueled (deny) cars. This means that the relevant attribute value is "I am an electric car." The claim needs to be proven, so that the entrance control system can ensure that it only grants access to an electric cars. The disclosure of the related attribute for the aim of inner city access control should be realized in a privacy-preserving means, using, for example, anonymous credentials [46].

11.3.3 *Credential management and end-to-end design*

Each entity in an IoT system needs to have credentials required to prove its own claims to the authorization components. This means that all participating systems need some form of keys and mutual trust associated with them. For example, would the smart city trust the car manufacturer to vouch for the "I am an electric car" claim?* Or would the car need to be regularly inspected and be issued with a token issued by a state-trusted institution? No matter how it is implemented, there is a need for secure key management and secure key distribution. If they are not in place, attackers might disguise their identities and credentials for their benefit. For example, an attacker can present their hybrid car on demand to appear as an electric car and freely cruise the aforementioned Area C of Milan.

Depending on the involved systems and communication links, transmitted data in the communication system can be protected by different means. One option is to protect the transport link, the other is to protect every message separately. Both options have serious disadvantages, suggesting that a layered approach is required to thwart intrusions. To make the differences obvious, let us consider confidentiality protection by encryption. Hop-to-hop, or so-called link-level, protection, encrypts data between neighboring network nodes. In end-to-end security, the confidentiality and the integrity protection is between the endpoints of the communication. As such, authentication (and finally the authorization) can be performed between the endpoints as well [39].

For example, to logically authenticate a specific car, one needs to be sure that what you technically authenticate is affixed to that specific car, so that it cannot be easily removed and placed into another car. Achieving end-to-end protection means that the need to trust the intermediate systems is removed. While this is preferable, it cannot always be achieved due to layered approaches and independent subsystems.

11.3.4 *Privacy for data-in-transit and data-at-rest*

The citywide communication networks of smart cities are very hard to physically secure against unauthorized access. In the IoT domain, the local network access is predominantly wireless and is therefore prone to eavesdropping. To protect citizens from any kind of hidden loss of personal information when accessing public resources, the communication also needs to be protected against traffic analysis [39,41,42].

Consider, for example, the case of encrypted voice-over-Internet protocol communication through a piece of software, e.g., Microsoft Skype. It sounds like eavesdropping on such communication would not be possible without the knowledge of proprietary (secret)

* The interested reader can consult the cases of the car manufacturers emission scandals Dieselgate in 2015 and 2017 regarding the level of trust that can be put on car manufactures: http://www.bbc.com/news/business-34324772 https://www.forbes.com/sites/bertelschmitt/2017/06/10/dieselgate-2-0-porsche-and-audi-caught-using-sophisticated-defeat-devices

communications protocols, access to cryptography keys, and capabilities to decrypt the conversations. Despite the strong encryption, isolated phonemes can be classified and given sentences identified from vectors of packet sizes of Skype traffic. The reported accuracy can reach more than 80% under specific conditions [47].

Smart energy meters provide automatic meter readings in intervals defined by the electricity provider. The meter readings should remain confidential; highly frequent measurements can reveal detailed information on household activities (e.g., pattern of use for specific electric appliances) [48]. Network traffic containing meter readings can be mapped to communication partners and traced to a specific meter and household. Even if traffic is encrypted, natural changes in traffic patterns and volumes can reveal precious information about the household operations, rendering encryption obsolete. An overview of privacy-preserving data aggregation techniques is provided by Erkin et al. [49]. The need to further anonymize meter readings that are stored by the energy providers, once they are received by them, is discussed by Efthymiou and Kalogridis [50].

There is no reproducible public interest to leave citizens traceable and facilitate continuous surveillance. Nevertheless do many smart city applications heavily depend on citizen location information to provide personalized services (e.g., location-based services). The traditional communication security goals are unable to protect location information. This kind of information leaks from metadata and can be extracted by traffic analysis with little effort. The protection of metadata requires a different approach provided by so-called privacy-enhancing technologies (PETs). However, potentially suitable protection mechanisms do still suffer from a huge overhead in terms of resources.

All citizens of a smart city together form a giant sensor, continuously contributing datapoints from the public space. This includes e-tickets for public transport, automated payment of tolls for car commuters, traffic navigation through smart city, and even presence in public spaces while carrying digital devices. Omnipresent smart cameras can also record entrance into public spaces. Even innocent-looking waste bins can assist in serving targeted advertisements to passing citizens, as demonstrated in London, United Kingdom.*

It is not an exaggeration that citizens are constantly producing datapoints in a smart city. While this can significantly improve their quality of life, such data are long lasting. They are continuously collected, processed, and stored for long times or even indefinitely. These data-at-rest are the new gold for numerous stakeholders. Privacy-preserving techniques, algorithms, and mechanisms are more than necessary to defend against misuse of stored information.

As an example, in 2016, a healthcare organization in Johannesburg, in South Africa, went paperless.† This included systems for deploying digital media health records to improve record keeping and as well patient care‡. There is a high risk that sensitive medical information will leak at some point and be processed in unintended ways, most probably not for the benefit of citizens. On the one hand, it may be very useful for benevolent local governments to hold information about the medical conditions of its citizens and dedicate appropriate budgets to benefit all. On the other hand, governments are not all benevolent, and individuals can be targeted by dishonest (state) employees or external contractors. For a beginning, strict access control and logging, privacy-preserving data processing, and data minimization are essential first steps. Still, if the loss of highly confidential data happens in the most secure environments, such as the National Security

* https://www.cnet.com/news/london-tosses-out-wi-fi-sniffing-smart-bins

† http://www.htxt.co.za/2016/06/08/joburg-clinics-say-goodbye-to-paper-at-digital-ehealth-system-launch

‡ http://ehealthnews.co.za/joburg-invests-in-ehealth-to-benefit-patients

Agency of the United States, it is hard to imagine that any (smart) city will be in better position to defend its own citizens. In the context of highly confidential health data and the economic interests of medical insurance companies, the interested reader is referred to Xu and Cremers [51].

11.3.5 Citizen considerations and concerns

Security experts raised concerns early on regarding the security and privacy of Internet-connected smart-home devices [52–54]. They put significant effort into analyzing existing systems, including privacy leakage through pairing and discovery protocols [55], insecure communication protocols [56], and vulnerabilities that allow remote spying on residents in smart homes [57–59]. Even simple IoT devices, such as smart locks and light bulbs, have been shown to be vulnerable [37,60,61]. The trustworthiness of an IoT application is impacted by the implemented privacy and security practices [62].

Although an independent concept, the smart home is the most widespread example of how humans interact with the smart city environment as technology users. Consequently, privacy and security concerns of citizens and consumers become more evident and can drive the adoption of new technologies [58,63–67]. Security and privacy concerns evolve over time and with the audience, from the preprocurement phase of smart-home IoT products and services to the deep integration of those products and services in daily life.

The European Union (EU) FP7 project "RERUM"* conducted a survey to explore the views of future consumers of potential smart city IoT products and services coming out of the project. As depicted in Figure 11.1, the respondents were most interested in smart-home services in domains of home automation and remote control, while less priority was given for smart transportation and e-health applications. The least priority was on smart-home security and surveillance services. Respondents were concerned for their privacy, citing photos and video streams that are transmitted over public Internet to third-party operators for further processing. They fully trust neither the technical means used to secure the communications nor the human operators whom they fear are constantly monitoring their private space with malicious intentions. This finding is in accordance with Lee and Kobsa [68,69], which also report that privacy considerations affect people's concerns about IoT services such as device tracking. Furthermore, Lee and Kobsa [68,69] highlight that the entity that monitors and collects personal information plays a crucial role in their decision whether they use an IoT service. Regarding the RERUM project findings, as depicted in Figure 11.2, respondents ranked the service price highest among their concerns when considering purchasing a smart-home service, along with the security and privacy, which are considerably more important factors compared to other criteria, such as the diversity of offered applications.

Differences between people living in the same smart home (e.g., technology enthusiasts and passive users—or teens and parents) may result in tensions about technology use and privacy considerations [70,71]. Privacy and security concerns are raised by consumers when they start to use smart-home devices and are in the novelty phase [72,73]. Even when people have lived in a smart home for many months and have had daily interaction with smart-home devices, it may be that their threat models is generally naive, aligned with the sophistication of their technical mental models. Consequently, in this knowledge state, their technology choices are likely to be based

* https://ict-rerum.eu

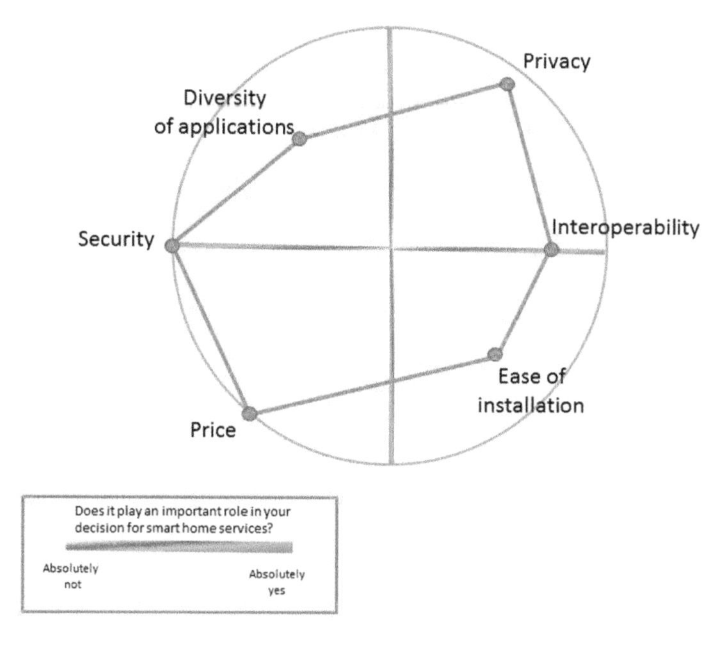

Figure 11.1 Importance of IoT services. (Source: RERUM project.)

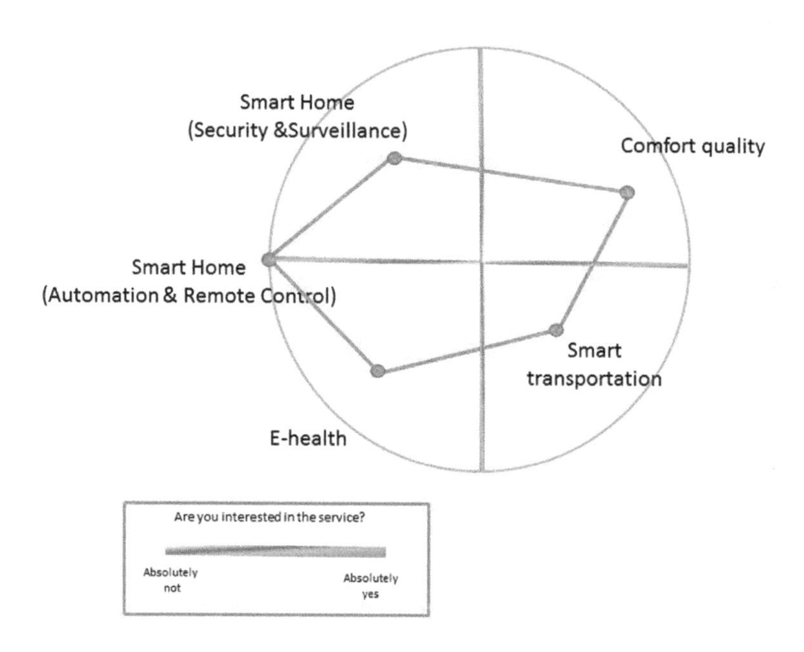

Figure 11.2 Import aspects of smart-home service. (Source: RERUM project.)

on requirements that may conflict with security and privacy (e.g., cost and interoperability) [74]. The comfort level of individuals with their IoT devices varies in different IoT data collection scenarios. The comfort level with IoT devices and device security is profoundly influenced by their perception on how their data are being used to be beneficial [64,75].

11.4 Recommendations

At first glance, IoT security challenges in the smart city domain (as for any IT system) could be considered already solved by applying good security practices, such as encrypting data end to end by default when communicating over the network. However, they become challenging to solve in large-scale distributed systems. IoT services may seem simple from the user point of view since the user experiences only the upper layer, i.e., how the information under consideration are displayed and utilized. However, the real picture is totally different. If we ask where we should expend efforts to enhance security in the IoT and, specifically, in the smart cities, the answer is everywhere! Because a system fails first at its weakest link [39].

11.4.1 Network security in constrained environments

As it was presented in the introduction of this chapter, IoT infrastructures that are typically deployed in a smart city integrate smart devices or smart things as the basic and fundamental components of the system. These devices provide sensing and actuation mechanisms that are the connecting links between the cyberworld and the physical world. They enable new services aiming to improve the lives of citizens.

Within the scope of smart city environments, a great diversity of use cases have been designed and implemented during the last decade. Each one of them targets different actors and users, establishes different objectives and requirements, and culminates in the design of different architectures and the integration of different technologies.

Despite this heterogeneity, the identification of a set of common requirements is possible [76]:

- Deployment of a huge number of devices to allow provisioning services to the majority of the citizens. For instance, more than 15,000 sensors were installed as part of the EU FP7 project "SmartSantander" to develop a first catalog of small applications in a relatively small city considering population and urban area.*
- Most devices must communicate data wireless due to their location and reduce installation complexity and costs. Some devices may be attached to mobile elements (e.g., vehicles).
- Distances between the different devices may be long, from hundreds of meters to several kilometers.
- Devices typically exchange small amounts of data over long periods.
- In many cases, devices are powered using batteries and may use energy harvesting. Thus, an efficient use of available resources is an essential nonfunctional requirement to optimize power consumption and extend the working life without supervision.
- Typical constraints also impose additional restrictions to the size and cost of each device.

Hybrid ICT architectures derived from these common requirements often lead to the deployment of low-power and lossy networks, which interconnect constrained devices with a variety of links, such as IEEE 802.15.4 or low-power Wi-Fi to backend infrastructures [35]. These devices are characterized by severe constraints on the central processing unit, memory, storage capacity, and communication bandwidth [77]. An example case is

* http://www.smartsantander.eu

the Zolertia Re-Mote platform*, which was developed in the aforementioned RERUM project. It is based on an ARM Cortex-M3 processor running at 32 MHz; it includes 512 KB of flash memory and 32 KB of random-access memory (aka RAM).

11.4.2 Security architecture recommendations

It is essential for information transfer and exchange to use constrained IoT systems. Smart cities must recognize this, since such communications of the IoT devices can control the behavior of city-critical systems such as water dams, traffic lights, or the power distribution network. At the same time, it is also essential, at a minimum, to ensure system resilience against most common attacks (e.g., packet replay attacks and denial-of-service attacks). Realizing security mechanisms for smart city IoT systems must also consider the impact on energy consumption. Security solutions that exist in enterprise environments (e.g., strong cryptography, control-flow integrity, and deep packet inspection for content filtering and intrusion detection) cannot be readily applied in citywide distributed low-powered devices that are in most cases physically hard to approach due to their special characteristics and constraints, e.g., strongly limited hardware resources, lack of advanced user interfaces, massive amount of them covering large areas, or even installed in moving or difficult-to-access locations.

All these goals combined result in very challenging requirements: mechanisms or algorithms that enforce privacy and security tend to be computationally intensive and consequently cost energy. It can be expected that processing times and energy consumption overheads for privacy and security are actually higher than the resources needed to process and transmit the data. Moreover, the integration of features to enforce security and privacy also increases the code size of the firmware, which also could be an important problem considering scarce memory resources.

Therefore, when designing the IoT architecture and to ensure an efficient and flexible implementation of authorization mechanisms, it is recommended that a distributed approach as proposed by the ACE (Authentication and Authorization for Constrained Environments) Working Group of the Internet Engineering Task Force be considered. In this approach, constrained nodes delegate authorization mechanisms to more powerful entities that are part of the overall IoT platform [78]. Constrained devices are known as resource servers, which host one or several resources, for instance, a sensor that measures the external temperature or humidity. The client represents an actor that tries to get access to the information or resources that are exposed by the resource server. We could think about a citizen that wants to retrieve the last measurements from his/her laptop or smartphone or even about another constrained device. Both actors, client and resource server, rely on more powerful entities for the authorization process: the authorization server and the client authorization server, respectively.

The following recommendations are proposed in order to ensure that security and privacy are preserved during the authorization process:

1. Secure protocols must be chosen for communications between clients and authorization server (e.g., implemented using transport layer security over transmission control protocol).
2. If the characteristics of the resource server permit it, secure protocols will be used for communications between clients and resource servers (e.g., end-to-end encryption with datagram transport layer security (DTLS) over user datagram protocol (UDP) [53]).

* http://zolertia.io/product/hardware/re-mote

3. If the characteristics of the resource server do not enable the use of secure communication protocols due to energy consumption constraints or limited computing capabilities, additional recommendations are needed:

    a. Tokens generated by authorization servers will be composed of two parts: one destined to the client and another to be attached to each request sent to the resource server. The former one must never be sent through an unsecured channel (e.g., constrained application protocol [53,79,80]).
    b. Resource servers shall be able to prevent typical attacks and to validate the integrity of the access token using mechanisms such proof of possession [81].
    c. Confidentiality of exchanged data shall be protected applying lightweight encryption algorithms (e.g., pseudorandom functions based on ChaCha/Poly1305 [82]) or with built-in hardware support if unavoidable.

11.4.3 *Secure over-the-air programming*

The firmware embedded in IoT devices must be updated to ensure safe and reliable operation. The integration of over-the-air (OTA) programming mechanisms as part of the IoT ecosystem allows bugs to be solved or new functionalities to be added in deployed devices. Moreover, OTA is essential in order to ensure that existing infrastructures will be able to address future security flaws without requiring the substitution of massive amounts of old units or manual flashing procedures. In fact, the application of continuous development and continuous integration techniques that make use of underlying OTA technologies may result in mitigating and preventing some of threats and vulnerabilities previously described [83].

While OTA is a crucial requirement for massive IoT deployments in smart cities, it cannot be denied that it also constitutes a potential vulnerability:

- The size of the firmware grows with the increasing number of functionalities that are implemented in IoT devices. This is a problem since OTA may require using a relevant percentage of the network bandwidth and of the resources of the device to execute a noncore task that must not affect the availability and performance of the device from the client's perspective. Denial-of service attacks may also try to exploit unprotected resources that are exposed to receive the updates.
- OTA could be used to gain control of devices by replacing the existing firmware or installing malware applications to sniff data, to monitor the device and network activity, to corrupt or modify the provided data, or just to completely disable victims.

To avoid these threats and safety flaws that may be used to attack OTA vulnerabilities, the following recommendations are encouraged [84]:

1. The IoT platform must implement communication mechanisms and interfaces between the version control component and the gateway that ensure privacy and security and that enforce appropriate authorization policies (i.e., only gateways that are part of the system will be allowed to retrieve updates, and gateways will receive information only about nodes that are connected to them).

2. The version control component of the IoT platform must sign the firmware images to be deployed to devices (e.g., RSA with SHA256 [85]). The gateway will be able to validate the origin and integrity of updates before sending them.
3. Safe communication protocols must be chosen to send updates from gateways to devices, providing privacy and data integrity (e.g., DTLS [53]).
4. Authorization policies must be enforced by the target device, i.e., only gateways will have permissions to send updates. Access tokens and authorization servers may be used to implement this measure.
5. Error detection techniques will be applied to ensure reliable transmission of firmware updates. Cyclic redundancy check codes or cryptographic hash functions may be used depending on the available computing resources of the target devices.
6. Devices must store the new firmware in a separate memory space to preserve the integrity of the currently used version.
7. Mechanisms to recover the previous state must be implemented as a backup solution, if something fails during the startup of the new image.

11.4.4 Integration of privacy

Although requirements and technologies stand on their own, to make an IoT system or generally any ICT system privacy enhanced, one needs to understand how privacy can be integrated in the design and planning of that system.

In security, security development life cycles are a well-understood way of conducting a systematic approach for ensuring security in software engineering. A privacy development life cycle should be defined the same way: to systematically introduce a privacy methodology in system engineering. However, security and privacy development life cycles have significant differences.

To make the definition of a privacy development life cycle tangible, we look at popular security development life cycles for software development and try to derive goals that can be used for privacy: the Microsoft Security Development Lifecycle (SDL) [86] and the Open Web Application Security Project Software Assurance Maturity Model [87]. Beneficial synergies form both approaches can be recognized [88,89], and the following set of guidelines for the steps of both security life cycle frameworks can be defined.

In general, the following can be stated:

1. Train personnel or ensure that personnel is qualified.
2. Identify threats, evaluate risks (acceptable vs. mitigable threats), and elicit requirements.
3. Design the system according to the requirements.
4. Implement the system, fulfilling all requirements.
5. Verify that the system fulfills the requirements.
6. Deploy the system while making sure the requirements will still apply in the deployment environment.
7. Keep the system developers ready to respond to any conflicting or emerging situation.

It should be noted that phases may differ from the SDL counterpart, as system engineers may adapt the content and the focus of each step according to their needs. The reader

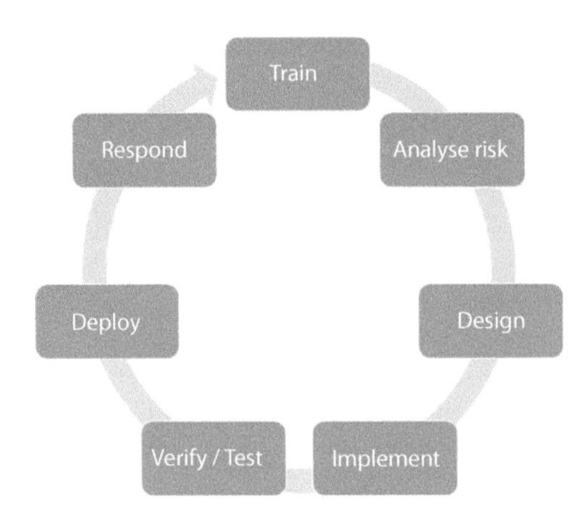

Figure 11.3 Processes of a privacy development life cycle.

is referred to Suppan [88,89] for further details. Figure 11.3 summarizes the life cycle (also note that the life cycle consists of living, continuous processes).

In the following sections, we will consider each step to support the definition of a privacy development life cycle.

11.4.4.1 *Privacy trainings for system developers*

Precondition: Training developers in privacy topics is as essential as training them in security. Although all team members should understand why privacy protection is fundamental and be familiar with the adequate guiding rules (e.g., the EU Data Protection Rules or the applicable guidelines), we recommend that one person in the team should be responsible for privacy, while everyone on the team should be knowledgeable about it.

The responsible person should be well trained in the technical aspects of designing and implementing privacy friendly systems. We call this person the "privacy expert" of the team. The privacy expert should know a privacy engineering framework such as PRIPARE (Preparing Industry to Privacy-by-design by supporting its Application in Research) [90]. The most critical condition for achieving a privacy-friendly product is the presence of at least one, but preferably several, security and privacy experts in the team.

The expert is also responsible for data protection expertise in the development team and should be a participant in every phase of the life cycle. The expert also brings the knowledge as to where mature PETs and best practices can be found. The life cycle itself does not focus on developing new technologies, which could cost a considerable amount of time, research, and technical expertise, but on using existing building blocks and suitable PETs, such as those we gave a brief overview of in previous sections.

A privacy expert needs to be continuously enhancing its technical skills and state-of-the-art knowledge, in particular for IoT with new developments in PETs. Legal support may be needed to resolve privacy-related incidents. In these situations, a privacy expert should be as well aware when legal support is required.

11.4.4.2 *Purpose definition and data minimization*

The first phase of the privacy life cycle development is the specification of requirements for the system. Here, functional requirements of the system are analyzed by posing the following questions, which follow one of the most basic principles in privacy, the principle of purpose. To be more precise, what personal data do the system collect, what is their specific purpose, and can the system reach its desired functionality with less personal data?

The following process is iterated stepwise to obtain answers to these questions and more concrete and operational privacy requirements (or PETs):

- Obtain or define the system data flow and the functional goals and requirements the system.
- Determine which personal data iareneeded to achieve the functional goals of the system.
- Analyze the functional requirements and determine if existing PETs can help minimize the data that are needed for the system.
- Define limits on data usage and data retention in the system.
- The privacy expert analyzes the proposed solution and suggests new possible technologies to reduce data usage in the system.

The reader is referred to Suppan [88,89] for further details on this process.

11.4.4.3 *Evaluation of privacy threats and risks*

After the definition of the required personal data in the system, privacy requirements, and privacy goals, this phase is used for privacy threat analysis. Several frameworks for privacy threat analysis have been proposed, such as LINDDUN (Linkability, Identifiability, Non-repudiation, Detectability, information Disclosure, content Unawareness, and policy and consent Non-compliance) [91], PriS [92] and FPFSD (Framework for Privacy-Friendly System Design) [93].

LINDDUN is especially well suited for the integration of a privacy development life cycle as it is based on STRIDE, a popular risk analysis methodology for security life cycle; see Microsoft Corporation [86]. System developers trained in an SDL and STRIDE should be able to learn the LINDDUN method easily, reuse existing system models (particularly data flow diagrams or DFDs) for their systems, and see synergies or problems of both security and privacy goals. LINDDUN follows STRIDE in defining six steps. The first three cover the "problem space," focusing on the problems, identifying privacy threats and defining requirements of the system. The last three steps cover the "solution space," which aim at fulfilling the requirements; see Microsoft Corporation [86]:

1. Define DFDs of the system (a graphical representation of the system).
2. Map privacy threats to elements of the flow diagrams.
3. Identify threat scenarios according to privacy threats.
4. Prioritize threats/risk analysis.
5. Elicit mitigation strategies.
6. Select PETs that support the mitigation strategies.

The last step is specially challenging for IoT environments. Existing PETs may need to be adapted for constrained and lossy environments or need to support a vast amount of fluctuating participants (e.g., due to devices that frequently change positions).

11.4.4.4 *Privacy by design*

The design phase develops strategies for implementation, verification, release, and response. Since this phase is also functional, security and privacy requirements are adjusted to one another. For example, functional requirements might need change to respect policies, security procedures might need to be adapted to support unlinkability, and privacy requirements might turn out impractical due to core functional requirements and need to be reshaped.

Conflicts might appear between goals, and therefore, best practices can be useful. Best practices are strategies that have been employed by others with good results. For example, Hoepman [94] has defined eight design strategies for privacy, which can be realized using privacy patterns (i.e., best practice solutions), namely, the following:

1. *Minimize* states that the amount of personal data that are processed should be restricted to the minimum amount possible (most basic privacy design strategy).
2. *Hide* states that any personal data, and their interrelationships, should be hidden from plain view.
3. *Separate* states that personal data should be processed in a distributed fashion, in separate compartments whenever possible.
4. *Aggregate* states that personal data should be processed at the highest level of aggregation and with the least possible detail in which it is (still) useful.
5. *Inform* corresponds to the important notion of transparency: Data subjects should be adequately informed whenever personal data are processed.
6. *Control* states that data subjects should be provided agency over the processing of their personal data.
7. *Enforce* a privacy policy, compatible with legal requirements, should be in place and should be enforced.
8. *Demonstrate* requires a data controller to be able to demonstrate compliance with the privacy policy and any applicable legal requirements.

Hoepman provides procedures to address each of the strategy elements listed earlier to support their technical engineering. As some strategies have rarely been used before, Hoepman points out that new patterns are needed. The reader is referred to the privacy patterns database [95].

At this point, we would like to add the following principles to the ones just mentioned:

- Early application of policies and filtering: The processing of personal data should take place in devices under the control of the data subject or, if that is not feasible, as close to the point of origin as possible. This strategy takes advantage of increased processing power in personal devices.
- Process data in a distributed fashion: Hoepman describes a separation strategy to process data in a distributed fashion, but, e.g., storing data in separated databases is not enough, if it can be relinked across databases. This strategy helps avoid such cases by establishing a mechanism that actively checks for possible identifiers and allows proper separation without the possibility to reaggregate the data.
- Ensuring the usability in privacy controls: This has several objectives, such as making privacy controls usable for a variety of users with different skill levels, integrating privacy controls seamlessly into the system, and making the users understand what they are seeing and how they can affect it with the controls provided to them.

11.4.4.5 Implementation and verification

Proper documentation and by-default configuration are key elements in the implementation and verification phase. Users must be able to perform informed decisions in a manner that preserves their privacy easily. The system should be "privacy friendly" from the start. This is called "privacy-by-default" and is one of the most important fair information and privacy-by-design principles, as the majority of users will interact with a system in its lifetime with the default settings, as pointed out by Willis [96].

Secure coding procedures will be needed to avoid privacy issues, which could otherwise become visible later. PETs need to be securely implemented in the same way as security mechanisms, e.g., by coding experts, and frequently verified with code reviews. Implementation strategies, as defined in the design phase, help assure that the implementation effort is controllable and on time and reaches the desired quality. Software developers and privacy experts should work closely together in this phase to avoid problems such as an improper choice of libraries with unwanted effects (such as the use of logging of data including personal information or the presence of vulnerabilities or leaks).

It remains unclear if testing procedures, such as penetration testing and fuzzing, can be used for privacy purposes. Nevertheless, the methods used in code reviews also offer valuable information about data and information flow in programs and about the presence and enforcement of privacy-enhancing mechanisms.

11.4.4.6 Release of system and education of stakeholders

The release of system and education of stakeholders phase is used to develop strategies in case vulnerabilities are discovered on release, and these strategies are carried over to the next phase. Strategies cover the assignment of responsibilities, emergency response methods and emergency assessments, technical actions, and communication strategies. Privacy cannot be protected simply by technical components; this holds for security as well.

The education of system stakeholders takes a significant role in this phase. Stakeholders are system administrators, operators, and system end users. Operation stakeholders need to know which data are processed by the system and what kinds of implications this might have for users. Technical protection might be useless in certain scenarios that might seem unlikely, yet operators should know them to be able to react in case they occur.

Data subjects need to be informed about how their data are processed and which tools they are provided to exercise their privacy rights. The released system should be accompanied by an according privacy disclosure that describes the use of personal data of the system; documentation, which details the tools that the system provides; and user communication tools like a brief explanation to addresses likely user questions. System administrators, users, and other personnel who may interact with the system need to be informed about how their actions affect the privacy of others and which actions can lead to privacy violations.

The Microsoft SDL proposes to validate the privacy standards of the system by a privacy advisor or a privacy seal of quality prior to release. A legal privacy expert should review the documents and overview the release process.

11.4.4.7 Technology and organizational response

The last phase is one of the most significant in the life cycle. It carries over the results from the release phase for rapid response strategies. Breaches might have a significant impact, and the team must be prepared to respond efficiently and timely to them, as they can occur unexpectedly. A response team must therefore develop a response plan that includes preparations for potential postrelease emergencies. The Canadian Office of the Privacy Commissioner (OPC) [97] proposes four steps for this phase:

- *Breach containment and preliminary assessment*: In this step, immediate actions to stop the breach are carried out and the investigation leader and a response team are assigned. Legal action against the attackers is suggested as well.
- *Evaluation* of the risks associated with the breach: In this step, the risks associated with the breach are evaluated and first actions are triggered. In case of compromised data, for example, sensitive data in plaintext will trigger different actions than encrypted data. Assessments can help identify the root cause, the foreseen harm, and the individuals affected and to find adequate mitigation strategies.
- *Notification*: Here the users are notified of possible consequences the breach might have. The notification should be as soon as (reasonably) possible and personal, by phone, e-mail, etc. In this step, additional organizations can also be informed, such as cyberdefense centers and credit card companies (if credit card data were stolen).
- *Prevention*: A prevention plan is defined at this point. The OPC suggests that the level of effort should reflect whether it was a systemic breach or an isolated instance. This step aims at fast communication and support strategies between companies and users. They help users to understand what possible consequences the breach may have and give them a transparent view of the emergency response strategies form the company.

Legal support should be expected to be needed to handle consequences and to initiate legal actions against the attackers. A root cause mitigation team could be engaged to investigate why a breach was possible and develop a mitigation plan that can be realized rapidly. This is an especially crucial step in battling cybercrime.

The life cycle, once implemented, will change with every iteration according to the needs of organization. With a proper privacy engineering organization in place, a solid privacy-by-design approach can be reached in system design as well as the operation of the IoT system. For smart cities, a proper privacy life cycle is a quality differentiator, one that needs just a first step to start and considerable practice to mature.

Acknowledgments

Ralf C. Staudemeyer was partly supported by the research pool of the Faculty of Computer Science and Mathematics of the University of Passau. Artemios G. Voyiatzis was partly supported by the Christian Doppler Forschungsgesellschaft through Josef Ressel Center project u'smile. Daniel Calvo was partly supported by the Internet of Everything Laboratory and the Energy and Transport Market of Atos Research and Innovation. Ralf C. Staudemeyer, George Moldovan, Santiago R. Suppan, Athanasios Lioumpas, and Daniel Calvo were partially supported by the European Union 7th Framework Programme under grant agreement number 609094 (RERUM).

References

1. C. Cerrudo, An Emerging US (and World) Threat: Cities Wide Open to Cyber Attacks, Technical report, 10Active Labs, 2015.
2. T. Raaijen and M. Daneva, Depicting the smarter cities of the future: A systematic literature review & field study, in Smart Cities Symposium Prague (SCSP'17), pp. 1–10, 2017.
3. G. Hackmann, W. Guo, G. Yan, Z. Sun, C. Lu, and S. Dyke, Cyber-physical codesign of distributed structural health monitoring with wireless sensor networks, Trans. on Parallel Distrib. Syst., vol. 25, no. 1, pp. 63–72, 2014.

4. T. Nagayama and B. F. Spencer, Jr, Structural Health Monitoring Using Smart Sensors, University of Illinois at Urbana-Champaign, Champaign, IL, Technical report, 2007.
5. G. Asimakopoulos, S. Christodoulou, A. Gizas, V. Triantafillou, G. Tzimas, J. Gialelis, A. Voyiatzis, D. Karadimas, and A. Papalambrou, Architecture and implementation issues, toward a dynamic waste collection management system, in Proceedings of the 26th International Conference on World Wide Web Companion (WWW '15), pp. 1383–1388, 2015.
6. A. G. Voyiatzis, J. Gialelis, and D. Karadimas, Dynamic cargo routing on-the-go: The case of urban solid waste collection, in Proceedings of the 10th International Conference on Wireless and Mobile Computing, Networking and Communications (WiMob'14), pp. 58–65, 2014.
7. C. Gogos, P. Alefragis, and E. Housos, Application of heuristics, genetic algorithms and integer programming at a public enterprise water pump scheduling system, in Proceedings of the 11th Panhellenic Conference on Informatics, pp. 579–589, 2007.
8. P. Jiang, H. Xia, Z. He, and Z. Wang, Design of a water environment monitoring system based on wireless sensor networks, Sensors, vol. 9, no. 8, pp. 6411–6434, 2009.
9. N. Castell, M. Kobernus, H.-Y. Liu, P. Schneider, W. Lahoz, A. J. Berre, and J. Noll, Mobile technologies and services for environmental monitoring: The Citi-Sense-MOB approach, Urban Clim., vol. 14, pp. 370–382, Dec. 2015.
10. S. Fang, L. Da Xu, Y. Zhu, J. Ahati, H. Pei, J. Yan, and Z. Liu, An integrated system for regional environmental monitoring and management based on internet of things, Trans. Ind. Informatics, vol. 10, no. 2, pp. 1596–1605, 2014.
11. S. Djahel, R. Doolan, G.-M. Muntean, and J. Murphy, A communications-oriented perspective on traffic management systems for smart cities: Challenges and innovative approaches, IEEE Comm. Surv. Tutorials, vol. 17, no. 1, pp. 125–151, 2015.
12. P. Rizwan, K. Suresh, and M. R. Babu, Real-time smart traffic management system for smart cities by using Internet of Things and big data, in 10th International Conference on Emerging Technological Trends (ICETT'16), pp. 1–7, 2016.
13. K. Nellore and G. Hancke, A survey on urban traffic management system using wireless sensor networks, Sensors, vol. 16, no. 2, p. 157, 2016.
14. S. Darby, Smart metering: What potential for householder engagement? Build. Res. Inf., vol. 38, no. 5, pp. 442–457, 2010.
15. S. S. S. R. Depuru, L. Wang, V. Devabhaktuni, and N. Gudi, Smart meters for power grid: Challenges, issues, advantages and status, Renew. Sustain. Energy Rev., vol. 15, no. 6, pp. 2736–2742, 2011.
16. I. Kunold, M. Kuller, J. Bauer, and N. Karaoglan, A system concept of an energy information system in flats using wireless technologies and smart metering devices, in Proceedings of the 6th International Conference on Intelligent Data Acquisition and Advanced Computing Systems (IDAACS'11), vol. 2, pp. 812–816, 2011.
17. A. Rial and G. Danezis, Privacy-preserving smart metering, in Proceedings of the 10th Annual Workshop on Privacy in the Electronic Society (WPES'11), pp. 49–60, 2011.
18. J. Lukkien and R. Verhoeven, The case of dynamic street lighting an exploration of long-term data collection, in Proceedings of the 20th Conference on Emerging Technologies & Factory Automation (ETFA'15), pp. 1–8, 2015.
19. R. Müllner and A. Riener, An energy efficient pedestrian aware Smart Street Lighting system, International J. Pervasive Comput. Commun., vol. 7, no. 2, pp. 147–161, 2011.
20. T. Lin, H. Rivano, and F. Le Mouel, A survey of smart parking solutions, in Transactions on Intelligent Transportation Systems, pp. 1–25, 2017.
21. H. Qin, Z. Li, Y. Wang, X. Lu, W. Zhang, and G. Wang, An integrated network of roadside sensors and vehicles for driving safety: Concept, design and experiments, in Proceedings of the International Conference on Pervasive Computing and Communications (PerCom'10), 2010, no. March 2010, pp. 79–87.
22. A. Khanna and R. Anand, IoT based smart parking system, in International Conference on Internet of Things and Applications (IOTA'16), pp. 266–270, 2016.
23. P. Ramaswamy, IoT smart parking system for reducing green house gas emission, in International Conference on Recent Trends in Information Technology (ICRTIT'16), pp. 1–6, 2016.

24. A. K. Debnath, H. C. Chin, M. M. Haque, and B. Yuen, A methodological framework for benchmarking smart transport cities, Cities, vol. 37, pp. 47–56, 2014.
25. R. Szabo, K. Farkas, M. Ispany, A. A. Benczur, N. Batfai, P. Jeszenszky, S. Laki et al. Framework for smart city applications based on participatory sensing, in Proceedings of the 4th International Conference on Cognitive Infocommunications (CogInfoCom'13), pp. 295–300, 2013.
26. P. Lin, M. Swimmer, A. Urano, S. Hilt, and R. Vosseler, Securing Smart Cities—Moving toward Utopia with Security in Mind, Trend Micro, Tokyo, 2017.
27. US Department of Homeland Security–National Protection and Programs Directorate (NPPD)–Office of Cyber and Infrastructure Analysis (OCIA), The Future of Smart Cities: Cyber-Physical Infrastructure Risk, Technical report, US Department of Homeland Security, Washington, DC, 2015.
28. A. Kirilenko, A. S. Kyle, M. Samadi, and T. Tuzun, The Flash Crash: The impact of high frequency trading on an electronic market, Available at SSRN, Vol. 1686004, p. 64, 2011.
29. A. Kirilenko, A. S. Kyle, M. Samadi, and T. Tuzun, The Flash Crash: High-frequency trading in an electronic market, J. Financ., vol. 72, no. 3, pp. 96–998, 2017.
30. US–Canada Power System Outage Task Force, Final Report on the August 14, 2003 Blackout in the United States and Canada: Causes and Recommendations, U.S. Department of Energy and Natural Resources Canada, p. 238, 2004.
31. B. Ghena, W. Beyer, A. Hillaker, J. Pevarnek, and J. A. Halderman, Green lights forever: Analyzing the security of traffic infrastructure, in Proceedings of the 8th USENIX Workshop on Offensive Technologies (WOOT'14), pp. 1–10, 2014.
32. Symantec Corporation, Internet Security Threat Report, Symantec Corporation, Mountain View, CA, 2016.
33. M. Guri, A. Kachlon, O. Hasson, G. Kedma, Y. Mirsky, and Y. Elovici, GSMem: Data exfiltration from air-gapped computers over GSM frequencies, in Proceedings of the 24th USENIX Security Symposium. (USENIX Security 15), pp. 849–864, 2015.
34. R. Langner, Stuxnet: Dissecting a cyberwarfare weapon, Secur. Priv., vol. 9, no. 3, pp. 49–51, 2011.
35. J. P. Vasseur, RFC7102—Terms Used in Routing for Low-Power and Lossy Networks, no. 7102. RFC Editor, Jan 2014.
36. A. Judmayer, N. Stifter, K. Krombholz, and E. Weippl, Blocks and chains: Introduction to Bitcoin, cryptocurrencies, and their consensus mechanisms, vol. 9, no. 1. Morgan & Claypool Publishers, San Rafael, CA, 2017.
37. E. Ronen, A. Shamir, A.-O. Weingarten, and C. O'Flynn, IoT goes nuclear: Creating a ZigBee chain reaction, in Symposium on Security and Privacy (SP'17), pp. 195–212, 2017.
38. J. Bauer, J. Cuellar, A. Fragkiadakis, B. Petschkuhn, H. C. Pöhls, D. Ruiz, R. C. Staudemeyer et al., Privacy Enhancing Techniques in Smart City Applications, (R. C. Staudemeyer and H. C. Pöhls, Eds.). Technical report. p. 292, University of Passau, 2015.
39. R. C. Staudemeyer, H. C. Pöhls, and B. W. Watson, Security and privacy for the Internet-of-Things communication in the SmartCity, in Designing, Developing, and Facilitating Smart Cities: Urban Design to IoT Solutions, Springer International Publishing, Cham, pp. 109–137, 2017.
40. M. Haus, M. Waqas, A. Y. Ding, Y. Li, S. Tarkoma, and J. Ott, Security and privacy in device-to-device (D2D) communication: A review, Commun. Surv. Tutorials, vol. 19, no. 2, pp. 1054–1079, 2017.
41. R. C. Staudemeyer, D. Umuhoza, and C. W. Omlin, Attacker models, traffic analysis and privacy threats in IP networks, in Proceedings of the 12th International Conference on Telecommunications (ICT'05), 2005, p. 7.
42. G. Danezis and R. Clayton, Introducing traffic analysis, in Digital Privacy: Theory, Technologies, and Practices, Auerbach Publications, pp. 95–117, 2007.
43. R. Shirey, RFC 4949—Internet Security Glossary. Version 2, IETF, 2007.
44. D. Gollmann, What do we mean by entity authentication? in Proceedings of the 1996 Symposium on Security and Privacy, pp. 46–54, 1996.
45. M. Martuzzi, C. Galassi, B. Ostro, F. Forastiere, and R. Bertollini, Health Impact Assessment of Air Pollution in the Eight Major Italian Cities, World Health Organisation Europe, p. 65, 2013.

46. J. Camenisch, M. Dubovitskaya, K. Haralambiev, and M. Kohlweiss, Composable and modular anonymous credentials: Definitions and practical constructions, in Lecture Notes in Computer Science, 2015, vol. 9453, pp. 262–288.
47. B. Dupasquier, S. Burschka, K. McLaughlin, and S. Sezer, Analysis of information leakage from encrypted Skype conversations, International J. Inf. Secur., vol. 9, no. 5, pp. 313–325, Jul. 2010.
48. A. Molina-Markham, P. Shenoy, K. Fu, E. Cecchet, and D. Irwin, Private memoirs of a smart meter, in Proceedings of the 2nd Workshop on Embedded Sensing Systems for Energy-Efficiency in Building (BuildSys '10), p. 61, 2010.
49. Z. Erkin, J. R. Troncoso-Pastoriza, R. L. Lagendijk, and F. Perez-Gonzalez, Privacy-preserving data aggregation in smart metering systems: An overview, Signal Process. Mag., vol. 30, no. 2, pp. 75–86, 2013.
50. C. Efthymiou and G. Kalogridis, Smart grid privacy via anonymization of smart metering data, in Proceedings of the International Conference on Smart Grid Communications (SmartGridComm'10), pp. 238–243, 2010.
51. L. Xu and A. B. Cremers, Patients' privacy protection against insurance companies in eHealth systems, in Proceedings of the Nordic Conference on Secure IT Systems (NordSec'14), pp. 247–260, 2014.
52 O. Arias, J. Wurm, K. Hoang, and Y. Jin, Privacy and security in Internet of things and wearable devices, Trans. Multi-Scale Comput. Syst., vol. 1, no. 2, pp. 99–109, Apr. 2015.
53. J. Granjal, E. Monteiro, and J. S. Silva, Security for the Internet of Things: A survey of existing protocols and open research issues, Commun. Surv. Tutorials, vol. 17, no. 3, pp. 1294–1312, 2015.
54. S. Sicari, A. Rizzardi, L. A. Grieco, and A. Coen-Porisini, Security, privacy and trust in Internet of Things: The road ahead, Comput. Networks, vol. 76, pp. 146–164, Jan. 2015.
55. D. J. Wu, A. Taly, A. Shankar, and D. Boneh, Privacy, discovery, and authentication for the internet of things, in European Symposium on Research in Computer Security (ESORICS'16), pp. 301–319, 2016.
56. A. Cui and S. J. Stolfo, A quantitative analysis of the insecurity of embedded network devices: Results of a wide-area scan, in Proceedings of the 26th Annual Computer Security Applications Conference (ACSAC '10), pp. 97–106, 2010.
57. T. Denning, T. Kohno, and H. M. Levy, Computer security and the modern home, Commun. ACM, vol. 56, no. 1, pp. 94–103, 2013.
58. T. Denning, C. Matuszek, K. Koscher, J. R. Smith, and T. Kohno, A spotlight on security and privacy risks with future household robots, in Proceedings of the 11th International Conference on Ubiquitous Computing (Ubicomp '09), 2009, pp. 105–114.
59. E. Fernandes, J. Jung, and A. Prakash, Security analysis of emerging smart home applications, in Symposium on Security and Privacy (SP 2016), pp. 636–654, 2016.
60. G. Ho, D. Leung, P. Mishra, A. Hosseini, D. Song, and D. Wagner, Smart locks: Lessons for securing commodity internet of things devices, in Proceedings of the 11th Conference on Computer and Communications Security, pp. 461–472, 2016.
61. P. Morgner, S. Mattejat, and Z. Benenson, All your bulbs are belong to us: Investigating the current state of security in connected lighting systems, arXiv Prepr. arXiv:1608.03732, Aug. 2016.
62. I. D. Addo, S. I. Ahamed, S. S. Yau, and A. Buduru, A reference architecture for improving security and privacy in Internet-of-Things applications, in Proceedings of the 3rd International Conference on Mobile Services (MS 2014), 2014, pp. 108–115.
63. D. J. Butler, J. Huang, F. Roesner, and M. Cakmak, The privacy-utility tradeoff for remotely teleoperated robots, in Proceedings of the 10th Annual International Conference on Human-Robot Interaction (HRI '15), 2015, pp. 27–34.
64. K. Courtney, G. Demiris, M. Rantz, and M. Skubic, Needing smart home technologies: The perspectives of older adults in continuing care retirement communities, J. Innov. Heal. Informatics, vol. 16, no. 3, pp. 195–201, 2008.
65. M. L. Mazurek, B. Salmon, R. Shay, K. Vaniea, L. Bauer, L. F. Cranor, G. R. Ganger et al., Access control for home data sharing, in Proceedings of the 28th International Conference on Human Factors in Computing Systems (CHI '10), p. 645, 2010.

66. D. Townsend, F. Knoefel, and R. Goubran, Privacy versus autonomy: A tradeoff model for smart home monitoring technologies, in Proceedings of the 33rd Annual International Conference of the Engineering in Medicine and Biology Society (IEEE EMBS), 2011, pp. 4749–4752.
67. J. Woo and Y. Lim, User experience in do-it-yourself-style smart homes, in Proceedings of the 2015 ACM International Joint Conference on Pervasive and Ubiquitous Computing (UbiComp '15), pp. 779–790, 2015.
68. H. Lee and A. Kobsa, Understanding user privacy in Internet of Things environments, in Proceedings of the 3rd World Forum on Internet of Things (WF-IoT'16), pp. 407–412, 2016.
69. H. Lee and A. Kobsa, Privacy preference modeling and prediction in a simulated campuswide IoT environment, in International Conference on Pervasive Computing and Communications (PerCom'17), 2017, pp. 276–285.
70. S. Mennicken and E. M. Huang, Hacking the natural habitat: An in-the-wild study of smart homes, their development, and the people who live in them, in International Conference Pervasive Comput., pp. 143–160, 2012.
71. B. Ur, J. Jung, and S. Schechter, Intruders versus intrusiveness: Teens' and parents' perspectives on home-entryway surveillance, in Proceedings of the International Joint Conference on Pervasive and Ubiquitous Computing (UbiComp '14), pp. 129–139, 2014.
72. A. J. B. Brush, B. Lee, R. Mahajan, S. Agarwal, S. Saroiu, and C. Dixon, Home automation in the wild, in Proceedings of the 2011 Annual Conference on Human Factors in Computing Systems (CHI '11), pp. 2115–2124, 2011.
73. E. K. Choe, S. Consolvo, J. Jung, B. Harrison, S. N. Patel, and J. A. Kientz, Investigating receptiveness to sensing and inference in the home using sensor proxies, in Proceedings of the Conference on Ubiquitous Computing, pp. 61–70, 2012.
74. E. Zeng, S. Mare, and F. Roesner, End User Security & Privacy Concerns with Smart Homes, in Symposium on Usable Privacy and Security (SOUPS), 2017.
75. P. E. Naeini, S. Bhagavatula, H. Habib, M. Degeling, L. Bauer, L. Cranor, and N. Sadeh, Privacy Expectations and Preferences in an {IoT} World, in Proceedings of the 13th Symposium on Usable Privacy and Security (SOUPS'17), pp. 399–412, 2017.
76. T. Watteyne, T. Winter, D. Barthel, and M. Dohler, RFC5548—Routing Requirements for Urban Low-Power and Lossy Networks, no. 5548. RFC Editor, May 2009.
77. C. Bormann, M. Ersue, A. Keranen, Z. Shelby, K. Hartke, and C. Bormann, RFC7228—Terminology for Constrained-Node Networks, Internet Engineering Task Force, May 2014.
78. L. Seitz, G. Selander, E. Wahlstroem, S. Erdtman, and H. Tschofenig, Authentication and Authorization for Constrained Environments (ACE), no. draft-ietf-ace-oauth-authz-06, Internet Engineering Task Force, Mar 2017.
79. C. Bormann, A. P. Castellani, and Z. Shelby, CoAP: An application protocol for billions of tiny internet nodes, IEEE Internet Comput., vol. 16, no. 2, pp. 62–67, 2012.
80. A. Capossele, V. Cervo, G. De Cicco, and C. Petrioli, Security as a CoAP resource: An optimized DTLS implementation for the IoT, in International Conference on Communications, pp. 549–554, 2015.
81. N. Asokan, V. Niemi, and P. Laitinen, On the usefulness of proof-of-posession, in Proceedings 2nd Annual PKI Res. Work., pp. 122–127, 2003.
82. Y. Nir and A. Langley, RFC7539—ChaCha20 and Poly1305 for IETF Protocols, no. 7539. RFC Editor, May 2015.
83. B. Greene, Agile methods applied to embedded firmware development, in Proceedings of the Agile Development Conference, pp. 71–77, 2004.
84. B. Candaele, D. Soudris, and I. Anagnostopoulos, Eds., Trusted Computing for Embedded Systems, 1st ed. Cham: Springer International Publishing, 2015.
85. Q. H. Dang, Secure Hash Standard, FIBS 180-4 Publ., vol. 4, no. August, p. 36, Jul. 2015.
86. Microsoft Corporation (Hrsg.), Security Development Lifecycle—SDL Process Guidance, Version 5.2. 5.2-1, Microsoft Corporation, Redmond, WA, May 2012. [Online]. Available: https://msdn.microsoft.com/en-us/library/windows/desktop/cc307748.aspx. [Accessed: Oct. 5, 2017].
87. P. Chandra and S. Deleersnyder, OWASP Software Assurance Maturity Model. Creative Commons (CC) Attribution-Share Alike, OWASP, 2012.

88. S. R. Suppan, Data Protection for the Internet of Things, Doctorate Thesis. University of Regensburg, Regensburg, 2017.
89. S. R. Suppan and J. C. Cuéllar, Privacy in the Internet of Things, in Engineering Secure Internet of Things Systems, vol. 2, Institution of Engineering and Technology, Stevenage, pp. 33–56, 2016.
90. N. Notario, A. Crespo, Y.-S. Martin, J. M. Del Alamo, D. Le Metayer, T. Antignac, A. Kung, I. Kroener, and D. Wright, PRIPARE: Integrating privacy best practices into a privacy engineering methodology, in Security and Privacy Workshops, pp. 151–158, 2015.
91. K. Wuyts, R. Scandariato, and W. Joosen, LINDDUN Privacy Threat Modeling, KU Leuven, Leuven, Jul 2015. [Online]. Available: https://distrinet.cs.kuleuven.be/software/linddun/. [Accessed: Oct. 7, 2017].
92. C. Kalloniatis, E. Kavakli, and S. Gritzalis, Addressing privacy requirements in system design: The PriS method, Requir. Eng., vol. 13, no. 3, pp. 241–255, Sep. 2008.
93. S. Spiekermann and L. F. Cranor, Engineering privacy, IEEE Trans. Softw. Eng., vol. 35, no. 1, pp. 67–82, 2009.
94. J.-H. Hoepman, Privacy Design Strategies, in ICT Systems Security and Privacy Protection, vol. 428, pp. 446–459, 2014.
95. N. P. Doty, M. Gupta, J. Zych, and R. McDonald, Privacy Dashboard, A Privacy Pattern, Jul 2015. [Online]. Available: http://privacypatterns.org/patterns/Privacy-dashboard. [Accessed: Oct. 5, 2017].
96. L. E. Willis, Why not privacy by default? Berkeley Technol. Law J., vol. 29, no. 1, pp. 61–134, 2014.
97. Office of the Privacy Commissioner of Canada, Key Steps for Organizations in Responding to Privacy Breaches, Office of the Privacy Commissioner of Canada, Gatineau, QC, p. 7, 2007.

chapter twelve

Securing supervisory control and data acquisition control systems

Nathan Lau, Hao Wang, Ryan Gerdes, and Chee-Wooi Ten

Contents

12.1 Security of critical infrastructures

National physical infrastructures are often operated by industrial control systems (ICSs) that heavily rely on information and communication technology (ICT). Specifically, industrial control systems of many sectors are built around the supervisory control and data acquisition (SCADA) architecture, consisted of communication network connecting servers, clients and embedded computer devices for automation, and operators to monitor and control the physical equipment. For example, water treatment, petrochemical, agriculture, and critical manufacturing facilities all adopt SCADA technology [1–3]. SCADA systems contribute to efficient operations of many critical infrastructure sectors. For example, the smart grid has increased prediction accuracy in loading shedding (i.e., matching generation to demand), which reduces 10% costs for utilities and consumers [4].

The significance of SCADA systems in operating critical infrastructures has made them prime targets of cyberattacks to inflict major disruptions on our society. As of 2015, the Repository of Industrial Security Incidents database contained over 250 SCADA incidents in the past two decades globally [5]. Table 12.1 [6–10] outlines the most prominent attacks on SCADA systems highlighting some worrisome cybersecurity trends. First, these prominent cyberattacks inflicted severe physical damages and service disruptions, indicating that malware are no longer only targeting traditional information technology systems. Second, the attacks targeted diverse infrastructure sectors, suggesting that all major systems are likely vulnerable to cyberintrusions. Third, the malware were unique and sophisticated, highlighting the skills of the attackers. For example, the most infamous

Table 12.1 Prominent cyberattacks on SCADA components in critical infrastructure sectors across the globe

Target organization	Time	Delivery method	Malware or exploits	Target device	Impact
Austrailian Maroochy Shire Sewage System [6]	Feb.–Apr. 2000	Insider with identical radio system	Insider	Pumping stations: RTUs	Overflow of over 800,000 liters of untreated sewage
Davis–Besse Nuclear Power Plant	Aug. 2003	E-mail attachments	Slammer worm: SQL server worm	Network hosts	(DOS) Disable monitoring system and control network communications
Baku–Tbilisi–Ceyhan (BTC) Pipeline [7]	Aug. 2008; 2014	Suspected insider	Unknown malware exploiting camera communication software vulnerability	Valve stations: RTUS and PLCs	Pipeline rupture and explosion and disruption in service
Iranian Natanz nuclear facility	?–Jul. 2010	Removable drives (USB), suspected insiders	Stuxnet	Gas centrifuges: PLCs	Destroyed gas centrifuges in Iranian nuclear facilities
Saudi Aramco Oil Refinery	Aug. 2012	N/A	Shamoon virus/ Ransomware	Oil refinery workstations	(DOS) Affecting global energy sector, such as Saudi Aramco (oil refiner)
US Power Utility intrusion	Oct. 2012	Removable drives (USB)	Mariposa virus	N/A	(DOS) Plants shut down for 3 weeks
German Steel Mill cyberattack [8]	Dec. 2014	Phishing with e-mail attachments	Unknown	Work station PLCs and plant network	Blast furnace explosions
Ukrainian power grid cyberattack [8]	Dec. 2015	Phishing with e-mail attachments	BlackEnergy 3 malware	Workstation RTUs	(DOS) Disconnect seven substations for three hours
Ukraine ransomware attack [10]	Jun. 2017	Unknown	Petya/NotPetya Malware	Monitoring system, plant network	(DOS) Disconnect multiple state-owned enterprises including national bank, airport, and nuclear power plant

Note: DOS, denial of service.

malware is Stuxnet, which employed four unique vulnerabilities to inflict damage on the gas centrifuges in the Iranian uranium enrichment facility [11].

The perpetrators for most cyberattacks on critical infrastructures are nation states motivated by complex, geopolitical objectives. This stands in contrast to traditional cybercriminals interested in financial incentives only. For example, the 2015 Ukrainian power grid cyberattack that caused power outages affecting over 225,000 customers [8] was followed by the 2017 Petya ransomware attacks, which affected multiple state-owned enterprises including the Boryspil International airport and the Chernobyl nuclear power plant [9]. Russia was the alleged attacker reacting to the ousting of former President Viktor Yanukovych (who favors ties with Russia) and the possibility of joining the European Union and North Atlantic Treaty Organization [12,13]. Further, as part of their military strategies, many governments operate state-of-the-art facilities and employ highly trained professionals to create and stockpile zero-day exploits—undisclosed vulnerabilities exploited by attackers for system access and control. For example, the 2017 global ransomware attack was allegedly enabled by stealing exploits created by the US National Security Agency [9]. Nation states can also employ other methods, including insider and supply chain attacks. These attacks can be difficult to defend due to the number of components and personnel involved in the design, operations, and maintenance of SCADA.

The cybersecurity of our critical infrastructures requires sociotechnical solutions to defend against highly sophisticated cyberattacks on SCADA systems. human–computer interaction (HCI) plays a central role in these sociotechnical solutions. This book chapter provides an overview of research and challenges in the human factors of SCADA cybersecurity. The remainder of this chapter is organized as follows: The next section begins with a description of SCADA technology pervasive in critical infrastructures and highlights attack targets and security issues that put functions of many critical infrastructures at risk. Following the next section, HCI issues are identified by examining the seven phases of the cyberkill chain, the process by which adversaries stage their attacks. Lastly, we conclude with potential research directions in HCI that can strengthen the defense of SCADA platforms.

12.2 SCADA systems

SCADA is a control architecture adopted by most industrial control systems that enable human and automation in monitoring and controlling industrial processes [14,15]. SCADA systems usually contain three segments of ICT—field devices, the SCADA/control network, and the corporate network—connected as in Figure 12.1 [16].

Referring to Figure 12.1, *networking devices*, such as switches, cables, and wireless radios, provide communication services to the SCADA components. These networking devices are mainly Ethernet-based [17], which supports the communication protocols commonly adopted by field devices. Common protocols implemented in SCADA include distributed network protocol (DNP3), modbus, and fieldbus that are not adopted by the corporate network or the World Wide Web. *Computer servers and clients* host supervisory control applications, a historian, and the human–machine interface (HMI). *Supervisory control applications* process the sensor data and configure control loops executed by the field devices. The *historian* logs all process data that can provide a basis for plant engineers to diagnose anomalies and develop new control algorithms. Finally, the *HMI* provides the means for control room operators to monitor the industrial processes and assume manual control as necessary.

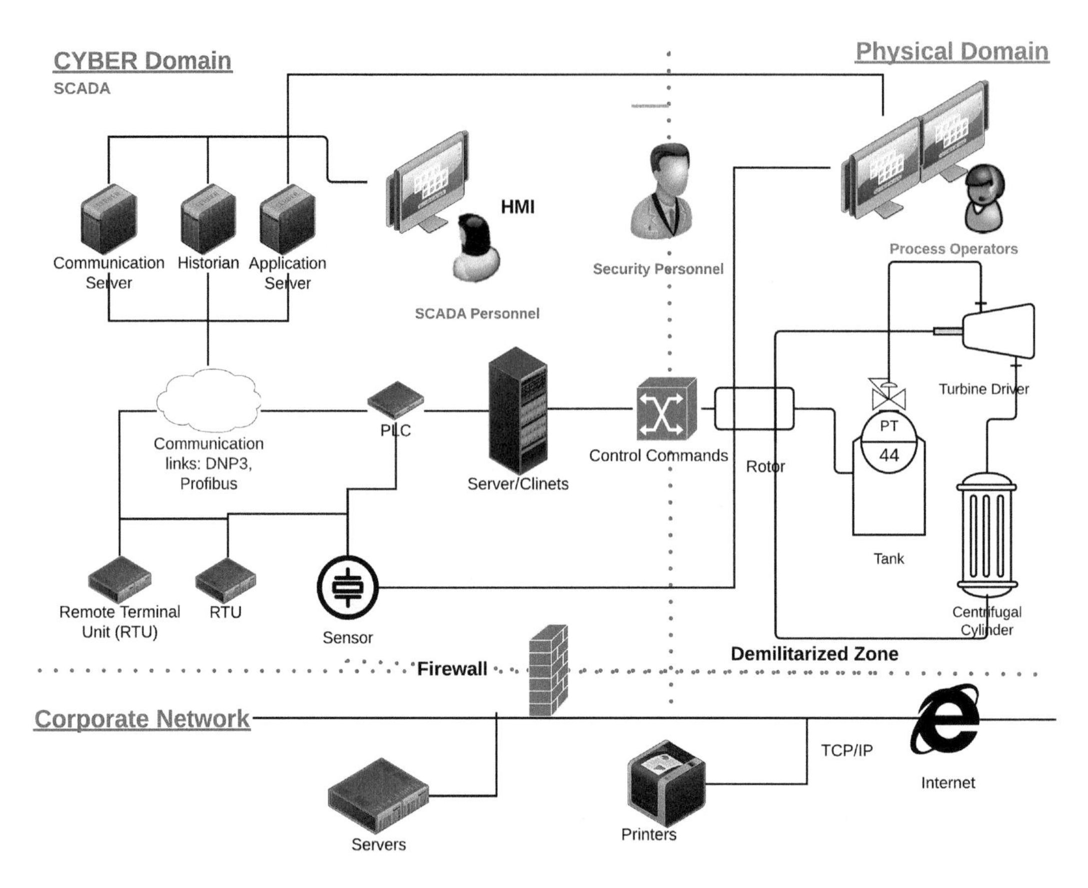

Figure 12.1 Generic implementation of SCADA.

Connected to the SCADA network are field devices that include remote terminal units (RTUs), programmable logic controllers (PLCs), and intelligent electronic devices (IEDs) for processing sensor data and issuing control commands [18]. PLCs and RTUs are digital computers customized for executing control and safety functions of industrial processes. IEDs are sensors and actuators with microprocessors for collecting plant data and changing equipment settings, respectively. The common communication protocols are DNP3 (IEEE 802.3) and modbus [14,19]. In other words, PLCs and RTUs process sensor data to issue control commands according to control loops configured by process engineers through supervisory control applications hosted by SCADA network computers. Further, PLCs and RTUs relay sensor data and actuator settings to computers hosting the historian and HMI via fieldbus protocols specified by IEEE 7-4.3.2 [20,21].

The corporate network is a common information technology (IT) network found in most businesses. Corporate networks typically consist of Ethernet-based network devices, servers, and desktop computers for carrying out business and engineering activities such as marketing, e-mails, engineering design, data analysis, and cloud-/Internet-based services. Traditionally, the corporate and SCADA networks are completely isolated from each other; however, recent implementation mediates the connection between the two networks through a demilitarized zone, firewalls, and/or additional routers [22]. This mediated network connection enables users to access data or update the configuration of SCADA components while performing other tasks only available on the corporate network (e.g., e-mail services). The corporate network adopts the more common communication protocols (e.g., user datagram protocol) and connects to the Internet.

12.2.1 Nature of SCADA cybervulnerabilities

Cybersecurity is essential for maintaining operations of our critical infrastructures. However, SCADA implementations in most critical infrastructures are extremely difficult to defend due to legacy, functional, and system design reasons [16].

SCADA technology evolves from the gradual deployment of embedded computers and microprocessors primarily designed to enhance the efficiency of supervisory control. Modern SCADA implementations increase connectivity among internal components and to other networks for maximum automatic and distributed process control [23], dramatically increasing the likelihood of intrusions [22]. Meanwhile, many SCADA technologies with virtually no security foundation in their design have become open standards and heavily deployed for their supervisory control functions. For example, commonly used industrial protocols such as modbus and DNP3 often do not support information integrity checking or authentication mechanisms between master and slave devices, resulting in potential risks in the corruption of control commands. Publicly known information of SCADA systems also allows attackers to gain in-depth system knowledge and access to technology [24]. Typically, critical infrastructure sectors are capital intensive and long-term investments and are likely to operate with either older equipment built with minimal security features or new equipment built on a relatively poor security foundation.

Besides legacy design issues, functional requirements place additional constraints on security. The control of physical processes can be highly sensitive to time lags prohibiting computationally intensive encryption or other mechanisms. For example, the nuclear industry gives high priority to the response time of safety systems [25]. SCADA systems also tend to run on out-of-date software that contains known, or probable, vulnerabilities. Prior to any software patching, validation activities must be conducted to ensure all control and safety functions must remain intact. Otherwise, the security updates would introduce undue risk into operations of the physical process [26,27]. This delay in patching software extends the time for attackers to deploy known exploits.

Finally, general system design characteristics may also constrain the cybersecurity of SCADA. These systems are inherently complex sociotechnical systems relying on many different types of components and professionals working internally and externally to their organizations [28]. Consequently, the attack surface for all system is immense. The individual components as well as the interfaces between the wide spectrum of technologies need to be secured. Also, connectivity is likely to increase for cloud-based services [29]. For example, smart grid systems will likely operate in a more decentralized paradigm, in which distribution network operators will need remote access to many geographically distributed and small-scale power generators, such as solar and wind [30]. As a result, many workers, or simply users such as homeowners, are necessary to operate and maintain SCADA components, thereby increasing the possibility of credentials being leaked through access control mismanagement (i.e., careless handling of passwords) or phishing (i.e., fraudulent e-mails resulting in malware infection that steals credentials). There is also the prospect of insider attacks, in which employees with system access can circumvent technological preventions [31].

The weak security foundation has resulted in SCADA systems being the prime targets for cyberattackers to inflict damages on physical infrastructures without any use of force [32]. Security measures guarding these cyberthreats must account for the two unique aspects of cybersecurity in SCADA systems. First, security efforts must consider how SCADA devices for controlling physical processes/equipment are designed, configured, operated, and maintained differently from conventional computer network systems. Equally important, cyberattacks on SCADA systems are best characterized as a part of warfare given their geopolitical motivations, as opposed to cybercrimes driven by financial incentives. These two essential aspects suggest that a military perspective on disrupting the cyberattack process would be appropriate for identifying and developing security mechanisms for the design, implementation, operation, and maintenance of SCADA devices. The next section examines the security mechanisms and HCI issues pertinent to thwarting adversaries' attack process, commonly known as the cyberkill chain [33]. We also relate each phase of the cyberkill chain to the infamous Stuxnet cyberattack on the Iranian enrichment facility to exemplify this approach for identifying HCI issues.

12.3 HCI issues in the cyberkill chain

The cyberkill chain refers to "a systematic process to target and engage an adversary to create desired effects." [33, p. 4]. The cyberkill chain consists of seven phases: reconnaissance, weaponization, delivery, exploitation, installation, command and control (C2), and actions on objectives [34].

The first phase is *reconnaissance*, during which adversaries identify, select, and profile targets through legal and illegal means. In other words, this is an information-gathering activity about the targets for developing custom cyberthreats. Many of these activities can be legal, such as gathering information on specific employees for targeted phishing [35] or studying the vulnerabilities of SCADA devices [36], while others, such as industrial espionage, may be illegal [37]. Adversaries may automate part of this process with web crawlers that generally do not pose any serious security threats [38]. For example, the Stuxnet attack in 2011 likely involved some form of espionage. It is now surmised that the alleged attackers were the American and Israeli governments, interested in obstructing Iran's progress toward developing nuclear weapons [39]. During this phase, the attackers became knowledgeable about personnel and details of the SCADA system configurations in Iranian nuclear facilities.

Thwarting reconnaissance, then, like that described earlier against Iran, clearly requires tight information control to prevent adversaries from acquiring personnel and technical information to devise cyberattacks [5]. However, tight information control presents major issues for employees striving against deadlines to complete their work or achieve other system goals. For example, roles and contacts of key employees (or their relatives and friends) are meant to be communicated freely, but this information is sufficient for adversaries to identify phishing targets for spreading malware to obtain credentials for access to SCADA networks [40]. Technical and configuration information about the SCADA system such as control application software is equally difficult to conceal, as it may be sufficient to be used by adversaries to infer system vulnerabilities [22]. Therefore, the central HCI issues would be how technical and personnel information can be communicated among authorized persons in a manner that would hinder reconnaissance while not affecting work efficiency.

The second phase is *weaponization*, during which adversaries employ information from reconnaissance to design and develop a malware or the payload—a binding software application with remote access that supports C2 by the attackers. The attackers set up clients and servers to support C2 and develop malware relying on exploits in the operating systems, software applications, communication protocols, and firmware of SCADA devices. The most infamous cyberweapon is Stuxnet, a malware that is capitalized on four zero-day exploits to self-install and self-destruct, replicate across removable drive and networks, communicate to external servers, and manipulate supervisory control software while bypassing any detection.

Weaponization technically does not involve any interaction with SCADA components or personnel, so thwarting this phase requires overall security improvements in the design, implementation, and operations of SCADA systems such that "hacking" becomes more difficult. Unfortunately, limited funds or research efforts are committed to secure SCADA-oriented software development, although hackers frequently take advantage of bugs in existing software for intrusion [41,42]. SCADA systems often employ commercial off-the-shelf (COTS) hardware and software to be cost-effective; however, COTS components permits adversaries to access the SCADA technology and discover exploits more easily than they can with custom "obscure" components [43]. Risk analysis is needed to clearly justify the return on the required investment For this reason, the main HCI issue concerning weaponization is developing "usable" cyberrisk analysis that can communicate to the decision makers about the investments needed to integrate upgrades to cybersecurity as part of the system design rather than as after thoughts [44,45].

The third phase is *delivery*, during which the cyberweapon is transmitted to the target network or computer system. Interestingly, cyberweapon delivery to SCADA relies on common and traditional methods: e-mail attachments [46], forced download [47], removable media [e.g., universal serial bus (USB) drives, [48]], and domain name system spoofing [49]. These delivery methods mostly intrude the corporate network that turns out to be a convenient stepping stone into the SCADA network. Delivery can be much simpler, with insiders who have sufficient privilege to install the malware or assist with the delivery in some other way. Insider delivery may also occur in the SCADA system supply chain, involving component manufacturers or transporters [50]. Stuxnet is infamous for deploying the payload on USB drives inserted into some SCADA system computers and spread to other computers through self-replication in the network of the system [51]. Although some types of generic malware can spread on its own through a network and removable drives, the Stuxnet attack was suspected to have involved insiders because the deep knowledge of the SCADA systems in the Natanz facility appeared necessary in order for this attack to have been carried out so successfully [52].

Thwarting delivery can be achieved by isolating system components, networks, and users to prevent exposure to cyberweapons. This approach has become the dominant strategy to cybersecurity. Specifically, cybersecurity has focused on perimeter-based solutions [53], including firewall, antivirus, access policy (e.g., disabling all USB ports), and authentication to block or identify suspicious traffic from entering into the system. Similar to preventing reconnaissance, isolation easily conflicts with productivity or other business objectives [53]. Application-oriented access control, including sandboxes, is also fraught with usability issues [54]. HCI issues in thwarting delivery involve developing perimeter-based solutions that are inherently flexible to accommodate a wide range of worker activities involving frequent communication with external networks or systems that can contain threats. For example, operations of many companies rely on cloud computing and the Internet of things that potentially expose the corporate and thus SCADA networks to much more malware. Further, these perimeter-based solutions must include user friendly interfaces for supporting appropriate configurations with considerations to cyberrisks.

The fourth phase is *exploitation*, which refers to leveraging the exploits in the software, communication protocols, or even workers to trigger the execution of malware code without detection. For example, a PDF attachment in a phishing e-mail that appears to be sent from a friend may be sufficient to elicit a click to initiate the installation of malware [55]. Further, sophisticated cyberattacks often employ multiple exploits to trigger installation (i.e., both human and software/protocol). Besides efforts in their discovery during weaponization, exploits engineered into the cyberweapon at the time of the development must remain effective against the operating systems and intrusion detection software after delivery. Otherwise, malware installation would be detected leading to removal or some cyberresponse. Stuxnet likely exploited both human and software vulnerabilities in that workers were successfully tempted to plug USB flash drives of unknown origin into a SCADA computer, and the infected USB drives took advantage of Windows "autorun" vulnerabilities (i.e., LNK and autorun.inf) to initiate self-install.

Mitigating exploits requires timely actions that eliminate human and machine vulnerabilities on which cyberweapons capitalize [56]. In general, software patching and updating can thwart some exploitations of the technology, but extensive functional testing is required to ensure safe and efficient operations of physical equipment [36]. Meanwhile, human vulnerabilities can be reduced with training and cognitive aids (e.g., spam or malware alerts). HCI issues related to human vulnerabilities and solutions are extensively examined for IT systems (see Iachello and Hong [57] and Sasse et al. [58]); however, both training and cognitive aids must be augmented to include materials specific to SCADA components and configurations. For example, SCADA employees must be trained to guard against latest infection methods that could jump the "air gap" (i.e., not connected to the Internet) such as removable drive infection after the forensics of Stuxnet was published. In general, essential HCI research includes effective training methods on information privacy and security and intuitive cognitive aids to reduce the likelihood of human exploitation [59]. Further, developing usable tools for testing software patches and tracking vulnerabilities for threats related to patching delays is needed.

The fifth phase is the *installation* of the malware that inject codes into the operating systems and software applications for adversaries to gain remote access and persistence (including some functional controls of the software) inside the target environment. This installation phase is intricately linked to exploitation (previous) and C2 (next). Installation is thus very complex because the exploits must target software applications specific to controlling SCADA components and the operating systems, to temporarily bypass security software and system authority. Then, installation needs to reoccur during C2 when attacker servers discretely update the malware. For Stuxnet, after the initial exploitation, the malware first gathered host information [e.g., operating system (OS) version and antivirus) to match targets. The actions are followed by either self-removal or installation/code injection to leverage the necessary exploits for authority escalation. Stuxnet has a two-part code injection. Common to most cyberattacks, the first part of code injection targeted background processes of the Windows OS that could obfuscate its traces, infect removable drives, and communicate with other resources in the network. Unique to attacks on SCADA systems, the second part of code injection targeted the supervisory control application (i.e., Siemens Step 7) to gain access and control of PLCs that gather sensor data and issue a command to centrifuges for enriching the uranium.

Thwarting the malware installation and code injection essentially requires the same security mechanisms for mitigating exploitation that provides authority escalation and vector obfuscation. In other words, mitigating exploitation with software updates or worker aids can directly hamper installation. When systems fail to prevent exploitation and installation, security would need to rely on scanning the network for malicious code and testing software for integrity. Although useful, both mechanisms cannot guarantee security, as signatures of malicious codes are often well hidden [60] and software integrity is testable only for well-defined criteria, such as safety-critical events that should never occur [61]. Further, the characterization of detected malicious codes or software compromises in the system is necessary to support security and operations personnel in assessing potential cyberimpacts and physical impacts to inform intervention (e.g., degraded operations versus shutdown). Hence, HCI issues concern developing tools that help workers identify and characterize compromises to support decision-making. Further, HCI research should support creating effective forensics tools for detecting points of delivery and exploitation that would require monitoring and subsequent elimination [62] and developing usable alarm systems for prioritizing attention toward high-risk areas.

The sixth phase is *C2*, during which the malware establish a communication channel to attacker servers to transmit information, receive updates, and execute commands. C2 can be particularly lengthy because adversaries must time their strikes based on the dynamics of geopolitics. An unnecessary strike could result in an extremely undesirable effect, including unwanted retaliation against the attackers. Forensics of the Stuxnet indicated a lengthy C2 phase as suggested by the registration of the very first server to be five years prior to action on objectives or physical damage to the centrifuges. During those five years, more C2 servers were registered to gather information on the operating system, the SCADA software configuration and network traffic [11].

The C2 phase can be foiled with advanced network intrusion and traffic monitoring that can aid workers in tracking and acting on malware infection. C2 can be like a lengthy game between attackers and defenders. Attackers can gather information and update the malware, while defenders can detect foreign intrusion and neutralize cyberweapons. HCI issues include developing visualization and expertise that help identify suspicious network traffic, with consideration to the geopolitical climate. Further, HCI research should help develope usable tools for the manual inspection of communication paths between SCADA network and untrusted areas.

The final phase is *actions on objectives*, when data exfiltration, network spreading, and system disruption occur. Attacks on SCADA typically begin with data exfiltration to acquire additional information on targets that could lead to malware updates for more targeted code injections to exact the physical sabotage of interest. Thus far, the phases of the kill chain are presented as sequential for simplicity. In reality, attackers persistently advance all phases to maintain access in the target environment because system updates or security responses could thwart one part of the kill chain and thereby neutralize the cyberweapon. This phase interacts with earlier phases through network communication and malware updates/installations. The alleged attackers of American and Israeli governments began developing Stuxnet in the early 2000s. American–Israeli attackers acted on their objective to obstruct the development of nuclear weapons when a series of International Atomic Energy Agency reports presented evidence of Iran's progress toward uranium enrichment in late 2010s [63–65]. To inflict physical damage on Iran's enrichment facility, Stuxnet injected codes into the supervisory control software (i.e., replaced the s7otbxdx.dll) that rewrote the control algorithm of the PLCs to issue commands of erratic rotational speeds to centrifuges and transmit misleading speed information to the operator displays.

Actions on the objectives of an adversary are not preventable if a defender cannot thwart earlier phases of the kill chain; however, the impact of cyberattacks can be reduced with resilient systems operation design. That is, the workers should be ready to respond to cyberevents by executing alternative procedures or adaptive work-arounds to operate through compromises or to shut down safely. HCI issues in this area require workers to engage in creative problem-solving supported by training and effective user interface design (e.g., Cone et al. [66] and Wei et al. [67]). Further, the response to cyberthreats aimed at inflicting damages to physical process also means that teamwork between security and operations personnel is essential [68]. Consequently, another HCI issue is developing intuitive collaboration user interfaces that illustrate how the compromised SCADA or cyber-components interact with equipment to affect the physical process. Finally, additional workforce may need to mobilize depending on geopolitical situations like how airports or other public areas increase physical security.

12.4 Promising research and design directions

As illustrated through the cyberkill chain and exemplified by the Stuxnet attack, the complex and evolving nature of cyberattacks on SCADA components present increasing challenges in the design, implementation, operations, and maintenance of current and future SCADA systems. SCADA systems must have the quality of preventing and adapting to successful cyberattacks. Thus, generally, human workers with better adaptive capabilities than machines are central to sociotechnical solutions for securing SCADA systems. However, HCI research on SCADA technology, let alone the cybersecurity challenge, is limited in the literature. The examination of the cyberkill chain, nevertheless, alludes to a multitude of interdisciplinary security approaches to thwarting cyberattacks. This section presents promising and emerging HCI research for securing SCADA, ranging from the perspectives of educating individual workers to shifting the asymmetric nature of cybersecurity.

12.4.1 Individual users and components

Much of current cybersecurity attention has been aimed at mitigating flaws in individual users and SCADA components. Common users are typically uneducated about cybersecurity for both IT and SCADA systems. They tend to place blind trust on the security features of the product to thwart malicious materials that could lead to cyberintrusion [69]. For example, researchers have been examining human susceptibility pointing out that males are more susceptible to phishing [70]. In addition, research has also discovered vigilance increments driven by incentives and diligence with appropriate trust on security functions [70,71]. Effective methods of minimizing common and naive human exploits include embedded training; contextual training; and online education against fraudulent e-mails, website, or attachments [72]. HCI efforts have been applied to studying different aspects of user authentication, such as password compositions [73] and two-factor authentication [74,75]. Regarding SCADA network security, research has shown narrative-based training to be more effective than tool-based training or other combinations of training methods [76]. Additional research and development on training tools is particularly necessary to address the risk associated with cloud services in SCADA technology. In general, much of the research and education effort focuses on the corporate network, neglecting the risk of SCADA components being compromised and potential impact on the physical processes.

An impending research area is a human-centric perspective of cybersituation awareness (cyber SA). Current research on cyber SA mostly adopts a data modeling perspective that focuses on what and how data in the system can be fused to identify activities of attacks. The literature contains virtually no empirical studies examining the cognitive work of how security analysts achieve cyber SA or acquire knowledge about attacks. Similarly, research formulates several key measurement dimensions of cyber SA: how well systems hypothesize true attack tracks (confidence), quality of evidence for the hypothesized tracks (purity), cost associated with signaling true and false positives, timeliness in detecting an attack for meaningful response, and the effectiveness in supporting cyberresponse [77]. Although theoretically applicable, measurements of these SA dimensions have not been investigated for assessing security personnel. The literature on cyber SA of the human workers is in paucity, representing a significant knowledge gap for designing optimal training programs and user interfaces.

Research on advancing average users in security skills and knowledge is comparatively limited with respect to the effort on technical solutions in securing current SCADA components or system processes. Although not necessarily driven by a user-centric development process, these technical solutions are briefly described to provide the context for the subsequent discussion on HCI associated with those solutions. Technical solutions under investigation and deployment include the following:

1. *Protocol security advancement* enables authentication and encryption that were neglected in the initial development of communication protocol. The development of advanced protocols incorporates evolving security technologies such as hash-based message authentication and cryptographic primitives [78].
2. *Perimeter-based access control* regulates data/communication traffic to prevent unauthenticated data or restrict authenticated data into a selected part of the network. Recent advances include software-defined networking [79] that relies on centralized traffic controllers rather than configuring network policy for individual networking devices.
3. *Network intrusion monitoring* observes and records network data including packet signature, time stamps, origin and destination Internet protocol addresses of network attempts, types of network event, and frequency of attempts/events [80]. Since heuristics based on past events are typically ineffective against zero-day attacks, network intrusion monitoring shifts toward a signature-less approach, relying on machine learning to detect traffic anomalies [81].
4. *Cyberdenial and deception* employs decoys to collect data on behaviors and induces poor decision-making by the adversaries. The most established technical implementation is honeypot, which serves as a decoy network intended to be attacked and closely monitored to obtain valuable attacker information. Honeypot can produce an early warning call while slowing down an external cyberattack [82]. A SCADA version of honeynet is being developed to protect critical infrastructures (see Lukas. Rist [83]).

12.4.2 User interface design

Realizing the potentials of various security features requires that technical innovators give attention to HCI, particularly user interface design. User interfaces can play a vital role in configuring perimeter-based solutions and understanding the impacts (or controls) on traffic to enable effective communication between, and preventing intrusion into, SCADA components (e.g., Wool [84]). More importantly, user interfaces visualizing network traffics in relation to the physical process can enhance cyber SA required for the resilient response to cyberthreats. However, traffic monitoring systems require substantially more attention in research and development to support human intelligence in confirming the detection and evolution of suspicious traffic. Advanced methods relying on machine learning for design and evaluation of user interfaces are warranted [85]. Simple but mature cognitive aides, such as spam filters and warnings of a suspicious attachment, should not be forgotten as they can improve user comprehension of potential risks of common intrusion vectors into IT systems or the corporate network [86].

HCI research is only recently attending to cyberdenial and deception. For example, One military exercise indicates the importance of adaptive problem-solving in human analysts when using honeynet for cyberdenial and deception [87]. The exercise also suggests that automation is currently incapable of sophisticated cyberdenial and deception, where the careful inference of adversarial behaviors, intents, and weaknesses is required. Thus, HCI research can play an important role in providing effective user interfaces for deploying advanced tools in SCADA systems to counterdeceive adversaries.

12.4.3 *Teamwork and resilience*

The technical and human response against SCADA attacks requires collective and coordinated efforts of employees with diverse interdisciplinary background to cover the vast attack surface of SCADA systems. The corporate and SCADA network staff must collaborate for system setup and threat diagnosis, as the two types of technology are converging. For example, many modern day control devices can likely connect to the Internet for cloud services to be offered by vendors [88]. Such trends present a trade-off between functionality and exposure that should be jointly examined by corporate and SCADA network personnel. Given confirmed intrusions, SCADA cyberresponders must collaborate with corporate administrators and control engineers to disconnect certain service paths for the prevention of further damage and review the restoration plans [8].

Cyberresiliency demands adaptive responses from many coordinated workers in SCADA [89,90]. For example, the financial cost and social impacts due to a temporary shutdown of a power plant must be considered with respect to the concerns in operational safety and compromises in security. That is, an effective cyberresponse not only demands security and operations staff to manage different technologies, but also challenges managers with financial and ethical dilemmas. There are many components to teamwork, as suggested a team effectiveness model for nuclear power plant cybersecurity [91] featuring communication, collective problem-solving, trust, shared knowledge of expertise, and adaptation. Further, government agencies can provide regular support by highlighting security solutions and alerts during times of high cyberthreats (cf., physical security level of public places). Each teamwork component can be enhanced with collaboration tools or computer-supported cooperative work. In the cyberdomain, large teams and massive information flow may also have drawbacks due to complexity with coordinating a larger number of people and the potential of propagating problems in the system [92]. In brief, research on multidisciplinary teamwork or collaboration between different personnel is pivotal to the robustness and resiliency of SCADA [93].

12.4.4 *Risk assessment and resource allocation*

The cybersecurity of critical infrastructures heavily depends on the commitment of upper management or leaders of organizations, similar to how safety must "begin from the top" [94]. Investment decisions or resource commitments are driven by the business case based on risk analysis rather than anecdotes of high-profile attacks and potential implications of SCADA attacks. For this reason, there has been substantial research in cyberrisk analysis [95], especially in quantitative modeling. Exemplary models include hierarchical models [87], attack trees or attack countermeasure trees [96], and graph theory-based models [97].

Much of the quantitative modeling work is built on machine-learning that is labor intensive with limited opportunities for empirical validation.

The complex nature of risk assessment methods can challenge practitioners who are continually performing analysis in response to constantly evolving threats. HCI research may ease the challenge by improving the usability of analysis tools and methods. For example, triage analyses (filtering for data of suspicious activities) requires more focused attention than escalation analyses [98]. This finding indicates that visualizations should differ between the two types of analysis. Meanwhile, incomprehensible risk assessment can challenge organizations in allocating resources and devising strategies (e.g., training) to defend different SCADA segments [99, pp. 41–80]. Risk analysis should thus be compatible with human reasoning or information processing. Recent work on work domain analysis presents one approach to building a psychologically relevant and physically faithful model of SCADA systems to highlight the risk of physical sabotage due to inadequate cybersecurity measures [100].

12.4.5 *Attacker attribution, intelligence coordination, and deterrence (geopolitics)*

The nature of cyberrisk and the business case for both defenders and attackers likely change with reliable attribution and retribution of the attackers. Attack attribution is defined as "determining the identity or location of an attacker or an attackers intermediary" [101]. Exposed identities can result in unwanted public spotlighting of the attackers, so other critical infrastructures sectors can become aware of the cyberrisks and begin tracking adversarial activities. Common attribution techniques mainly involves data traffic monitoring [101]. However, attackers are increasingly creative in distorting their traffic patterns with intermediate hosts/servers (i.e., "hops" [102]). Recently, security experts discovered merits in monitoring specific human behaviors in attributing attacks. Boebert [103] identifies critical human-related attribution characteristics including keystroke intervals; misspellings of command names; time of day references; and duration of intrusion to infer demographics of the adversaries, such as their language [104]. Web browsing behaviors can provide insights into an attacker's behavioral patterns [105]. To enhance cybersecurity, HCI research can support the attribution of attacks in three ways—investigating how the attacker's interaction behaviors with SCADA systems could potentially expose their identity, how the higher likelihood of identity exposure might alter the attacker's behaviors (e.g., likelihood of a strike on SCADA systems), and how decision support systems can help cyberdefenders identify attackers.

Attribution enables retribution for attacks if nation states and international organizations are willing to pursue perpetrators. Retribution may shift the asymmetry of cybersecurity in that, without any retribution strategies, defenders bear most of the cost and consequence, while the attackers reap most of the rewards [106]. However, this asymmetry is reduced when consequences and monetary recuperation can be imposed on perpetrators who are caught. Legal ramifications, or simply retaliation postures, may deter attackers while financial recuperations may change investment decisions. Financial recuperations may drive new product markets in addition to the current cyberinsurance market, which provides financial security in a manner that draw investment away from permanent technical solutions [107,108]. However, details on retributions for cyberattacks are limited, as only security and government agencies are privy to details of such knowledge. Future research in HCI can contribute to cybersecurity by studying how different rules of engagement and retribution methods can influence the security investment, the design and operations of different SCADA components, as well as the behaviors of attackers. Perhaps it is only a possibility in the distant future can retribution be a significant factor in cybersecurity symmetry.

12.5 Conclusions

This chapter provides an extensive review of current and emerging SCADA security research related to human factors. A generalized SCADA architecture is presented along with the description of various components that can be targeted in cyberattacks. SCADA systems can suffer from security risks due to legacy design and functional complexity. The general IT solutions, such as vulnerability patching, are not always immediately applicable to securing SCADA/control network because some security mechanisms could interrupt real-time operations or degrade the reliability of the physical processes. The modern SCADA network often has firewall-mediated connection to the corporate network for exchanging information or accessing cloud services, presenting another risk factor. HCI research can enhance SCADA security through user interface design that supports workers in the effective configuration of security tools and acquisition of cyber SA. Further, research must begin to focus on teamwork starting with staff in security and operations for coordinating response to cyberattacks. Finally, HCI could examine how the attribution and retribution of attackers may shift the asymmetry of cybersecurity in favor of the defenders.

References

1. M. A. Grigoras, A. Popescu, E. Merce, and F. Arion, Scada system-tool and partner for sustainable agriculture development, *Bulletin of University of Agricultural Sciences and Veterinary Medicine Cluj-Napoca. Horticulture*, vol. 64, no. 1–2, pp. 703–708, 2008.
2. A. Creery and E. Byres, Industrial cybersecurity for power system and SCADA networks, in *2005 Industry Applications Society 52nd Annual Petroleum and Chemical Industry Conference*, pp. 303–309: Piscataway, NJ: Institute of Electrical and Electronics Engineers, 2005.
3. V. Rajeswari, L. P. Suresh, and Y. Rajeshwari, Water storage and distribution system for pharmaceuticals using PLC and SCADA, in *2013 International Conference on Circuits, Power and Computing Technologies (ICCPCT)*, pp. 79–86 Piscataway, NJ: Institute of Electrical and Electronics Engineers, 2013.
4. Y. Mo et al., Cyber–physical security of a smart grid infrastructure, *Proceedings of the IEEE*, vol. 100, no. 1, pp. 195–209, 2012.
5. RISI. *RISI Online Incident Database* [Online]. Available: http://www.risidata.com/Database/P240 (accessed Sept. 27, 2017), 2017.
6. M. Abrams and J. Weiss, *Malicious Control System Cyber Security Attack Case Study—Maroochy Water Services, Australia*: McLean, VA: MITRE Corporation, 2008.
7. R. M. Lee, M. J. Assante, and T. Conway, *ICS CP/PE (Cyber-to-Physical or Process Effects) Case Study Paper—Media Report of the Baku-Tbilisi-Ceyhan (BTC) Pipeline Cyber Attack*. SANS Institute, ed: SANS Industrial Control Systems, 2014.
8. R. M. Lee, M. J. Assante, and T. Conway, German steel mill cyber attack, *Industrial Control Systems*, vol. 30, 2014.
9. R. M. Lee, M. J. Assante, and T. Conway, *Analysis of the Cyber Attack on the Ukrainian Power Grid*: SANS Industrial Control Systems, 2016.
10. T. Fox-Brewster, *NotPetya Ransomeware Hackers "Took Down Ukrainian Power Grid"*: Jersey City, NJ: Forbes Media, July 3, 2017. [Online]. Available: https://www.forbes.com/sites/thomasbrewster/2017/07/03/russia-suspect-in-ransomware-attacks-says-ukraine/#5ed1f5406b89 (accessed Sept. 27, 2017).
11. N. Falliere, L. O. Murchu, and E. Chien, W32: Stuxnet Dossier, White paper, Symantec Corp., *Security Response*, vol. 5, no. 6, 2011.
12. A. Greensberg, *How an Entire Nation Became Russia's Test Lab for Cyberwar*, 2017.

13. R. M. Lee, *Potential Sample of Malware from the Ukrainian Cyber Attack Uncovered*: North Bethesda, MD: SANS Industrial Control Systems Security Blog, Jan. 1, 2016. [Online]. Available: https://ics.sans.org/blog/2016/01/01/potential-sample-of-malware-from-the-ukrainian-cyber-attack-uncovered (accessed Sept. 27, 2017).
14. S. Hong and M. Lee, Challenges and direction toward secure communication in the SCADA system, in *2010 Eighth Annual Communication Networks and Services Research Conference (CNSR)*, pp. 381–386: Piscataway, NJ: Institute of Electrical and Electronics Engineers, 2010.
15. T. Brown, Security in SCADA systems: How to handle the growing menace to process automation, *Computing & Control Engineering Journal*, vol. 16, no. 3, pp. 42–47, 2005.
16. A. Nicholson, S. Webber, S. Dyer, T. Patel, and H. Janicke, SCADA security in the light of Cyber-Warfare, *Computers & Security*, vol. 31, no. 4, pp. 418–436, 2012.
17. Institute of Electrical and Electronics Engineers. *IEEE 802.3 Ethernet Working Group*: Piscataway, NJ: Institute of Electrical and Electronics Engineers, 2005. [Online]. Available: http://www.ieee802.org/.
18. S. Sridhar and G. Manimaran, Data integrity attacks and their impacts on SCADA control system, in *2010 IEEE Power and Energy Society General Meeting*, pp. 1–6: Piscataway, NJ: Institute of Electrical and Electronics Engineers, 2010.
19. IEC (International Electrotechnical Commission). *IEC 60870-5-104: Telecontrol Equipment and Systems*: Geneva: IEC, June 2006. [Online]. Available: https://webstore.iec.ch/preview/info_iec60870-5-104%7Bed2.0%7Den_d.pdf (accessed Sept. 27, 2017).
20. H. S. Kim, J. M. Lee, T. Park, and W. H. Kwon, Design of networks for distributed digital control systems in nuclear power plants, in *International Topical Meeting on Nuclear Plant Instrumentation, Controls, and Human-Machine Interface Technologies (NPIC&HMIT 2000)*, 2000.
21. Institute of Electrical and Electronics Engineers. *Standard Criteria for Programmable Digital Devices in Safety Systems of Nuclear Power Generating Stations*: Piscataway, NJ: Institute of Electrical and Electronics Engineers, 2016. [Online]. Available: https://standards.ieee.org/findstds/standard/7-4.3.2-2016.html (accessed Sept. 27, 2017).
22. V. M. Igure, S. A. Laughter, and R. D. Williams, Security issues in SCADA networks, *Computers & Security*, vol. 25, no. 7, pp. 498–506, 2006.
23. S. Karnouskos, Cyber-physical systems in the smartgrid, in *2011 9th IEEE International Conference on Industrial Informatics (INDIN)*, pp. 20–23: Piscataway, NJ: Institute of Electrical and Electronics Engineers, 2011.
24. A. A. Cárdenas, S. Amin, and S. Sastry, Research challenges for the security of control systems, in *HotSec*, 2008.
25. J. She and J. Jiang, On the speed of response of an FPGA-based shutdown system in CANDU nuclear power plants, *Nuclear Engineering and Design*, vol. 241, no. 6, pp. 2280–2287, 2011.
26. A. Cardenas, S. Amin, B. Sinopoli, A. Giani, A. Perrig, and S. Sastry, Challenges for securing cyber physical systems, in *Workshop on Future Directions in Cyber-Physical Systems Security*, 2009, p. 5.
27. B. Babu, T. Ijyas, P. Muneer, and J. Varghese, Security issues in SCADA based industrial control systems, in *2017 2nd International Conference on Anti-Cyber Crimes (ICACC)*, pp. 47–51: Piscataway, NJ: Institute of Electrical and Electronics Engineers, 2017.
28. E. M. Frazzon, J. Hartmann, T. Makuschewitz, and B. Scholz-Reiter, Toward socio-cyber-physical systems in production networks, *Procedia CIRP*, vol. 7, pp. 49–54, 2013.
29. A. Sajid, H. Abbas, and K. Saleem, Cloud-assisted IoT-based SCADA systems security: A review of the state of the art and future challenges, *IEEE Access*, vol. 4, pp. 1375–1384, 2016.
30. S. A. Boyer, *SCADA: Supervisory Control and Data Acquisition*: Research Triangle Park, NC: International Society of Automation, 2009.
31. D. J. John, R. W. Smith, W. H. Turkett, D. A. Cañas, and E. W. Fulp, Evolutionary based moving target cyber defense, in *Proceedings of the Companion Publication of the 2014 Annual Conference on Genetic and Evolutionary Computation*, pp. 1261–1268: New York: Association for Computing Machinery, 2014.
32. B. Zhu, A. Joseph, and S. Sastry, A taxonomy of cyber attacks on SCADA systems, in *2011 International Conference on Internet of Things and 4th International Conference on Cyber, Physical and Social computing (iThings/CPSCom)*, pp. 380–388: Piscataway, NJ: Institute of Electrical and Electronics Engineers, 2011.

33. E. M. Hutchins, M. J. Cloppert, and R. M. Amin, Intelligence-driven computer network defense informed by analysis of adversary campaigns and intrusion kill chains, *Leading Issues in Information Warfare & Security Research*, vol. 1, no. 1, p. 80, 2011.
34. T. Yadav and A. M. Rao, Technical aspects of cyber kill chain, in *International Symposium on Security in Computing and Communication*, pp. 438–452: Berlin: Springer, 2015.
35. D. D. Caputo, S. L. Pfleeger, J. D. Freeman, and M. E. Johnson, Going spear phishing: Exploring embedded training and awareness, *IEEE Security & Privacy*, vol. 12, no. 1, pp. 28–38, 2014.
36. G. M. Coates, K. M. Hopkinson, S. R. Graham, and S. H. Kurkowski, A trust system architecture for SCADA network security, *IEEE Transactions on Power Delivery*, vol. 25, no. 1, pp. 158–169, 2010.
37. O. Thonnard, L. Bilge, G. O'Gorman, S. Kiernan, and M. Lee, Industrial espionage and targeted attacks: Understanding the characteristics of an escalating threat, *Research in Attacks, Intrusions, and Defenses*, pp. 64–85, 2012.
38. Y. Y. Haimes and C. G. Chittester, A roadmap for quantifying the efficacy of risk management of information security and interdependent SCADA systems, *Journal of Homeland Security and Emergency Management*, vol. 2, no. 2, 2005.
39. J. P. Farwell and R. Rohozinski, Stuxnet and the future of cyber war, *Survival*, vol. 53, no. 1, pp. 23–40, 2011.
40. M. Workman, Wisecrackers: A theory-grounded investigation of phishing and pretext social engineering threats to information security, *Journal of the Association for Information Science and Technology*, vol. 59, no. 4, pp. 662–674, 2008.
41. K.-K. R. Choo, Cloud computing: Challenges and future directions, *Trends and Issues in Crime and Criminal Justice*, no. 400, p. 1, 2010.
42. A. Hildick-Smith, Security for critical infrastructure scada systems, *SANS Reading Room, GSEC Practical Assignment, Version*, vol. 1, pp. 498–506, 2005.
43. K.-K. R. Choo, The cyber threat landscape: Challenges and future research directions, *Computers & Security*, vol. 30, no. 8, pp. 719–731, 2011.
44. C. Colwill, Human factors in information security: The insider threat–Who can you trust these days? *Information Security Technical Report*, vol. 14, no. 4, pp. 186–196, 2009.
45. B. Jerman-Blažič, An economic modelling approach to information security risk management, *International Journal of Information Management*, vol. 28, no. 5, pp. 413–422, 2008.
46. R. C. Dodge and A. J. Ferguson, Using Phishing for User E-mail Security Awareness, in *SEC*, pp. 454–459: Berlin: Springer, 2006.
47. T. Moore and R. Clayton, Examining the impact of website take-down on phishing, in *Proceedings of the Anti-Phishing Working Groups 2nd Annual eCrime Researchers Summit*, pp. 1–13: New York: Association for Computing Machinery, 2007.
48. S. Karnouskos, Stuxnet worm impact on industrial cyber-physical system security, in *IECON 2011—37th Annual Conference on IEEE Industrial Electronics Society*, pp. 4490–4494: Piscataway, NJ: Institute of Electrical and Electronics Engineers, 2011.
49. B. Yan, B. Fang, B. Li, and Y. Wang, Detection and defence of DNS spoofing attack, *Jisuanji Gongcheng/Computer Engineering*, vol. 32, no. 21, pp. 130–132, 2006.
50. J. Slay and M. Miller, Lessons learned from the Maroochy water breach, in *Critical Infrastructure Protection: ICCIP 2007*, pp. 73–82: Boston, MA: Springer, 2007.
51. J. Clark, S. Leblanc, and S. Knight, Risks associated with usb hardware trojan devices used by insiders, in *2011 IEEE International on Systems Conference (SysCon)*, pp. 201–208: Piscataway, NJ: Institute of Electrical and Electronics Engineers, 2011.
52. T. M. Chen and S. Abu-Nimeh, Lessons from stuxnet, *Computer*, vol. 44, no. 4, pp. 91–93, 2011.
53. P. S. M. Pires and L. A. H. Oliveira, Security aspects of SCADA and corporate network interconnection: An Overview, in *International Conference on Dependability of Computer Systems, 2006. DepCos-RELCOMEX'06*, pp. 127–134: Piscataway, NJ: Institute of Electrical and Electronics Engineers, 2006.
54. Z. C. Schreuders, T. McGill, and C. Payne, The state of the art of application restrictions and sandboxes: A survey of application-oriented access controls and their shortfalls, *Computers & Security*, vol. 32, pp. 219–241, 2013.

55. M. Chandrasekaran, K. Narayanan, and S. Upadhyaya, Phishing e-mail detection based on structural properties, in *NYS Cyber Security Conference*, vol. 3, 2006.
56. S. Kraemer, P. Carayon, and J. Clem, Human and organizational factors in computer and information security: Pathways to vulnerabilities, *Computers & Security*, vol. 28, no. 7, pp. 509–520, 2009.
57. G. Iachello and J. Hong, End-user privacy in human–computer interaction, *Foundations and Trends® in Human–Computer Interaction*, vol. 1, no. 1, pp. 1–137, 2007.
58. M. A. Sasse, S. Brostoff, and D. Weirich, Transforming the 'weakest link'—A human/computer interaction approach to usable and effective security, *BT technology Journal*, vol. 19, no. 3, pp. 122–131, 2001.
59. C. B. Mayhorn and P. G. Nyeste, Training users to counteract phishing, *Work*, vol. 41, no. Supplement 1, pp. 3549–3552, 2012.
60. U. Bayer, A. Moser, C. Kruegel, and E. Kirda, Dynamic analysis of malicious code, *Journal in Computer Virology*, vol. 2, no. 1, pp. 67–77, 2006.
61. F. B. Schneider, Enforceable security policies, *ACM Transactions on Information and System Security (TISSEC)*, vol. 3, no. 1, pp. 30–50, 2000.
62. J. R. Nurse, S. Creese, M. Goldsmith, and K. Lamberts, Guidelines for usable cybersecurity: Past and present, in *2011 Third International Workshop on Cyberspace Safety and Security (CSS)*, pp. 21–26: Piscataway, NJ: Institute of Electrical and Electronics Engineers, 2011.
63. F. Izadi and H. Saghaye-Biria, A discourse analysis of elite American newspaper editorials: The case of Iran's nuclear program, *Journal of Communication Inquiry*, vol. 31, no. 2, pp. 140–165, 2007.
64. International Atomic Energy Agency. *IAEA Annual Report for 2008*: Vienna: International Atomic Energy Agency, 2008. [Online]. Available: https://www.iaea.org/publications/reports/annual-report-2008 (accessed Sept. 27, 2017).
65. International Atomic Energy Agency, *IAEA Annual Report for 2009*: Vienna: International Atomic Energy Agency, 2009.
66. B. D. Cone, C. E. Irvine, M. F. Thompson, and T. D. Nguyen, A video game for cyber security training and awareness, *Computers & Security*, vol. 26, no. 1, pp. 63–72, 2007.
67. D. Wei, Y. Lu, M. Jafari, P. Skare, and K. Rohde, An integrated security system of protecting smart grid against cyber attacks, in *Innovative Smart Grid Technologies (ISGT), 2010*, pp. 1–7: Piscataway, NJ: Institute of Electrical and Electronics Engineers, 2010.
68. P. Rajivan, M. Champion, N. J. Cooke, S. Jariwala, G. Dube, and V. Buchanan, Effects of teamwork versus group work on signal detection in cyber defense teams, in *International Conference on Augmented Cognition*, pp. 172–180: Berlin: Springer, 2013.
69. A. Acquisti, L. Brandimarte, and G. Loewenstein, Privacy and human behavior in the age of information, *Science*, vol. 347, no. 6221, pp. 509–514, 2015.
70. S. L. Pfleeger and D. D. Caputo, Leveraging behavioral science to mitigate cyber security risk, *Computers & Security*, vol. 31, no. 4, pp. 597–611, 2012.
71. S. Sheng, M. Holbrook, P. Kumaraguru, L. F. Cranor, and J. Downs, Who falls for phish? A demographic analysis of phishing susceptibility and effectiveness of interventions, in *Proceedings of the SIGCHI Conference on Human Factors in Computing Systems*, pp. 373–382: New York: Association for Computing Machinery, 2010.
72. P. Kumaraguru, S. Sheng, A. Acquisti, L. F. Cranor, and J. Hong, Teaching Johnny not to fall for phish, *ACM Transactions on Internet Technology (TOIT)*, vol. 10, no. 2, p. 7, 2010.
73. Y.-Y. Choong, A cognitive-behavioral framework of user password management lifecycle, in *International Conference on Human Aspects of Information Security, Privacy, and Trust*, pp. 127–137: Berlin: Springer, 2014.
74. P. Hoonakker, N. Bornoe, and P. Carayon, Password authentication from a human factors perspective: Results of a survey among end-users, in *Proceedings of the Human Factors and Ergonomics Society Annual Meeting*, vol. 53, no. 6, pp. 459–463: Los Angeles, CA: SAGE Publications, 2009.
75. N. Gunson, D. Marshall, H. Morton, and M. Jack, User perceptions of security and usability of single-factor and two-factor authentication in automated telephone banking, *Computers & Security*, vol. 30, no. 4, pp. 208–220, 2011.

76. S. Stevens-Adams et al., Enhanced training for cyber situational awareness, in *International Conference on Augmented Cognition*, pp. 90–99: Berlin: Springer, 2013.
77. G. Tadda, J. J. Salerno, D. Boulware, M. Hinman, and S. Gorton, Realizing situation awareness within a cyber environment, in *Multisensor, Multisource Information Fusion: Architectures, Algorithms, and Applications 2006*, vol. 6242, p. 624204: Bellingham WA: International Society for Optics and Photonics, 2006.
78. M. Puys, M.-L. Potet, and P. Lafourcade, Formal analysis of security properties on the OPC-UA SCADA protocol, in *International Conference on Computer Safety, Reliability, and Security*, pp. 67–75: Berlin: Springer, 2016.
79. I. Ahmad, S. Namal, M. Ylianttila, and A. Gurtov, Security in software defined networks: A survey, *IEEE Communications Surveys & Tutorials*, vol. 17, no. 4, pp. 2317–2346, 2015.
80. P. Oman and M. Phillips, Intrusion detection and event monitoring in SCADA networks, *Critical Infrastructure Protection: ICCIP 2007*, pp. 161–173: Boston, MA: Springer, 2007.
81. L. Koc, T. A. Mazzuchi, and S. Sarkani, A network intrusion detection system based on a hidden naive Bayes multiclass classifier, *Expert Systems with Applications*, vol. 39, no. 18, pp. 13492–13500, 2012.
82. M. B. Salem, S. Hershkop, and S. J. Stolfo, A survey of insider attack detection research, in *Insider Attack and Cyber Security*, S. J. Stolfo, S. M. Bellovin, A. D. Keromytis, S. Smith S. W. Hershkop, and S. Sinclair (eds), pp. 69–90: Boston, MA: Springer, 2008.
83. L. Rist, *The Honeynet Project*: Ann Arbor, MI: The Honeynet Project. [Online]. Available: https://www.honeynet.org/blog/201 (accessed Sept. 27, 2017).
84. A. Wool, The use and usability of direction-based filtering in firewalls, *Computers & Security*, vol. 23, no. 6, pp. 459–468, 2004.
85. H. V. Singh and Q. H. Mahmoud, Vidaq: A framework for monitoring human machine interfaces, in *2017 IEEE 20th International Symposium on Real-Time Distributed Computing (ISORC)*, pp. 141–149: Piscataway, NJ: Institute of Electrical and Electronics Engineers, 2017.
86. A. Patrick and S. Kenny, From privacy legislation to interface design: Implementing information privacy in human-computer interactions, in *Privacy Enhancing Technologies*, pp. 107–124: Berlin: Springer, 2003.
87. F. Baiardi, C. Telmon, and D. Sgandurra, Hierarchical, model-based risk management of critical infrastructures, *Reliability Engineering & System Safety*, vol. 94, no. 9, pp. 1403–1415, 2009.
88. D. S. Markovic, D. Zivkovic, I. Branovic, R. Popovic, and D. Cvetkovic, Smart power grid and cloud computing, *Renewable and Sustainable Energy Reviews*, vol. 24, pp. 566–577, 2013.
89. S. Young, Incident response and SCADA, in *Handbook of SCADA/Control Systems Security*, p. 159, 2013.
90. L. I. Millett, B. Fischhoff, and P. J. Weinberger, *Foundational Cybersecurity Research: Improving Science, Engineering, and Institutions*: Washington, DC: National Academy of Sciences, 2017. [Online]. Available: http://nap.edu/24676.
91. J. Steinke et al., Improving cybersecurity incident response team effectiveness using teams-based research, *IEEE Security & Privacy*, vol. 13, no. 4, pp. 20–29, 2015.
92. J. L. Marble, W. F. Lawless, R. Mittu, J. Coyne, M. Abramson, and C. Sibley, The human factor in cybersecurity: Robust & intelligent defense, in *Cyber Warfare*, pp. 173–206: Berlin: Springer, 2015.
93. W. L. Heinrichs, P. Youngblood, P. M. Harter, and P. Dev, Simulation for team training and assessment: Case studies of online training with virtual worlds, *World Journal of Surgery*, vol. 32, no. 2, pp. 161–170, 2008.
94. Q. Hu, T. Dinev, P. Hart, and D. Cooke, Managing employee compliance with information security policies: The critical role of top management and organizational culture, *Decision Sciences*, vol. 43, no. 4, pp. 615–660, 2012.
95. Y. Cherdantseva et al., A review of cyber security risk assessment methods for SCADA systems, *Computers & Security*, vol. 56, pp. 1–27, 2016.
96. A. Roy, D. S. Kim, and K. S. Trivedi, Attack countermeasure trees (ACT): Toward unifying the constructs of attack and defense trees, *Security and Communication Networks*, vol. 5, no. 8, pp. 929–943, 2012.

97. J. Guan, J. H. Graham, and J. L. Hieb, A digraph model for risk identification and mangement in SCADA systems, in *2011 IEEE International Conference on Intelligence and Security Informatics (ISI)*, pp. 150–155: Piscataway, NJ: Institute of Electrical and Electronics Engineers, 2011.
98. A. D'Amico and M. Kocka, Information assurance visualizations for specific stages of situational awareness and intended uses: Lessons learned, in *IEEE Workshop on Visualization for Computer Security 2005 (VizSEC 05)*, pp. 107–112: Piscataway, NJ: Institute of Electrical and Electronics Engineers, 2005.
99. W. Boone, Governance and compliance, in *Handbook of SCADA/Control Systems Security*, p. 201, 2016.
100. H. Wang, N. Lau, and R. Gerdes, Application of Work Domain Analysis for Cybersecurity, in *International Conference on Human Aspects of Information Security, Privacy, and Trust*, pp. 384–395: Berlin: Springer, 2017.
101. D. A. Wheeler and G. N. Larsen, *Techniques for Cyber Attack Attribution*: Alexandria, VA: Institute For Defense Analyses, 2003.
102. J. Hunker, B. Hutchinson, and J. Margulies, Role and Challenges for Sufficient Cyber-Attack Attribution, pp. 5–10: Institute for Information Infrastructure Protection, 2008.
103. W. E. Boebert, A survey of challenges in attribution, in *Proceedings of a workshop on Deterring CyberAttacks*, pp. 41–54, 2010.
104. D. G. Brizan, A. Goodkind, P. Koch, K. Balagani, V. V. Phoha, and A. Rosenberg, Utilizing linguistically enhanced keystroke dynamics to predict typist cognition and demographics, *International Journal of Human-Computer Studies*, vol. 82, pp. 57–68, 2015.
105. M. Abramson and D. W. Aha, User Authentication from Web Browsing Behavior, in *FLAIRS conference*, pp. 268–273, 2013.
106. K. Geers, The challenge of cyber attack deterrence, *Computer Law & Security Review*, vol. 26, no. 3, pp. 298–303, 2010.
107. N. Shetty, G. Schwartz, M. Felegyhazi, and J. Walrand, Competitive cyber-insurance and Internet security, in *Economics of Information Security and Privacy*, N. Shetty, G. Schwartz, M. Felegyhazi, and J. Walrand (eds), pp. 229–247: Boston, MA: Springer, 2010.
108. R. Pal, L. Golubchik, K. Psounis, and P. Hui, Will cyber-insurance improve network security? A market analysis, in *2014 Proceedings of IEEE INFOCOM*, pp. 235–243: Piscataway, NJ: Institute of Electrical and Electronics Engineers, 2014.

chapter thirteen

Healthcare information security and assurance

Ulku Yaylacicegi Clark and Jeffrey G. Baltezegar

Contents

13.1 Introduction

Healthcare represents a significant segment of the US economy. In 2016, total health expenditures reached $3.3 trillion and are projected to grow to 5.6% per year until 2025 (CMS, 2016). Per Centers for Medicare and Medicaid Services statistics, the 2016 figures translate into $10,348 per person or 17.9% of the gross domestic product of the nation. The use of in healthcare organizations is believed to alleviate the expenditure while increasing the overall quality of patient care. HIT services involve the use of technology to provide healthcare as well as to enable the comprehensive exchange the digital health information (Office of National Coordinator for Health Information Technology, 2015). This chapter introduces and discusses some issues related to the digitization of health sector. Section 13.1 summarizes regulations Health Insurance Portability and Accountability Act (HIPAA) and Health Information Technology for Economic and Clinical Health (HITECH) that started the digitization. Section 13.2 discusses the three-step evolution of digital records: adoption of electronic health records (EHRs), EHR-to-EHR information exchange, and EHR–personal health record (PHR) information exchange. The challenges and opportunities introduced by mobile devices (smartphones, tables, etc.) are presented in Sections 13.3 [mobile health (mHealth) applications] and 13.4 [mobile device management (MDM) and bring your own device (BYOD)]. Section 13.5 reviews the ransomware attacks and disaster recovery plans (DRPs) as a countermeasure. The user perspective of digitized health industry is reviewed in Section 13.6.

13.1.1 HIPAA

The HIPAA of 1996 (HIPAA, Public Law 104-191) is intended to provide continuous health insurance coverage for workers who lose or change their job and to reduce the administrative burdens and cost of healthcare by standardizing the electronic transmission of administrative and financial transactions. The US Department of Health and Human Services (HHS) set guidelines by the HITECH Act in 2009 and expanded it with the HIPAA omnibus rule in 2013. Out of five titles of HIPAA, title II detailing administrative simplification addresses how electronic healthcare transactions are transmitted and stored. The HIPAA security rule mandated by title II establishes a standard for the security of electronic protected health information (aka ePHI) (Table 13.1).

The HIPAA security rule intends to safeguard the confidentiality, integrity, and availability of all ePHI by securing any individually identifiable health information during electronic or digital storage, processing, or transmission. The rule applies to covered entities (CEs) and business associates. CEs include healthcare providers, health plans, healthcare clearinghouses, and certain business associates. A business associate is any organization or person working in association with or providing services to a CE who handles or discloses personal health information (PHI) or PHRs. CEs choose the appropriate technology and controls for their own unique environment taking into consideration their size and capabilities, their technical infrastructure, the cost of the security measures, and the probability of risk. HHS Office of Civil Rights (OCR) authority is responsible for investigating violations and enforcing the security rule, and the fines for noncompliance are up to $1,500,000 per violation per year. The HIPAA security rule details administrative, physical, and technical safeguards.

The HIPAA privacy rule establishes national standards to protect individuals' medical records and other PHI and applies to health plans, healthcare clearinghouses, and those healthcare providers that conduct certain healthcare transactions electronically. The rule requires appropriate safeguards to protect the privacy of PHI and sets limits and conditions on the uses and disclosures that may be made of such information without patient authorization. The rule also gives patients rights over their health information,

Table 13.1 Five sections of HIPAA

Title I: HIPAA Health Insurance Reform	Title I protects health insurance coverage for individuals who change or lose jobs and prohibits group health plans and issuers from denying coverage to individuals with specific diseases and preexisting conditions.
Title II: HIPAA Administrative Simplification	Title II entails the HHS to establish national standards for processing electronic healthcare transactions. It also directs healthcare organizations to implement secure electronic access to health data and to remain in compliance with privacy regulations set by HHS.
Title III: HIPAA Tax-Related Health Provisions	Title III includes tax-related provisions for healthcare.
Title IV: Application and Enforcement of Group Health Plan Requirements	Title IV provides further details on provisions for individuals with preexisting conditions and those seeking continued coverage.
Title V: Revenue Offsets	Title V includes provisions on company-owned life insurance and the treatment of those who lose their US citizenship for income tax purposes.

Source: (https://www.hhs.gov/hipaa/for-professionals/index.html)

including rights to examine and obtain a copy of their health records and to request corrections.

The HIPAA enforcement rule contains provisions relating to compliance and investigations, the imposition of civil money penalties for violations of the HIPAA administrative simplification rules, and procedures for hearings.

In 2013, HHS announced the omnibus rule, which included modifications to the HIPAA privacy, security, enforcement, and breach notification rules under the HITECH Act, the Genetic Information Nondiscrimination Act, and other modifications to HIPAA rules.

13.2 EHR, health information exchanges, nationwide health information network, meaningful use, fast healthcare interoperability resources standard, and PHR

Currently, one of the HIT services is the EHR system, which is an electronic record of patient health information generated by one or more encounters in any care delivery setting (PCORI, 2017). The mandated digitization of health records by the American Recovery and Reinvestment Act of 2009 resulted in 93% adoption of EHR systems by the end of 2014. Majority of the health organizations adopting EHR systems demonstrated meaningful use of EHR, which refers to the use of EHR to improve quality and safety, improve care coordination, and maintain security and privacy of ePHI. The complete picture of the patient's health history presented by the EHR provide essential information to physicians to make the most informed decision while cutting the costs of redundant tests and examinations. EHR implementations have been shown to increase the quality of healthcare delivery and reduce the associated costs (Bourgeois and Yaylacicegi, 2010).

One of the primary goals behind the initiative of the government for encouraging the adoption of EHRs is to increase health information exchanges (HIEs) and eventually maintain a nationwide health information network (NHIN). The NHIN aims to provide a secure and interoperable health information infrastructure that allows stakeholders, such as physicians, hospitals, payors, state and regional HIEs, federal agencies, and other networks, to exchange health information electronically (Cline, 2012). The NHIN will significantly help reduce healthcare spending in the United States while improving the patient care quality.

As EHR adoption and meaningful use increased, health professionals quickly realized a multitude of interoperability problems between the EHRs of different vendors. HL7, an organization that develops standards for the exchange, integration, sharing, and retrieval of EHR, drafted the first healthcare interoperability resources guide (FHIR) in 2004 in an attempt to address integration issues with EHRs, portals, HIEs, mobile phone applications, cloud communications, and other health information technology (IT) systems (Mandel et al., 2016). FHIR is expected to have full standard status by 2017.

The logical next step is the EHR-to-patient information exchange. Many providers permit patients to access their PHRs via patient portals (Schleyer et al., 2016). In addition, patients are collecting/recording data regarding their health using various mobile applications such as MyFitnessPal. In the 10-year vision of the Office of the National Coordinator for Health Information Technology, the empowerment of users/patients to manage their own PHR and consequently get timely individualized diagnosis and treatment with the help of real-time data shared by the providers are emphasized (ONC, 2015).

13.3 mHealth applications

mHealth is used to denote how mobile and wireless technologies can be used to improve health-related services. The field of mHealth has undergone rapid changes and continues to move up the healthcare agenda (Istepanian and Xhang, 2012; Sebelius, 2011; Varshney, 2011). Seventy-seven percent of the US population now have a smartphone (Pew Research Center, 2017), and these phones continue to develop new features and see improvements in computing power. Smartphones can now be used to track, manage, and improve health (Landau, 2012a, 2012b; Powell et al., 2014). Perhaps the most visible element of mHealth is the profusion of phone applications (apps), especially the ones related to fitness and wellness. A simple search in application stores shows the presence of a large number of such applications. There were more than 90,000 iOS mHealth applications available in 2015—with a more than 100% increase compared to 43,000 iOS mHealth applications available in 2013. In addition, global health application downloads increased from 1.7 billion in 2013 to 3.7 billion 2017 (Statista, 2018). In 2020, the global mHealth market value is estimated to be $58.8 billion.

mHealth "apps" are widely used by consumers and medical professionals (e.g., patients, doctors, pharmacists, and others). The main categories of mHealth apps that are in use are reference apps (such as WebMD), wellness applications (such as MyFitnessPal), social media apps (such as PatientsLikeMe), and apps designed to access EHRs (such as Care360) and PHI (such as Microsoft HealthVault) (mHIMMS, 2015). In another report, commissioned by Royal Philips Electronics, it is reported that a growing number of mobile users are turning to—and trusting—mHealth applications (RPE, 2012). One in 10 Americans surveyed in the study believe that if it were not for web-based health information, "they might already be dead or severely incapacitated." A quarter of those surveyed use symptom checker websites or home-based diagnosis technology as much as they visit the doctor, while another 27% use these interactive applications instead of going to the doctor. While only 66% of the patients would be willing to fill a prescription from their physician, 90% of the patients are willing to use an application prescribed by their doctor (mHIMMS, 2015).

Although, there is currently little evidence-based research that can directly support the health benefits of mHealth applications, there is good reason to believe that these applications have potential to significantly benefit overall health. Perhaps the most benefit potential lies in applications designed to access EHRs and PHIs; however, both patients and doctors need to know that the privacy, security, and safety of these applications are adequately addressed before mHealth can be successfully integrated into the healthcare system. mHealth apps allow patients to take control of their own health, especially in areas of healthy eating, managing chronic disease, and quitting smoking (Varshney, 2011). Additionally, PHRs, which include medical history, laboratory health results, and insurance information, help people manage their lives and actively participate in their own healthcare (Davis et al., 2017). For doctors, mHealth can help provide point-of-care resources and aid in managing their practices. For patients, mHealth can improve the convenience, cost, and quality of their healthcare. mHealth is an important tool in the healthcare arena and its significance and success or failure will be determined from how it integrates with health systems and allows for better care of patients.

Populations that currently use mHealth technologies have the most benefit from the use of this technology. Patients with chronic health conditions, as well as people who want to maintain good health, would benefit from the implementation of mHealth (Varshney, 2011). As more patients become aware of the health benefits of mHealth, they are anticipated to increase subscriptions to mobile technologies and health applications. Even though there are applications that help manage a specific condition, applications that integrate and consolidate the data are still in the development phase.

For doctors, mHealth apps can help provide point of care resources and aid in managing their practices. Doctor's list privacy and security as concerns as the leading barriers to greater use of mHealth (Gagnon, 2016). Patients using mHealth applications need to have confidence that the products they are using are safe, secure, and accurate. Data security, access control, policy, and confidentiality are the main issues that must be addressed in order for mHealth to continue to flourish and deliver safe healthcare benefits.

13.4 MDM systems and BYOD

Many healthcare organizations utilize mobile devices in the workplace to allow providers to travel as needed. While these devices are a great asset for the providers, they have also presented many problems to the IT staff responsible for managing them. With the increased adoption of these devices for technology such as telemedicine, the need for secure management capabilities is increasing. There are several MDM systems (MDMSs) on the market that are designed to make it easier to manage smartphones, tablets, and laptops that are in use in the workplace. These systems typically offer a centralized interface for deploying and configuring agents on the mobile devices.

The challenge with managing these mobile devices is that a healthcare organization must protect patient data at rest, in transit, and in use on a device that may not be designed to be natively managed by an enterprise. To make matters worse, many users take their devices with them while travelling off site and outside of the protected corporate network of the organization. While travelling, new potential issues arise such as data interception, loss or theft of the device, and unauthorized access by third parties. To address these threats, a MDMS must be able to add a layer of encryption by utilizing a virtual private network. The MDMS enable users to remotely access and wipe the device on demand and ensure that only authorized users are able to access the data on the device. While some vendors have created management systems that are capable of providing these services, the IT department must also be able to effectively implement and enforce these controls on a daily basis.

In addition to configuring the device for MDM, users must be educated on how to use the MDM application, what the limitations of the MDM application are, and what their role is in supplementing the MDM application to ensure the security of their device. For instance, an IT department may choose not to enforce automatic updates to mobile devices, to help minimize downtime, or to prevent potential problems from new software revisions. The user is then responsible for updating their mobile device to ensure that it has the latest security patches in place. When considering mobile device vulnerabilities disclosed recently, patching mobile devices is critical to maintaining confidentiality of data in use on these devices. Users who are not aware of the importance of updating their mobile device will place their data at a much greater risk of exposure.

Some of the common capabilities found in MDM applications are device encryption, screen locks, encrypted backups, and remote wiping. Mobile device users must understand the purpose of these features and how they work to ensure the security of their device and the data stored on it. From a mobile device user's perspective, the MDM application may even be a hindrance rather than a tool that can greatly increase the security of their data. For instance, screen locks using a personal identification number (PIN) or password may be inconvenient, but they help prevent unauthorized parties from accessing the mobile device. Some organizations may even enforce data destruction if the PIN or password is incorrectly entered more than five times. This protection measure greatly increases security while making the user more accountable for access to their device. Leaving a device in a bag or allowing a child to play with it could potentially cause the phone to be reset without the user's knowledge.

Alternative authentication measures such as fingerprint scanning or facial recognition are also possible. Using a fingerprint scan is already possible with several iPhone and Android devices, but is still not proven to be completely secure. According an article in the *New York Times*, researchers at the New York University and the Michigan State University "were able to develop a set of artificial 'MasterPrints' that could match real prints similar to those used by phones as much as 65 percent of the time" (Goel, 2017). Facial recognition is also available in some smartphones but, in some cases, has been defeated simply by placing a photo of the authorized user in front of the device's camera (Amadeo, 2017).

What does all this mean for the end user? The safest built-in authentication method available for mobile devices is still the PIN or password. If the user is following password best practices, however, it is likely that they must keep up with as many as 20 different passwords for various accounts. The result is that many users will resort to using unsafe password management practices, such as reusing passwords for multiple accounts or using passwords that are easily guessed by an attacker. When considering the sensitive nature of data on mobile devices in the healthcare industry, strong authentication measures on mobile devices are a requirement. To supplement weak authentication measures, a two-factor authentication (2FA) mechanism such as a token (Yubikey, Duo, etc.) or time-based one time password (Google Authenticator, Duo, etc.) could be implemented. With growing support and integration into existing login portals, 2FA offers improved authentication security and may be easier to get users to "buy in" to secure processes. Simply entering an additional code into a login portal after typing in a password may be more easily adopted by a user than continually increasing the complexity of the password.

The work does not end when the device is delivered to the employee; regular auditing of its compliance status is critical but can easily be overlooked. In larger organizations, it is possible that some mobile devices are lost or stolen for extended periods before the IT staff has been notified. This scenario creates even more work for the IT staff, since they must then begin the process of reviewing log files for the time that the device was lost to identify whether a data breach occurred. Other poorly designed or implemented MDM platforms could allow a user to circumvent or altogether remove the MDM agent from the device. The device would then be unmanaged and therefore unprotected by the corporate controls that were put in place. Failure to properly configure just one mobile device could cause a major breach if the device becomes lost or stolen. According the HSS Breach Portal, there were approximately 824,324 records exposed due to a lost or stolen laptop or other portable electronic device from 2015 to 2017 (see Table 13.2). According to the data provided, 279,233 records were exposed due to lost or stolen laptops. Another 114,458 records were exposed due to lost or stolen devices identified as "other portable electronic devices." Of the breaches identified in this table, the two largest breaches—due to a lost or stolen laptop—account for over 74% of the records exposed during this time frame.

As the workplace evolves, many organizations are allowing their employees, contractors, and associates to bring their own personal devices into the workplace for use with corporate data systems. The immediate challenge is that the organization must come up with an effective method of management for these devices that meets organizational

Table 13.2 Reported number of records breached involving a mobile device from 2015 to 2017

Spokane VA Medical Center	3,275	Laptop
W. W. Grainger, Inc.	1,594	Laptop
Indiana Health Centers, Inc.	1,697	Desktop computer and laptop
South Bend Orthopedic Associates Inc	1,272	Laptop
Mercy Family Medicine	2,069	Other portable electronic device
Spectrum Health System	902	Other portable electronic device
California Pacific Orthopedics and Sports Medicine	2,263	Laptop and paper/films
Little River Healthcare	542	Laptop
Bay Area Pain and Wellness Center	548	Laptop
Southwest Community Health Center	6,000	Desktop computer and laptop
Durango Family Medicine, PC	18,790	Other portable electronic device
LKM Enterprises, Inc.	3,400	Desktop computer and laptop
Pacific Ocean Pediatrics	18,637	Other portable electronic device
LSU Healthcare Network	2,200	Other portable electronic device
Nova Southeastern University	1,086	Other portable electronic device
Michigan Facial Aesthetic Surgeons d/b/a University Physician Group	3,467	Laptop
Spine Specialist	600	Laptop
Lifespan Corporation	20,431	Laptop
Western Health Screening	15,326	Other portable electronic device
Specialty Dental Partners of Philadelphia, PLLC–DBA Rich Orthodontics	960	Desktop computer and laptop
Local 693 Plumbers & Pipefitters Health & Welfare Fund	1,291	Other portable electronic device
Denton Heart Group–Affiliate of HealthTexas Provider Network	21,665	Other portable electronic device
Sharp Memorial Hospital	754	Laptop and other portable electronic device
Wonderful Center For Health Innovation	3,358	Laptop
Children's Hospital of Los Angeles	3,594	Laptop
Managed Health Services	5,500	E-mail and laptop
Kinetorehab Physical Therapy, PLLC	665	Laptop
MGA Home Healthcare Colorado, Inc.	3,119	Laptop
Gibson Insurance Agency, Inc.	7,242	Laptop
Fred's Stores of Tennessee, Incorporated	9,624	Laptop
StarCare Speciality Health System	2,844	Laptop and paper/films
Kaiser Permanente Northern California	1,136	Other portable electronic device
California Correctional Health Care Services	400,000	Laptop
Quarles & Brady, LLP	1,032	Laptop
OptumRx, Inc.	6,229	Laptop
W. Christopher Bryant DDS PC	2,200	Other portable electronic device
Premier Healthcare, LLC	205,748	Laptop
Centers Plan for Healthy Living	6,893	Laptop
St. Luke's Cornwall Hospital	29,156	Other portable electronic device
Total	824,324	

Source: HHS, *Breach Portal: Notice to the Secretary of HHS Breach of Unsecured Protected Health Information*, HHS, Washington, DC, https://ocrportal.hhs.gov/ocr/breach/breach_report.jsf.

policy but allows the user to retain some control over their own device—some users may be reluctant to hand over control of their personal devices to their employer. To further complicate matters, there is a wide range of smartphones and tablets that must be considered when considering a strategy to manage them.

While there are some MDMSs that integrate both Android and iOS managements, there is not one perfect solution that integrates natively across all platforms. The result is that the IT department must integrate third-party systems with their enterprise network, while protecting the connections between them. The increased risk created by implementing new software increases an organization's attack surface. Hence, new software configurations and remote access requirements can cause an increase in the number of headaches for the IT staff who implement it. While the recent push for cloud-based technologies has been a welcome shift for some organizations, other problems may arise as a result of transmitting and storing data with a cloud provider. For example, some MDM platforms may require access to the internal network to issue certificates to mobile devices for authentication to wireless networks, creating more exposure of internal services to the Internet. It is also possible that smaller organizations may be unable to implement an MDMS because of the cost and/or complexity of the system. This only increases the likelihood that they will experience a breach because of an unprotected mobile device being used to access or store patient data.

Even in an IT department with staff dedicated to managing mobile devices, there is still a possibility that an incorrect configuration could allow employees to access company data without restrictions. For example, an organization could adopt a very secure MDM application and deploy it to all company-owned mobile devices, but fail to prevent users from accessing their corporate e-mail account on a personal device that does not have the MDM application installed. With the availability and popularity of cloud-based e-mail systems such as Office 365, it is possible for a user to connect their own smart devices with their e-mail accounts with little or no trouble at all. According to Trend Micro (2015), 69% of employees say that they use their smartphones for work, while their IT staff believed that only 34% of them do.

Although MDMSs can provide organizations with great control over their devices, issues still arise with the implementation of these systems. For instance, the design of the Apple iOS does not allow for integration into a Microsoft Windows domain environment, preventing the devices from utilizing device certificates. The IT staff must then create user accounts to allow the iOS devices to utilize user certificates, which results in a new problem—orphaned user accounts that will exist within the active directory until the devices have been retired.

If mobile device manufacturers could work together to agree on an open standard for developing devices that could be managed in a corporate setting, it is very likely that the security of patient data could be greatly improved. Given the rapid adoption of mobile devices in the workplace, it is easy to see that this trend will continue and will increase the need for a unified management system for these mobile devices. Creating a common framework for the integration of devices into an lightweight directory access protocol environment could allow for these devices to be integrated into corporate networks that are relying on them more and more every day, allowing for native management of the configuration of the devices as well as ensuring the secure configuration of the devices is correctly implemented and enforced. User experience would also be greatly improved by using a common security baseline in devices from all vendors, eliminating vendor-specific training and the use of third party applications for management.

13.5 Ransomware attacks and the role of the DRP

Although careful planning and implementation of security systems is critical to the mitigation of risk, not all attacks and breaches can be prevented. Healthcare organizations must therefore carefully consider the actions to take when security incidents do occur so that they can create an effective action plan to be executed when needed. This type of plan is known as a business continuity plan, or BCP. A smaller component of the BCP is the DRP, which focuses on a specific department or service. For example, a BCP will identify processes and procedures to be carried out by an organization in the event of a disaster so that the overall business can return to normal operations as quickly as possible. A DRP created by the IT department in an organization would focus on identifying procedures that restore systems and services critical to IT department operations as quickly as possible.

While BCPs have always been a requirement of the HIPAA, the threats that organizations must prepare for have evolved. Twenty years ago, an organization may have considered a physical intrusion on company property to be more likely to occur than a remote attack by a hacker. Today, however, remote attacks occur at an alarming rate thanks to the evolution of both hardware and software, allowing attackers to gain access to remote systems more quickly and more easily than ever before. Twenty years ago, an organization might not have considered the possibility that an employee could walk off site with a copy of company data in their pocket. Today, flash drives and portable hard drives have increasingly large storage capabilities and can be transported in a coat pocket. Some organizations may even find that all their sensitive data could be stored in a single portable storage device.

In recent years, a new type of malware, called ransomware, has created a need for organizations to review their existing BCP and DRP documents to ensure that they are effective in stopping it. Ransomware typically infects a system through an e-mail attachment, by clicking a link in a phishing e-mail or by plugging in an infected universal serial bus drive. Once the ransomware has infected the system, it begins to encrypt all data that does not belong to the host operating system. After the encryption process is complete, the user is notified that their data are inaccessible, and a fee will need to be paid to regain access. The encryption used by ransomware is typically unbreakable, leveraging public key cryptography that is commonly used to protect confidential data while communicating with websites over the World Wide Web. To increase the pressure to pay the ransom as quickly as possible, the attackers may only give the victim a few days to pay the ransom, at which point the attacker destroys the decryption key and access to the stolen data are lost.

Until recently, the most effective approach to recovering from a ransomware attack was to create an out-of-band backup of the data of the organization so that it could be restored as quickly as possible when needed. Some recent ransomware attacks, however, have taken things a step further; the attackers threaten to release the encrypted data to the public if a ransom is not paid. Now, not only is the IT staff dealing with an infection that is impacting day-to-day operations, they must consider that possibility that their sensitive data will be exposed if they do not pay the ransom. While the backup provides them with a way to recover quickly, they must still pay the ransom to prevent their data from further exposure. The interruption of day-to-day operations and the cost of recovery from the infection can greatly affect an organization as well. Unfortunately, simply having an out-of-band backup is not sufficient anymore.

To be prepared for a ransomware infection, healthcare organizations must have a plan in place to quickly respond to an incident, isolate the infected machine, remove the infection, and restore the systems back into production state. Simply purchasing a standard template or hiring a consultant to create the DRP is not enough, however. Every organization must regularly test its DRP for effectiveness so that when needed, the organization is able to follow the predetermined procedures with minimal delay. One effective method of testing the DRP is the "table top exercise," or "TTX," which facilitates the review of procedures in the DRP by having key personnel meet to respond to a mock disruptive event and discuss the response plan and how it would apply. Any potential problems could then be identified, and adjustments would be made to the DRP to ensure that it remains effective. Regular training should also be provided to all key personnel so that they are aware of the procedures to follow during a disruptive event, such as a ransomware infection.

A DRP should include a section that identifies all personnel who will be part of the computer incident response team, or CIRT. This team should include key members of the IT department, including system admins, network admins, technicians, as well as anyone in management. All members of the CIRT should be aware of their role as part of the team, as well as their responsibilities if called upon to respond to an incident. In the event of an emergency, an organization cannot afford to waste time deciding on which employees are responsible for each task in the recovery process. When considering the capability of recent ransomware variants such as WannaCry and Petya, it is apparent that a quick, efficient action is needed to isolate the infection before it spreads throughout a network, compromising other devices in the process. Without this critical step, a healthcare organization could potentially spend weeks recovering after a ransomware infection hits their network.

Communication is a critical component of the disaster recovery process. Once an incident occurs, it is imperative that an organization has clearly identified procedures for notifying key personnel so that they can execute the DRP. Having a documented path for the escalation of a potential incident could mean the difference between effectively stopping a ransomware infection from spreading and cleaning it up after it infects a network of devices. Once the disaster recovery procedures have begun, the communication between key personnel is critical to ensure that the plan is carried out effectively. Time spent waiting for a return phone call for authorization, for example, could cripple response efforts and allow valuable time to slip by. Recording contact information and alternate contact methods for key personnel in the DRP is a good step toward facilitating communication during this process. Notifying users of expected downtime is another critical step in communicating during a ransomware infection. Whether the effects are limited to a single server or the whole data center, the inability to access company resources during this time can have a significant impact on the employee's ability to perform their job. If they have not already been trained on how to continue operations during a disruptive event such as this, make sure they are provided specific instructions for continuing to function in their role within the organization until services can be fully restored. Finally, the communication between the organization and the patient is another critical step in the disaster recovery process. Notifying them of any potential breach and/or outage will ensure compliance with industry requirements and may help mitigate public relations problems down the road.

Given the serious nature of a ransomware attack in healthcare industry, it is imperative that organizations have a DRP in place to respond as quickly and efficiently as possible. As ransomware attacks evolve, health organizations must adapt to these threats proactively or face potentially catastrophic circumstances.

13.6 User experience

One important aspect of information security in any industry—especially healthcare—is the user's experience while interacting with the controls put into place. The oldest and simplest authentication method, the password, is becoming more of a problem as technological advances in computing have made simple passwords extremely easy to defeat. Many security experts will agree that the recommended minimum length of a password should be 12 characters, but most organizations would find it difficult to enforce this requirement, if possible at all. To make matters worse, many users have multiple accounts online. In the healthcare industry, it is possible that healthcare staff members must remember passwords for the EHR system, e-mail account, computer login, mobile device login, voice-over-Internet protocol (VOIP) voicemail, and even time clock. From a password management standpoint, this could mean up to six different passwords remembered for various accounts. Many users will resort to using unsafe methods of remembering their passwords, such as writing down their password on notes attached to their monitor or keyboard, defeating the protection provided by requiring a strong password for authentication.

Another challenge that arises in the healthcare field is that devices are sometimes shared by multiple employees. For example, one laptop placed on a rolling cart could be used by several nurses while they make their rounds to check on patients and distribute medication. A user who forgets to logoff when they finish using the laptop might return to find that someone else has entered data under the wrong username. In some cases, passwords could be shared with users who forget theirs or who cannot access their account for some reason. As a result, the organization will not be able to get an accurate accounting record of what actions were taken by a user on the shared laptop. Another reason users may be required to share a laptop is that many healthcare facilities operate 24 hours a day. Staff who work rotating shifts will often share devices with other employees in their department while they are off the clock. To make this work, some devices—such as tablets—must be set up with a shared user account so that multiple users can access a single device. In this case, it is very likely that the password to the account is found written down somewhere close by.

To support their healthcare staff, IT departments must strive to make their systems as user friendly as possible while maintaining the security of patient data. In demanding work environments such as a healthcare facility, the stress of having to remember a unique password to login to an EHR system may create inefficiency in the workflow process. One way to improve upon using a unique password for every business system is to implement a single sign on, or SSO. The SSO is a login method whereby a user authenticates to one server, who then validates the user's identity to multiple other systems. Many software vendors already support some form of SSO based on the security assertion markup language (SAML), which is an open standard for authentication between systems. By utilizing SAML in all business systems, a healthcare organization could require their users to create one strong password that protects their user account, which is then used to authenticate them within the organization. For instance, a user could then login one time to their company web portal, which would verify the user's identify. That authentication server would then automatically provide SAML identity verification to any other system (e-mail, EHR, VOIP phones, etc.) the user attempted to login to, without the user having to enter any additional passwords. With only one strong password to remember, the user would have a much better experience when authenticating to a business system and security would be greatly improved.

One of the major hurdles that healthcare organizations and software vendors must overcome is that healthcare staff often go through training and certification just to enter the field and must then maintain licensure and accreditation through on-going training and recertification. The result is that healthcare employees must spend most of their time learning and honing their skills as healthcare professionals, which does not always include technological training. Asking providers to take time away from their patients to improve their security awareness is not easily done and can negatively impact their ability to provide care to their patients. As is the case in many other industries, the organization must design a plan that can educate staff on information security fundamentals such as password strength and mobile device security in a way that can be understood and adopted by healthcare professionals.

Finally, healthcare organizations must also consider their patients when implementing EHR systems. The HITECH Act of 2009 requires that healthcare organizations provide patients with a way to access and download their electronic information. Providing external access to protected information presents a unique set of challenges to information security and must therefore be carefully considered before implementation. Many EHR systems provide a patient portal to meet the HITECH requirement and mitigate some of the risk faced by the healthcare organization. To access their data, patients typically go through a setup process where they access the portal to prove their identity and validate the personal information stored about them. Once this process is complete, the patient will be able to access their health records on demand. While this new technology is a great improvement, an extra burden is placed on the organization's staff to maintain the publicly accessible system through updates as well as continually monitor it for security events.

References

Amadeo, R. (2017, March 31). Galaxy S8 face recognition already defeated with a simple picture. *Ars Technica*, https://arstechnica.com/gadgets/2017/03/video-shows-galaxy-s8-face-recognition-can-be-defeated-with-a-picture/.

Bourgeois, S., and Yaylacicegi. U. (2010). Electronic health records: Improving patient safety and quality of care in Texas acute care hospitals. *International Journal of Healthcare Information Systems and Informatics*. 5(3), 1–13.

Cline, S. (2012). *About Health IT in North Carolina*. NDoHaH Services, North Carolina Department of Health and Human Services, Raleigh, NC.

CMS (Centers for Medicare & Medicaid Services) (2016). *NHE Fact Sheet*. CMS, Baltimore, MD. https://www.cms.gov/research-statistics-data-and-systems/statistics-trends-and-reports/nationalhealthexpenddata/nhe-fact-sheet.html (last accessed: 1/15/2018).

Davis, S., Roudsari, A., Raworth, R., Courtney, K. L., and MacKay, L. (2017, July 1). Shared decision-making using personal health record technology: A scoping review at the crossroads. *Journal of the American Medical Informatics Association*. 24(4), 857–866, https://doi.org/10.1093/jamia/ocw172.

Gagnon, M.-P., Ngangue, P., Payne-Gagnon, J., and Desmartis, M (2016). m-Health adoption by healthcare professionals: A systematic review. *Journal of the American Medical Informatics Association*. 23(1), 212–220, https://doi.org/10.1093/jamia/ocv052.

Goel, V. (2017, April 10). That fingerprint sensor on your phone is not as safe as you think. *New York Times*. https://www.nytimes.com/2017/04/10/technology/fingerprint-security-smartphones-apple-google-samsung.html.

HIPAA (US Department of Health and Human Services). HIPAA for Professionals. HHS, Washington, DC. https://www.hhs.gov/hipaa/for-professionals/index.html.

HHS (US Department of Health and Human Services). *Breach Portal: Notice to the Secretary of HHS Breach of Unsecured Protected Health Information*. HHS, Washington, DC. https://ocrportal.hhs.gov/ocr/breach/breach_report.jsf.

Istepanian, R., and Xhang, Y.-T. (2012). Guest editorial introduction to the special section: 4G health—The long-term evolution of m-health. *IEEE Trans on IT in BioMedicine*. 16(1), 1–5.

Landau, E. (2012a). Smartphone apps become "surrogate therapists." *CNN*. http://www.cnn.com/2012/09/27/health/mental-health-apps.

Landau, E. (2012b). Tracking your body with technology. *CNN*. http://www.cnn.com/2012/09/21/health/quantifiedself-data/index.html.

Mandel, J., Kreda, D., Mandl, K., Kohane, I., and Ramoni, R. (2016). SMART on FHIR: A standards-based, interoperable apps platform for electronic health records. *Journal of the American Medical Informatics Association*. 23(5), 899–908.

mHIMMS (2015). *mHealth App Essentials: Patient Engagement, Considerations, and Implementation*. Healthcare Information and Management Systems Society, Chicago, IL. http://www.himss.org/mhealth-app-essentials-patient-engagement-considerations-and-implementation.

ONC (2015). *Update On The Adoption Of Health Information Technology And Related Efforts To Facilitate The Electronic Use And Exchange Of Health Information*, Washington, DC. https://www.healthit.gov/sites/default/files/Attachment_1_-_2-26-16_RTC_Health_IT_Progress.pdf (last accessed: 5/23/2018).

PCORI (Patient-Centered Outcomes Research Institute) (2017). *Users' Guide to Integrating Patient-Reported Outcomes in Electronic Health Records*. PCORI, Washington, DC. https://www.pcori.org/document/users-guide-integrating-patient-reported-outcomes-electronic-health-records (last accessed: 1/24/2018).

Pew Research Center (2017). *Record Shares of Americans Now Own Smartphones, Have Home Broadband*. Pew Research Center, Washington, DC. http://www.pewresearch.org/fact-tank/2017/01/12/evolution-of-technology/, (last accessed: 1/24/2018).

Powell, A. C., Landman, A. B., and Bates, D. W (2014). In search of a few good apps. *JAMA*. 311(18), 1851–1852. https://doi:10.1001/jama.2014.2564.

RPE (Royal Philips Electronics) (2012). *Philips Survey Reveals One in 10 Americans Believe Online Health Information Saved Their Life*. RPE, Andover, MA. http://www.newscenter.philips.com/us_en/standard/news/press/2012/20121212_Philips_Survey_Health_Info_Tech.wpd#.UZVUN7Xrz7Q.

Schleyer, T., King, Z., and Miled, Z. B. (2016). A novel conceptual architecture for person-centered health records. *AMIA Annual Symposium Proceedings*, 1090–1099.

Sebelius, K. (2011). Keynote Address at mHealth Summit, Washington, DC. http://www.hhs.gov/secretary/about/speeches/sp20111205.html.

Statista (2018). *Number of mHealth App Downloads Worldwide from 2013 to 2017*. Statista, Hamburg. https://www.statista.com/statistics/625034/mobile-health-app-downloads/.

Trend Micro (2015). *Implementing BYOD: How Lost or Stolen Devices Endanger Companies*. Trend Micro, Shibuya. https://www.trendmicro.com/vinfo/ph/security/news/mobile-safety/how-lost-or-stolen-devices-endanger-companies.

Varshney, U. (2011). *Pervasive Healthcare Computing: EMR/EHR, Wireless and Health Monitoring*. Springer, New York.

section five

Governance

chapter fourteen

US cybersecurity and privacy regulations

Mark Schertler and Jill Bronfman

Contents

14.1 *Cybersecurity law compliance basics*

14.1.1 *Episodes of* Homeland

People have made a joke about the television show *The Walking Dead*: every episode has the same plot. We need supplies! The supplies are over there! But there are zombies over there! The problem inherent in this dilemma is obvious, and there will be casualties.

This same scenario plays out in nearly every episode of *Homeland*, a television series available on the Showtime network that dramatizes US data security and privacy issues. The name of the show itself is a reference to the actual Department of Homeland Security (DHS), a federal government agency tasked with protecting the US from terrorism and other threats to the security of the nation (US DHS, 2017).

On *Homeland*, Carrie is the central character in a series of dramatic episodes involving national security, including physical security and data security. On any given episode, Carrie needs some data, aka electronic documents. They are in a secure facility/server/behind a firewall with a password. But if she goes and gets them, she will get killed! Why? Everything the average person does is tracked online.

- Financial data: If Carrie uses a credit card to travel, that information is stored as data on a server. If anyone tracks where she spends money, she will be found.
- Healthcare information: If someone gets shot, they cannot go to a regular hospital, because in order to get medical care, you need to give out an extraordinary amount of personal and health data, and you can be found.
- Telecommunications and communications transmissions: Even pre-Internet, or really pre-web, most of the developed world was connected via the landline telephone network. Location data has always been available for landline calls, which enables the 911 network. E911 identifies your location on your cell.

Yet, it is a television show, and poor beleaguered Carrie has a limited amount of time to save the day. So what Carrie would often do is persuade someone to let her in (password), wear a disguise (use someone else's account), or carry a gun (hack/brute force). Nevertheless, they always find her and sometimes she gets several significant scratches on her nose.

14.1.2 *Introduction to cybersecurity*

Why do we need to know about cybersecurity in order to save the day? What are the risks we are trying to minimize? A myriad of new jobs are available in the cybersecurity field, mostly created in an attempt to answer these two questions. There are newly focused cybersecurity-specific professions, including audit, compliance, and risk management. In addition, there are entirely new job categories for privacy professionals, security experts, and privacy-trained product counsel. Let us say you are assigned one of these roles in your new job. What can you learn in order to shine in your new position and/or profession?

There are big picture issues that you can apply to various scenarios. Corporate compliance may demand competing requirements. When you are directly dealing with a business unit, what is the goal? Look to the legal and information technology (IT) departments to create direction and training for entry-level employees. Creating a corporate compliance structure for privacy and security is a more elaborate exercise and requires a different skill set than would have previously been sufficient for advising a chief executive officer.

The risks associated with data security and privacy law failures are myriad. The most obvious and familiar is data breach, and its associated financial and legal liabilities, including state government-required reporting. The Federal Trade Commission (FTC) keeps tabs on companies who act in unfair and deceptive ways toward the public, including breaches and substandard security practices. The payment card industry (PCI) imposes additional standards for the use of credit card data. Business to business (B2B) and business to consumer (B2C) contracts may impose obligations on companies to maintain certain cybersecurity minimums standards and/or leave an open obligation to comply with industry standards, generally or with regard to specific named standards. The liability for negligence outside of contracts, and in addition to them, concerns many companies who fear a moving target standard of care for how much security will be expected and whether hindsight is required after an unforeseen, but perhaps not unforeseeable, breach. Shareholder derivative lawsuits loom in the shadows, with some of the same concerns mentioned for general liability about rapidly updating technology applied in the cybersecurity field and whether a company, especially a small company with limited resources, will be expected to keep up.

14.1.3 *Players and playbooks: Who controls corporate privacy and data security*

14.1.3.1 *Legal standards*

Cybersecurity law in the United States is a patchwork quilt of regulations, including some antiquated regulations that govern a small sector of federal government agencies, such as the Privacy Act of 1974 (US Department of Justice, 2017). This so-called Privacy Act followed a Watergate-era concern with government intrusion into the personal lives of average citizens. In actuality, the scope of this Privacy Act could not imagine the extent to which the National Security Administration and other federal and state government agencies would have access to citizens' data. This expansion in scope and breadth to the collection of data is enabled by, first, the expansion of hardware and software technologies that can interact with humans and, second, by the types of data we voluntarily, and sometimes involuntarily, share with other individuals, corporations, and the government.

14.1.3.2 *Corporate culture*

Corporations vary in their cultural responses to the primacy of data collection, processing, and storage in the twenty-first century. In nearly all cases, corporations are collecting data, although they often outsource the processing and storage of data. Increasingly, this processing of data is done by subcontracted specialists, and the storage of data is often off site, remote, and possibly beyond the reaches of the US Government to access the data by subpoena or other legal process.

Corporations are attempting to keep pace with rapid changes in technology, and the ability of the legal profession to reflect changes in technology is often limited to revising B2B contracts that are only reviewed and revised every 1–3 years. Further, products and services are designed, implemented, and thrust upon the market periodically, but without full understanding of the consequences of their ability to collect and store personal data.

14.1.3.3 *Process improvements*

Most companies begin their cybersecurity assessments with an audit of data location and an evaluation of existing processes. Obvious process improvements center upon a more consistent review and audit of data collection procedures and processes. Once a year as a standard protocol may no longer be enough for a privacy impact assessment (PIA) or data mapping exercise. Education and training supporting these procedures are crucial.

Less obvious, and perhaps more onerous, suggestions include reformulating how we do business around data collection. Should legal and policy procedures for the collection of personal data continue to be largely left to the private sector? Or should the inroads made by the sectoral regulation of certain industries, described in the next section, be made universal, in order to protect data privacy across industries and across the United States?

Further, ideally, the audit or assessment is not the first time a company has considered how its actions affect its customers, employees, and vendors. Privacy by design is a movement to imbed privacy considerations into products and services at the onset, rather than as an add-on layer after the design is completed or even after lawsuits or government prosecution yields a consent decree mandating privacy and security be added.

14.1.4 *Sectoral regulation overview: Case law, statutes, and agency regulation*

14.1.4.1 *Financial regulations: Gramm–Leach–Bliley*

The Gramm–Leach–Bliley Act (GLBA) (US FTC, 2002) was enacted to require financial institutions, according to the FTC, to disclose their data security and privacy practices to their customers and to create security protocols for their customer's personal data. Financial institutions are defined broadly by the GLBA, including not only banks that hold personal funds, but also institutions that loan money or offer insurance. The regulation covers those who render financial advice as well, in keeping with the understanding of the responsibility for private financial data at the time in which the legislation was drafted. This law created not only an industry of tools to comply with the law, but also an army of personnel associated with compliance. Further, this law, along with healthcare privacy and security regulations described below, kicked off a movement in the legal, compliance, and IT fields toward outward-facing compliance and information sharing among companies rather than just establishing privacy and security standards on a case-by-case, company-by-company basis.

14.1.4.2 *Healthcare: Health Insurance Portability and Accountability Act*

The Health Insurance Portability and Accountability Act (HIPAA) of 1996 began with an acknowledgement that medical records were increasingly becoming electronic and needed to be accessible by medical professionals. Arguably, the security and privacy requirements, while a "side effect" of this medicine, may have had the most lasting impact on the public. Covered entities, the healthcare providers, must adhere to security and privacy protective requirements for the public. The goal is to protect personal health information (PHI) now unleashed from the locked file cabinet in the doctor's office onto the world of electronic storage.

The act covered both privacy and security issues and defined for many the distinction between the two. The security requirement reads as follows: "The HIPAA Security Rule establishes national standards to protect individuals' electronic personal health information that is created, received, used, or maintained by a covered entity. The Security Rule requires appropriate administrative, physical and technical safeguards to ensure the confidentiality, integrity, and security of electronic protected health information" (US Department of Health and Human Services, 2017a). Security was focused on the safety of the data, whereas privacy rules protected individuals. The privacy rule adds that "the HIPAA Privacy Rule establishes national standards to protect individuals' medical records and other personal health information and applies to health plans, healthcare clearinghouses, and those healthcare providers that conduct certain healthcare transactions electronically. The Rule requires appropriate safeguards to protect the privacy of personal health information, and sets limits and conditions on the uses and disclosures that may be made of such information without patient authorization" (US Department of Health and Human Services, 2017b).

The HIPAA-covered entities will contract with companies who carry this information forward, by storing it or processing it, and these companies, called business associates under the law, have similar obligations to protect the covered entities' patients. In addition, the breach notification rule also requires HIPAA-covered entities and their business associates to notify those affected by a breach of their protected health information.

14.1.4.3 Telecom: Federal Communications Commission regulation

The Federal Communications Commission (FCC) ventured into the field of data privacy regulation with its Customer Proprietary Network Information (CPNI) (US FTC, 2002) restrictions. CPNI regulation was intended to limit the ability of telecommunications provider to take customer information gathered to provide one service and use it to market another service to that customer (US Government Printing Office, 2011). These laws became more and more relevant as companies began mergers, acquisitions, and joint ventures in a deregulated marketplace in the late 1990s and into the first decade of the twenty-first century. CPNI regulations attempt to create a bridge between the commercial efficiencies of using personal information gathered about existing customers and the privacy interests of customers. Several exceptions have evolved to this general principle in the law, including exceptions for emergencies and consent of the customer. In an acknowledgement of the specially situated regulated position of telecommunications providers, providers are allowed to use depersonalized, aggregated information, but only if the providers also allow others to use the same aggregated information, for a reasonable fee. Into the following decade, the technological ability to collect and share personal data has grown into a substantial industry, and it will be interesting to see how regulated versus unregulated companies deal with personal data over the next few years.

14.1.4.4 Utility and technology rulings: Federal Trade Commission regulation

The Federal Trade Commission (FTC) has gone head to head with the FCC over primary federal agency jurisdiction over security and privacy issues. The FTC relies on Section 5 of the FTC Act for its authority over these issue, but it is limited by the scope of the FCC's authority over some carriers. Section 5 of the FTC Act is general rather than specifically related to either security or privacy in that it prohibits unfair or deceptive practices by companies in their business practices. On its website, the FTC lists a broad variety of privacy-related acts to which it may enforce, including the Truth in Lending Act, the CAN-SPAM Act (for e-mail practices), the Children's Online Privacy Protection Act (protecting the personal information of children under 13), the Equal Credit Opportunity Act, the Fair Credit Reporting Act, the Fair Debt Collection Practices Act, and the Telemarketing and Consumer Fraud and Abuse Prevention Act (US FTC, 2015). As a result, the FTC has had several enforcement actions against companies to call out violations of these acts, and to, under its more general Section 5 authority, notify companies of insufficient privacy and data security practices.

14.1.4.5 Federal: FISMA and Federal Risk and Authorization Management Program

Recognizing the need to address cybersecurity incidents that affect the US Government, the US Congress has passed two different acts both with the acronym "FISMA" to improve the information security posture of systems used by the US Government for information processing. The first is the Federal Information Security Management Act of 2002 (FISMA 2002), which "requires each federal agency to develop, document, and implement an agency-wide program to provide information security for the information and systems that support the operations and assets of the agency, including those provided or managed by another agency, contractor, or other sources" (US NIST, 2017).

The follow-up is the Federal Information Security Modernization Act of 2014 (FISMA 2014), addressing the lessons learned and increased technological complexity and risk that surfaced in the intervening 12 years, "amends the Federal Information Security Management Act of 2002 (FISMA) provides several modifications that modernize federal security practices to address evolving security concerns. These changes result in less overall reporting, strengthens the use of continuous monitoring in systems, increased focus on the agencies for compliance, and reporting that is more focused on the issues caused by security incidents" (US NIST, 2017). The evolving FISMA acts promotes a risk-based approach to the implementation of cybersecurity policies and controls at an agency level so that each individual agency can address the risks to its mission in an appropriate manner versus a one–size-fits all solution for the entire US Government. There is centralized control that is mandated including the development of security standards, guidelines and minimal requirements by the US National Institute of Standards and Technology (NIST), development of appropriate government-wide information security policies by the Office of Management and Budget, and development of government-wide security incident reporting and agency annual reporting requirements by the DHS.

Federal Risk and Authorization Management Program (FedRAMP) is a US federal government-wide program for the assessment and authorization of cloud services providers. It follows FISMA guidelines and uses NIST 800-53 controls to define its cybersecurity requirements. FedRAMP applies to software-as-a-service, platform–as-a-service, and infrastructure-as-a-service providers. It provides a "do once, use many time" framework (US FedRAMP, 2017) for both cloud service providers and US agency buyers. Cloud service providers can be accessed against FedRAMP requirements and achieve an authority to operate (ATO). All US agency buyers when looking for cloud services can then review the ATO and know that the service provider has been assessed and meets the required cybersecurity controls to protect US Government data.

14.1.5 State regulation overview

14.1.5.1 State data breach/privacy regulations

Due to the failure to pass comprehensive US federal cybersecurity and privacy regulations, states have stepped in and started passing regulations governing aspects of cybersecurity themselves. As of this writing, 48 states, the District of Columbia, Guam, Puerto Rico, and the Virgin Islands have passed personal identifiable information (PII) breach disclosure laws. These state laws differ, in some cases significantly, as to what qualifies as PII. The definition of PII of most states is a combination of basic identification information such as name, social security number, state identification (ID) (driver license information), and financial account data. Various states include additional information in their PII definition including biometric information, DNA (as of this writing, only Wisconsin codifies this in law), electronic signature (as of this writing only, North Dakota codifies this in law), medical information, date of birth, employee ID, mother's maiden name, health insurance information, and tax information. In addition to differences in what constitutes PII, what constitutes a breach, who must be notified, and when and what type of disclosure is required also differ from state to state. Further, some states specify the content of the breach notification. These state breach laws are constantly being amended, and new related regulations are becoming law. The National Conference of State Legislatures (2017) has resources for tracking the constantly changing landscape of state cybersecurity and breach notification legislation.

While it is a positive step forward for citizens of states that have breach notification regulations, it is a burden on organizations that do business across state lines as differing requirements of each state can overlap or contradict each other creating a complex breach notification environment. It would seem that a unified federal breach notification law would be advantageous, but several states are worried that a federal regulation would offer their citizens less protections than their current state laws (National Law Review, 2017). It is difficult to pass federal legislation that would preempt all state regulations due to political constraints, and therefore, any federal legislation would simply add another overlapping set of standards.

14.1.5.2 New York state cybersecurity requirements for financial services companies

Another example of states stepping in due to the lack of US federal cybersecurity regulations is the New York state cybersecurity requirements for financial services companies, which went into effect on March 1, 2017. The law was passed by the New York state legislature to address increasing threats posed by cyberattacks on institutions, particularly financial institutions, and is the first state regulation to address these cybersecurity threats. "This regulation requires each company to assess its specific risk profile and design a program that addresses its risks in a robust fashion" (New York State Department of Financial Services, 2017). The law applies to any person or nongovernment entity, defined as a covered entity, operating under the banking, insurance, or financial services law in the state of New York.

The law requires each covered entity to develop and maintain a cybersecurity program based on a risk assessment of its internal and external cybersecurity risks. Specific components of a cybersecurity program that are required include the following:

1. Cybersecurity policy—The regulation specifies and outlines a policy or policies that must be written, maintained, and approved by a senior officer, the board of directors, or an appropriate governing body.
2. Chief Information Security Officer (CISO)—A CISO must be designated, and that person is responsible for the development, implementation, and enforcement of the cybersecurity program.
3. Penetration testing and vulnerability assessments—The regulation requires monitoring and testing to evaluate the effectiveness of the cybersecurity program. It states that monitoring should be continuous, or penetration testing is required at least annually and vulnerability assessments are required at least biannually.
4. Audit trails—Cybersecurity audit trails that allow detection and response to cybersecurity events must be implemented and records maintained for at least 3 years.

5. Access privileges—Covered entities shall limit access privileges and regularly review existing privileges.
6. Application security—A cybersecurity program must put in place policies, guidelines, and procedures for all internally developed application software to ensure that secure development practices are followed, assessments are performed, and testing on all externally developed application software is performed to determine the risk third party software may introduce.
7. Risk assessment—The risk assessment previously mentioned must be conducted periodically, and the cybersecurity program must be able to adapt to the new and evolving risk identified in the assessment.
8. Cybersecurity personnel and intelligence—The covered entity must utilize qualified personnel and must ensure that these personnel receive adequate training, including training on the changing cybersecurity threat landscape.
9. Third-party service provider security policy—Similar to the internal security policies, a covered entity must also implement, maintain, and enforce security policies that apply to third-party service providers that have access to covered entities' systems or nonpublic information.
10. Multifactor authentication—Covered entities must implement multifactor authentication or similar controls for access to systems and nonpublic data. In particular, access to internal networks from external networks is called out.
11. Data retention—The regulation requires policies addressing the retention and secure disposal of nonpublic data.
12. Training and monitoring—In addition to the training required for cybersecurity personnel, all authorized users must be monitored and trained as well.
13. Encryption of nonpublic information—Encryption of nonpublic information over external networks and at rest on covered entity systems is required. If encryption is not feasible, compensating controls must be implemented, and these controls must be annually reviewed by the CISO to determine if they are still effective and if encryption has become feasible since the last review.
14. Incident response plan—The covered entity shall have a written incident response plan to address cybersecurity events.

In addition, the law requires covered entities to notify the superintendent of the New York State Department of Financial Service if a cybersecurity event occurs that must be reported to other government agencies under other laws or may materially harm any part of the covered entity's operations. Covered entities must also annually provide written certification that it is in compliance with this regulation. To encourage the required information sharing, the regulation does exempt covered entities from the disclosure of information under other New York state and federal laws.

The New York state cybersecurity requirements for financial services companies is the first cybersecurity regulation of its type in the United States, and being just months old at the time of this writing, it will be interesting to see the effect it has on cybersecurity regulations at the state and federal levels in the United States. In general, companies who operate nationally often set their company policies according to the strictest security standards in place in states in which they operate. As a result, one state can set industry standards on a national level, and a small number of companies who operate nationally can model industry standards for many regional and smaller companies.

14.2 Industry standards and the role of corporations

14.2.1 Creation of industry standards

Again, due to the lack of US federal cybersecurity regulations and a myriad of state actions, industry players have developed and advanced standards for security and privacy. A couple of the best known are the Payment Card Industry Data Security Standard (PCI DSS) and the Health Information Trust Alliance (HITRUST) common security framework (CSF). In addition, the NIST cybersecurity framework, while developed by the US Government as a nonbinding framework for critical infrastructure, is slowly being adopted by many different industries as a standard framework to follow to assure that adequate risk-based cybersecurity programs are implemented. Industry standards function to provide consistency for companies to plan their budgets, hire personnel, and shield themselves from liability and government scrutiny.

14.2.1.1 PCI DSS

The PCI Security Standards Council is a global organization founded by some of the largest payment card processing companies—American Express, Discover Financial Services, JCB International, MasterCard, and Visa Inc.—in the world. The Security Standards Council develops and promotes payment card industry standards for the protection of cardholder data. All five founding members have agreed to abide by the standards developed and require their vendors and processors to do so as well. To validate that their standards are being properly followed, the council also has a program to train, test, and accredit assessors to validate that those who process and/or store payment card information are properly following the council's standards.

The most well-known standard developed by the council is the PCI DSS. PCI DSS is an evolving standard and is currently at version 3.2. The PCI DSS defines technical and operational requirements for the protection of payment card account data. "PCI DSS applies to all entities involved in payment card processing including merchants, processors, acquirers, issuers, and service providers. PCI DSS also applies to all other entities that store, process or transmit cardholder data (CHD) and/or sensitive authentication data (SAD)" (PCI Security Standards Council, 2016). Cardholder data consists of primary account number, cardholder name, expiration data, and service code. Sensitive authentication data includes the full track data on the magnetic stripe or chip equivalent, card validation value, and PIN data. PCI DSS covers the following six high-level areas:

1. Build and maintain a secure network and systems
2. Protect cardholder data
3. Maintain a vulnerability management program
4. Implement strong access control measures
5. Regularly monitor and test networks
6. Maintain an information security policy

14.2.1.2 HITRUST CSF

The HIPAA regulations, as outlined earlier, and, specifically for our discussion in this section, the security and privacy rules were written to apply to a wide range of healthcare organizations, for example, everything from a small doctor's office to a large healthcare organization. As a regulation should be, it identifies requirements for protecting electronic

PHI (ePHI), but makes no recommendations on what controls to use or how to meet those requirements. This led to the requirements being generic, making it is very subjective as to what is required to be HIPAA compliant. The HITRUST is a healthcare industry-focused nonprofit that has developed a CSF for protection protecting healthcare information and ePHI. The HITRUST CSF was developed to provide a risk-based and compliance-based approach and provide a prescriptive framework of controls.

The HITRUST CSF:

- Includes, harmonizes and cross-references existing, globally recognized standards, regulations and business requirements, including ISO, NIST, PCI, HIPAA, and State laws
- Scales controls according to type, size, and complexity of an organization
- Provides prescriptive requirements to ensure clarity
- Follows a risk-based approach offering multiple levels of implementation requirements determined by risks and thresholds
- Allows for the adoption of alternate controls when necessary
- Evolves according to user input and changing conditions in the industry and regulatory environment on an annual basis
- Provides an industry-wide approach for managing Business Associate compliance" (HITRUST Alliance)

14.2.1.3 NIST cybersecurity framework

While the US Government has no general-purpose cybersecurity or privacy regulations at present, it has directed the creation of a cybersecurity framework. On February 12, 2013, then President Obama signed an executive order (EO), which, among other things, including improving cybersecurity information sharing with the US private sector, directed the US NIST to create a "Baseline Framework to Reduce Cyber Risk to Critical Infrastructure." The EO directed NIST to create a cybersecurity framework that would "enhance the security and resilience of the Nation's critical infrastructure and to maintain a cyber environment that encourages efficiency, innovation, and economic prosperity while promoting safety, security, business confidentiality, privacy, and civil liberties" (White House, 2013). The cybersecurity framework, officially titled the "Framework for Improving Critical Infrastructure Cybersecurity" was completed and officially released by NIST 1 year later on February 12, 2014.

The cybersecurity framework promotes a risk-based approach to cybersecurity. It defines five parallel and continuous cybersecurity functions—identify, protect, detect, respond and recover—that should be evaluated to "help identify and prioritize actions for reducing cybersecurity risk" (US NIST, 2014). For each function, the cybersecurity framework defines categories and subcategories of activities and outcomes pertinent to the function to help evaluate needs and risks. Informative references for each function identify already developed NIST and industry standards, guidelines, and best practices that provide further detail for the function. These functions and related information are called the Framework Core.

Next, the cybersecurity framework defines tiers to help an organization understand their level of sophistication related to the implementation of cybersecurity and risk management. Tiers are based on an "organization's current risk management practices, threat environment, legal and regulatory requirements, information sharing practices, business/mission objectives, cyber supply chain risk management needs, and organizational constraints" (US NIST, 2014). The tiers defined in the cybersecurity framework are the following:

- Practical (tier 1)
- Risk Informed (tier 2)
- Repeatable (tier 3)
- Adaptive (tier 4)

The cybersecurity framework points out that not every organization needs to be at tier 4 but should rather strive to obtain the tier that reduces their cybersecurity risk and is most cost-effective to obtain.

Finally, the cybersecurity framework defines a profile for aligning the Framework Core functions, categories, and subcategories with the business requirements, risk tolerance, and resources of the organization. The cybersecurity framework recommends developing a current profile describing the current state of cybersecurity activities within the organization and a target profile for defining the desired state the organization would like to work toward. The gaps between the current and target profiles will identify the action items that organization needs to take on to improve its risk management and cybersecurity posture.

The cybersecurity framework is intended to be a living document changing with the cybertechnology and risk landscape. To ensure this, on December 18, 2014, the US Congress passed the Cybersecurity Enhancement Act of 2014, "to provide for an ongoing, voluntary public-private partnership to improve cybersecurity, and to strengthen cybersecurity research and development, workforce development and education, and public awareness and preparedness, and for other purposes" (US Congress, 2014). The Act directed NIST to "on an ongoing basis, facilitate and support the development of a voluntary, consensus-based, industry-led set of standards, guidelines, best practices, methodologies, procedures, and processes to cost-effectively reduce cyber risks to critical infrastructure" (US Congress, 2014). NIST is, as of this writing, working on version 1.1 of cybersecurity framework. Draft version 1.1 was released on January 10, 2017, based on feedback received on version 1.0 and comments received at an NIST-sponsored workshop in April 2016. The main updates to version 1.0 are the following:

- The addition of a section on measuring cybersecurity effectiveness
- An expansion on the use of the cybersecurity framework for cybersupply chain risk management
- Improvements in the access control category related to authentication, authorization, and identity proofing
- Improved explanation of the relationship between tiers and profiles

NIST intends to publish the final of version 1.1 in the fall of 2017.

The cybersecurity framework is a voluntary framework available to everyone within and outside the United States. Among its benefits, it provides a common language and framework for cybersecurity discussion that will hopefully foster communication between the public and private sector as well as among all organizations interested in improving cybersecurity. The cybersecurity framework is in use by organizations around the world and may become the de facto standard for measuring and improving the cybersecurity posture of organizations.

To promote the usage of the cybersecurity framework within the US Government, on May 11, 2017, President Trump issued an EO directing US Government executive branch agencies to use the cybersecurity framework "to manage the agency's cybersecurity risk" (White House, 2017). This EO requires agencies to provide a written action plan to the executive branch on how the agency will implement the cybersecurity framework and align the agency's policy, guidelines, and standards with the cybersecurity framework. The action plans are not required until after the deadline for this book, but it is hoped that having the entire US Government executive branch using the cybersecurity framework to implement its cybersecurity risk management and cybersecurity plans will provide a significant quantity of real-life actionable lessons learned and feedback on the cybersecurity framework. NIST can, per its authority under the Cybersecurity Enhancement Act of 2014, use this data to further evolve and improve the cybersecurity framework to address the ever-evolving threat landscape. Such an effort can only benefit the cybersecurity community at large.

So while the US Government has not passed any general-purpose cybersecurity and privacy laws or regulations, EOs directing the creation and usage of the cybersecurity framework and the Cybersecurity Enhancement Act of 2014, which directs NIST to continuously facilitate and support cybersecurity efforts, have produced an actionable and evolving framework that will hopefully help any type of organization implement, measure progress, and improve their own cybersecurity efforts.

14.2.2 Compliance/audits

Even if well-defined general-purpose US cybersecurity regulations were in place, organizations would still want assurances that their service providers are taking due care in safeguarding the sensitive, business critical information that the service provider is processing and/or storing for the organization. To provide this assurance, several industry and technical groups have come up with compliance standards and defined audit regimes so that service providers can provide third-party attestation that they take due care with the information/data entrusted to them.

Following is an overview of some of the more well-known and utilized compliance and audit regimes. This is by no means an exhaustive list and service providers should listen to their customers to determine the specific regime(s) that a service provider's customers would like them to follow. These compliance mechanisms may be built into a company's protocols via company policy, via contractual commitment with a customer or other third party, or more informally done as needed. Each standard involves reporting, internal or external as required, and ideally, follow-up actions to assure continuing compliance.

14.2.2.1 American Institute of Certified Public Accountants system and organization controls

The American Institute of Certified Public Accountants has defined a set of system and organization controls (SOC) service offerings that allow the system-level controls of a service organization or the entity-level controls of other organizations to be audited by certified public accountants (CPAs). The reports that an organization can have a CPA firm provide are the following:

- SOC 1—SOC for services organizations: internal controls over financial reporting (ICFR)

  A SOC 1 report covers ICFR, which are controls that the service provider has in place to protect their customer's data that would have an effect on the customer's own controls for their financial reporting. The service provider's customer and the customer's CPA use these reports to evaluate the service provider's controls that affect the customer's financial reports.

- SOC 2—SOC for services organizations: trust services criteria
  SOC 2 reports cover the security, availability, and integrity of the systems the service provider uses to process their customer's data and the confidentiality and privacy of the data processed by these systems. These reports can be used by the service provider to monitor and measure progress on governance and risk management programs.
- SOC 3—SOC for services organizations: trust services criteria for general use report
  A SOC 3 report covers the same controls and scope as the SOC 2 report. They are written to be more widely disseminated and therefore do not contain the level of detail found in a SOC 2 report.

14.2.2.2 ISO/IEC 27000 series

The International Organization for Standardization (ISO) and the International Electrotechnical Commission (IEC) are international organizations that work together through a joint committee to develop technical standards for IT. The ISO/IEC 27000 series (2017) provides background, common terminology, principles, techniques, and guidance on information security management systems (ISMSs). The main document in the ISO/IEC 27000 series is ISO/IEC 27001, officially titled "Information technology—Security techniques—Information security management systems—Requirements," was updated and rereleased in 2013. It describes what is required to implement an ISMS including understanding organizational context, required leadership support, risk assessment, and continuous improvement. The annex provides a list of security controls, grouped into 14 categories, that may, but are not required to, be used to implement the ISMS based on the risk assessment. Not requiring specific controls enables organizations to use a risk assessment methodology that makes the most sense for their business and still achieve compliance with ISO/IEC 27001.

ISO/IEC does not certify organizations as compliant with its standards. Certification is done by independent third-party organizations. Certifications are valid for 3 years after which a full renewal audit is required. The normal process is for an organization to have documentation and certification audits at the beginning of the 3-year cycle to ensure the ISMS is in place and properly documented followed by annual surveillance audits to check ongoing progress and operations at the beginning of years 2 and 3.

14.2.2.3 European Union's global data protection regulation

Last but not least in our discussion of cybersecurity standards is a look at the future of worldwide compliance. The European Union (EU)'s global data protection regulation (GDPR) (Trunomi, 2017) will take effect in 2018 and will have a significant effect on US businesses that operate in the EU, touch personal information of EU citizens, or simply do business on an international basis. While a full analysis of the potential impact of GDPR regulations on the security industry are beyond the scope of this chapter, suffice it to say that it will provide several cautionary tales for companies due to its high EU-based privacy standards, its comprehensive regulations, and its focus on monetary fines.

At this point, we return to the focus offered at the beginning of this chapter on the actual actors accomplishing the tasks laid out over the last few sections. Who are the people who implement each of these compliance goals and how do they interact with one another?

14.2.3 Navigating corporate structure, roles, and conflicts

The following positions reflect roles of individuals, their interactions with the corporate boards that govern companies, and potential conflicts as their responsibilities begin to overlap.

1. Executives: The role of those in the executive suite has been increased in recent years with a growing assignment of responsibility upward. In the past, IT was IT, and the technology side of the business stopped at the highest level with at most the CTO. Now the entire C-suite and executive-level personnel are expected to both understand and execute security protocols.
2. Cybersecurity lawyers, including in-house and outside counsel: Both inside and outside counsels for companies need to understand the basics of HIPAA, GLBA, and the GDPR, which is looming, at the time of this writing, for those organizations with EU customers in 2018.
3. IT personnel: Traditionally tasked with nearly all the work in this chapter save the most esoteric legal review, IT should now expand beyond educating its own to educating the C-level executives on what could and most likely will happen in the event of a major data breach. See, herein, the discussion of tabletop exercises, which give these parties an opportunity to interact and share knowledge on these issues.
4. Security and privacy professionals: A growing profession of security and privacy professionals has come up the ranks from a variety of backgrounds—IT, legal, employment/human relations (HR), and compliance professionals. Their role is to cross-pollinate among the other groups mentioned in this section, to make sure each group understands its role in cybersecurity. These are the individuals that design and run the tabletop exercises and prepare internal and external policies with input from the other stakeholders.
5. Marketing: Marketing and product design personnel get involved in privacy and data security at the very earliest stages of the process via a concept called privacy by design or security by design. These processes consider the regulations and laws that may apply to a new product or a new method of marketing the product and "bake in" the security and privacy protections rather than adding them as an overlay after the product is complete, or even launched. Marketing people also have a role to play in matching expectations about privacy and security with the reality of an Internet-connected device, including suggesting user-accessible privacy protections such as data input minimization and security self-help such as superior password design.
6. Investor relations: Reaching out to the investor community is the function of the investor relations team. Good security is a good business practice and may increase the value of the company or a particular product line with differentiated privacy protection.
7. Government relations: In addition to the rather specific function of working with government entities in a data breach scenario for notifications purposes, government relations personnel have a larger function to understand the motivations and expectations of a regulatory agency to prevent prosecutions or other negative attention on a more ongoing basis.
8. Customer care and call center managers: The first line of defense in a data breach or general concern about privacy and data security is the call center. The ability to scale up the customer response center in a disaster situation caused by humans (a hacking scenario) or by natural disasters (a flood taking out a data center) is key to security management.

9. HR management: As discussed in the education and training section, it is difficult to create a good security system in a company without creating a culture of security to support it. HR personnel are responsible for recruiting and background checks to decrease the likelihood of an insider attack and responsible for ongoing training about new regulations, new technologies, and newly creative ways of accessing the employer's network. Human-centered hacking, such as pretending to be someone with credentials or other methodologies that take advantage of human error or vulnerabilities are a particular focus of this team's expertise and creative efforts to protect data belonging to the company and to individual customers.
10. Cyberinsurance and risk management: Cyberinsurance is also discussed in this chapter, but here we focus on the individual personnel associated with choosing and implementing insurance and risk management techniques in the process. This group can enhance security by running through checklists embedded in risk management to avoid data loss and secure perimeters. Physical security, such as locking doors and checking badges, may be housed in the HR and/or risk management group.
11. Law enforcement and government contacts: Law enforcement, from the neighborhood police force to state attorneys general and federal agencies, not only have a specific role to play in the notification and investigation of a criminal hacking instance, but can also be brought in to educate employees or consumers about protecting privacy, avoiding fraud, and sidestepping identity theft bandits.
12. Auditors: Auditing has, for every intent and purpose, a financial meaning, a security meaning, and, increasingly, a privacy meaning as well. It can be used in internal policy practice to prevent security incidents and to assess the damage once a security incident has occurred. It may also appear in each or any of these contexts in the text of a B2B contract, with associated consequences for the companies in the event of a data breach.

14.2.4 *Compliance activities conducted by personnel*

14.2.4.1 *Security/PIAs*

PIAs and, in the EU, data DPIAs are, in a nutshell, a data audit of your company. Where are the data and how are they stored, transmitted, and processed? It is common to evaluate these issues internally and assign red, yellow, or green colors to each existing practice to evaluate whether they are dangerous, questionable, or just fine as is. The next step is to research industry standards and get each category up to speed. See the section in this chapter on security standards for details on how to accomplish this task. While security audits and evaluations are, as discussed herein, quite detailed and standardized, the privacy industry has more variance in its ability to assess how much privacy is available to individuals. PIAs can incorporate security standards and build upon them to provide assurances to business partners, consumers, and government officials that privacy is an important part of the protocols of a company.

14.2.4.2 *Tabletop exercises*

A tabletop exercise is exactly what it sounds like—there is a group of people sitting around a table, and an exercise, or rather a game, is played out over a period. Tabletop exercises get to the idea that in order to be prepared for a disaster, or even a minor dilemma, it is best to run through various scenarios that may occur in advance of day zero. There is a variety of security incidents that could occur, from cables cut by vandals to national cybersecurity meltdowns, but here we will go into some detail about two tabletop exercises that are worth spending some time outlining, and ideally, actually running.

1. Data breach hypothetical(s): The data breach hypothetical tabletop exercise requires a bit of imagination to posit some basic facts: Who or what caused this imaginary breach? How did we find out about it? What was the reaction of the media, the government and the individuals affected? Ultimately, the goal of the exercise is to create a checklist of items to review systems and procedures to avoid such a scenario from actually occurring.
2. Responses to class action lawsuits: A little less imagination is required here. One of the well-known class-action firms, or perhaps someone new to the game, files a suit against your company. What is the first thing that you do? Notify legal, but then what? Your tabletop exercise should run through the process of gathering a team to respond, including IT, legal, policy, investigations, media and customer relations personnel, and the head of any department involved in the security snafu. Investigate cautiously. Preserve necessary documents and processes under any litigation hold as instructed by counsel. Respond to media inquiries via a centralized point of contact and with prepared messaging.

14.2.4.3 *Creating a corporate privacy culture*

1. Internal policies: Internal policies for each company should include privacy policies (in addition to any legally required external privacy policy for customers) and security policies. They should be drafted within the parameters of industry standards discussed in this chapter and, when feasible, go beyond legal requirements to address forward-thinking security practices. The law is slow to incorporate new technologies, and even when legislators leap on a new technology, it takes some time for their ideas to become law.
2. Education and training: It does not do a company much good to have state-of-the-art security policies if no one is aware of them or follows them. Education designed to explain the meaning of each protocol, including the "whys" as well as the "shoulds," will disperse the knowledge in the IT and legal departments throughout the company. Training is the "how," i.e., once you have the requisite knowledge about company security practices, how does this apply to an individual employee's job and how can he/she pass on this knowledge to his/her reports and successors?
3. Resources for professionals: The International Association of Privacy Professionals, the Cloud Security Alliance, RSA, and Information Systems Audit and Control Association are several of the trade associations that offer resources for new and experienced privacy and data security professionals. Each of these organizations has a website with available resources, some of which are behind a paywall, and articles addressing current topics in privacy and security for professionals. They also offer live conferences and meetings during which individuals can confer on problems and solutions in the industry.

14.2.4.4 *Contracts and security clauses*

When designing a contract for the provision of tech services, the parties should begin with prevention and creating a secure interconnected network. This preliminary should create a structure of words that supports a company in avoiding a data breach or security incident and mitigate the effects of a data breach. The contract is that framework, and the lawyers, IT professionals, and security personnel are each a piece of the puzzle with support and information gathering from engineering and then sharing/training after the contract is signed.

Following are factors to examine when evaluating the efficacy of security and data breach clauses in protecting each company involved in a transaction from not only the likelihood of breach, but also the resulting rather enormous liability consequences of breach. First, consider the relative sophistication of parties. Are the parties equally situated with regard to technological sophistication and experience? In some situations, the answer may well be yes, the parties are both technology providers, as in a transaction between a software manufacturer and an enterprise-level telecommunications company. In other instances, the companies may be divergent, for example, a cloud storage company serving a small chain of pizza restaurants. Next, evaluate the control of data by each company. Again, there may be widely different levels of ability to control data based on the type of services being offered in the contract. In a true outsourcing arrangement, one party controls nearly all the other party's data. In many cases, IT personnel will need to explain to management whether they truly have access to the data versus the mere possibility of access, from an accidental or intentional breach of access protocols. Also, the types of use of data should be factored in, including data storage or data processing, each of which entails different risk protocols when the data are in transit (often encrypted) and then when the data are at rest, preferably behind a network security firewall and physical security protections. Finally, does one or more parties have data security or breach insurance and/or access to availability of insurance? Some parties will sidestep the issue of insurance by representing that they have self-insurance, i.e., deep pockets.

14.2.4.5 Legal negotiation strategy for data breach and data security

There are several factors that can throw a wrench in negotiations, even if the parties agree on the basic plan to increase mutual data security and avoid data breaches if possible:

1. Definitions: Most contracts begin with a definitions section. One tricky definition to create is the definition of a data breach. For example, most security professionals will acknowledge that security incidents are subbreach. Some level of security incident occurs all the time, and it matters quite a bit when they roll over into a "data breach" as defined by the contract or the law. Several state laws have minimum requirements for data breach by defining the number of affected individuals or state citizens and have another trigger for encrypted versus unencrypted data.
2. Legislation: New legislation is always being proposed, especially in the media noteworthy areas of data breach, individual privacy, and national security as it applies to private companies. Many corporations would kill for comprehensive data breach legislation as long as they can write it themselves. As a matter of practicality, most legislators do not have either the expertise or the time to write highly technical bills and often rely on industry experts to draft portions of the legislation that relate to data security. As a result, the draft usually winds up touting best practices or industry standards or another somewhat vague standard if you do not have the context of interpretative case law or contracts in hand with detailed attachments outlining security requirements. This situation leads to full employment for lawyers to interpret and litigate over the enacted laws and regulation. Further complicating the legislative issue is the multiple jurisdictions that may be jockeying for position on the issue of data security. This also requires legal interpretation, both on a general level

and specific to any given deal. For example, a law firm may provide not just ad hoc advice, but an interactive tool to juxtapose regulations in multiple jurisdictions. If at some point after this writing we see an enacted federal legislation on the issue of data security, the law will likely, in order to avoid an unending battle of wills, still exempt state laws. That leaves 51 potential jurisdictions, including Puerto Rico, Washington DC, Guam, as well as the financial and healthcare sectoral regulations, to push around the parameters of the requirements.

3. Standing: Standing is the legal concept that whoever sues another must have a legal right to do so. Several recent cases in the privacy and data security area have begun looking at harm done as the threshold for standing to sue. The court decisions mention a requirement that there not just be hurt feelings as the result of a privacy violation, but that there must also be proof of some sort of financial harm to the affected individual in order for him/her to have standing to sue. It is difficult to prove an increased risk of identity theft, but identity theft is nearly the only harm discussed in many of the cases. Identity theft protection is a marketed service, so there are damages associated with needing such protection as the result of having your personal data leaked to the public. Nevertheless, this is a small ticket item in comparison to the difficulties many individuals suffer as the result of being a victim of a data breach. In order to definitely provide compensation for such damages, some states may find a need for statutory damages, i.e., a set amount of compensation delivered to each victim regardless of proof of actual damages.
4. Departmental silos: The image of a grain or water silo tower standing tall and alone in the field is a vivid one, and it accurately represents the idea that large companies have different departments for each functionality, and the departments often do not communicate with each other often or well. There may be different legal departments even for privacy and for security issues, or security operations may be housed in IT while privacy compliance is settled in with the regulatory group. While litigation seems like one department, issues that arise as the result of a company being targeted by class action lawsuit may find a different response than a privacy compliance matter that arises from an FTC enforcement action. Small companies may have no department to deal with privacy and security, and all issues must be defaulted to busy C-suite executives. Better business practices would be to designate a liaison to communicate policies among these departments. Also, as part and parcel of the training ideas espoused herein, all-hands calls and meetings can be used to educate each employee about his/her role in the company and its relation to the other roles in other departments. With these practices, silos may not be eliminated entirely, but at least everyone will be on the same page when an incident occurs.
5. Transparency: There is quite a bit of variation in how much is disclosed to the public about the internal security and privacy policies of a company. Privacy policies are generally public, whereas security policies are probably only public to the extent they are discussed in privacy policy or are offered as generalities about securing personal data to consumers. On a B2B level, companies will often allow highly sensitive data to be released under a higher level nondisclosure agreement. Highly sensitive data for a company would include trade secrets, network configuration maps, or proprietary and patent information. The ultimate transparency issue is an involuntary one—should we have a back door for encryption to enhance national security or would it decrease security because it opens up vulnerabilities?

14.2.4.6 Cyberinsurance

Cyberinsurance has enormously grown as an industry during the last few years. Still, it is not a mature complete product, and the statisticians are still working out the kinks. What can you legally and commercially obtain coverage for? The following discussion explores what is available and what to look for in a cyberinsurance policy. Starting with the obvious, a policy should cover basic security privacy and liability. Coverage amounts are still the big question. How much liability coverage would you need for a data breach? We know that California and other states have created or at least mirrored an industry standard for 12 months of identity theft protection for each affected individual/customer, plus the cost of customer and government notices and a wide variety of media interactions. Beyond data breach, consider insurance for a host of breach-related losses associated with business interruption, data recovery, regulatory procedure participation and compliance fines, and crisis management costs. A rising army of professionals can deal with each of these issues rather than having to divert internal operations to this crisis, but each item outsourced will have associated legal and business fees.

There are several cautionary tales associated with the insurance evaluation. It is worth noting that regular insurance ("commercial general liability" or CGL) will not cover this specialized loss. In addition, policies have exclusions for war or terrorism. Note that if you say that this is the reason for your breach or data loss, this may undermine your insurance coverage. The policy may have exclusions for vendors. The now infamous Target breach was likely the heating, ventilation, and air-conditioning vendor's fault, not a technology issue. The lesson here is to look at your coverage and investigate the source of the leak or breach before making a public statement.

What happens after you sign up for a cyberinsurance policy? Your work is not done. Next up is to flow down to subcontractors your security policies, review their policies and practices, or at least get representations and warranties from vendors. Contract language should limit vendors' access to data to a need-to-know basis, among other security precautions. Also, the legal counsel should lock in a flow down requirement to get cyberinsurance to any vendors/subcontractors on the deal, so the correct/liable party will be insured. As with any legal specialty, there are insurance-specific lawyers with expertise in these areas, and coverage lawyers can advise on types of insurance. Overall, consider any liability limitations in the policy, especially caps on policy. For example, even a rather generous $90 million policy may still not cover you if have $248 million in credit card losses as a big box retailer would suffer. Costs of cover for a data breach may include anything from simple customer notifications to elaborate data reconstruction/recreation. There are ancillary and consequential costs to the data breach, including post-contract monitoring, reporting, and auditing. All in all, be not overwhelmed by the choices and buy cyberinsurance or self-insure.

In the coming years, we will all become Carrie from *Homeland*. We will need to address cybersecurity in our daily lives, for our work and to protect our children. A basic understanding of how security and privacy are handled in the law, in technology, and in government policy will be useful for thriving in the coming cybercentury.

Bibliography

American Institute of CPAs, 2017, *System and Organization Controls—SOC Suite of Services*, http://www.aicpa.org/InterestAreas/FRC/AssuranceAdvisoryServices/Pages/SORHome.aspx.

Cornell Law School Legal Information Institute, 2017, 47 US Code § 222—Privacy of customer information, https://www.law.cornell.edu/uscode/text/47/222.

Harvard Law School Forum, 2017, *New York Cybersecurity Regulations for Financial Institutions Enter Into Effect*, https://corpgov.law.harvard.edu/2017/03/25/new-york-cybersecurity-regulations-for-financial-institutions-enter-into-effect/.
National Council of State Legislatures, 2017, *Security and Privacy*, http://www.ncsl.org/research/telecommunications-and-information-technology/privacy-and-security.aspx.
Payment Card Industry Security Standards Council, 2017, Homepage, https://www.pcisecuritystandards.org/.
US Congress, 1999, S.900—*Gramm–Leach–Bliley Act*, https://www.congress.gov/bill/106th-congress/senate-bill/00900.
US Congress, 2014, S.2521—*Federal Information Security Modernization Act of 2014*, https://www.congress.gov/bill/113th-congress/senate-bill/2521.
US Federal Trade Commission, 2017, *Consumer Privacy*, https://www.ftc.gov/tips-advice/business-center/privacy-and-security/consumer-privacy.
US National Institute of Standards and Technology, 2017, *NIST News*—NIST releases update to cybersecurity framework, https://www.nist.gov/news-events/news/2017/01/nist-releases-update-cybersecurity-framework.

References

HITRUST Alliance, 2017, *Understanding and Leveraging the CSF*, https://hitrustalliance.net/understanding-leveraging-csf/.
International Organization for Standardization, 2017, ISO/IEC 27000 Family—Information Security Management Systems, https://www.iso.org/isoiec-27001-information-security.html.
National Conference of State Legislatures, 2017, *States Laws Related to Internet Privacy*, http://www.ncsl.org/research/telecommunications-and-information-technology/state-laws-related-to-internet-privacy.aspx.
National Law Review, 2017, *State Data Breach Notification Laws*, http://www.natlawreview.com/article/state-data-breach-notification-laws-february-2017-privacy-update.
New York State Department of Financial Services, 2017, *Cybersecurity Requirement for Financial Services Companies*, http://www.dfs.ny.gov/legal/regulations/adoptions/dfsrf500txt.pdf.
Payment Card Industry Security Standards Council, 2016, *PCI Data Security Standard Requirements and Security Assessment Procedures Version 3.2*, https://www.pcisecuritystandards.org/documents/PCI_DSS_v3-2.pdf [Must Accept PCI DSS Agreement to view document].
Trunomi, 2017, *EU GDPR Portal*, http://www.eugdpr.org/.
US Congress, 2014, *Cybersecurity Enhancement Act of 2014*, https://www.congress.gov/113/plaws/publ274/PLAW-113publ274.pdf.
US Department of Health and Human Services, 2017a, *The Security Rule*, https://www.hhs.gov/hipaa/for-professionals/security/index.html.
US Department of Health and Human Services, 2017b, *The Privacy Rule*, https://www.hhs.gov/hipaa/for-professionals/privacy/index.html.
US Department of Justice, 2017, *Privacy Act of 1974*, https://www.justice.gov/opcl/privacy-act-1974.
US DHS (US Department of Homeland Security), 2017, *About DHS*, See https://www.dhs.gov/about-dhs.
US FedRAMP, 2017, Program Overview, https://www.fedramp.gov/.
US FTC (US Federal Trade Commission), 2002, *How To Comply with the Privacy of Consumer Financial Information Rule of the Gramm-Leach-Bliley Act*, https://www.ftc.gov/tips-advice/business-center/guidance/how-comply-privacy-consumer-financial-information-rule-gramm.
US FTC, 2015, *Privacy and Security Update*, https://www.ftc.gov/reports/privacy-data-security-update-2015.
US Government Printing Office, 2011, 47 US Code § 222—Privacy of customer information, https://www.gpo.gov/fdsys/granule/USCODE-2011-title47/USCODE-2011-title47-chap5-subchapII-partI-sec222/content-detail.html.

US NIST (US National Institute of Standards and Technology), 2014, *Framework for Improving Critical Infrastructure Cybersecurity*, https://www.nist.gov/sites/default/files/documents/cyberframework/cybersecurity-framework-021214.pdf.
US NIST, 2017, *FISMA Background*, http://csrc.nist.gov/groups/SMA/fisma/overview.html.
White House, 2013, Executive Order—Improving Critical Infrastructure Cybersecurity, https://obamawhitehouse.archives.gov/the-press-office/2013/02/12/executive-order-improving-critical-infrastructure-cybersecurity.
White House, 2017, *Presidential Executive Order on Strengthening the Cybersecurity of Federal Networks and Critical Infrastructure*, https://www.whitehouse.gov/the-press-office/2017/05/11/presidential-executive-order-strengthening-cybersecurity-federal.

chapter fifteen

Impact of recent legislative developments in the European Union on information security

Gerald Quirchmayr

Contents

15.1 *Privacy and critical information infrastructure protection regulation: Background*

Privacy and information security are some of the core concerns in the design, development, and operation of IT systems. With recently published solid evidence from Europol [2016 and 2017; Internet Organised Crime Threat Assessment (IOCTA)], the size and intensity of the problem facing Europe is well documented. It was in May and June 2017 that waves of serious attacks based on exploits leaked after the intrusion of secret service systems (EternalBlue 2017) have again shown the need for a concerted action against these now very dangerous attacks (National Audit Office 2017). With new legislation in both areas, privacy protection [General Data Protection Regulation (GDPR), Regulation (European Union [EU]) 2016/679] (European Parliament and Council of the EU 2016a), and critical infrastructure security [NIS Directive, Directive (EU) 2016/1148] (European Parliament and Council of the EU 2016b), the EU is now countering the growing danger on a strategic level. These two pieces of legislation are a direct consequence of the European cybersecurity strategy (EU 2013), which paved the way for a now far more integrated approach.

This new legislation introduces a much needed basis for information sharing about cyberattacks and reporting obligations on incidents and updates the by now partially obsolete European privacy legislation that came into effect in 1995 (Directive 95/46/EC) (European Parliament and Council of the EU 1995). One of the very significant differences is that the new privacy legislation, which will become effective in 2018, is now a directly applicable law, a major step forward from previous guidelines that were aimed only at the harmonization of laws in member states of the EU. For system developers and system operators, the resulting competitive advantage is that from May 2018, they will have only one central privacy law to be compliant with for all member states in the EU, the world's largest single market. The much younger European network and information security legislation still takes form in a directive (Directive (EU) 2016/1148). Considering the many different interests and approaches of member states in the fields of national security and national defense, a joint European strategy and umbrella legislation in the domain of cybersecurity is therefore also a major improvement that only a few decades ago would have been unthinkable. With national interests and economic needs of member states and of the EU as a whole being reflected in the new legislation, a successful implementation of the legislation can be expected. This chapter looks at the two central pieces of legislation, the GDPR [Regulation (EU) 2016/679] and the NIS Directive [Directive (EU) 2016/1148] primarily from a system development and information technology (IT) operations perspective.

15.2 GDPR [Regulation (EU) 2016/679]

The GDPR was introduced as the successor of the European privacy legislation introduced in 1994 (Directive 95/46/EC) to address new technological developments, such as cloud computing, Internet of things (IoT), and smartphones, and to establish a new adequate basis for technology applications such as social media that were not yet on the horizon in 1994 but are a major economic and societal factor today. As a compromise is reached between privacy advocates, government agencies, and industry, the regulation is primarily aimed at providing a balanced and workable solution and does consequently receive continuing criticism from privacy advocates (Fielder 2017).

The EU data protection reform was adopted by the European Parliament and the European Council on April 27, 2016. The European Data Protection Regulation will be applicable as of May 25, 2018, and replace the Data Protection Directive. The regulation is grouped (EU Info 2017) into introductory chapters (Chapters 1 and 2), covering general provisions and principles, and Chapters 3–5 covering the rights of the data subject, obligations of controller and processor, and the transfers of personal data to third countries or international organizations.

Independent supervisory authorities, cooperation and consistency; remedies, liability, and penalties; provisions relating to specific processing situations; delegated acts and implementing acts; and final provisions are regulated in Chapters 6–11.

Within these chapters, the rights of the data subject; obligations of controller and processor; transfers of personal data to third countries or international organizations; and remedies, liability, and penalties have received special attention.

Obligations regarding the safeguarding of personal information are specified in Chapter 4—Controller and Processor—Section 2—Security of Personal Data. Article 32—Security of Processing—contains a list of obligatory organizational and technical measures to be taken. As this article contains the primary obligations for system developers and operators, the full text is shown in Box 15.1.

A risk-based approach and adequate encryption and pseudonimization of personal data are the major new obligations introduced by Article 32. The requirement of regular testing, regular assessment, and evaluation of security measures is now made explicit.

While already well-established rights of individuals, such as the confidentiality of personal data, continue to be protected, the right to erasure ("right to be forgotten") in Article 17 and the right to data portability in Article 20 represent significant new challenges for system developers and operators.

It is, however, Article 25, Data Protection by Design and by Default, which has introduced the necessity to rethink system design and development whenever personal data are involved.

BOX 15.1 ARTICLE 32: SECURITY OF PROCESSING

1. Taking into account the state of the art, the costs of implementation and the nature, scope, context and purposes of processing as well as the risk of varying likelihood and severity for the rights and freedoms of natural persons, the controller and the processor shall implement appropriate technical and organizational measures to ensure a level of security appropriate to the risk, including inter alia as appropriate:
   a. the pseudonymisation and encryption of personal data;
   b. the ability to ensure the ongoing confidentiality, integrity, availability and resilience of processing systems and services;
   c. the ability to restore the availability and access to personal data in a timely manner in the event of a physical or technical incident;
   d. a process for regularly testing, assessing and evaluating the effectiveness of technical and organizational measures for ensuring the security of the processing.
2. In assessing the appropriate level of security account shall be taken in particular of the risks that are presented by processing, in particular from accidental or unlawful destruction, loss, alteration, unauthorised disclosure of, or access to personal data transmitted, stored or otherwise processed.
3. Adherence to an approved code of conduct as referred to in Article 40 or an approved certification mechanism as referred to in Article 42 may be used as an element by which to demonstrate compliance with the requirements set out in paragraph 1 of this Article.
4. The controller and processor shall take steps to ensure that any natural person acting under the authority of the controller or the processor who has access to personal data does not process them except on instructions from the controller, unless he or she is required to do so by Union or Member State law.

While in the past, it was rather typical to retrofit IT security safeguards, they now have to be considered as a core requirement guiding the system design and development process.

As can be directly derived from the preceding text, the newly introduced legal obligations will lead to a very significant increase in terms of additional requirements and system design principles. Apart from design and development, Article 25 will also change the way in which systems dealing with personal data operate. (The controller shall implement appropriate technical and organizational measures for ensuring that by default, only personal data, which are necessary for each specific purpose of the processing, are processed.)

The obligatory data breach notification comes in two forms, Article 33—Notification of a Personal Data Breach to the Supervisory Authority—and Article 34—Communication of a Personal Data Breach to the Data Subject. A personal data breach means a breach of security leading to the destruction of, loss of, alteration of, unauthorized disclosure of, or access to personal data. This means that a breach is more than just losing personal data. Notifying the relevant supervisory authority of a breach becomes obligatory where such a breach is likely to result in a risk to the rights and freedoms of individuals. If unaddressed, such a breach is likely to have a significant detrimental effect on individuals—for example, result in discrimination, damage to reputation, financial loss, loss of confidentiality, or any other significant economic or social disadvantage [see ICO (2017)].

Article 35—Data Protection Impact Assessment—will have a significant consequence in the form of a privacy risk assessment having to be performed in the feasibility study stage of every IT project. Especially considering agile programming techniques, this will result in a major change of processes in terms of slowing down system development. Together with the principle of data protection by design and by default (Article 25), this legal requirement should ensure that privacy protection is adequately taken care of from the early stages of a project on.

Regulations governing the transfer of personal data to third countries or international organizations are laid out in Articles 44–50. The general principle for transfers sets the basis, followed by rules regulating transfers on the basis of an adequacy decision, transfers subject to appropriate safeguards, binding corporate rules, transfers or disclosures not authorized by Union law, derogations for specific situations, and international cooperation for the protection of personal data.

Penalties being imposed for violations of the guideline were discussed widely, because consequences can now be quite severe, since especially Article 83-4 ("Infringements of the following provisions shall, in accordance with paragraph 2, be subject to administrative fines up to 10,000,000 EUR, or in the case of an undertaking, up to 2% of the total worldwide annual turnover of the preceding financial year, whichever is higher") introduces a very credible deterrent against violations.

This new privacy legislation has already led to a rethinking of privacy protection as an essential aspect of system design and development. While introducing new requirements (e.g., privacy by default) that certainly lead to higher development costs, it also leads to a higher level of system security. Given the recent sophisticated attacks, these obligations do reflect, however, only the growing threats and might soon be viewed as a very welcome guide to counter them. When applied by developers to system design beyond the protection of personal data, they might in a very positive way contribute to finally achieving the goal of designed-in, built-in security by default. How much such a general redesign of system development is already in the context of critical information infrastructures is addressed in the following description of the NIS Directive.

15.3 Directive on security of network and information systems (Directive (EU) 2016/1148)

With the Cybersecurity Strategy of the European Union (EU 2013) setting the stage for a better coordinated and more integrated approach to cybersecurity, the second strategic legislation to be passed was the NIS Directive in 2016 [Directive (EU) 2016/1148]. "Recognizing that network and information systems and services play a vital role in society. Their reliability and security are essential to economic and societal activities, and in particular to the functioning of the internal market," the directive aims at providing a harmonized approach "responding effectively to the challenges of the security of network and information systems therefore requires a global approach at Union level covering common minimum capacity building and planning requirements, exchange of information, cooperation and common security requirements for operators of essential services and digital service providers," setting a minimum standard across the EU. Higher levels of protection than required by the directive are of course highly welcomed by the document.

In (7) of the preamble, the applicability of the directive is stated as follows:

> To cover all relevant incidents and risks, this Directive should apply to both operators of essential services and digital service providers.

> However, the obligations on operators of essential services and digital service providers should not apply to undertakings providing public communication networks or publicly available electronic communication services within the meaning of Directive 2002/21/EC of the European Parliament and of the Council (1), which are subject to the specific security and integrity requirements laid down in that Directive, nor should they apply to trust service providers within the meaning of Regulation (EU) No 910/2014 of the European Parliament and of the Council (2), which are subject to the security requirements laid down in that Regulation.

The governing principles are laid out in Article 1—Subject Matter and Scope (Section 2).

> 2. To that end, this Directive:
>   a. lays down obligations for all Member States to adopt a national strategy on the security of network and information systems;
>   b. creates a Cooperation Group in order to support and facilitate strategic cooperation and the exchange of information among Member States and to develop trust and confidence amongst them;
>   c. creates a computer security incident response teams network ("CSIRTs network") in order to contribute to the development of trust and confidence between Member States and to promote swift and effective operational cooperation;
>   d. establishes security and notification requirements for operators of essential services and for digital service providers;
>   e. lays down obligations for Member States to designate national competent authorities, single points of contact and CSIRTs with tasks related to the security of network and information systems.

It is Article 1, Section 2 (c), which implements the vision of giving computer security incident response teams (CSIRTs)/computer emergency response teams (CERTs) a central role in the fight against cyber threats, as envisaged in EU (2013) (Figure 15.1):

The security and notification requirements for operators of essential services and for digital service providers established in (d) should assure a common minimal level of protection and information sharing regarding critical information infrastructures.

Article 2—Processing of Personal Data—references to the relevant EU legislation. Article 3—Minimum Harmonization—introduces the principle of a minimal level of security that should be achieved in all member states of the EU. A guideline for the "identification of operators of essential services" is provided by Article 5 of the directive.

Chapter II—National Frameworks on the Security of Network and Information Systems—defines the obligation of member states, respectively:

The high importance of ENISA, the European Network and Information Security Agency, as hub for knowledge and information exchange, is underlined in the preamble and in several articles of the directive and especially in Article 12.

Chapter IV—Security of the Network and Information Systems of Operators of Essential Services—introduces the minimal obligations to be fulfilled, with Article 14—Security Requirements and Incident Notification—being the central point of reference

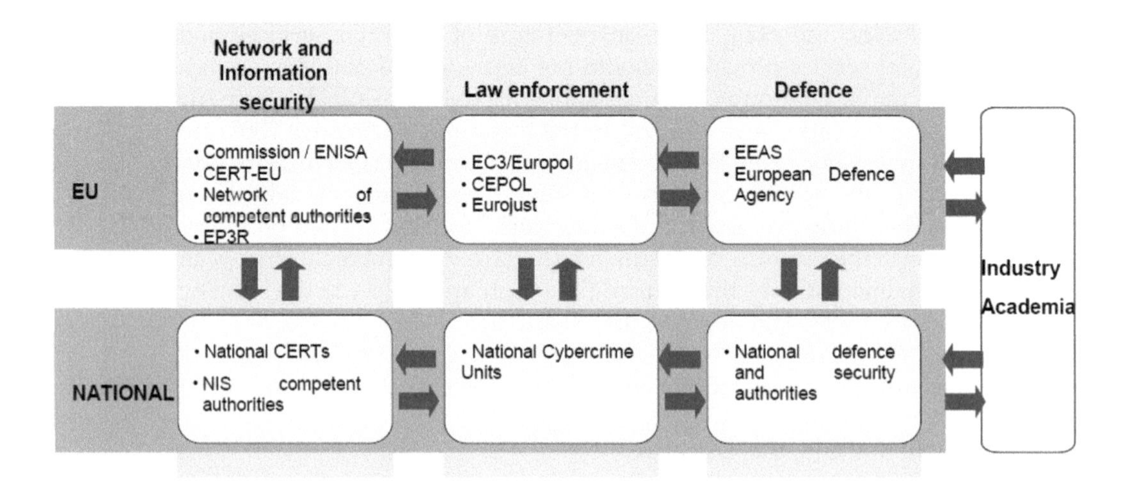

Figure 15.1 Role of CERTs/CSIRTs. CEPOL: European Union Agency for Law Enforcement Training; CERT-EU: Computer Emergency Readiness Team for the EU institutions, agencies and bodies; EC3: European Cybercrime Centre; EEAS: European External Action Service. (Extracted from EU, Joint Communication to the European Parliament, the Council, the European Economic and Social Committee and the Committee of the regions: *Cybersecurity Strategy of the European Union: An Open, Safe and Secure Cyberspace,* http://ec.europa.eu/information_society/newsroom/cf//document.cfm?doc_id=1667, 2013.)

for the safeguards and measures to be implemented. Article 15—Implementation and Enforcement—defines who Article 14 is to be acted on.

From an IT technology point of view, Chapter V—Security of the Network and Information Systems of Digital Service Providers—should be used as a core guidance. While not being surprising in its content, Article 16 introduces quite challenging requirements for system operators (Box 15.2).

The full implementation of the preceding requirements can be expected to result in substantial costs for system development and operations. With Article 16 stating the obligations in rather general terms, a more detailed implementation guideline will be needed. This guideline will most probably come in the form of national NIS legislation.

Incidents leading to reporting duties are defined in Article 16, Section 4:

> 4. In order to determine whether the impact of an incident is substantial, the following parameters in particular shall be taken into account:
>   a. the number of users affected by the incident, in particular users relying on the service for the provision of their own services;
>   b. the duration of the incident;
>   c. the geographical spread with regard to the area affected by the incident;
>   d. the extent of the disruption of the functioning of the service;

BOX 15.2 ARTICLE 14: SECURITY REQUIREMENTS AND INCIDENT NOTIFICATION

1. Member States shall ensure that digital service providers identify and take appropriate and proportionate technical and organizational measures to manage the risks posed to the security of network and information systems which they use in the context of offering services referred to in Annex III within the Union. Having regard to the state of the art, those measures shall ensure a level of security of network and information systems appropriate to the risk posed, and shall take into account the following elements:
   a. the security of systems and facilities;
   b. incident handling;
   c. business continuity management;
   d. monitoring, auditing and testing;
   e. compliance with international standards.

e. the extent of the impact on economic and societal activities. The obligation to notify an incident shall only apply where the digital service provider has access to the information needed to assess the impact of an incident against the parameters referred to in the first subparagraph.

Article 16, Section 5, regulates notification duties in the case of outsourced services:

> Where an operator of essential services relies on a third-party digital service provider for the provision of a service which is essential for the maintenance of critical societal and economic activities, any significant impact on the continuity of the essential services due to an incident affecting the digital service provider shall be notified by that operator.

Chapter VI—Standardisation and Voluntary Notification—is aimed at achieving a standardized communication and information exchange and provides a legal basis for nonobligatory information sharing (voluntary notification). This voluntary notification will be especially helpful in unclear situations where new emerging cyberthreats have to be dealt with and in the early stages of a massive cyberattack, when first indicators of compromise become visible, but the damage resulting from an attack is still below the threshold that would lead to an obligatory notification.

With creating a network of trusted CSIRTs and national hubs, establishing standardized information exchange, and strengthening the position of ENISA, the NIS Directive is an essential step toward securing the European cyberspace. It finally gives government agencies, CSIRTs, and law enforcement the long overdue legal basis for a much needed EU-wide closer and better structured cooperation. Focusing on critical infrastructures this directive also adheres to the principle of concentrating available European resource on essential efforts.

15.4 *Expected impact on information security and privacy management*

The expected impact of the introduction of the GDPR [Regulation (EU) 2016/679] is a more comprehensive harmonization of privacy protection across the EU. New and emerging technologies as well as new business models should be covered by the GDPR. While it fully satisfies all desires neither of privacy activists nor of industry, it is a solid compromise that gives society and economy a good basis to address the needs of privacy-related governance issues in cloud computing; social media; smart mobile devices; smart infrastructures, such as the smart power grid and smart metering; and the abundance of privacy issues related to IoT technology.

The NIS Directive [Directive (EU) 2016/1148] is aimed at establishing a common minimal level of critical information infrastructure protection in all member states of the EU. With the central role attributed to CSIRTs and the establishment of dedicated national hubs that serve as focal points for information exchange and cooperation in the case of a major cyberattack against Europe, a new infrastructure for more effectively countering cyberthreats will be introduced. The obligation to report major incidents is expected to lead to a much better and much faster coordinated reaction in case an attack starts spreading across the EU. Improved situation awareness, better and earlier information about developing threats and attacks, and the ability for a coordinated reaction is aimed at leading to a safer European cyberenvironment. The new European privacy and security legislation can be considered as a milestone for securing critical infrastructures, services, and processes across the EU while at the same time updating privacy legislation to enable it to cope with new digital business models and new technologies. The legislative process started by Directive 95/46/EC has now entered a next level, acknowledging the continuously increasing importance and growing dependence of the economy and of society as a whole on information and communication technology-based (cyber-) infrastructures.

References

Burgess, M. (2007) Everything you need to know about EternalBlue – the NSA exploit linked to Petya, WIRED, 28 June 2017, https://www.wired.co.uk/article/what-is-eternal-blue-exploit-vulnerability-patch.

ENISA. https://www.enisa.europa.eu/; accessed 2018-05-23.

EU (2013) Joint Communication to the European Parliament, the Council, the European Economic and Social Committee and the Committee of the regions: *Cybersecurity Strategy of the European Union: An Open, Safe and Secure Cyberspace*, http://ec.europa.eu/information_society/newsroom/cf//document.cfm?doc_id=1667.

EU Info (2017) *Structured Overview of General Data Protection Regulation GDPR*, https://gdpr-info.eu/.

European Parliament and Council of the EU (1995) Directive 95/46/EC of the European Parliament and Council of the EU of 24 October 1995, on the protection of individuals with regard to the processing of personal data and on the free movement of such data, Official Journal L 281, 23/11/1995 P. 0031–0050.

European Parliament and Council of the EU (2016a) Regulation (EU) 2016/679 of the European Parliament and of the Council of 27 April 2016 on the protection of natural persons with regard to the processing of personal data and on the free movement of such data, and repealing Directive 95/46/EC (General Data Protection Regulation).

European Parliament and Council of the EU (2016b) The Directive on security of network and information systems (NIS Directive), https://ec.europa.eu/digital-single-market/en/network-and-information-security-nis-directive.

Europol (2016) *The 2016 Internet Organised Crime Threat Assessment (IOCTA)*, European Police Office, 2016, ISBN 978-92-95200-75-3; ISSN 2363-1627, available on https://www.europol.europa.eu/activities-services/main-reports/internet-organised-crime-threat-assessment-iocta-2016.

Europol (2017) *The 2017 Internet Organised Crime Threat Assessment (IOCTA)*, European Police Office, 2017, available on https://www.europol.europa.eu/sites/default/files/.../iocta2017.pdf.

Fielder, A. (2017) *New EU Data Protection Laws: Ok, but a Tremendous Missed Opportunity with Possible Threats Looming*, https://www.privacyinternational.org/node/689.

ICO (2017) *Guide to the General Data Protection Regulation (GDPR)*, https://ico.org.uk/for-organisations/data-protection-reform/overview-of-the-gdpr/.

National Audit Office (2017). National Audit Office, Report by the Comptroller and Auditor General Department of Health Investigation: WannaCry cyber attack and the NHS, HC 414 SESSION 2017–2019 25 APRIL 2018, published 24 October 2017.

chapter sixteen

Privacy and security in the IoT—Legal issues

Rolf H. Weber

Contents

16.1 Introduction

The legal environment in technology-exposed areas is not easy to develop and to frame as experience has shown in many cases. The rule-making processes are confronted by fast technological changes; in addition, rules must be based on the technological designs of the concerned devices and software. The case to be discussed in this chapter is the Internet of things (IoT) applications.

The Internet Society (2013, 12) defined the term *IoT*, coined by the British technology engineer Kevin Ashton, in 1999, as the development of item identifications, devices, and sensor technologies that enable everyday items to interact with the environment. The IoT adds the dimension of "any thing" to information and communications technologies that already feature "any time" and "any place" aspects of functionality (ITU 2012, 2). Thus, the IoT is "a global infrastructure for the information society, enabling advanced services by interconnecting (physical and virtual) things based on existing and evolving interoperable information and communication technologies" (ITU 2012, 2). A substantively wide but shortly worded definition qualifies the IoT as a network encompassing a broad spectrum of device forms used in many varying settings (Weber 2015, 618).

In the IoT field, the most commonly used technology is radio frequency identification device (RFID). RFID aims at preventing the disappearance of goods and maintaining their quality through the shipment process. Tracking parts in manufacturing processes and measuring variables (temperature and humidity) in storage facilities are other common IoT applications today.

For private use purposes, the IoT helps increase household efficiency by allowing devices to communicate and take action such as ordering items to refill the fridge or starting a washing machine. Particularly in this context, the collection of device activity data can be sensitive; consequently, data privacy becomes an issue.

A particular challenge in this respect concerns the interaction between IoT-generated data and customer data. As such, machine-produced activity data are not subject to data protection regulations, but the combination with personal data can lead to the application of data privacy laws. This consequence mainly occurs if big data analytics are applied.

Furthermore, experience shows that the effects of malfunctions (e.g., in the case of placing of an order) created by corrupted data can be substantial: If devices and/or software are not working properly, an entire production process can be interrupted or damages can be caused to a private household. On the one hand, the traditional liability regimes are applicable, but, on the other, the simultaneous disclosure of data to third persons might also cause noncompliance with data protection laws.

As mentioned, technological elements provide a framework for designing the conditions for rule-making processes; in particular, the following key elements related to different IoT applications must be taken into account when seeking to regulate this environment (Weber 2015, 618):

- The *technology* has to be "global" in order to make the same technical processes applicable all over the world; the respective industry standards should ensure interoperability and data security.
- *Ubiquity* describes the extension (scope) of the technological environment: The IoT regulatory framework must be designed to ubiquitously encompass persons, things, plants, and animals.

- *Verticality* means the potential durability of the technical environment; an IoT application needs to function long enough to enable its use in the supply chain until it reaches the final customer.
- *Technicity* is an important basis for the development of rules protecting the data; thereby, the complexity of the techniques (active and passive, rewritable, processing, and sensor-provided products) as well as the complexity of background devices must be taken into account.

In a nutshell, IoT applications can add value to individuals as well as businesses, but they also cause risks. In order to protect against such risks, the law must understand the technological features and set rules accordingly.

This contribution begins with a discussion of the normative framework applicable to IoT by taking into account the suitability of regulatory measures as well as the current state of regulation and the strategies being proposed for future regulation. Next, we discuss the privacy and security challenges in light of data protection rights and obligations before addressing specific issues in the context of IoT. These issues are based on the nature of the technology and thus require for the most part technological solutions. In conclusion, an outlook on future developments is presented.

16.2 Normative framework

16.2.1 Legal suitability and systematic structure

The normative framework governing the IoT should apply globally; it should be applicable to every device on earth. The present lack of international rules and the improbability of reaching a respective multilateral agreement require stronger leadership by the industries in establishing the relevant standards for the applications and devices. Such an approach should avoid causing a disruption between many potentially varying data protection rules across states. Harmonization processes based on standardizations of industry organizations could be a first move into the direction of legal stability.

Since IoT applications and new technological opportunities have organizational, social, and cultural implications, a simple legal framework is not easily developed. Also, different types of information used in the context of the IoT increase the difficulty of identifying single factors. Only through combinations of approaches and analytical methods will it be possible to develop a stable legal environment. Since the collection of data by IoT applications is carried out in an automated manner, the risk of being noncompliant with the applicable laws must be addressed in their design.

The establishment and implementation of an appropriate legal framework enshrining effective rules calls for a systematic approach. Thereby, a systematization of legal problems potentially occurring should be done by coordination along certain technical axes. Reference points can, for example, be the already mentioned technological elements of globality, ubiquity, verticality, and technicity. The normative challenges of data privacy and security need to be reflected in a qualitative classification. In this context, the question that must be addressed is "how much privacy is society prepared to surrender to increase security?" Solutions should permit the understanding of privacy and security not as opposites, but as principles affecting each other.

16.2.2 Regulatory environment

16.2.2.1 Governmental initiatives

1. A multilateral agreement, similar, for example, to the World Trade Organization agreements governing international trade, does not exist in the field of the IoT. In addition, the negotiation of such an agreement is also unlikely to happen during the next few years. But even if the respective efforts were undertaken, it appears to be doubtful whether an adequate legal framework would be possible in view of the fast developing technologies and different legal regimes around the world.
2. On a regional level, the European Commission started relatively early looking into the regulatory challenges caused by the IoT. With the support of an expert group, the Commission published a detailed questionnaire that provoked many valuable inputs (European Commission 2013). Nevertheless, the Commission, which was invited by the addressees of the questionnaire to implement a multistakeholder initiative for the establishment of IoT guidelines and recommendations, has withdrawn these activities from the political agenda [for further details see Weber (2016, 29/30)].

   In March 2015, the European Commission initiated the creation of the Alliance for Internet of Things Innovation (AIOTI). This organization was invited to prepare a European IoT roadmap toward the year 2020. In October 2015, the AIOTI published 12 reports, which set forth the "Recommendations for future collaborative work in the context of the Internet of Things Focus Area in Horizon 2010." The reports address IoT applications, innovation ecosystems, IoT standards, policy issues, the smart living environment, the smart farming and food safety, wearables, smart cities, the smart mobility, and the smart manufacturing.
   In addition to Article 29—Data Protection Working Party of the EU (WP29)—consisting of representatives of the national data protection authorities, the European Data Protection Supervisor and the European Commission published its Opinion on the Internet of Things in September 2014 (WP29, Opinion 8/2014) (European Commission 2014). In this Opinion, the WP29 alerts both businesses and customers to the challenges and risks arising from the use of the IoT technologies (quantified self, home automation, and wearable computing) and proposes measures that could enhance and secure data privacy (e.g., privacy impact assessment, quality control by device manufacturers and application developers, and improvement of standardization). Finally, research groups (such as the European Research Cluster on the Internet of Things and the Dynamic Coalition on the Internet of Things) tackle normative IoT issues; however, their political impact is quite remote [see Weber (2016, 30/31)].

3. The Federal Trade Commission (FTC) in the United States has also begun to look into privacy issues in the IoT environment. The FTC in a staff report (2015) declared that IoT-specific legislation at this time would be "premature" and instead encouraged the development of self-regulatory programs for industry sectors in order to improve privacy and security issues. In May 2016, a policy paper for further research with the title "Developing and Growing the Internet of Things Act" ("DIGIT") was presented. As far as privacy issues are concerned, for obvious reasons, different levels of data protection in the United States and in the European Union (EU) create challenges in coming to a common understanding.

16.2.2.2 Self-regulatory initiatives

In view of the difficulty to develop a genuine normative environment for the IoT, as mentioned, rule-making processes should start at the technological designs of the IoT. The present lack of a global regulatory environment makes it necessary that industries producing and using the IoT devices are self-regulating by adhering to the state-of-the-art standards of the industry for such devices.

Therefore, notwithstanding the fact that the IoT will be a discussion topic for many years, the present regulatory model is still based on self-regulation through many business standards, beginning from technical guidelines and extending to fair information practices. Indeed, under the given circumstances, it appears to be appropriate that standard setting by the industry itself should be encouraged as long as this model meets the demand of the market and offers parties engaged with the IoT a choice as to the level of privacy protection they wish.

Self-regulation realizes the principle of subsidiarity, meaning that the participants of a specific community try to find suitable solutions (structures, behaviors) themselves. The legitimacy of self-regulation is to be seen in the fact that private incentives lead to a need-driven rule-making process. In addition, self-regulation is usually less costly and more flexible than governmental rules see also Schmid (2008, 199)].

Related to the concept of self-regulation, legal doctrine has developed the notion of "soft law." This term did not yet gain a clear scope or reliable content; often, it is used in parallel to self-regulation. But the word *soft law* shows the neighborhood to law usually covering certain forms of expected and acceptable codes of conduct (Weber 2010, 27/28).

The specific problem about data privacy and security consists of the acknowledgment that the applied principles are not identical in the different regions of the world, which makes the application of general principles difficult in cross border business activities.

16.2.3 Regulatory strategies and agenda

The legal framework for data privacy and data security issues of the IoT could be based on five different strategies (Weber 2010, 29; Schmid 2008, 208):

- *Right-to-know legislation:* This approach envisages keeping the customer informed about the applied IoT scenarios, i.e., the customer should know which data are collected and should have the possibility to deactivate the tags after a transaction.
- *Prohibition legislation:* This concept, corresponding to traditional atate legislation, envisages to forbid or at least to restrict the use of IoT applications in certain scenarios. The mentioned self-regulatory mechanisms, however, rather tend to introduce incentives (if at all) instead of prohibitions.
- *Information technology (IT) security legislation:* This model develops initiatives that demand the establishment of certain IT security standards; usually, respective rules are developed by the concerned market participants, but state intervention remains possible. The respective standards could develop, for example, a new-generation framework of data protection protocols allowing the setting up of stringent safeguards as to reporting and frequent audits of implemented measures.
- *Utilization legislation:* This approach intends to support the use of IoT applications under certain conditions; such a normative concept must fine-tune an appropriate balance between prohibited and utilizable approaches.
- *Task force regulation:* This model covers legal provisions supporting the technical community in investing into the research of the legal challenges of the IoT.

The aforementioned approaches can also be combined; in principle, however, in the IoT environment, the IT security legislation should become the driver of a regulatory framework. The other approaches appear to be less suitable due to the limited scope and/or the unclear technological perception.

The regulatory agenda must consider the requirements of technological designs and applications [see also Nappinai (2017, 40/41)]. This awareness is crucial for the development of appropriate new rules. Therefore, based on the outlined general assessment of the normative framework governing data protection and data security issues in the IoT field, the following considerations start with a discussion of the technological topics, in particular, the devices and software requirements, the privacy and security challenges, and the privacy enhancing technologies, followed by specific new issues related to the IoT technology. For these reasons, the legal issues caused by the pertaining privacy and security threats caused by the IoT are to be analyzed hereinafter.

16.3 Specific privacy and security issues

16.3.1 Device and software requirements

During the last few years, microchips are constantly becoming cheaper to produce resulting in IoT sensor prices dropping below 50 cents per unit. The smaller devices such as RFID will be a main driver of growth in this area. In addition, cellular devices can provide an access point and a gateway for other (low-power) technologies (Weber 2015, 621).

At this moment, the technological challenge consists of the limited storage space, particularly on a simple passive RFID. An ongoing flow of information cannot be saved on an RFID tag. Therefore, the supply of the collected information must be made available through linking and cross-linking with the help of an object naming service (ONS). The ONS has some similarities with the well-known domain name system (DNS) of Internet Corporation for Assigned Names and Numbers but is not identical; nevertheless, the ONS is also authoritative in the way that the entity having change control over the information about the electronic product code (EPC) is the same entity having assigned the EPC to the concerned item (EPC Global 2008).

Depending on the IoT device, access by a third party can be made possible at any point in the technological chain. But since aggregated data are of value at the access point (router or cellular device), hackers and other interested third parties can also use the access point as main entry to the stored information. Usually, data are not encrypted at this location; encryption and anonymization are only done at a later stage on the cloud server. Consequently, privacy risks cannot be overlooked.

Recently, new standards have been designed to better deal with the various types of data. Mostly, an open-access approach is chosen. In using this method, the industry also enables criminals to profit from the accessible information. In order to avoid negative consequences, uniform security standards should be developed in order to ensure the safety of the data at every step from its collection to the processing (for the identification of criminal, see, e.g., Wynzard Group).

16.3.2 Privacy and security challenges

Obviously, the IoT devices collect a large amount of information; consequently, these devices also carry a substantial potential of privacy risks in relation to the use of the data and its access. As IoT devices are increasingly used in all fields of daily life, the identification of an individual and his/her behavioral patterns become a growing concern.

16.3.2.1 *General disclosure risks*

With the widespread implementation of the new technologies, better-designed safeguards for privacy and data integrity must be created. The potential of IoT for daily life needs to be balanced against the risks of undue information disclosures. For example, with the help of big data analytics, the accumulated raw data can become highly valuable (particularly in the health sector) since specific patterns can be extracted, but the privacy risks inherent are not to be neglected as the IoT data would allow the identification of an individual and his/her (health) condition [for further details on security risks, see Xiao et al. (2000)].

IoT devices often collect certain data that are aggregated with other data and sent through a router to a communication device. Such a device is then able to transfer the data to a cloud server for processing. During this procedure, various protocols and compression technologies are employed, since—as mentioned—the storage space on devices is limited. In addition, the devices often are unable to cope with the big headers used for the Internet protocol IPv6.

Technologically and practically, the interconnection between the devices and infrastructures has not yet reached the appropriate level allowing for its seamless implementation into daily life. With an increased number of services offered based on IoT technology, however, this limitation will lose importance over the next couple of years. Hence, the location points of interconnection need to be designed in a secure way.

16.3.2.2 *Requirements of technical architecture*

The technical design of the IoT is not without impact on the privacy and security of the involved individuals. Privacy includes the concealment of personal information as well as the treatment of the data. Many stakeholders are interested in the data; for example, private actors such as marketing enterprises, national security services, and public utility operators. Therefore, the degree of reliability must be high (Weber 2010, 24).

The following privacy and security requirements are relevant as criteria for achieving the desired goals (Fabian and Günther 2007):

- *Resilience to a tag:* The system must avoid single points of failure and should adjust itself to node failures.
- *Data authentication:* Retrieved address and object information should generally be authenticated.
- *Access control:* Information providers must be able to implement access control on the data provided.
- *Client privacy:* Only the information provider should be able to infer from observing the use of the lookup system related to a specific customer.

These requirements are to be integrated into the risk management concept of private enterprises and government agencies. A good IT governance approach considers the concerned business activities and limits exposure.

16.3.2.3 *Cybersecurity risks*

The general security risks in the IoT have a further exposure in the context of cybersecurity causing new and unique challenges. In fact, recent examples such as the hacking of baby monitors have shown the vulnerability of IoT devices. A key issue consists of the increase of overall attack surface for malicious attacks, as compared to isolated (i.e., nonconnected) systems [for further details, see Weber and Studer (2016, 719/20)].

A 2015 study by Hewlett–Packard (2015) showed that 70% of IoT devices contain serious vulnerabilities stemming from the following:

- *Lack of transport encryption:* Many IoT devices are simple "unit-taskers" and have cost, size, and processing constraints, i.e., most devices will not support the processing power required for strong security measures.
- *Insufficient authentication and authorization:* The weakness is due to poor password requirements, careless use of passwords, lack of periodic password resets, and failure to require reauthentication for sensitive data.
- *Insecure web interface:* Issues in this respect include persistent cross site scripting, poor session management, and weak or plain default credentials.
- *Insecure software and firmware:* Most IoT devices are designed without the ability to accommodate software or firmware updates making vulnerability patching difficult.

In addition, digital attacks on connected devices not only pose risks in the online world, but they also create physical risks to the devices themselves and, even more critically, safety risks for IoT users [for further details, see Weber and Studer (2016, 720/721)].

16.3.3 Privacy-enhancing technologies

A number of technologies have been developed in the past couple of years to balance IT risks against information privacy goals. The best-known measures are based on the so-called privacy-enhancing technologies (PETs); the following techniques are often implemented (Fabian and Günther 2009, 124/25; Weber 2010, 24/25):

- *Virtual private networks* as extranets established by close groups of business partners
- *Transport layer security* based on an appropriate global trust structure, thereby improving confidentiality and integrity of the IoT
- *DNS security extensions* (DNSSEC) making use of public key cryptography for the signature of resource records to guarantee origin and integrity of delivered information;
- *Onion routing* encrypting and mixing Internet traffic from many different sources and channels;
- *Private information retrieval* systems assessing the customers' interest in a specific information.

Increased privacy and security can also be achieved by *peer-to-peer* (P2P) systems. These designs can achieve good scalability and performance.

Another relatively new approach addresses technical possibilities being able to integrate the data privacy safeguards into the IT devices and applications. This new notion, called "privacy by design" (PbD), requires the adherence to seven basic principles: (1) a proactive approach to protection measures, (2) privacy as default setting, (3) privacy embedded into the design of the technology, (4) full functionality, (5) end-to-end security spanning the life cycle of the device, (6) visibility and transparency allowing the stakeholders to verify the privacy claims made, and (7) respect for user privacy (Cavoukian 2009). In the meantime, PbD has become a processor's obligation according to the new EU General Data Protection Regulation (GDPR) of 2016 entering into force in late May 2018 (Article 25 GDPR).

A similar new approach is called "privacy engineering." Privacy is considered in this approach to be a nonfunctional requirement or quality attribute. Such an attribute needs to be broken down into components that can be architected and evaluated. In this concept, the collection, processing, and storage of personal information becomes a business requirement that makes it necessary to rethink topics such as privacy policies and notices [for further details, see Finneran et al. (2014, 73–226)]. Therefore, new elements structuring data and privacy governance concepts gain importance.

16.3.4 *Privacy through deletion*

Privacy can be realized through the diversion or the destruction of an RFID tag or the deletion of data.

1. As far as the lifetime of RFID tags is concerned, individual tags can be disabled if it is decided that an alternative use of the tag would be preferable. The disabling can be done by putting the tag in a protective mesh of foil known as a "Faraday cage," which is impenetrable by radio signals of certain frequencies or by "killing" the tag, i.e., removing and destroying it (Eschet 2005, 317/18).

   Nevertheless, it cannot be overlooked that both approaches have certain disadvantages: Possibly, some tags are overlooked or left with the individual. In addition, the "kill" command related to a tag still leaves room for the possibility of reactivation (Weber 2010, 25). Finally, businesses may be inclined to offer clients certain incentives for not destroying tags or secretly giving them the tags.
2. Deletion rights and automatic data deletion are also an important aspect of ensuring privacy as the amount of data collected are growing exponentially [see Schwartz and Solove (2011, 1819)]. Data saved in various scattered databases make it very complicated for an individual to not only theoretically retain but also enforce the right to have the data deleted after a certain period or upon request. Looking from a general perspective, data deletion challenges merit much higher attention, exceeding the notion of the human "right to be forgotten" (Google Spain) recently acknowledged by the European Court of Justice and encompassing new technical solution for the improvement of data privacy.

16.3.5 *Challenges through technological innovations*

16.3.5.1 *Ongoing technological developments*

The regulatory framework is usually relatively slow, while technological innovations move very fast. Therefore, efforts to adjust the legal rules are often effectively nullified by new innovations. Recent technologies encompass, for example, location-based services, sensor networks, delay-tolerant networks, and the smart grids. Some legislators still try to adapt traditional telecommunications rules without taking into account that information exchange moves to other infrastructures.

All these new technologies have caused additional risks to privacy and security. In order to minimize such risks, data collection devices should be designed to include basic privacy protection features from the beginning. In fact, the new so-called G2 RFID technology allows the user to hide a part or all the memory of the tag and the ability to read or alter the data depending on the proximity to the tag. Thus, such new technologies will ensure that the control over the data is given back to the data subject (Weber 2015, 623).

However, more and more devices used in daily life create additional challenges. These emerging risks include (i) automatically generating data that are not necessary for providing a service and its collection could potentially have severe privacy implications, (ii) private data scattered across large distributed systems leads to a loss of control, and (iii) deanonymization results from the linking of data collected across an ever-growing number of devices.

The issues of (i) the quality of data and (ii) the quality of context add a new critical dimension of data privacy and security discussions. These phenomena play an increasingly important role in IoT debates and, therefore, are discussed in more detail in the following sections.

16.3.5.2 *Quality of data*

The quality of the data collected is increasing with any further information added. Location and environment information leads to a better usability of the data. Thus, the aggregation of data is a quality-improving activity.

The collected data are in most cases not encrypted, as its face value is low. The risks emerge when the automatically collected unencrypted data from various sources are combined and aggregated into one database. Therefore, automated processes should be implemented in IoT devices to ensure privacy by encrypting and anonymizing data.

The quality of the data can be achieved by taking the environment in which it is collected into account. A context attribute may be unknown as long there is no information about it. It may also be ambiguous and/or imprecise when the reported information is correct but not provided with a sufficient degree of precision (Weber 2015, 623).

16.3.5.3 *Quality of context*

The quality of context raises new issues of confidentiality. The quality of context refers to the embedment of information into the life sphere of an individual and not to the process or the hardware components that possibly provide the information. Up to now, quality of context issues have not been sufficiently discussed even as these phenomena play an increasingly important role in privacy debates [see, for example, Machara Marquez et al. (2013) and de Montjoye et al. (2013)].

A context data may be unknown if there is no information available about it, leading to incomplete context information and wrong interpretations (Weber 2015, 624). It may also be ambiguous or imprecise in case of contradictory information from different context sources. As context data are by nature dynamic and heterogeneous, they tend to be erroneous (i.e., an exact reflection of the real state of the modeled situation is not given).

However, even low-quality context data are useful for data mining algorithms. In addition, the reliable quality of context information is improving the efficiency of such algorithms. Legally, the quality of context data encompasses sensitive information since its value and the potential effects of such data relate to individuals. The combination of parameter values may be used to infer what the operating system of remote machines is because different operating systems, and different versions of the same operating system, set different default values for these parameters (Weber 2015, 625).

16.4 *Specific challenges of the IoT for privacy and security*

16.4.1 *Types of privacy infringements*

Data privacy can be infringed in the IoT context at various stages (Miorandi et al. 2012): The first stage concerns the access to the collected data by third parties; the second stage is the use and distribution of data by the data collector; and the third stage encompasses the risks that the data is combined with other data. Individuals are especially unaware of the third case when using IoT devices supplying the data. Such kind of combinations with other data allows the creation of new information about a person or a situation being potentially of high commercial value.

As mentioned, a particular issue is the quality of context. Information surrounding the collection of the primary data such as the status and attributes of the data-collecting device can lead to a personality profile. The aggregation of a large amount of data (for example, power usage of a household and travel patterns from mobile phone location data) causes the formation of sensitive data collections. Such a development is not compliant with the data minimization principle of data protection laws requiring the limitation of data collection to the furthest extent possible (Weber 2015, 624).

The automated data collection does not necessarily lead to a higher level of trust than in the case of manually collected human data. Therefore, adverse judgments affecting a person based on such data collected by whatever devices should be prevented not only through appropriate technical safeguards but also through regulatory restrictions on the data use (Weber 2015, 624).

16.4.2 *Enforcement of the transparency and the data minimization principle*

Data collection and data storage must be transparent. Individual rights can be effectively exercised only if the concerned person is aware of the data processing entity and the contents of the databases. Therefore, awareness should be generally created in the society as to the many privacy implications IoT devices can have on an individual.

Transparency tools have been addressed for a few years. Usually, these tools intend to improve the users' understanding and control of their data profiles. Transparency was particularly acknowledged as an important element of privacy in the "Mauritius Declaration on the Internet of Things," proclaimed by the Data Protection and Privacy Commissioners (2014) of more than 100 countries on October 14, 2014: "Transparency is key: those who offer Internet of Things devices should be clear about what data they collect, for what purposes and how long this data is retained."

Four basic characteristics that transparency tools should possess appear to be important in the long run (Weber 2015, 625):

- Provide information about the intended collection, storage, and/or data processing
- Provide an overview of what personal data have been disclosed to what data controller under what policies
- Provide online access to the personal data and how they have been processed by the data controller
- Provide counterprofiling capabilities helping the users to anticipate how their data matches relevant group profiles, which may affect future opportunities and threats

Transparency in the privacy context is also important with respect to technical safeguards. Only if people understand the functions of PETs can they use them efficiently or decide on the implementation in an automated fashion. Transparency might even gain importance in view of the next IoT-based distributed systems compared to the present web-based ubiquitous applications, since users can no longer control the data coming from terminals with which they directly interact (laptop, smartphone, etc.). However, they will have to handle the control of the data automatically produced by the connected devices (Weber 2015, 626).

In the future, data could be scattered across a large distributed system while facing issues such as heterogeneity and scalability of processes. So far, neither experts nor regulators have undertaken much research in this field notwithstanding the importance of the arising privacy-related topics.

Apart from transparency, the general principle of data minimization is enshrined in practically all data protection laws, such as, in Article 5 of the EU GDPR 2016. This principle aims at limiting the collection and processing of personal data to the amount necessary for the concerned businesses. Again, the legal requirement of data minimization needs to be supported by technical measures. Possible implementation approaches are (Weber 2015, 626) as follows:

- *Encrypted aggregation techniques:* Data are collected only to the extent to which they add value for the intended use.
- *Perturbation:* Data gets systematically altered using a perturbation function (e.g., adding random numbers).
- *Obfuscation:* This concept replaces a certain percentage of data by random values (e.g., change of code in order to make reverse engineering difficult).

In particular, the data minimization principle aims at deleting data from the IoT device or supporting systems when the data are no longer relevant and when continued storage is not justified. Nevertheless, in the IoT context, the data minimization principle must be balanced against the demands of civil society and businesses for more functionality. Thus, the problem (to be solved) arises to what extent and under what circumstances technical possibilities exist for putting back "deleted" data in its original state, if at a later stage, a new need to read the information should occur.

16.4.3 Confidentiality and anonymity challenges

Confidentiality is a legal notion that has been relevant in public and private law for centuries. Under certain circumstances, information is only disclosed to a small circle of persons. Its content should not be available to persons other than the specific addressees. For example, in governments and administrative agencies, certain data are classified as confidential information; typically, this is the case for federal and state investigating authority inquiries (e.g., the Federal Bureau of Investigation, the Department of Justice, and the police). In the private law, similar protections exist, for example, the attorney–client privilege or the medical doctor–patient secrecy in the healthcare sector.

The new technologies allow an improvement of the framework for confidentiality interests. Particularly, the PETs (cf. Section 16.3.3) can protect data from unauthorized access by third persons. Privacy measures can support confidentiality requirements by providing solutions for anonymizing the collected data (including the communications).

The "anonymity of data" [see Weber and Heinrich 2012)] can be indexed by cryptographic measures. Such measures are designed to reflect properties such as (i) unlinkability (two information items or two actions of the same person cannot be related), (ii) undetectability (a third person is not able to ascertain whether an information item exists), (iii) unobservability (it is not possible to detect whether a system is being visited by a certain user), and (iv) communications content confidentiality (Weber 2015, 622).

During recent years, new technologies have been developed allowing a disclosure of information to a third person but protecting the anonymity of the person from whom the data were collected while still retaining its value. The best-known model is called *k*-anonymity, which is aimed at reducing the risk of reidentification by linking datasets. The *k*-anonymity model addresses the problem of directly matching externally available data and claims that an individual cannot be identified within a set of k users. Therefore, protection is provided if the information for each person is not distinguishable from at least $k - 1$ individuals whose information is also contained in a given data set (Sweeney 2002). Thus, this approach requires a structuring of the data, which protects against the identification of an individual.

However, the *k*-anonymity model is susceptible to background knowledge and homogeneity attacks (Machanavajjhala et al. 2006). As a consequence, a further refined variant has been developed, namely, the L-diversity model; a block of data is L-diverse if it contains at least L well-represented values for the sensitive attribute S (Wang and Wang 2013). So far, this approach has mainly been effective in cases of static data, but due to increased research efforts, the L-diversity method is now also applicable to incremental data disclosure (Wang and Wang 2013).

Normative frameworks from competent regulatory authorities come into play in this context, since legal rules must define under which circumstances the noncompliance with anonymity requirements constitutes a data breach. The relevant term is called *differential privacy*, which is aimed at providing means to maximize the accuracy of queries from statistical databases while minimizing the risks of identifying its records (Dwork and Roth 2014).

Anonymity in communications has the objective of protecting traffic data by avoiding disclosure of who talks to whom. Even if the content of a communication is kept confidential (or anonymous), sensitive information can still be gained by analyzing the leaked traffic data, namely, the respective data can include locations and identities of the parties in the communication, time, frequency, and the volume of the information exchange (Weber 2015, 623). Therefore, privacy laws also need to protect such kind of data.

16.4.4 Cybersecurity regulations in particular

As mentioned, cybersecurity issues play a role in the data security context of the IoT. Apart from the quite outdated and not globally applicable Cybercrime Convention (CETS 185) of the Council of Europe (Budapest Convention) of November 2001 [for further details, see Weber and Studer (2016, 722/23) and Nappinai (2017, 43)], only a recent regional legal instrument is available, namely, the Network and Information Security (NIS) Directive of the EU of May 2016. Based on the preparatory work of the 2004 established European Network and Information Security Agency, the NIS Directive has the objective of implementing a culture of network and information security for the benefit of citizens, consumers, businesses, and public sector organizations in the EU [for further details Weber and Studer (2016, 723)]. The NIS concept encompasses the ability of networks and information systems to resist, at a given level of confidence, any action that compromises the availability, authenticity, integrity, or confidentiality of stored or transmitted or processed data or the related services [for the United States, see Shackelford et al. (2016)].

The NIS Directive sets forth the following main topics and measures to realize the desired level of information security:

- *Improved national cybersecurity capabilities:* A national cybersecurity strategy should be developed and adopted designing a policy and a regulatory environment for information security as well as establishing adequate institutional capacities.
- *Improved EU-level cooperation:* Strategic cooperation and exchange of information must be secured.

- *Security and incident notification requirements:* The respective requirements are differently designed for operators of essential services and digital services providers [for further details, see Weber and Studer (2016, 714)].

The NIS Directive has been subjected to critical assessments; mainly, the weakening of certain provisions during the legislative process (relatively low level of "minimum" harmonization) is debated. Furthermore, some requirements imposed on digital operators are subject to an open interpretation. Nevertheless, this new legal instrument has at least the chance to adapt the regulatory environment to new cybersecurity needs by obliging a wide range of industries and other players to pay more attention to the challenging issue of information security.

16.4.5 Interoperability and connectivity requirements

With the increased availability of IoT applications, challenges in respect to the interoperability and connectivity of the new services also increase. An added value for users (civil society and businesses) depends on the possibilities to bring different networks and services "together" as was done in the case of different Internet protocols some 30 years ago. Interoperability and connectivity requirements have been known for decades from telecommunications markets; in the IoT field, the problems are not substantially deviating but technically often more complicated.

The Open Interconnect Consortium (OIC) was founded in July 2014 to improve interoperability and connectivity. The efforts of the OIC go in the direction of defining a common communications framework based on industry-standard technologies to wirelessly connect and intelligently manage the data flow among emerging IoT devices, regardless of form factor, operating system or service provider. The OIC assembles leaders from a broad range of industry vertical segments, from smart home and office solutions to automotive and more. As an objective, OIC specifications and open-source implementations should support businesses in the design of products that intelligently, reliably, and securely manage and exchange data under changing conditions, power, and bandwidth, even in the case of lack of an Internet connection (Broadcom 2014).

At first instance, interoperability and connectivity facilitate the execution of business transactions. But the respective technical requirements can also uphold privacy standards; if the respective instruments are in place and enable the smooth "transfer" from one system to another system, technically encrypted, and thereby, privacy-protecting data do not risk being disclosed during a transmission chain.

16.5 Outlook

The IoT opens up a world of opportunities because it is applicable to a wide range of sectors and markets, including logistics and transportation management, connected furniture and appliances, agricultural monitoring systems, smart clothes and accessories, toys, entertainment, and art. The steadily growing number of IoT products and devices will make life easier for individuals while creating privacy risks by targeting individuals unknowingly at the same time.

Nevertheless, based on the technical and regulatory complexity of IoT, the future of digital privacy will strongly depend on the willingness of the IoT industry to implement its own standards. These standards must consider the nature and context of the collected data and offer tailored solutions as part of the technological backbone.

Furthermore, efforts must be made to negotiate international standards which are based on an international agreement and have binding effect, providing both enforceable rights against the suppliers of IoT technology as well as the providers of the ancillary software environment. So far industry standards as to security have been very effective. However, data protection and privacy require a much broader foundation that is based on both technical standards and an appropriate legal framework.

The ultimate solution to complex issues raised by IoT may be the combination of law and technology together with a universally agreed industry standard. Any such legal–tech solution must allow for both an efficient technical process and a data protection policy conform use of the collected IoT data. In doing so, such a toolset could also address many of the issues that create uncertainty in the market such as the application of the controller or processor definition to IoT device manufacturers. If they are part of a broader solution, they will not need to rely on ambiguous interpretations of the GDPR to be compliant. Furthermore, such tools would facilitate the documentation of the collected data and the use by third parties, thereby allowing for more transparent processing and facilitating of the enforcement of data subject rights. Such a PbD approach would mitigate most of the data protection and security issues that arise in most IoT settings.

In summary, many challenges lie ahead: security and privacy issues, product energy and maintenance needs, new product–person relationship models, product–user–manufacturer relationships, as well as new business models reflecting this duality. To facilitate the seamless implementation process of both privacy and security frameworks, a coordinated approach on both the international and national levels is necessary and warranted.

References

Broadcom (2014), *Industry Leaders to Establish Open Interconnect Consortium to Advance Interoperability for Internet of Things*, Press Release, July 9, 2014, Broadcom, San Jose, CA, available at http://www.broadcom.com/press/release.php?id=s858114.

Cavoukian A. (2009), *Privacy by Design, 7 Foundational Principles*, available at https://www.privacybydesign.ca/index.php/about-pdb/7-foundational-principles/.

Data Protection and Privacy Commissioners (2014), *Mauritius Declaration on the Internet of Things*, 36th International Conference of Data Protection and Privacy Commissioners, available at http://www.privacyconference2014.org/madia/16596/Mauritius-Declaration.pdf.

de Montjoye Y., Hidalgo C.A., Verleysen M., and Blondel V.D. (2013), Unique in the crowd: The privacy bounds of human mobility, *Nature Scientific Reports* 3, 1376.

Dwork C., and Roth A. (2014), The algorithmic foundations of differential privacy, *Theoretical Computer Science* 9 (2014), 211–407.

EPC Global (2008), *Object Naming Service (ONS) Version 1.0.1*, EPC Global, Lawrenceville, NJ, available at http://www.gs1.org/sites/default/files/docs/epc/ons_1_0_1-standard-20080529.pdf.

Eschet G. (2005), Protecting privacy in the web of radio frequency identification, *Jurimetrics* 45 (2005), 301–322.

European Commission (2013), *Public Consultation on the IoT Governance*, European Commission, Brussels, available at https://ec.europa.eu/digital-single-market/en/news/conclusions-internet-things-public-consultation.

European Commission (2014), *WP29 (Article 29 Data Protection Working Party): Opinion 8/2014 on the Recent Developments on the Internet of Things* (September 16, 2014), European Commission, Brussels available at http://ec.europa.eu/justice/data-protection/article-29/documentation/opinion-recommendation/files/2014/wp223_en.pdf.

Fabian B., and Günther O. (2007), Distrubted ONS and its impact on irivacy, *IEEE Communications 2007*, Institute of Electrical and Electronics Engineers, Piscataway, NJ, available at http://ieeexplore.ieee.org/document/4288878.

Fabian B., and Günther O. (2009), Security challenges of the EPC Global Network, *Communications of the ACM* 52 (2009), 121–125.

Finneran Dennedy M., Fox J., and Finneran T.R. (2014), *The Privacy Engineer's Manifesto*, Apress Media, New York.

FTC (Federal Trade Commission) (2015, January) *Staff Report, Internet of Things: Privacy and Security in a Connected World*, FTC, Washington, DC, available at https://www.ftc.gov/system/files/documents/reprts/federal-trade-commission-staff-report-november-2013-workshop-entitled-internet-of-things-privacy/150127iotrpt.pdf.

Hewlett–Packard (2015), *Hewlett–Packard Internet of Things Research Study*, Report 2015, Palo Alto, CA, available at http://www.hp.com/h20195/V2/GetPDF.aspx/4AA5-4759ENW.pdf.

Internet Society (2015), *The Internet of Things: An Overview*, Internet Society, Reston, VA, available at https://www.internetsociety.org/sites/default/files/ISOC-IoT-Overview-20151014_0.pdf.

ITU (International Telecommunications Union) (2012), *ITU-T Recommendation Y.2060, Overview of the Internet of Things (6/2012)*, ITU, Geneva, available at http://www.itu.int/ITU-T/recommendations/rec.aspx?rec=11559.

Machanavajjhala A., Gehrke J., and Kifer D. (2006), L-Diversity: Privacy beyond k-anonymity, *Proceedings of the 22nd International Conference on Data Engineering, Los Alamitos*, IEEE Computer Society, Washington, DC, 24–35.

Machara Marquez S., Chabridon S., and Taconet C. (2013), Trust-based Context Contract Models for the Internet of Things, UIC/ATC 2013, 557–562, available at https://ieeexplore.ieee.org/stamp/stamp.jsp?tp=&arnumber=6726259.

Miorandi D., Sicari S., de Pellegrini F., and Chlamtac I. (2012), Internet of things: Vision, Applications and Research Challenges, *Ad Hoc Networks* 10 (2012), 1497–1516.

Nappinai N.S. (2017), Dark side of IoT, *CRi* 2 (2017), 39–45.

Schmid V. (2008), Radio frequency identification law beyond 2007, in Floerkemeier Ch., Langheinrich M., Fleisch E., Mattern F., and Sarma S.E. (eds.), *The Internet of Things*, Springer, Berlin, 196–213.

Schwartz P., and Solove D. (2011), The PII problem: Privacy and a new concept of personally identifiable information, *New York University Law Review* 86/6 (2011), 1814–1894.

Shackelford S.J., Russell S., and Hunt J. (2016), Bottoms up: A comparison of voluntary cybersecurity frameworks, *UC Davis Business Law Journal* 2016, 217–260.

Sweeney L. (2002), k-Anonymity: A model for protection of privacy, *International Journal on Uncertainty, Fuzziness and Knowledge-Based Systems* 10 (2002), 557–570.

Wang P., and Wang J. (2013), L-diversity algorithm for incremental data release, *Applied Mathematics & Information Services* 7 (2013), 2055–2060.

Weber R. H. (2010), Internet of things—New security and privacy challenges, *Computer Law & Security Review* 26 (2010), 23–30.

Weber R. H. (2015), Internet of things: Privacy issues revisited, *Computer Law & Security Review* 31 (2015), 618–627.

Weber R. H. (12016, Governance of the Internet of things—From infancy to first attempts of implementation? *MPDI Laws* 5 (2016), 28–39.

Weber R.H., and Heinrich U. 2012, *Anonymization*, Springer, London.

Weber R.H., and Studer E. (2016), Cybersecurity in the Internet of things: Legal aspects, *Computer Law & Security Review* 32 (2016), 715–728.

Wynzard Group, Advanced Crime Analytics, The fastest way to reveal actionable intelligence hidden in your data, available at https://wynyardgroup.com/crime_analytics.php.

Xiao Q., Gibbons T., and Lebrun H. (2000), *RFID Technology, Security Vulnerabilities, and Countermeasures*, 357, in Huo Y. (ed.). Supply Chain the Way to Flat Organisation seq., available at http://cdn.intechopen.com/pdfs-wm/6177.pdf.

chapter seventeen

US government and law enforcement

Greg A. Ruppert

Contents

17.1 Introduction

The complexity of cybersecurity in both scope and definition is evidenced in the real world by the myriad federal, state, local, and international agencies which play an integral role in the prevention, mitigation, and investigation of cyber-related incidents. Cyberadversaries range from nation states to organized crime groups to cyberhacktivists, all of which can vary their vectors of attack against a wide range of victims throughout the United States. Victims can include government sites, critical infrastructure, businesses, or individuals. Given the online and digital interactions between companies, systems, and customers, the ability to effectively fortify systems has become increasingly impossible. Adding to the complexity of defense are the complexity and size of the criminal adversaries. Nation states primarily engage in warfare, espionage, malicious attacks, and corporate theft. Organized crime groups engage in fraud, identity theft, ransomware, and spear phishing campaigns for financial gain. Cyberhacktivists 7pursue a social agenda through a variety of cyberattacks designed to gain attention for their cause. As a result, the governmental reaction to cybersecurity has been to create or empower a panoply of agencies with authorities or jurisdiction, which was designed to provide the necessary response expertise to the evolving threat driven by the Internet and cyber-enabled enhancements to our global environment. However, this approach over time has led to confusion for the victims of cyberattacks as well as jurisdictional fighting among the many agencies. Thus, numerous legislative actions and executive orders have been enacted to define "the lanes in the road" for lead roles and to facilitate the coordination, outreach, and liaison between the agencies. This chapter will discuss the government and law enforcement agencies involved in cybersecurity, their roles, governing structures, and coordinating entities.

17.2 US governmental landscape

Cybersecurity has multiple facets that range from net defense to investigation to proactive activities to mitigation postattack. The US response has been historically divided into law enforcement, defense, and intelligence agencies. These were all developed agencies prior to the increased focus on cybersecurity. Thus, several departments and agencies have assumed or been given new or enhanced jurisdiction. This has resulted in an often confusing patchwork of agencies being involved in an investigation depending on the actor, cyberactivity, or government response. Under the US governmental landscape, this chapter will discuss the executive branch oversight and the coordination approach as well as the most prevalent federal agencies involved in the cybersecurity response in the US.

17.3 Overarching executive branch policy governance for a "whole-of-government" approach

Traditionally, the National Security Council (NSC) is the main policy development and coordination arm of the administration. President Obama prioritized the need to institutionalize policies to facilitate stronger cybersecurity and better protect the United States against cyberthreats [1]. Depending on the administration, the NSC appoints a high-ranking official to oversee the administration of the federal government of cybersecurity through the many departments and agencies and oversee a central interagency coordinating body related to cybersecurity. This interagency group will meet on a regular basis and as needed to provide operational coordination.

17.3.1 Cyber Unified Coordination Group

One of the more significant acts undertaken by the federal government was to outline the lanes in the road related to the federal government response to a cyberattack. This was accomplished through the creation of a Cyber Unified Coordination Group to coordinate the response to a significant cyberincident. Significant cyberincidents were defined as those that affect critical infrastructure owners and operators or cyberincidents that could have catastrophic regional or national effects on public health or safety, economic security, or national security.

17.4 *Department of Homeland Security*

The Department of Homeland Security (DHS) has a number of agencies and offices responsible for multiple areas of cybersecurity. Under the National Protection and Programs Directorate, the DHS executes on multiple missions from the Office of Cybersecurity and Communications, Office of Cyber Infrastructure and Analysis, and the Office of Infrastructure and Protection. As a result of the combined effort, dozens of cybersecurity alerts to the private sector and general public related to cyberthreats are issued daily.

17.4.1 *Office of Cybersecurity and Communications*

The Office of Cybersecurity and Communications (CS&C), within the National Protection and Programs Directorate [2], is responsible for enhancing the security, resilience, and reliability of the cyberinfrastructure and communications infrastructure of the nation. CS&C works to prevent or minimize disruptions to critical information infrastructure in order to protect the public, the economy, and government services. CS&C leads efforts to protect the federal ".gov" domain of civilian government networks and to collaborate with the private sector—the ".com" domain—to increase the security of critical networks [3].

17.4.2 *National Cybersecurity and Communications Integration Center*

Additionally within CS&C, there is the National Cybersecurity and Communications Integration Center (NCCIC), which serves as a 24/7 cybermonitoring, incident response, and management center and as a national point of cyberincident and communications incident integration [4]. The NCCIC (pronounced "n-kick") is composed of four branches: NCCIC Operations and Integration, United States Computer Emergency Readiness Team (US-CERT) [5], Industrial Control Systems Cyber Emergency Response Team, and National Coordinating Center for Communications [4].

17.4.3 *Office of Cyber and Infrastructure Analysis*

The Office of Cyber and Infrastructure Analysis provides consolidated all-hazards consequence analysis, ensuring that there is an understanding and awareness of cybercritical and physical critical infrastructure interdependencies and the impact of a cyberthreat or incident to the critical infrastructure of the nation [6].

17.4.4 *Office of Infrastructure Protection*

The Office of Infrastructure Protection (leads the coordinated national effort to reduce the risk to US critical infrastructure posed by acts of terrorism. In doing so, the level of preparedness and the ability to respond and quickly recover in the event of an attack, natural disaster, or other emergency of the nation is increased [7].

17.4.5 *US Cyber Emergency Response Team*

As mentioned earlier, the DHS has the responsibility for the US Cyber Emergency Response Team. US-CERT provides a broad range of reporting and analysis to the government and private sector. US-CERT also exchanges information across a global Computer Security Incident Response Team community to improve the security of the critical infrastructure and the systems and assets of the nation on which Americans depend. Partners with which US-CERT may share information include US federal agencies; private sector organizations; the research community; state, local, tribal and territorial governments; and international entities [8]. US-CERT is a member of the Forum for Incident Response and Security Teams [9]. US-CERT also provides a robust collection of reporting, analysis, and alerts valuable to the ongoing defenses needed for a robust cybersecurity program under their Cybersecurity Awareness System. As noted on their website, individuals can sign up to receive their four products that provide a range of information for users with varied technical or general expertise [10]. Those with more technical interest can read the *Alerts*, *Current Activity*, or *Bulletins*, while users looking for more general-interest pieces can read the *Tips* [10].

17.5 US law enforcement

In the United States, there are several different federal law enforcement agencies with overlapping jurisdictions to include the wide-ranging array of illegal activities connected to cybercrime. Cybercrimes include complex computer intrusions by nation states, hacktivism, and cyber-enabled criminal activity related to fraud and other financial scams. Also, the Internet and electronic communications are used in a multitude of other federal crimes such as terrorism, human trafficking, drug trafficking, and child pornography. In addition, there are state, local, tribal, and territorial law enforcement agencies which also have jurisdiction over cybercriminal activity. In 2016, the Presidential Policy Directive-41 [11] on US Cyber Incident Coordination Policy was released, setting forth principles governing the response of the federal government to cyberincidents. The policy designates certain federal agencies to take the lead in three different response areas—threat response, asset response, and intelligence support. Those agencies and their roles are as follows:

- The Department of Justice, acting through the Federal Bureau of Investigation (FBI) and the National Cyber Investigative Joint Task Force (NCIJTF), will be taking the lead on threat response activities.
- The DHS, acting through the NCCIC, will be the lead agency for asset response activities.
- The Office of the Director of National Intelligence, through its Cyber Threat Intelligence Integration Center, will be the lead agency for intelligence support and related activities [12].

17.5.1 National Cyber Investigative Joint Task Force

The NCIJTF was created in 2008 to provide coordination among law enforcement and intelligence communities. The NCIJTF is composed of over 20 partnering agencies and has representatives who are colocated and jointly work to accomplish the mission of the organization from a whole-of-government perspective. As a unique multiagency cybercenter, the NCIJTF has the primary responsibility to coordinate, integrate, and share information to support cyberthreat investigations, supply and support intelligence analysis for community decision-makers, and provide value to other ongoing efforts in the fight against the cyberthreat to the nation [13]. The NCIJTF is managed by the FBI and led by an FBI senior executive as the director of the task force. Recent congressional legislation mandated the increased sharing of cyberthreat information, and the response of the government highlighted the role the NCIJTF.

17.5.2 FBI

The FBI is the lead federal agency for investigating cyberattacks by criminals, overseas adversaries, and terrorists. In fact, the majority of cybercrime is investigated by one of the FBI's 56 field offices' cyber task forces. These investigations are usually prioritized for larger intrusions or major attacks. They typically include massive distributed denial-of-service attacks, as well as Botnets and malware investigations. Efforts of the cyber task forces are overseen by the Cyber Division of the FBI within FBI headquarters. Unless the Internet merely facilitates the commission of the criminal activity (such as bank fraud or child sex trafficking), matters are assigned to the cyber task forces. Given the interplay between cyber-enabled crimes and the underlying type of crimes, often FBI squads will collaborate against a specific target [14]. The FBI Cyber Division in addition to overseeing and supporting field office efforts also manages several specialty units to facilitate the more robust engagement with the private sector, other governmental entities, and victims of cybercrimes.

17.5.2.1 FBI Internet Crime Complaint Center

The FBI Cyber Division also runs the Internet Crime Complaint Center (IC3) so the public can report information to the FBI regarding Internet-facilitated criminal activity. Reports are collected and disseminated to a variety law enforcement agencies for investigative and intelligence purposes. Reports are also created for public awareness [15]. Along with partner agencies, the IC3 participates in multiple initiatives targeting the following types of frauds:

- Charitable contributions fraud
- Counterfeit check fraud
- Identity theft task force
- International fraud
- Investment fraud
- Online pharmaceutical fraud
- Phishing
- Work-at-home scams [16]

17.5.2.2 FBI InfraGard

A partnership with members of the private sector, called InfraGard, is also overseen by the Cyber Division of the FBI. InfraGard is locally run by the FBI field offices. It has 84 chapters with more than 46,000 members nationwide, helping to protect and defend critical infrastructures [17]. The InfraGard program was designed to facilitate public–private collaboration between the private sector and the government [17].

17.5.3 DHS–US Secret Service

The Secret Service also engages in the investigation of cybercrime. It has developed an electronic crimes special agent program and established network of 46 financial crimes task forces and 39 electronic crimes task forces (ECTFs).

17.5.3.1 Electronic crimes task force

In 2001, the US PATRIOT Act mandated that the Secret Service establish a national network of ECTFs to prevent, detect, and investigate various forms of electronic crimes including cybercrime [18]. Similar to the InraGard of the FBI, the ECTF model relies on partnerships between the law enforcement, the private sector, and academic community to share information and intelligence [18]. These ECTFs are a strategic alliance of over 4000 private sector partners; over 2500 international, federal, state, and local law enforcement partners; and over 350 academic partners [18].

17.6 State and local law enforcement efforts

The United States consists of tens of thousands of state-level and local-level law enforcement agencies. These agencies have differing geographic as well as state and local law jurisdictions. In addition, the agencies can greatly vary in size such as major large city departments with over 40,000 sworn officers to small departments consisting of as few as two deputies. As a result, the cyber-related expertise greatly varies. While the FBI has management responsibilities over several national databases such as fingerprint records* or the National Crime Information Center,† there are no standardized cyber-related databases. Additionally, many state and local departments do not possess the equipment, expertise, or related resources to conduct cyber-enabled or cyber-related investigations. Some states also vary with the degrees of criminal legislation related to cyberattacks. Additionally, most agencies lack the ability to efficiently investigate criminal activity, which crosses state lines, let alone national borders. Thus, given the speed and distance in which cybercrimes occur, the methods of state and local law enforcement investigation, which was designed around protecting the local community from criminal activity occurring in that area by actors also located in the same area, have not kept pace with technological advancements.

17.6.1 FBI regional computer forensic laboratories

In an effort to assist state and local law enforcements, the FBI established regional cyber forensic laboratories (RCFLs) to provide technical assistance, develope new digital

* In July 1999, the fingerprint identification function was automated in the Integrated Automated Fingerprint Identification System (IAFIS). This national, computerized system for storing, comparing, and exchanging fingerprint data in a digital format permits comparisons of fingerprints in a faster and more accurate manner. It is located in, and operated by, the Criminal Justice Information Services Division of the FBI in Clarksburg, West Virginia. IAFIS provides three major services to its customers. First, it is a repository of criminal history information, fingerprints, criminal subject photographs, as well as information regarding military and civilian federal employees and other individuals as authorized by Congress. Second, it provides positive identification of individuals based on fingerprint submissions (both through 10 print fingerprints and latent fingerprints). Third, it provides tentative identification of individuals based on descriptive information such as a name, date of birth, distinctive body markings, and identification numbers. The primary function of IAFIS is to provide the FBI a fully automated fingerprint identification and criminal history reporting system. Additionally, IAFIS has made several other accomplishments. It has improved latent fingerprint identification services to the law enforcement community, and it has helped develop uniform biometric standards. These improvements have eliminated the need to process and retain paper fingerprint cards and has thereby accelerated the identification process. Another benefit has been the development of improved digital image quality. See FBI [19].

† The National Crime Information Center, or NCIC, has been called the lifeline of law enforcement—an electronic clearinghouse of crime data that can be tapped into by virtually every criminal justice agency nationwide, 24 hours a day, 365 days a year. It helps criminal justice professionals apprehend fugitives, locate missing persons, recover stolen property, and identify terrorists. It also assists law enforcement officers in performing their duties more safely and provides information necessary to protect the public. NCIC was launched on January 27, 1967 with five files and 356,784 records. By the end of 2015, NCIC contained 12 million active records in 21 files. During 2015, NCIC averaged 12.6 million transactions per day. See FBI [20].

evidence forensics tools, and create training programs for digital evidence examiners and law enforcement officers [21].

As a result, the FBI has created 15 RCFLs to serve as a forensics laboratory and training center for the examination of digital evidence in support of all types of criminal investigations. A typical midsize RCFL consists of 15 people whose responsibilities range from crime scene-based digital evidence collection to computer evidence examinations as well as related testimony before a court or grand jury. These RCFLs provide much needed expertise to smaller departments throughout the United States [21].

17.7 Conclusion

The speed of which cybercrime has grown from a mechanism to cause mayhem and engage in small cyber-enabled fraud into a worldwide epidemic has not allowed law enforcement to keep pace with the threats. The international scope of the infrastructure used and the speed at which an attack can be carried out make it nearly impossible to prevent criminal, nation state, or terrorism attacks, and the established mechanisms built over centuries to investigate and bring criminals to justice are oftentimes efforts of futility. Adding to the complexity is that the victims are often private sector companies with no interest in coming forward to report a massive intrusion as such a pronouncement will result in damage to their brand, civil lawsuits, executive dismissals, and loss of revenues. As a result, the cooperation between the primary victim of the crime is filtered through law firms and often delayed. The ultimate victims are often the clients or customers of the company that suffered the initial cyberattack and their interests are sometimes only a secondary concern to the companies they once entrusted with their information. Further complicating law enforcement efforts is the traditional methodology of investigating crimes where they occur through regional offices connected to the community. The ability for cybercrimes and attacks, which primarily emanate from overseas locations and strike multiple US victims in differing locations in rapid succession, requires a centralized investigative division in order to collect all the threat intelligence and evidence and then conduct truly holistic investigations. A centralized program would serve as a primary hub of all collection and investigation, thereby possessing the ability to fully comprehend the attack and facilitate the full-range response actions and preventative measures which only the whole-of-government approach has to offer. Unfortunately, such a massive shift in established bureaucratic structures to accomplish this centralization would require a level of leadership in the political structures of Washington, DC, that is unprecedented in recent history.

References

1. *Fact Sheet: The Administration's Cybersecurity Accomplishments*, The White House, Washington, DC, https://obamawhitehouse.archives.gov/sites/default/files/fact_sheet-administration_cybersecurity_accomplishments.pdf, May 19, 2018.
2. *National Protection and Programs Directorate*, Department of Homeland Security, Washington, DC, https://www.dhs.gov/national-protection-and-programs-directorate, May 19, 2018.
3. *Office of Cybersecurity and Communications*, Department of Homeland Security, Washington, DC, https://www.dhs.gov/office-cybersecurity-and-communications, May 19, 2018.
4. *National Cybersecurity and Communications Integration Center*, Department of Homeland Security, Washington, DC, https://www.dhs.gov/national-cybersecurity-and-communications-integration-center, May 19, 2018.
5. *US-CERT*, https://www.us-cert.gov/, May 19, 2018.

6. *Office of Cyber and Infrastructure Analysis*, Department of Homeland Security, Washington, DC, https://www.dhs.gov/office-cyber-infrastructure-analysis, May 19, 2018.
7. *Office of Infrastructure Protection*, Department of Homeland Security, Washington, DC, https://www.dhs.gov/office-infrastructure-protection, May 19, 2018.
8. US-CERT (United States Computer Emergency Readiness Team). *About Us*, US-CERT, Washington, DC, https://www.us-cert.gov/about-us, May 19, 2018.
9. *FIRST*, https://www.first.org/, May 19, 2018.
10. *National Cyber Awareness System*, US-CERT, Washington, DC, https://www.us-cert.gov/ncas, May 19, 2018.
11. *Presidential Policy Directive—United States Cyber Incident Coordination*, The White House, Washington, DC, https://obamawhitehouse.archives.gov/the-press-office/2016/07/26/presidential-policy-directive-united-states-cyber-incident, May 19, 2018.
12. *Countering the Cyber Threat*, Federal Bureau of Investigation, Washington, DC, https://www.fbi.gov/news/stories/new-us-cyber-security-policy-codifies-agency-role, May 19, 2018.
13. *National Cyber Investigative Joint Task Force*, Federal Bureau of Investigation, Washington, DC, https://www.fbi.gov/investigate/cyber/national-cyber-investigative-joint-task-force, May 19, 2018.
14. *FBI-Cyber*, Federal Bureau of Investigation, Washington, DC, https://www.fbi.gov/investigate/cyber, May 19, 2018.
15. *FBI, IC3*, Internet Crime Complaint Center, Clarksburg, West Virginia, https://www.ic3.gov/about/default.aspx, May 19, 2018.
16. *FBI, IC3—Brochure*, Internet Crime Complaint Center, Clarksburg, West Virginia, https://www.ic3.gov/media/IC3-Brochure.pdf, May 19, 2018.
17. *FBI, InfraGard*, https://www.infragard.org, May 19, 2018.
18. *USSS—Cyber ECTF Brochure*, US Secret Service, Washington, DC, https://www.secretservice.gov/data/investigation/USSS-Cyber-Investigations-Flyer.pdf, May 19, 2018.
19. *Criminal Justice Information Services (CJIS)*, Federal Bureau of Investigation, Washington, DC, https://www.fbi.gov/services/cjis
20. *National Crime Information Center*, Federal Bureau of Investigation, Washington, DC, https://fbi.gov/services/cjis/ncic, May 19, 2018.
21. *Regional Computer Forensic Laboratory*, https://www.rcfl.gov/, May 19, 2018.

chapter eighteen

Enterprise solutions and technologies

Michael Cook

Contents

18.1 Introduction

In the enterprise, cybersecurity programs exist to protect the confidentiality, integrity, and availability of the information of the business. Some cybersecurity programs on the surface may be striving to maintain confidentiality of intellectual property, others focus on the availability of information systems, and others focus on maintaining compliance with privacy regulations in their given field. Regardless of its focus, however, the ultimate goal of any cybersecurity program is to allow the organization to effectively conduct operations while sufficiently reducing financial, reputational, or functional risk. Enterprise solutions involve a widespread array of technical tools including protections for hardware, software, databases, and physical security to reduce this risk, but the most important aspects of any cybersecurity program are the organic pieces between the seats and the keyboards, e.g., the users.

18.2 Challenges of securing the human

The employees an organization can make or break the cybersecurity program. There is very little information in this world which cannot be transmitted, copied, photographed, reproduced, recorded, or stored by both high-tech and low-tech methods. While the news is full of romantic reports of hackers in dimly lit rooms exploiting vulnerabilities in software to get their hands on private data, the reality is that the most frequent causes of exposure within organizations today are inadvertent or naive actions from within. Each year, IBM conducts a Cybersecurity Intelligence Index, which increasingly cites employee error as the source of a breach. The 2014 report in particular cites that as many as 95% of the incidents referenced in the study involve human error [1].

18.2.1 Phishing

As discussed in our chapter on social engineering, phishing is the practice of a malicious third party sending fraudulent communications to employees of the enterprise in an attempt to coerce them into revealing their system passwords or to introduce malware to the systems of the enterprise. Phishing messages often take the form of mimicking an information technology (IT) department indicating that an account is about to expire or a mailbox is full or posing as a real person the individual knows. These messages often demand a username and password to ensure that service continues uninterrupted or provides a link to download a file containing malware. These sorts of attacks work, based largely on volume, and the numbers are alarming. Phishing expert Cofense published the results of a 13-month eight million message phishing campaign across 23 industries in the United States, which did not practice phishing awareness training. The results on average indicated that the campaigns produced a 20% response rate [2], an alarming number considering that it only takes one point of entry to cause a substantial security breach.

The individuals behind these attacks are very good at what they do. Real websites and logos are duplicated, and domain names similar to the those of the organization are registered, making it extremely challenging to detect a fraudulent message. Attackers research their targets on social media and corporate websites, and they write code that alters itself to avoid detection. More advanced phishing techniques, sometimes referred to as whaling, mimic the e-mail accounts of a chief executive officer (CEO) or Chief Financial Officer requesting a wire transfer. Other attacks include phoning into accounts payable departments posing as a vendor to be paid or asking for a bank account update and more.

The following example was received by the chief financial officer of San Jose State University (SJSU) in January 2017 from a malicious third party posing as the university president:

> From: **President's Correct Name and Work E-Mail Address**
> Date: Fri, Feb 10, 2017 at 10:25 AM
> Subject: Re: REQUEST
> To: Chief Financial Officer
>
> Kindly process a domestic wire transfer now on the beneficiary details below and reply with payment confirmation when completed.
>
> Beneficiary Name: J & D Express
> Account Name: 111810823
> Routing Number: 325070760
> Bank Address: 921 Alder Ave Sumner WA 98390
> Beneficiary Address: 24843 45th Ave South R302 Kent, WA 98032
> Amount: $67,678.00 USD
>
> Best regards
> Sent from my iPad

The shock factor of receiving a message from the president of the university is intended to cause the recipient to ignore warning signs, such as an incorrect reply-to address, and complete the transaction. While this particular attempt was caught without incident, Coastal Carolina University was tricked into wiring payment to a false bank account [3].

Some of the largest reported incidents in history can be traced to phishing. In 2014, Sony Pictures incurred an estimated $100,000,000 in damages when targeted e-mails were sent to employees tricking them into trusting the wrong people exposing the credit card numbers of thousands of customers. A few months later, another group of hackers gained access to as many as 80 million medical records belonging to Anthem, one of the largest health insurance providers in the United States. The breach occurred when employees were tricked into installing malware on trusted enterprise systems [4].

What can be done about phishing? The ability to discern a fraudulent message, website, or link from a legitimate source is critical to combat phishing. Enterprises should invest in training and awareness programs which bring attention to common mistakes made by malicious third parties including missing or incorrect secure sockets layer certificates and poor spelling/grammar, and utilize URLs which are not part of the domain of the organization. Reporting phishing attempts should be easy for the employees as well. Products such as Cofense allow cybersecurity teams to send messages to their own employees in a safe controlled environment, providing employees with learning opportunities. For less than $1 per employee per year, there is a strong return on investment argument to be made when a single breach could cost a company hundreds of thousands.

In addition to nontechnical campaigns, cybersecurity programs should take a layered approach to combat phishing. Multifactor authentication (MFA) tools ensure that disclosed usernames and passwords alone do not harm the organization by requiring a one-time key, token, fob, fingerprint, cell phone app, or another "factor" aside from a username and password to sign in. While cumbersome, MFA is becoming easier to live with every day, but many organizations still do not have this technology deployed widely due to cost and complexity of support. Antivirus and antimalware applications prevent known malicious applications from installing or spreading, but often, the individuals behind these attacks are writing code faster than antivirus vendors can write definitions to detect it. Today's advanced firewalls and e-mail filters can prevent data from being sent to known malicious addresses and can even detect and prevent new threats. Encryption technology also helps prevent the impact of a breach if an exposure were to occur. On the extreme side of the spectrum, some organizations prevent direct electronic communication outside the organization or only permit access to e-mail from known trusted networks and devices. The key is security in layers. Assess the risk to phishing, the business needs of the organization, and the budget available to deploy as many tools as practical.

18.2.2 E-mail accidents

The act of not reviewing who is in the "to" field of an e-mail or not double checking the content of a message is one of the most common forms of information disclosure within organizations. This particular challenge is exacerbated by modern e-mail clients, cell phones, and tablets, which often hide previously read portions of messages, try to predict message content, and auto-fill in recipients. In 2016, the City of Calgary investigated an incident resulting in a breach of over 3700 employees' personal data due to a member of the organization forwarding sensitive content as part of a technical support request [5].

How does an organization reduce e-mail accidents? The inadvertent transmittal of information to third parties can only be prevented through education and awareness programs. Google has recently begun integrating into their Gmail suite a reminder whenever an e-mail address outside the organization is included on a message, but this feature has yet to see widespread adoption and could easily be overlooked (Figure 18.1). In situations where communications with outside entities are required, a strong awareness campaign is really the only option.

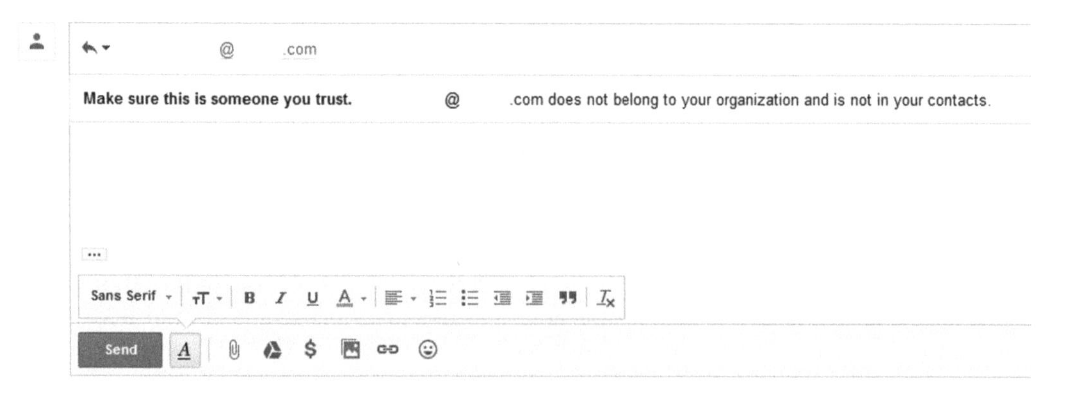

Figure 18.1 Example of an e-mail application that identifies and warns users when responding to an unknown user outside your organization.

18.2.3 Physical security

The 2014 US State of Cybercrime Survey, a joint effort by PricewaterhouseCooper, Carnegie Mellon University, CSO magazine, and the US Secret Service, reported that "only 49% of companies have a plan to address and respond to insider security threats—even though 32% of the same companies agree that crimes perpetrated by insiders are more costly and damaging than those committed by outsiders" [6]. If insiders can walk into your data center and grab a removable hard drive, they have no need to remotely break into your servers, yet physical security is often an overlooked function of an enterprise cybersecurity program [7].

Surprisingly, the Montana Department of Administration and even the Internal Revenue Service received audit findings for inadequate visitor and contractor safeguards [8]. While no two organizations are identical in the aspect of physical security the concept of open lobbies, relying on humans to be the gatekeepers to secured spaces, inadequate locking mechanisms, and unmonitored facilities are all too common, especially for small businesses and public entities.

How do companies improve physical security? Do not overlook the importance of physical security. A strong cybersecurity program should not only include extensive controls, procedures, incident response programs, and audit mechanisms for sensitive areas, but should also include controls for general-purpose spaces. Require name badges for visitors, empower employees to approach unknown parties in their workspaces, encourage clean desk policies, lock up sensitive information, and remind employees of their role in physical security at regular intervals.

18.2.4 Record retention and disposal

A lesser discussed aspect of cybersecurity is the proper retention and disposal of information. *Psychology Today* cites that there is anxiety associated with the disposition of records, especially at work, as it is all too easy to think of a scenario where a particular document type may be needed as reference in the future. This contrasts with the organizational need to destroy records at the end of their retention period in order to reduce risk and improve employee productivity. Organizations not only need to establish schedules of how long each of the document types they keep on file must be stored, but also ensure that at the end of their retention period, documents are destroyed securely. No sensitive information should be stored without establishing first how long it should be stored for and how the documents will be identified and securely destroyed following their retention periods. A strong, widely publicized, and well-known record retention and disposal program will not only reduce the risk of an organization to data exposure or discovery in legal proceedings, but also the impact of any data disclosures.

What is the right way to address record retention? Retention policies should be set with legal and operational priorities in mind, and reminders should be sent to employees regularly including procedures and retention schedules. Managers should make it part of the procedures of their departments to complete review cycles and ensure compliance. On the technical side, some document management solutions such as Hyland OnBase and ImageNow include retention management software based on document type. Other areas, especially files stored on traditional file servers, offer little in terms of real document creation dates, document types, and other needed information. Departments should establish naming conventions and procedures to ensure compliance in these more challenging scenarios and consider utilizing a more advanced document management system.

18.2.5 Procurement

A 2013 study by Trustwave, one of the world's most prominent cybersecurity firms, revealed that an estimated 63% of data breaches occurred due to some form of third-party involvement [9]. Cybersecurity programs must develop procedures for evaluating and securing agreements with third parties including evaluations of technical solutions, contract verbiage, designation of liability, background checks for employees of third parties, compliance, and other necessities. Perhaps the most famous of these incidents relates to the inadequate cybersecurity practices of Target Stores in 2013. The private network of Target, which was intended to protect over 40 million customer credit card numbers, was compromised by a heating, ventilation, and air-conditioning (HVAC) vendor, who was credentialed to have access. The vendor's credentials were used to circumvent security resulting in as many as 1.1 million credit card numbers being disclosed in 2013. Following the resignations of both the chief information officer and CEO, the $162 million breach was publicly announced [10]. Adding complexity to the challenge is the need to address all methods of procurement including purchase orders, petty cash, procurement cards, personal reimbursements, and donations. Identifying which purchases require security evaluations can be challenging; hardware, software, and software as a service purchases may all require a security evaluation.

How can procurement be improved? With the ever-growing business of software-as-a-service applications and cloud computing, storage of or access to sensitive data should be reviewed to ensure compatibility, security, compliance, and appropriate language in agreements in order to reduce the liability of the organization to exposure, fines, and other damages. Procurement personnel and the individuals requesting purchases are unlikely to be cybersecurity professionals, so a cybersecurity program must work hand in hand with the procurement department(s) of the organization to ensure that requests are sufficiently evaluated prior to purchase. Evaluations are often a conversation between the requestor, procurement, and information security teams to gain a better understanding of what types of data will be involved and how that will be transmitted or stored in order to get a better sense of what security controls may be required. Organizations such as the Cloud Security Alliance have developed comprehensive questionnaires and certification processes, which set standards and provide guidelines by which a third party's security practices can be evaluated [11]. Factors such as the third party's geographical location of data centers, disaster recovery protocols, data encryption practices, and internal security training procedures should be evaluated to ensure that the third party is held to the same standards as is the enterprise itself. The role of the cybersecurity program in procurement should pose alternatives when a requested purchase does not meet standards, consider the risks and the data types involved with the third party, be collaborative, and ensure that constituents understand why or why not a purchase was approved.

18.2.6 Human resources

Human resources departments can be a strong ally for cybersecurity programs. Human resources is the gatekeeper for who is being trusted to be privy to sensitive information within the organization. Always ensure that human resources requires appropriate criminal background checks for employees not only on the job they will be performing, but also the physical access they will require to complete that job. The same rules should apply for volunteers, interns, contractors, and anyone else who will ultimately supplement the workforce in one way or another.

18.2.7 Employee awareness programs

With so much risk placed on the employees of an organization, cybersecurity programs have a strong emphasis on employee awareness programs. Legal departments, auditors, lawmakers, and sanctioning bodies alike have made their mark on information security awareness campaigns. Regulations such as the Health Insurance Portability and Accountability Act (HIPAA), Payment Card Industry Data Security Standard (PCI-DSS), and US Department of Justice require annual training for all employees with access to data. There is a temptation to utilize online training, training during orientation, and other easy-to-administer solutions to meet required awareness standards and satisfy the needs of auditors. This approach risks compromising the real goals of these awareness programs. The focus should be less on appeasing auditors and more on ensuring that each employee is aware of the unique aspects of their job in which information security may become relevant and know when and who to ask for help when an issue arises.

The timing and length of delivering training is critical for employees to retain the information needed to securely operate within the enterprise. The employee on-boarding process often includes information about medical benefits, signing of confidentiality agreements, emergency preparedness information, facility tours, and more. With so much information conveyed during an employee's first few days of work, security awareness may not be effectively retained. Furthermore, lengthy online trainings make security training a chore and more likely to be passively completed than other solutions. According an extensive 2007 study regarding the effectiveness of cybersecurity awareness programs in universities, factors such as cultural assumptions, regulatory agencies, beliefs, attitudes toward work, trust in leadership, and others determine the effectiveness of training. With so many variables and so many employees with differing feelings toward the work environment, there is no one-size-fits-all program approach to eliminate employee cybersecurity risks. For some employees, open forums may be successful, while for others, brief updates can be made in monthly staff meetings; for some, fliers and newsletters are effective, while for others, a brief training as part of the password reset process may be sufficient. No two organizations are alike, and no two individuals learn in the same way, so a strong cybersecurity training program needs to cover as many learning methods as possible. Instilling a culture of information security takes time; celebrate the victories, discuss thwarted phishing whaling attacks, learn from mistakes, and let the incidents of the past be an educational tool for the future. Real, relevant examples will always be a great ally. Be realistic about employees' knowledge of policies, laws, and standards [12].

18.2.8 Special considerations for software and web developers

Enterprises that develop their own applications are exposed to a whole new world of threats which must be mitigated. While the security of applications is a topic beyond the scope of this chapter, it is important to mention that there are real risks that must be addressed by programmers within the enterprise. SQL injection, where an attacker can inject malicious code into web-based applications, is perhaps the most popular of dozens of exploits which programmers need to be aware of. Heartland Payment Systems, which at the time provided credit card payment processing solutions for 175,000 customers, was fined $145,000,000 after a hacker was able to breach their security controls through SQL injection [10]. Cybersecurity programs should include special training elements for developers. Developers should be well versed in techniques attackers use to exploit software vulnerabilities including SQL injection, cross-site scripting, and web and application server configuration vulnerabilities.

The Open Web Application Security Project publishes a top 10 list of vulnerabilities, which all developers should be aware of [13]. Consider sending programmers to secure development training courses or bringing a trainer on site to keep programmers up to date on vulnerabilities. Cybersecurity and development managers should keep in mind that developing secure code is more work than developing code simply for functionality. With ever-increasing demands, it is easy for a developer to be so focused on meeting functionality goals that security is overlooked, so it is always helpful to integrate a peer code-review as part of the release management process. Make security code reusable, review it regularly, and make use of automated tools for testing. The concept of ethical hacking or penetration testing allows enterprises to test their applications using techniques developed by hackers. For environments where a live resource is not feasible, there are tools available as well, which can test for some of the most common vulnerabilities. Qualysguard Web Application Security, Trend Micro, Accunetics, TrustWave, VeriCode, and countless others offer testing applications and services to identify vulnerabilities.

18.3 Role of enterprise information technology teams

18.3.1 Making the right solutions easy and available

Enterprises today are facing the challenge of a world where technology touches every aspect of an employee's life, and technology is more obtainable than it ever has been before. Many of the employees within the enterprise are armed with technical knowledge and tools through their interactions with home computers, networks, and devices, and of course, this knowledge is often used to present ideas at work. This poses a challenge, as often technology designed for ease of use and the security demands of a home environment do not nearly meet the requirements of the enterprise. This puts IT teams in the position of having to listen more and explain why things are not as simple as they are at home.

With technology solutions at everyone's fingertips, IT teams must make the solutions of the enterprise as easy to obtain and as simple to work with as they are with their competition. Low and no-cost solutions can be readily found by any member of the organization armed with a web browser. Software-as-a-service, platform-as-a-service, and infrastructure-as-a-service providers are easy to obtain and frequently offer free trials or no-cost low transaction volume environments. Free and low-cost providers are often able to provide these services because they are accessing, sharing, and even selling the information stored in the systems which poses a significant risk to the enterprise. Even industry giant Google is utilizing information it collects from your accounts for other purposes:

> We use the information we collect from all of our services to provide, maintain, protect and improve them, to develop new ones, and to protect Google and our users. We also use this information to offer you tailored content—like giving you more relevant search results and ads.

Under the guise of tailored content, this privacy statement for Google's Gmail application is actually making their users aware that they are selling e-mail and browsing habits to the highest bidder to deliver customized advertisements. Furthermore, servers provided by third parties do not come by default with appropriate security controls such as antivirus, patch management, encryption, and strong password controls often required by enterprise cybersecurity programs. Amazon Web Services, for instance, clearly states the end user's responsibility for security in their Terms of Service:

> You are responsible for properly configuring and using the Service Offerings and otherwise taking appropriate action to secure, protect and backup your accounts and Your Content in a manner that will provide appropriate security and protection, which might include use of encryption to protect Your Content from unauthorized access and routinely archiving Your Content.

When individual contributors sign up for these services, security controls and system administration tasks are put into their hands. This means they can turn off or improperly administer firewalls, forget to install antivirus, not meet password complexity requirements, etc. In spring 2017, a server was created on Amazon Web Services for free using a personal account belonging to a SJSU faculty member. The server was used to store confidential Family Educational Rights and Privacy Act (FERPA)-protected data and compile 4-month-long projects for students in a class. When the improper configuration of the firewall of the server was combined with a weak administrator password, a malicious third party not only accessed the FERPA-protected records, but also installed ransomware encrypting the contents of the server and stopping the course dead in its tracks. The result was not only a violation of a federal law, but also a loss of the hard work of an entire semester. The underlying cause of this incident, deeper than the lack of controls, is that a problem needed to be solved, in this case, a server was needed, and the secure way to solve it was not easily obtainable. While data loss prevention (DLP) tools such as CloudLock can assist in identifying data in private clouds and tools such as Layer 7 decryption and inspection can block data from leaving a corporate network, these technologies lack the maturity to be totally effective. IT teams must therefore focus on making the right solutions easy to obtain, cost-effective, and available, or their users may be tempted to obtain their own solution. The cybersecurity program should work in concert with the IT teams to ensure that users are aware of the risks associated with finding their own solutions and know where to go in order to solve business problems in alignment with the security goals of the organization.

18.3.2 *Technology from the home to the office: Preparing for the Internet of things*

Among the challenges faced by IT professionals today is the expectation that the computing environment at work should be similar to that of the home. Employees today often expect to be able to wirelessly project from their laptop, expect to wirelessly print from their mobile tablets and bring any new gadget being sold online into the office. This introduces new complexities for IT professionals as not only do many of these tools rely on technologies which might not exist in the enterprise such as peer-to-peer communication or a lack of a virtual local area network segmentation to operate, but also that many of these technologies are designed for the home environment where confidential data is not stored on networks and where the risk of intrusion is far less than that in the workplace. Some organizations address these risks by enforcing policies which prevent the use of these technologies. This leaves service desk and desktop support personnel, often the primary technology interface for individuals within the organization, to explain complex security and compatibility issues. Unfortunately, the answer to "why doesn't my Apple Airplay work on the Wi-Fi?" involves explaining both why receiving an Internet protocol (IP) address in the same subnet as the AppleTV cannot be guaranteed and the intricacies of the Bonjour protocol. However, employees are typically not looking for a lecture on enterprise wireless, but rather simply want to share their screen with a projector without any wires. It is a lose–lose situation. The employee does not get what they want, and the IT team is perceived as less than helpful. Nevertheless, these situations can be turned into win–wins. IT personnel can address many of these security risks by finding secure alternative solutions. For example, in the case of the AppleTV, a Crestron screen-sharing device will meet the need or through microsegmentation, they can create environments for known nonsecure devices to function without impacting the security of the enterprise.

The concept of the microsegmentation of networks, involves breaking networks into small segments of devices with similar access needs. In this design, for example, network-enabled HVAC devices coexist in their own network, while the public connecting to the guest Wi-Fi are in another network, and developers needing to access intellectual property are in another isolated network. Microsegmentation, especially in Wi-Fi networks, is becoming a go-to solution to support untrusted devices and improve security within the enterprise.

Network access software can identify the type of device connecting to the wired or wireless network and place it in an appropriate area of the network of the organization where it can do what it needs to do (transmit video, access the Internet, etc.) without being able to access irrelevant and unneeded sensitive resources. This allows the IT team to accommodate the request without putting any sensitive assets at risk. The 802.1x solutions such as Cisco's Identity Services Engine and Aruba's Clearpass can ensure that personal laptops have the latest antivirus definitions and operating system patches through posture evaluation. They can even make security decisions based on who is connecting and what software is installed on the device and the last time it received updates using the 802.1X protocol.

18.4 Impact of controls and tools on the enterprise

Organizations widely vary by the policies and controls that they implement and must balance risk and controls with the need to efficiently operate. The nature of the business of an organization, value of its confidential assets, its leadership, culture, regulatory agencies, and governing bodies all influence what can and cannot be written in policy and implemented using technology. In the defense or banking industries, the value of confidential assets is extremely high. It makes perfect sense for these organizations to employ strict controls, blocking communication to/from the Internet, scanning and decrypting all network traffic with the strictest of policies, requiring multiple forms of authentication, and ensuring that the only systems which can access sensitive data are owned and secured by the organization. On the opposite side of the spectrum, in the education space, the concepts of intellectual freedom, open research, and experimentation dominate the business objectives that IT departments must deliver on. Cybersecurity controls are often a detriment to mobility and rarely serve to improve the user experience. Cybersecurity programs must therefore walk a fine line between risk mitigation and business outcomes.

18.4.1 Essentials

Some organizations have the ability to take extreme measures such as blocking e-mail from untrusted devices and inspecting all web traffic; others must learn to secure their environments with little restriction. No matter what the environment, there are some basic security practices that are easy to implement, affordable, and effective in just about any organization. No enterprise today should be operating without policies and tools to ensure basic patch management and antivirus for all endpoints, firewalls to secure networks, physical security policies, and policies to assess and retire or secure dated systems.

The Center for Internet Security collaborates with cybersecurity giants such as the SANS Institute and the US Department of Defense to publish the *Critical Security Controls for Effective Cyber Defense*. While some organizations may not be able to implement in entirety, this top 20 list of essential cybersecurity controls it is a strong starting point for any cybersecurity program (Table 18.1) [14].

Table 18.1 Center for Internet Security: 20 critical controls

Critical control	Enterprise system type	Examples
CSC 1: Inventory of authorized and unauthorized devices	Endpoint management applications, inventory applications, and dynamic host configuration protocol applications	Microsoft SCCM, Jamf Casper, IBM BigFix, manual inventories, active directory, TrackIt Isupport, and Infoblox
CSC 2: Inventory of authorized and unauthorized software	Application whitelisting, antivirus/antimalware, and software inventory tools	Lumension, AppLocker, removal of administrator rights, Sophos, Symantec, Intel/McAffee, and IBM BigFix
CSC 3: Secure configurations for hardware and software on mobile devices, laptops, workstations, and servers	Endpoint management applications	Microsoft SCCM, Jamf Casper, and IBM BigFix
CSC 4: Continuous vulnerability assessment and remediation	Vulnerability scanning	Qualysguard, Rapid 7, Symantec, and Sophos
CSC 5: Controlled use of administrative privileges	Desktop configuration and MFA	Microsoft Windows, Apple OSX, Microsoft Active Directory, DUO, and Okta
CSC 6: Maintenance, monitoring, and analysis of audit logs	Log management	Splunk, LogRhythm, and Intel/McAffee
CSC 7: E-mail and web browser protections	Mail and web filters and endpoint management application	Cisco Ironport, Palo Alto Networks, Gmail, Microsoft SCCM, Jamf Casper, and IBM BigFix
CSC 8: Malware defenses	Antivirus/antimalware	Symantec, Norton, Sophos, and Intel/McAfee
CSC 9: Limitation and control of network ports, protocols, and services	802.1x authentication, port security, and firewalls	Cisco ISE, Aruba Clearpass, Cisco, HP, Palo Alto Networks, and Juniper
CSC 10: Data recovery capability	Backup and replication tools	Symantec, Veeam, IBM Spectrum Protect, Netapp, and EMC
CSC 11: Secure configurations for network devices such as firewalls, routers, and switches	Switch, network and firewall administration tools	Vendor specific
CSC 12: Boundary defense	Next-generation firewalls	Cisco, HP, Palo Alto Networks, and Juniper
CSC 13: Data protection	Encryption and DLP	Microsoft BitLocker, Apple FileVault, Netapp, EMC, Cloudlock, and Identity Finder
CSC 14: Controlled access based on the need to know	Network segmentation	Cisco, HP, Palo Alto Networks, and Juniper
CSC 15: Wireless access control	802.1x authentication	Cisco ISE and Aruba Clearpass
CSC 16: Account monitoring and control	Directory and automation tools	Microsoft Active Directory, Okta, and Phisher Identity

(*Continued*)

Table 18.1 (Continued) Center for Internet Security: 20 critical controls

Critical control	Enterprise system type	Examples
CSC 17: Security skills assessment and appropriate training to fill gaps	Information security awareness program	Skillport, LawRoom, SANS, custom solutions, in-person trainings, and Cofense
CSC 18: Application software security	Penetration testing software and ethical hackers	BURP, Qualysguard, and Rapid 7
CSC 19: Incident response and management	Policies and procedures	
CSC 20: Penetration tests and red team exercises	Penetration testing software and ethical hackers	BURP, Qualysguard, and Rapid 7

18.4.2 BYOD

Bring your own device, or BYOD, is a relatively new concept in the cybersecurity discussion. As the availability and sophistication of private services (social media, online banking, web e-mail, etc.) increases and the proliferation of work e-mail on cell phones, text messages for business communications, virtual private networks (VPNs), and telecommute programs expands, the line between work and home is blurred. At the same time, portability, the ability to customize, and the cost of personal computing devices continue to improve for the consumer, leading many individuals, when given the opportunity, to opt to use their personal devices for work tasks. After all, why carry and manage two or three or four devices when a single device is capable of getting the job done?

There is, of course, a real risk involved in allowing personal devices onto enterprise networks. BYOD means that the individual employee owns, manages, and controls the security on the endpoint. The employee is responsible for ensuring that their accounts have appropriate permissions and run without administrator rights. They are also responsible for antivirus, in control of the password policies, and in control of patch management. Some organizations, especially those in high-risk environments, simply opt to disallow unknown devices from connecting to the networks, and truly, this is the most secure method of addressing the matter; but for other organizations, BYOD is a fact of life. Enterprises should consider employing and educating employees regarding minimum security requirements for personal computers (PCs) and implementing tools to reduce the risk.

The concepts of microsegmentation and posture assessment prove useful in the BYOD space. Through posture assessment, Wi-Fi and wired networks and VPN can identify when a computer does not have antivirus installed or is not current on patches. Networks can allow BYOD on to low-risk segments, while protecting protected information. Virtual desktop infrastructure allows BYOD devices to connect to a private, secured, and organization-controlled environment to access sensitive data. Technology has adapted to the times.

The management of organization-owned devices is changing as well. The concept of immediately wiping out the manufacturer-installed copy of Windows or OSX to replace with a locked down corporate version with no authorization to change the screen saver, let alone install software is changing. Tools such as Microsoft SCCM and Jamf Casper are developing new ways of configuring, deploying, and securing devices without significantly impacting the user experience. PC giant IBM has recently embraced this change in technology by deploying Jamf Casper to secure Apple computers as the standard-issue device for its employees. The shift has automated system deployment, while improving the security of the organization [15]. The tools today are ever changing, so before making a decision regarding BYOD, you must first understand your risk and do your homework on what tools are available to mitigate those risks while accomplishing the business objectives.

18.5 Special considerations for universities, schools, public entities, libraries, and enterprises supporting intellectual and academic freedom

Academic and other related enterprises have their own unique set of cybersecurity challenges. While in a corporate setting, it may be relatively straightforward to implement strict controls regarding device security and access to information, universities must balance risk avoidance with the concept of intellectual freedom. Intellectual freedom, according to the American Library Association, is "the right of every individual to both seek and receive information from all points of view without restriction." For cybersecurity professionals in a university setting, intellectual freedom means that many techniques, such as whitelisting, web traffic inspection, and e-mail monitoring, directly conflict with the university mission. Information should be able to be accessed from any device, from anywhere, without anybody watching. While primarily focusing on the university setting, the concepts in this chapter can easily be applied to cities, libraries, and other enterprises operating in an environment where risk avoidance controls may carry an intellectual freedom or political price tag.

As in other types of organizations, the biggest threats to cybersecurity in a university setting are the very people cybersecurity programs are designed to protect, the students, faculty, and staff. Universities operate much like small cities performing the work of a wide range of businesses. For example, universities house research institutes, libraries, day care centers, restaurants, construction companies, automotive repair facilities, power plants, police departments, hospitals, professional sports teams, custodial crews, Internet service providers, and computer software development teams and, of course, execute the primary mission of teaching and learning. Each area has its own set of cybersecurity challenges and regulations and unique set of users. With such a diverse environment, especially the diversity of technical skill sets belonging to individuals who have access to sensitive data, it is no surprise that the vast majority of information breaches in universities occur due to the actions of a person rather than a piece of equipment. A typical university responds to employee-created incidents vs. those from external sources at a rate of 8:1. This, coupled with the often conflicting objective of free and open learning, makes risk reduction uniquely challenging and requires extra attention to designing a multifaceted cybersecurity awareness program that appeals to all audiences, mediums, and technical skill sets.

Universities have an operational need to support intellectual freedom across an enormous number of domains and, hence, are often precluded from filtering or block content, ports, or IP addresses. There are legitimate use cases where individuals may need to access the content of questionable ethical or legal nature. Pornographic content, hacking, malware, and black-market content are all legitimate research subjects. While universities may be able to justify blocking some unwanted content such as known command and control servers and known illegal protocols (such as BitTorrents) using Layer 7 inspection on firewalls, chances are that sooner or later, an exception will need to be made if restrictions are placed where they will have widespread impact (such as on an Internet border firewall).

Universities serve as peoples' homes as well and must function essentially as an Internet service provider. The number one consumer of bandwidth in universities is not web traffic for classes, but is often video streaming demands from dormitories. Universities must be prepared to handle large volumes of streaming traffic and be prepared to respond to or block attempts at illegal downloading from the Recording Industry Association of America and more.

18.5.1 Compliance

All universities must secure student information, and many, depending on size, have health centers, hospitals, and police departments. All universities in the United States must comply with the Family Rights and Privacy Act to protect student information and must comply with the PCI-DSS for the processing of credit cards. As such, security in each area should be as unique as the business tasks that take place in them. Each set of regulations is large enough in scale to serve as its own area of expertise. While it is beyond the scope of this chapter to explain how to comply with each regulation, there are some common pitfalls that every organization should watch out for. When working with protected information, it is always advisable to seek counsel with industry experts, legal departments, audit departments, and make use of all the resources available at your disposal as there are significant consequences for noncompliance in all cases.

18.5.2 HIPAA

The HIPAA specifies strict controls for the protection of electronic and paper information for people. In a university setting, there are typically two areas where these rules may apply, in human resources, where employees are privy to and store information regarding medical insurance, and in hospitals/health centers, which may treat faculty, staff, and students. A university may have one or many human resources departments or healthcare providers, especially universities that operate auxiliary organizations. HIPAA requires annual training for staff with access to medical records, firewalls to segment all network traffic, specific requirements for data centers, and complex contractual language to be added to all agreements with third parties among dozens of other regulations.

It is not uncommon for departments on campus that focus on accessible education or provide other clinic-type services to operate under the impression that they are a HIPAA-covered entity, when in reality, they are not. While it never hurts to have end users wanting to go the extra mile to protect data, HIPAA breach reporting requirements can be quite strict, and a cybersecurity program should identify the appropriate incident response mechanism for each area. In fact, following the HIPAA incident response protocol, which includes notification to the local media, for non-HIPAA-covered entities would pose fairly significant public relations risk for the university and its reputation. HIPAA is a complex law with specific rules regarding the transaction types and transmission/storage methods that necessitate HIPAA compliance. It is strongly advised that legal counsel, the cybersecurity office, and the administrator of each potential HIPAA-covered entity within the organization work together to determine for sure if an area is considered part of HIPAA, especially when designing incident response protocols and deploying security tools.

18.5.3 Credit cards

Credit card processing is becoming more and more available to members within any organization, and there are two areas which need to be analyzed closely and regularly: (1) financial controls and (2) the PCI-DSS. Established by the Payment Card Industry Security Standards Counsel, and composed of some of the world's largest card issuers, the PCI-DSS provides security requirements that organizations must comply with. In the past, the cost and availability of credit card processing technology made obtaining it virtually infeasible for smaller departments and individuals. However today, services such as Paypal and Square offer low or no-cost credit card processing technology. Now the flower stand on campus during graduation, the food truck driving through during lunch hour, and the vendor selling hotdogs in the stands at a football game can accept credit cards. Department managers need to be aware of their responsibilities to comply with PCI-DSS and know where to go for help. Compliance requirements may turn up in unexpected places including student clubs, supplemental applications for graduate school, and one-time fund raisers. Real headline news has been made when poor financial control decisions were made by university administrators, staff, faculty, or students leading to the misplacement of university funds, embezzlement, and other abnormalities [16].

As with traditional enterprises, an exposure of credit card information or improper processing of cardholder data has resulted in some of the most public and heavily fined exposures. What makes universities different, however, is that they often operate hundreds of distinct units with their own cardholder data environments (CDEs), whereas corporations are more likely to have one or a few explicit and extensively secured environments. University cybersecurity and credit card processing authorities have a responsibility to remain aware of changes to PCI standards and ensure compliance wherever cards are accepted. Depending on resource availability, all enterprises are advised to seek the aid of outside qualified security assessors or train internal security assessors to ensure compliance with the PCI-DSS. PCI-DSS requires extensive segmentation of networks in certain scenarios, training for card handlers, penetration testing of the CDE, annual reporting, and more. In a world where each department has its own set of applications custom tailored to their business needs, it is critical to establish standards. Each time a new method of card processing is introduced, new training and compliance programs must be established. Establish standards for the campus and minimize the number of merchant identifications used for processing to reduce efforts needed to comply with this complex set of regulations.

18.5.4 FERPA

Unique to educational organizations is the FERPA of 1978. This US federal law protecting student information was passed decades before tools such as modern enterprise resource planning applications such as Oracle's Peoplesoft, SAP, and Workday were extensively integrated into university operations but, nonetheless, significantly impacts cybersecurity today. FERPA protects every aspect of student information up to and includes information such as a student's name, grades, and class roster. FERPA breaches are the most common form of exposure in many universities and can lead to real consequences including loss of federal funding and accreditation. For example, universities are required to establish contracts ensuring total ownership and control of student information with all third parties, limit information available to parents, and more. In the university setting, FERPA should be at the forefront of the information security awareness campaign.

- Department of Justice—The systems involved in university police operations have special considerations due to their role not only in the safety of lives on campus, but also in the sensitive nature of the data that can be accessed.
- Digital Millennium Copyright Act (DMCA)—The DMCA protects the intellectual property of digital works including, but not limited to, television, movies, and music. Universities should be aware of and prepared to handle inquiries from various entities attempting to locate those who have illegally obtained or reproduced material. A typical letter may demand the removal of the illegal content or threaten to impose fines:

> You are being contacted on behalf of NBC Universal and its affiliates ("NBC Universal") because your Internet account was identified as having been used recently to illegally copy and/or distribute the copyrighted NBC Universal motion picture(s) and/or television show(s) listed at the bottom of this letter. This notice provides you with the information you need in order to take immediate action that can prevent serious legal and other consequences. These actions include:
>
> 1. Stop downloading or uploading any motion pictures or TV shows owned or distributed by NBC Universal and/or its affiliates without authorization; and
> 2. Permanently delete from your computer(s) all unauthorized copies you may have already made of such films and/or TV shows.
>
> The illegal downloading and distribution of copyrighted works are serious offenses that carry the risk of substantial monetary damages and, in some cases, criminal prosecution.
>
> Copyright infringement also undoubtedly violates your school's policies governing acceptable use of campus network resources and could lead to serious disciplinary action.

University students, often for the first time, are being provided with commercial Internet speeds when connecting on campus, and the temptation to download illegally obtained content through BitTorrent or other services is high. Universities must require authentication for wired and wireless network ports and maintain adequate logs of IP address assignment to users. Often, the IP address and timestamp are the only pieces of information the third party will have available upon discovery of a breach to their systems. Universities must also consider blocking known ports, sites, applications, protocols, and IP addresses such as BitTorrent and PirateBay.

18.5.5 Divide and secure: Microsegmentation

The individual parts of the whole in a university setting have different business objectives. The concept of microsegmentation, or the segmenting of large networks into small, more manageable pieces that can be individually secured, is becoming increasingly popular.

By utilizing the strengths of modern multicontext firewalls, universities can start to think about security in a different way, in alignment with business needs: Most institutional research involves data which is not in any way sensitive or secret. The network traffic in dorms does not have any reason to communicate with the ERP system in human resources. The network connecting the point-of-sale systems in campus dining has to be on its own segment in order to maintain PCI compliance. Microsegmentation allows universities to implement a security strategy as unique as are the departments within the organization. Identify the areas of campus which have no need to communicate with sensitive systems, e.g., the dorms, the guest Wi-Fi, and research. Then, identify the areas of campus that have strict compliance rules, e.g., police departments and health centers, and for each area, create a unique network segment. The security of legacy HVAC, electrical control networks, and other devices with historically very poor security designs is now possible by allowing them to exist in a portion of the network unavailable to the outside world. Securing the university is not an insurmountable task but rather a process of collecting unique networks with their own unique business objectives and, as a result, security configurations.

18.5.6 BYOD for universities

Universities have little choice but to embrace and support BYOD with some estimating the impact of the Internet of things causing the number of devices on campus to double by 2020 [16]. Many universities rely on personally owned devices belonging to their faculty to support classes as well. As a result, Virtual Desktop Infrastructure, 802.1x, posture assessment when connecting to networks and strong policies and awareness programs, reminding individuals that they are responsible for their device meeting security standards, are all key pillars of a cybersecurity program. A strong responsible use policy reminding constituents that they are responsible for ensuring that personal devices meet security standards is a must as well. Universities often benefit from the generosity of their vendors. Antivirus, password vault, and patch management applications frequently offer free licenses for students or faculty use alongside the purchase of licensing for the enterprise-owned systems. Make getting the right tools into the hands of those who need them as simple as possible.

18.5.7 Physical security

An interesting aspect of universities, especially public universities, is the concept of open physical access to facilities. Libraries, classrooms, and administrative buildings are typically unlocked during regular business hours, are open for public use, and do not discriminate as to who is allowed to enter the facility. Most department offices on campuses, even police and healthcare departments, regularly employ student assistants and interns. The result is a highly open and collaborative environment for business operations, teaching, and learning. But the openness of university buildings and offices also results in an inherent information security risk. Where a typical company may employ security checkpoints, security guards, badges, biometrics, electronic access to facilities, facial recognition, and other controls, universities frequently choose not to, or cannot, implement these controls due to resource constraints or intellectual freedom. Adding to the challenge is that employees, especially student employees, are rarely required to wear name badges, so self-policing of physical parties in areas where confidential data (physical or electronic) are infrequent, and inventorying access to physical and electronic data, is even more rare. Solving this issue is primarily a matter of changing the culture. University cybersecurity programs must include a piece to empower employees to question visitors in secure spaces, encourage departments to require name badges, lock doors or drawers which control access to sensitive data, encourage the installation of electronic door locks in lieu of hard keys, and employ name badge programs for visitors. A discernable percentage of security incidents, even thefts, involve not hackers breaking into systems, but rather a lack of a culture of physical security.

18.5.8 Facilities and systems

Seldom talked about are the unique manners in which universities receive funding and their impact on their extremely diverse facilities. Universities operate power plants, police departments, residential buildings, commercial buildings, restaurants, and more. In many universities, the term *deferred maintenance* is the vernacular term for resource constraints which result in facilities having not been upgraded in years, sometimes decades. From a cybersecurity perspective, there are a couple things that have impact. The first is that often, buildings are constructed with one-time funding using building technology that was cutting edge at the time of construction. The result is many universities are running building management systems, supervisory control and data acquisition (SCADA) systems, door lock control systems, and security camera systems which are dated and possibly plagued with information security vulnerabilities. SCADA systems, used for power management within buildings, cogeneration facilities, and in power plants, are notoriously built on dated technologies and are thus easy to compromise. Security camera systems are often procured and installed by individual contributors and departments and as part of building projects. Due to the low cost and ease of implementation, it is not uncommon for departments to purchase and install these devices without the knowledge of requirements for video retention, regular functionality testing, or the ability for the cameras to be accessed by security personnel. Regardless of the system or environment, facilities pose a myriad of challenges where upgrades may be necessary or where, in some cases, microsegmentation is the only feasible solution; the key is to be inclusive with facilities services departments and have an open dialog and an accurate inventory of the devices that they manage so that risk can be assessed and remediated as needed.

The second aspect of facility management may not be so obvious, but it centers on strategizing and setting standards. Companies that manufacture technology for facilities are good at what they do and often practice rather rapid development. They are very good at sales as well. Great care needs to be taken to ensure that compatibility and security standards are established for not just things such as network, telephony, and Wi-Fi in facilities, but also the specific facility technology which runs the building. This will not only ensure that procured systems follow laws and policies, but it also avoids duplication and unwanted diversification of systems as well. It is not at all uncommon to find multiple incompatible building management systems, multiple door access control systems, or other related services of varying degrees of compatibility and security within universities. SJSU, for instance, has three building access control systems, administered by five different departments. A strong cybersecurity program includes inventory and security methodologies for facilities management and standards for future procurement so that when decisions are made to build new buildings or procure new technology, they are done so in a fashion that helps better align security, facilities management, and the business needs of the organization.

18.5.9 Procurement for universities

Procurement can be especially challenging in a university setting as universities often do not have a single procurement office, but may have several or dozens to serve the needs of student organizations, campus dining, facilities, athletics, research, and other operations of the campus, each with their own procurement procedures all equally contributing to risk. A strong cybersecurity program will address all procurement mechanisms used within the university including auxiliaries and individual members of the organization who may be taking advantage of reimbursements or other easy-to-use services. Irrespective of how technology is procured and what its purpose is, research or otherwise, it needs to be inventoried and properly secured by the IT staff.

18.5.10 Personal tools and making the right solution the easy solution

University employees can be quite resourceful. In the absence of availability, or barriers to make use of a service, it is not uncommon to find individuals contributing their own equipment, utilizing personal accounts for university business, or taking advantage of free programs offering services to universities. This can range from bringing in personal equipment, such as servers, to utilizing personal Hotmail accounts to communicate with students, to standing up servers in Microsoft Azure. While most organizations allow BYOD to occur and have policies requiring certain security measures, these policies are often forgotten or improperly implemented. Cybersecurity programs have an obligation to identify and address the risks associated with these practices.

The use of personal systems, personal e-mail accounts, and personal cloud services (i.e., Dropbox, Microsoft One Drive, etc.) for university business causes a number of concerns. When an individual without signing authority for the university signs an agreement for a cloud service, the security of the data stored there is immediately in jeopardy. As soon as university data make their way onto a system not controlled by the university, there could be a lack of security controls, intellectual property could be brought into question, Public Records Act queries cannot be fulfilled, administrative investigations cannot be completed, and more. Cybersecurity is not something that can always be accomplished quickly or easily and often the motivation needed to properly set up a service according to university standards is outweighed by the need to provide a solution. University IT teams must find ways to identify the types of services being provided personally, work with the individuals who have the needs, and find a way to make the right decisions easier than their less secure counterparts. Many universities have passed policy restricting the usage of personal e-mails, servers, and cloud services. Others provide no-cost virtual services to faculty. If the right service is not as easy to get as Amazon Web Services, for instance, the audience will continue to gravitate toward the easy-to-obtain solution.

IT teams do have some tools available to them to help address many situations, but they are no substitute for education, outreach, and making the right choices easy. VPN services, switches, and other network-enabling technologies now have the ability to block or permit access based on compliance. Services now exist, albeit often rather costly, that can scan cloud services within the scope of the university to perform DLP or identify sensitive data being stored in a cloud service that is not compliant with their standards. There is also a dramatic increase in cloud service providers, such as Dropbox and LucidChart, that are working with IT organizations to prevent the creation of unauthorized personal accounts using university e-mail addresses.

18.5.11 Centralization

Universities often struggle with challenges associated with decentralization and shadow IT. In some situations, departments have created their own technology teams; in others, hobbyists have set up services that become integral parts of the organization. Some universities have 1 active directory; others have 50, with hundreds of administrators. While the discussion of centralized vs. decentralized is still a topic of debate among university administrators, the fact is that things are easier for cybersecurity programs when a single team can be the authority for configuration, compliance, and product selection. In the case of decentralized IT, individual departments customize their technology to their specific needs, and cybersecurity programs may not have as much authority to specify solutions. In these environments, cybersecurity programs need to rely on setting achievable standards and providing incentives for compliance. By offering low or no-cost tools to decentralized teams, and being openly collaborative about decisions made in the technology departments, some decentralized departments will not want to have to reinvent the wheel and will come on board. Leverage audits help auditors see where decentralized controls do not meet security standards. Finally, employ at least annual internal risk assessments to identify issues and make sure that those issues are made visible to not only technical personnel but also relevant administrators as well.

18.6 Conclusion

Enterprise systems are an ever-growing and ever-changing world. There is no shortage of new tools and services to keep any cybersecurity team busy, but it is important to not get distracted and lose focus on the human side of the cybersecurity program. A cybersecurity program needs to spend as much time on educating and building solutions with the employees in mind as it does on technology. Cybersecurity is never an easy job and often demands the support and attention of professionals in enterprise environments. Remember to communicate with stakeholders early and often and gain buy-in from business leadership and technology partners alike as new controls are implemented. Celebrate the successes, learn from the mistakes, and always remember the systems will only ever be as secure as the people who use them.

References

1. International Business Machines. 2014. *IBM Threat Force Intelligence Index*. International Business Machines, New York, https://www.ibm.com/security/data-breach/threat-intelligence-index.html.
2. PhishMe.com. 2016. *Enterprise Phishing Susceptibility Report*. PhishMe.com, Leesburg, VA, https://phishme.com/project/enterprise-phishing-susceptibility-report/.
3. Brown, Jo. 2016. Coastal Carolina University scammed out of more than $1M. *WBTW News*, http://wbtw.com/2017/03/22/coastal-carolina-university-scammed-out-of-more-than-1m/
4. Sporck, Lauren. 2016. *8 of the Largest Data Breaches of All Time*. January 18. OPSWAT, San Francisco, CA.
5. Von Ogden, Jacqueline. 2016. *8 Examples of Internal-Caused Data Breaches*. October 18. Cimcor: Merrillvile, IN.
6. Pricewaterhouse Cooper, Carnegie Mellon University, CSO magazine, US Secret Service. 2014. *US State of Cybercrime Survey*. United States Department of Homeland Security, Washington, DC.
7. Covington, Robert. 215. *Physical security: The Overlooked Domain*. Computer World, Framingham, MA, http://www.computerworld.com/article/2939322/security0/physical-security-the-overlooked-domain.html.
8. Knapp, Kenneth; Denney, Gary; and Barner, Mark E. 2011. *Key Issues in Data Center Security: An Investigation of Government Audit Reports*. August 27. Elsevier: New York, NY.

9. Ashford, Warwick. 2013. Bad outsourcing decisions cause 63% of data breaches. February 15. Computer World, Framingham, MA, http://www.computerweekly.com/news/2240178104/Bad-outsourcing-decisions-cause-63-of-data-breaches.
10. Amerdlng, Taylor. 2017. *The 15 Worst Data Security Breaches of the 21st Century*. June 14. CSO, Framingham, MA, http://www.csoonline.com/article/2130877/data-protection/data-protection-the-15-worst-data-security-breaches-of-the-21st-century.html.
11. Cloud Security Alliance. 2017. *STAR Self Assessment*. Cloud Security Alliance, https://cloudsecurityalliance.org/star/self-assessment/. Seattle, WA.
12. Resgui, Yacine. 2007. *Information Security Awareness in Higher Education: An Exploratory Study*. December 18. Elsevier: New York, NY.
13. Open Web Application Security Project. 2017. *WASP Top Ten Project*. Open Web Application Security Project, https://www.owasp.org/index.php/Category:OWASP_Top_Ten_Project. Maryland, US.
14. Center for Internet Security. https://www.cisecurity.org/controls/. East Greenbush, NY. January 19, 2017.
15. Madden, Brian. 2016. *Jamf and IBM Are Showing How Mac in the Enterprise Is Changing*. October 25. http://www.brianmadden.com/opinion/Jamf-and-IBM-are-showing-how-Mac-in-the-enterprise-is-changing. Techtarget, Newton, MA.
16. American Library Association. 2017. *Intellectual Freedom and Censorship Q & A*. American Library Association, Chicago, IL, http://www.ala.org/advocacy/intfreedom/censorship/faq.
17. *News and Observer*. 2016. Former NCSU professor charged with embezzling from agriculture student groups. February 11. http://www.newsobserver.com/news/local/crime/article59740506.html.

section six

Perspective

chapter nineteen

Perspectives on the future of human factors in cybersecurity

Abbas Moallem

Contents

19.1 Introduction

Computer technology is rapidly changing, and the world will soon be connected in one way or another to the Internet. Personal, enterprise, and government information will be digitalized from its physical (paper documents and printed image) form. As this is already happening in the more industrialized countries, we can now predict that sooner or later, all countries, poor or rich, will be joined to this network.

We also can predict that many technological solutions will be found to increase the protection of the digital assets of individuals, groups, and organizations. However, with more people connecting to this huge network, more individual vulnerability will also result. In the field of cybersecurity, the prediction is that this is to be a growing field, and one might assign the same likelihood to the importance of human factors in cybersecurity.

In this chapter, after providing a summary of future perspectives on the future of human factors in cybersecurity, the perspective of each contributor is extracted from their chapter and presented as a separate subsection.

19.2 Summary

19.2.1 Authentication

- Authentication in the future: Authentication will no longer be dominated by passwords. Consequently, users should expect authentication with less pain and more ease in the future.
- Authentication will still use "something you have, something you know, and something you are."
- Biometric usage has the potential of greatly improving cybersecurity, because it relies on inherent biological traits rather than knowledge factors for authenticating or identifying users.
- Biometric verification will increase in use in combination with other methods of identity proofing and authentication.
- With the accumulation of different biometrics information, privacy will be the main future challenge.
- Biometric characteristics will likely become extensively used for biometric verification, but if the site implements effective attacks detection, then using stolen biometric data to authenticate might not work. However, the main privacy challenge arises from other usage. For example, facial images might be used to link a user's account to the user's activity on social networks or the user's visit to a physical store equipped with customer identification cameras.
- The keys and digital certificates are going to get more mature and be used to secure sensitive data, protect its integrity, and authenticate the participants in any digital interaction.

19.2.2 Trust and privacy

The need to elaborate new theories, methods, and research to understand how human and nonhuman team members work most effectively in cybersecurity context will grow.

- Disruptive innovations will have more impact on society with mass deployment.
- Worries will grow regarding privacy and exponential increases of mass data collection from all aspects of the human's life.
- Storage of data in the cloud, i.e., centralized repositories, and predictive analysis of collected data will be a big area of concern.
- Big data will be analyzed by machine learning techniques to provide detailed, personalized characteristics of an individual and prediction of individual future behavior.
- The control on data stored will be the main concern of all human societies, countries, organizations, and human communities.
- Predictions with 99% accuracy based on big data analysis will result in concerns for the 1% of predictions that are inaccurate and wrong. One percent of a population of 50 million is still 500,000; that is not a negligible number, and such error rates associated with big data analysis may be intolerable.
- People whose privacy is violated are vulnerable to manipulation. In dictatorships, people behave according to the dictator's wishes because of fear preventing them from doing something else. In western democracies, this issue will not be less dangerous.

- The needs for tools that facilitate human decisions and responses based on data will grow. Consequently, such tools will provide tremendous power for the cybersecurity professional.

19.2.3 Threats

- The insider threat might not be as frequent as malicious software, but its impact can be costly.
- Big data analytics and the incorporation of personality traits, psychological factors, psychosocial data, and motivations in profiling might be used to prevent insider events.
- Social engineering is going to be an evolving practice with many new perpetrators.
- Social engineers use well-known techniques and continually explore to find new ways to use human behavior to exploit the weakness in people unable to distinguish lies from truth to acquire information.
- As more consumers shop and pay with connected devices, and commerce increasingly migrates to digital channels, industries need to invest and invent new standards, technologies, and products to remove sensitive account data from the payment environment, putting it into a form that cannot be used by criminals for fraud. Products, services, and online platforms should develop built-in security and privacy features, thereby protecting both the product and the customer information from being hacked.

19.2.4 Smart networks and devices

- Security in home networking devices will improve first by making the settings and configurations of the home networking device more user-friendly and increasing user awareness on security.
- Ambient assisted living (AAL) applications for supporting the aging will still expand, particularly the usage of Internet of things (IoT) technologies.
- The needs for regulations on IoT are going to be growing.

19.2.5 Governance

- The need for more laws and regulations with regard to technology, cybercriminality, privacy, and particularly IoT will increase in all modern societies. The regulations will have a huge impact on our daily lives, affecting how we work and protect our children.
- The need to address cybersecurity in our daily lives, for our work, and to protect our children will grow.
- Everyone will need to have the basic understanding of how security and privacy are handled in the law, in technology, and in government.
- Legislative efforts on cybersecurity and privacy will continuously grow.

In the following section, the perspectives of the authors and experts in each area are provided.

19.3 Authentication and access management

In terms of authentication, according to Furnell (Chapter 1),

> one thing we can be sure about moving forward is that the authentication landscape will no longer be as dominated by passwords as it has been in the past. The use of alternative approaches (particularly biometrics) has already become commonplace, so users can expect a far more varied experience than they have previously been offered. Moreover, if the technology choices are appropriately judged and correctly matched to their needs, then users should be able to expect authentication to feel less of a barrier to legitimate use. The options and opportunities are there—but they still require manufacturers and service providers to take them.

Corella (Chapter 2) believes that

> cybersecurity depends on the integrity of computer systems and their operation by human users and administrators. The most dangerous cyber attacks are those that target both system and operational vulnerabilities, leveraging system exploits to compromise user accounts, or compromising systems by impersonating users or modifying their behavior using social engineering techniques. The science of human factors is concerned with the interaction between humans and computer systems, and therefore has key contributions to make towards strengthening cybersecurity. One of those contributions will no doubt be an acceleration of the ongoing research on biometrics.
>
> Operational vulnerabilities derive primarily today from the reliance on knowledge factors to identify and authenticate users, either knowledge of secret passwords or knowledge of private information used in automated knowledge-based verification or human-to-human interaction. Passwords are still the most common means of authenticating users on the Internet, despite many efforts to replace them, and knowledge-based verification is still the primary means of remote identity proofing.
>
> Biometrics is a disruptive technology that has the potential of greatly improving cybersecurity, because it relies on inherent biological traits rather than knowledge factors for authenticating or identifying users. Biometric authentication is not vulnerable to phishing attacks or breaches of backend databases. Even if an adversary breaches a database of user accounts containing facial images, the adversary will not be able to use one of those facial images to impersonate a user unless he or she is able to prove to a verifier that the image is that of his or her face.
>
> But the security gains promised by biometrics depend on the ability to thwart spoofing with effective presentation attack detection, which is difficult. Recent advances in deep neural network technology have yielded remarkable improvements in the accuracy of

facial image verification, and similar improvements may be achievable in other biometric modalities. But such accuracy improvements are consistently reproducible only in non-adversarial settings and have been shown to be vulnerable to adversarially crafted images in spoofing attacks. Both neural networks and classical biometric technologies have to contend with powerful digital attacks such as voice morphing that construct virtual realities in real time and may thereby be able to respond to anti-replay challenges.

In biometric verification there will always be an arms race between verifiers and impersonators. Therefore, biometric verification should be used in combination with other methods of identity proofing and authentication, so that the emergence of an unforeseen method of attack is not catastrophic for the verifiers. The old mantra of proving your identity with "something you have, something you know, and something you are" remains valid.

While spoofing attacks pose a security challenge to biometric verification, a severe privacy challenge arises from the fact that the biometric characteristics used for biometric verification may be used to link the user activities across both cyberspace and the physical world. An adversary who breaches a web site's user database containing facial images may not be able to impersonate users if the site implements effective presentation attack detection but may be able to use the facial images to link each user's account to the user's activities on social networks or the user's visits to physical stores equipped with customer identification cameras.

The privacy challenge may be addressed by two distinct avenues of research. One of them is continued research on revocable biometrics, also known as biometric cryptosystems, which rely on randomized and revocable helper data that is deemed to reveal no useful biometric information, in lieu of traditional biometric templates. The other is research on biometric architectures that do not make use of databases of biometric information. In one such architecture, commonly used today in mobile devices, biometric verification data is kept within the user's device, in local storage that provides some resistance to physical tampering and/or malware. In another architecture that dispenses with biometric databases, biometric verification data is included in a hybrid crypto-biometric credential issued by a certification authority.

All this means that biometrics will be a most active area of research for the foreseeable future, concerned with presentation attacks and their detection, revocable biometrics, biometric verification architectures, new biometric modalities, and improvements of neural network robustness in adversarial settings.

Nair (Chapter 3) believes that

keys and certificates are poised for explosive growth fueled in part by trends in virtualization, cloud, DevOps, and IoT. Yet they are also among the least understood concepts of cybersecurity, as evidenced by the low investment in good cryptography management practices. This is especially stark when compared to the billions of dollars we spend as an industry to protect older, less secure technology such as usernames and passwords.

It is because of this lack of awareness that malicious actors, whether it be private groups of individuals or nation states, target vulnerabilities in the implementation of cryptographic security—SSL/TLS attacks dwarf all other forms of network attacks today, and this difference will only continue to grow. The technology itself is mature and, when implemented correctly does what it is supposed to do, it secures sensitive data, protects its integrity, and authenticates the participants in any digital interaction. This robustness of the technology is now being utilized for malicious purposes, leading to an ongoing debate between the benefits afforded by security/confidentiality on one hand and the threats they pose to citizens from all walks of life on the other.

19.4 Trust and privacy

Cellary (Chapter 4) states that without the science and methods of human–computer interaction (HCI), new tools are not likely to support effective decisions. Consequently, HCI will also need to elaborate new theories, methods, and research to understand how human and nonhuman team members work most effectively in cybersecurity contexts.

> Thoughts regarding future perspectives of e-privacy: Disruptive innovations start to have impact on society only when they are deployed on a mass scale. Currently we are witnessing the mass collection of digital data in the clouds, i.e., centralized repositories, and predictive analysis of collected data. An individual is likely to be conscious of the collection of some data concerning him or her, but not all. He or she is conscious of data entered by him or her into computers to receive a required digital service. However, he or she is not conscious of data collected automatically by devices like cameras or sensors deployed in smart environments. Collected data, even if originally anonymous, may be quite easily attributed to identified individuals. Those data are then combined with data concerning millions of other individuals, products, services, situations, behaviors, reactions, etc., comprising big data. Big data may be analyzed by machine learning techniques to provide detailed, personalized characteristics of an individual and prediction of his or her future behavior. Given the state of the art of information technology briefly depicted above, we may now ask questions about the human and social consequences. It is clear that any institution (and its leaders), which has the possibility to use the above technology, can gain extreme power over individuals and societies. A question arises, who will these people and institutions be? Will society have control of such an institution? The question of control is very pertinent, as machine learning is not based on a cause-and-effect relationship, but on correlations among different datasets and analysis of a big number of training examples.

If one mathematician proves a theorem, another mathematician may check the proof and find errors, if any. If a human programmer writes a program, another human programmer may inspect program code and find errors, if any. However, if a neural network is trained by peta-bytes of data, nobody is able to check whether a particular prediction is correct or not. If big data analysis provides predictions with 99% accuracy, which is great, there are still 1% of the predictions that are wrong. Taking the United States, whose population is 320 million, as a case, wrong predictions would concern 3.2 million people; that is huge. If a wrong prediction concerns advertisement of a product being uninteresting for an individual, the consequences are probably negligible. However, if a wrong prediction concerns a medicine or a medical procedure, the consequences for a patient may be very severe. Wrongly applied predictions may undermine principles of the judicial system. Currently, an individual is guilty if he or she committed a crime in the past and that fact was proved in the court. Prediction based on big data analysis with 99% or even higher accuracy that an individual will commit a crime in the future is not sufficient to pronounce him or her guilty, because the crime must really happen first. An approach based on predictions is motivated by a will to protect the possible victims of prospective crime, but neglects the free will of an individual who finally decides to commit a crime or not. Predictive big data analysis may also corrupt democratic political systems. People whose privacy is violated are vulnerable to manipulation. In dictatorships, people behave according to the dictator's wishes because of the fear that prevents them from doing something else. In the case of the mass deployment of big data analysis, people will behave according to self-fulfilling predictions because they will not be able to imagine anything else.

The problem of e-privacy is so difficult to solve, and so important to study because there are no effective technical means at this time to protect e-privacy. In the past, similar problems have been solved by establishing proper laws forbidding actions that were inacceptable by a society concerned. However, the digital world is global, so a law established in one country is ineffective in other countries. Therefore, representatives of one country may violate e-privacy of citizens of other countries and manipulate them. To protect the privacy of people in a global digital world, a global law should be established and respected. This is unfortunately rather utopic.

Schuster and Alexander (Chapter 5) believe that the future of

cybersecurity is a challenging problem because of its massive scale and rapid rate of change. The pace of technology makes it hard to predict what cybersecurity will look like in even the near future. We think it is a bit easier to envision how innovation in HCI will play a role in cybersecurity solutions. In the future, HCI will include greater use of naturalistic interfaces and the use of big data. These interfaces will become disruptive when they become highly accurate and robust. Our tools will feel as though they behave intelligently; they will anticipate our needs and adapt to us and our situations. This vision of naturalistic communication and higher-level tasks performed by automation is called human-automation teaming. Future automation for cybersecurity will talk to us and listen to us. It will understand higher-level tasks and appreciate the context of problems. Cybersecurity work will continue to require automated tools, but we will interact with them almost as though they are human team members. The ability to adapt responses to context and collaborate on human terms will facilitate human decisions and responses based on data. We imagine such a tool will provide tremendous power for the cybersecurity professional. In our vision, there will continue to be a need for effective human decision-making, but the nature of this work will be dramatically different than seen today. This power would also be available to the attacker. Thus, we do not suggest that human-automation teaming alone would change the cat and mouse game between attackers and defenders.

However, the game might be disrupted if HCI innovation is combined with technological approaches to the problem. For example, an automated cyber defender teammate might allow analysts to rapidly identify novel attacks. Therefore, the future of cybersecurity depends on continued interdisciplinary science and practice, with HCI playing an important role. Without the science and methods of HCI, new tools are not likely to support effective decisions. HCI will also need new theories, methods, and research to understand how human and non-human team members work most effectively in cybersecurity contexts.

19.5 Threats

Papadaki and Shiaeles (Chapter 6) think that

> the insider threat might not be as frequent as malicious software, but its impact can be costly. Detecting insider threats is not a purely technical solution, and the human factor can play an important role. Recent research has recognized its importance and has incorporated personality traits, psychological and psychosocial data, as well as motivations and possible catalysts of insider events. Beyond prevention and detection, though, best practices and guidelines recognize that insider threat is a multifaceted problem, and the success of insider threat mitigation strategies depends on the cooperation of various groups within an organization. Specifically, specific emphasis is given on management, human resources, legal, physical security, data owners, IT, and software engineering.

On social engineering, Moallem (Chapter 7) believed that

> it is an evolving practice with many sources of new perpetrators. Social engineers use well-known techniques and continually explore

> to find new ways to use human behavior to exploit the weakness in people unable to distinguish lies from truth to acquire information. Until technology offers automatic solutions to help users in detecting the lies and protecting people from being victims of social engineers, users will be required to gain awareness and knowledge to protect themselves from deception techniques. Cybersecurity experts should be constantly evaluating and detecting tactics of social engineering and providing efficient warning and training to protect personal and organizational assets.

With regard to money laundering and black markets, Bayatmakou (Chapter 8) believes that

> as more consumers shop and pay with connected devices, and commerce increasingly migrates to digital channels, industries must invest in new standards, technologies, and products. One of the best defenses is removing sensitive account data from the payment environment, putting it into a form that cannot be used by criminals for fraud. Products, services, and online platforms should develop built-in security and privacy features, thereby protecting both the product and the customer information from being hacked.

19.6 Smart networks and devices

The home network will have a huge impact on the life of each individual in the future. The (Moallem, Chapter 9)

> ...router works as the front door to your digital data and due to the complexity of protecting systems, most home networks are very vulnerable. Security can be achieved first by making the settings and configurations of the home networking device more user friendly and intuitive to user needs. Secondly, there needs to be an increase in user awareness on security, along with basic trainings to how to properly manage a home network, at the very least to help users protect their system by enabling basic security settings.

Tragos (Chapter 10) thinks AAL applications for supporting the ageing will still expand particularly usage of IoT technologies. He believes that

> Human Computer Interaction and cybersecurity have a bi-directional relationship, which is very critical for future applications, especially in scenarios with people in the need. On the one hand, HCI needs cybersecurity mechanisms in order to ensure that the systems are secured and no unauthorized third parties can intervene and interact with computers in a malicious way. However, past research has worked a lot on identifying solutions to cybersecurity issues in HCI. Lately, the focus in the IoT world has shifted towards designing and developing trusted IoT systems. HCI can help on this by bringing the human factor closer to the IoT devices and assisting

> towards increasing the users' perception of trust in an IoT system. Most interfaces for management of IoT applications and systems are not user friendly and do not provide enough information about the status, any issues, emergencies or attacks. For an average home user, more user-friendly and simple interfaces should be developed, so that the user is able to easily understand if something is wrong in his environment and try to resolve the issue. Thus, HCI should also focus on improving the situational awareness of users with respect to cybersecurity issues in their environment.
>
> Another important thing for future research is how HCI can help improve cybersecurity. This becomes more important with respect to the privacy regulation of the European Commission that will take effect in May 2018 (GDPR).* According to this regulation that will become law in all EU countries, users should have full control over their data. HCI applications can provide significant assistance for the realization of interfaces so that technology-illiterate users will be able to understand in a simple, yet effective and interactive way the data they are sharing through all their devices and applications. Additionally, using simple interaction the users should be able to control their data, getting alarms for new requests to share data and managing the response to these requests. This is of utmost importance to ensure the requirements for privacy by design and privacy by default as described in the GDPR. Thus, HCI researchers should also work on this area of research, trying to work together with end-users to understand their requirements for designing user-friendly interfaces to improve the privacy of user sensitive information.

On attacks on smart cities, Staudemeyer (Chapter 11) believes that

> legal support should be expected to be needed to handle consequences and to initiate legal actions against the attackers. A root cause mitigation team could be engaged to investigate why a breach was possible and develop a mitigation plan that can be realized rapidly. This is an especially crucial step in battling cybercrime.
>
> *The life cycle, once implemented, will change with every iteration according to the needs of organization. With a proper privacy engineering organization in place, a solid privacy-by-design approach can be reached in system design as well as the operation of the IoT system. For smart cities, a proper privacy life cycle is a quality differentiator.*

Lau et al. (Chapter 12) believe that supervisory control and data acquisition (SCADA) will be still be a widely used control architecture for many industrial systems in the critical infrastructures. Thus, HCI research should try to enhance SCADA security

> through user interface design that supports workers in the effective configuration of security tools and acquisition of cyber SA. Further,

* GDPR: EU General Data Protection Regulation, http://ec.europa.eu/justice/data-protection/reform/index_en.htm.

> research must begin to focus on teamwork starting with staff in security and operations for coordinating response to cyberattacks. Finally, HCI could examine how the attribution and retribution of attackers may shift the asymmetry of cybersecurity in favor of the defenders.

19.7 Governance

Laws and regulations need to evolve to regulate technologies and issues that will include everything from criminality to data protection to protection of citizens' privacy.

Schertler (Chapter 14) thinks that

> in the coming years, we will all become Carrie from Homeland. We will need to address cybersecurity in our daily lives, for our work and to protect our children. A basic understanding of how security and privacy are handled in the law, in technology, and in government policy will be useful for thriving in the coming cybercentury.

And Jill Bronfman (Chapter 14) put it in more poetic terms:

> What of ourselves are we willing to sacrifice for comfort?
> For safety?
> For security?
> To see the future, we look past the past
> I'll miss you: hand-held remote control, power cord, steering wheel,
> Or not.
> The past is perhaps not only prologue
> But epilogue
> Surprise me

In terms of regulation, Quirchmayr (Chapter 15) emphasizes that

> recent and ongoing legislative efforts around the globe stress the continuously growing importance of cybersecurity and privacy. Security by design and security by default will be mandated by EU legislation from May 2018 with respect to the protection of personal data. The importance of HCI in cybersecurity will, therefore, increase significantly, among other reasons, because it will be a core factor in achieving legal compliance.

With the expansion of Internet of things Weber (Chapter 16) believes that

> cybersecurity plays an important role in the data processing/collecting context of the Internet of Things (IoT). Governmental regulations can build a certain normative framework, but human factors must also be aligned to the cybersecurity challenges. In the European Union (EU), the Network and Information Security (NIS) Directive of 2016 has the objective of implementing a "culture" of security elements for the benefits of citizens, consumers, businesses

> and public sector organizations. The NIS concept encompasses the ability of networks and information systems to resist, at a given level of confidence, any action that compromises the availability, authenticity, integrity or confidentiality of processed, stored or transmitted data or related services.
>
> The term *culture* highlights that human factors must be taken into account in connection with the main topics and measures that can realize the desired level of information security, namely the improved national cybersecurity capabilities, the improved cooperation between governmental agencies as well as the design of security and incident notification requirements. An extension of the NIS principles beyond the geographic area of the EU could become a valuable contribution to a global improvement of cybersecurity in the IoT context. In particular, the IoT industry having the most advanced practical experience and know how through its human resources should start implementing cybersecurity standards based on analyses of present weaknesses, to be later followed by the involvement of intergovernmental organizations.

Security agencies all over the world need to evolve in terms of organization and investigative method with (Ruppert, Chapter 17)

> a centralized program would serve as a primary hub of all collection and investigation, thereby possessing the ability to fully comprehend the attack and facilitate the full-range response actions and preventative measures which only the whole-of-government approach has to offer. Unfortunately, such a massive shift in established bureaucratic structures to accomplish this centralization would require a level of leadership in the political structures of Washington, DC, that is unprecedented in recent history.

Along with all these changes, the enterprise system needs to continuously evolve to conform with this changing world. There will be "no shortage of new tools and services to keep any cybersecurity team busy, but it is important to not get distracted and lose focus on the human side of the cybersecurity program."

We need to "celebrate the successes, learn from the mistakes, and always remember the systems will only ever be as secure as the people who use them" (Cook, Chapter 17).

As David Thaw (University of Pittsburgh) put it,

> humans are often cited as the "weak link" in cybersecurity. While perhaps true in a highly technical sense, this claim deeply misunderstands the fundamental goals of cybersecurity, and the nature and character of the threat vectors, which will dominate the future. Rather than being the "weak link"—individuals are the purpose for which cybersecurity exists. And thus, the antiquated goal of driving humans to comply with "better security practices" must be reconsidered in light of a comprehensive approach to ensuring that information systems are designed to enable human endeavor in a manner, which effectively manages risk.

Thaw, considering the example of password complexity, believes that

> for decades, conventional wisdom recommended using complex passwords for authentication purposes. This concept originated from a 1979 paper by Morris and Thompson which demonstrated that passwords using only single-case alpha characters (26 total) were substantially more vulnerable to "key search" attacks than were more complex passwords which required more "classes" of characters (e.g., uppercase, numeric, punctuation, etc.).* For nearly 40 years following the publication of this paper, this conclusion was adopted nearly unanimously among cybersecurity practitioners.†

For Thaw

> as it turns out, this "expert" advice was wrong. Florencio, Herley, and Coskun pointed out flaws in this advice in 2007,‡ but their conclusions went largely ignored in practice until 2017. In early 2017, the National Institute of Standards and Technology reversed its position in official guidance and removed much of its recommendations for use of complex passwords to increase the security of authentication interfaces. Human factors and the additional risk vectors created by password complexity requirements were a key element of this decision.§ The lead author behind NIST's original position recommending password complexity requirements subsequently publicly repudiated the original recommendation and expressed regret about its negative impacts.¶

Thaw questions,

> How did this happen? Security is a field often driven by emotion, and the desire to feel (or help others feel) "safe." But perfect safety is an illusion. Fortunately, it is [one illusion] that society is capable of coming to terms with—as anyone who has driven (or ridden in) a car knows. Unfortunately, when it comes to cybersecurity, it is not one for which we have yet reconciled our emotions with scientific reality. Morris and Thompson's original paper did not actually counsel practitioners to require extremely complex passwords as a solution for strengthening authentication practices. It rather recommended a series of measures—most of which remain valid today—including prohibiting the use of dictionary words and specific commonly used

* R. Morris and K. Thompson, Password security: A case history, *Communications of the ACM: Operating Systems*, Vol. 22, No. 11, pp. 594–597, 1979.

† D. Thaw, Cybersecurity stovepiping, *Nebraska Law Review*, Vol. 96, No. 2, pp. 901–925, 2017.

‡ D. Florencio, C. Herley, and B. Coskun, Do strong web passwords accomplish anything?, *HOTSEC'07 Proceedings of the 2nd USENIX Workshop on Hot Topics in Security*, No. 10, 2007.

§ P. Grassi et al., Digital identity fuidelines, *National Institute of Standards and Technology Special Publication 800-63B*, 2017.

¶ R. McMillan, About those online password rules...N3v$r M1#d!—Expert who touted mixing letters, digits, symbols now regrets it," *Wall Street Journal*, August 8, 2017, p. A1.

> passwords, and the implementation of system-level security measures including guess-rate limitation and cryptographic hashing of password storage with "salting" of the cryptographic hash functions used for this purpose. Yet the most common recommendation adopted from their paper was one they did not make—requiring users to adopt and frequently change extremely complex, lengthy passwords. In other words, to provide users with a "feeling" of "taking action" to preserve their security—classic "security theater."*†
>
> To move beyond these risks, we must accomplish two goals. First, we must understand that users (humans) must actually be able to use the systems we design. Second, we must move past an emotional desire for security theater and accept that, as with any complex system, modern information technologies carry risk. There is no perfect security. But we can effectively manage risk. Moving away from rigid checklists and towards risk management plans is not only superior but necessary to a functioning society increasingly dependent on integrated information technology systems.

We will conclude this chapter with what Mohd Anwar (North Carolina A&T State University) wrote us:

> With sophisticated cybersecurity solutions, it has increasingly become hard to hack into the computer systems. Rather, new strategies to exploit human vulnerabilities are contributing regularly to successful cyberattacks. Additionally, with the rapid growth of cyber physical systems (CPS), human factor issues will be the front and center of cybersecurity. No cybersecurity solution is adequate without sound judgement and proper efforts of the human actors in the operating environment. Many times, the human actors play critical roles to deploy and operationalize cybersecurity solutions. However, the cybersecurity tasks can be impractical, time consuming, and cognitively burdensome for the human actors to perform. As a result, the human elements of cybersecurity need to be fully investigated. In essence, the cybersecurity solutions need to be usable. The usable security research should focus on answering two critical questions: (a) How humans can be assisted in carrying out cybersecurity tasks? (b) If possible, how security solutions can obviate the need of human interventions? In addition to novel cybersecurity mechanisms, this research requires understanding of the strength and limitations of human cognition and behaviors.

* B. Schneier, Beyond security theater, *New Internationalist*, November 2009.

† S. Bellovin, *Thinking Security: Stopping Next Year's Hackers*, Addison-Wesley Professional, Boston, MA, 2016.

chapter twenty

Movies and media

Abbas Moallem

Contents

20.1 Introduction

In today's digital world, movies and media have a large influence on the population, in both informing and misinforming people. With the spread of mobile computers, one observes the increasing role of media in influencing younger populations. Youtube, online streaming, and on-demand media and television easily fill an individual's downtime, since everything is available to autonomously watch whenever one desires.

This chapter aims to provide a list of and commentary on some fictional movies and documentaries among the many cybersecurity-related films. The commentary provided is not a film critique, but information about the content from a cybersecurity point of view.

(Films are listed by date from the most recent in each category.)

20.2 Classic movies

The following are a few classic movies that are interesting to watch given their historical perspective on security. Watching these movies mainly illustrates how security issues have changed and how the future of security and privacy were envisioned in the past.

20.2.1 Nineteen Eighty-Four (1984)

The first movie that one should view before anything else is *Nineteen Eighty-Four*, based on George Orwell's novel by the same name. The story is about a man in a totalitarian system of a fictional country where everybody is watched, and history is rewritten as desired by a supreme leader, "Big Brother." Michael Radford directed this movie starring actors John Hurt, Richard Burton, and Suzanna Hamilton. When Orwell's book was originally published in 1949, the society that he imagined was almost impossible to conceive, yet now it is arguably our reality. With massive surveillance and tools that can transform any laptop or mobile device into a surveillance camera, this masterpiece is extremely relevant to the modern day.

20.2.2 Three Days of the Condor (1975)

Classic movies with their rich dialogue can be a very good source of information to put everything in perspective, and this is particularly true when it comes to cybersecurity. This movie follows the story of a Central Intelligence Agency (CIA) researcher who finds all his coworkers dead and must outwit those responsible until he figures out who he can trust. Directed by Sydney Pollack and based on a novel by James Grady, this film takes a different look at security agencies and how they operate. It received an Oscar nomination in 1976.

20.2.3 *Sebastian (1968)*

This is a movie that might not easily be found on video streaming services, but is especially interesting to watch. The movie is about Sebastian, a former Oxford professor, who in the late 1960s directed the all-female decoding office of the British Intelligence Agency. This is a British film, directed by David Greene, and is based on a story by Leo Marks and Gerald Vaughan-Hughes.

20.2.4 *Stolen Identity (1953)*

This is one of the first movies on identity theft. It tells the story of an Austrian taxi driver who dreams of going to the United States. He cannot, however, because he does not have proper papers or identification. One day, an American businessman is waiting for his cab, when another man kills him. The taxi driver grabs the dead man's papers and takes over his identity. The film depicts a classic example of old-style identity theft by possessing someone else's paper documentation, something that nowadays with more and more smart card is less probable.

20.2.5 *The Great Impersonation (1942)*

There are two versions of this movie: one from 1935 directed by John Rawlins and a 1942 version directed by Alan Crosland. Both are based on the novel written by E. Phillips Oppenheim, which was published in 1920. The story is about an unconscious man who is found in a boat, which has drifted to the landing of an isolated African outpost. There, Baron von Regenstein, an enemy agent, recognizes the man as his exact double, Sir Edward Dominey, with whom von Ragenstein went to school. He plots to kill Dominey and pose as the dead man. This is a fictional movie taking place in the early 1900s, but like *Stolen Identity* (preceding item), it provides an interesting perspective on how one could change their identity in the past.

20.3 *Fictional action and drama movies*

During the last two decades, several Hollywood-style movies and TV series were released with plots focused on cybersecurity, although many of these are entertainment films, not necessarily cinematically valuable. Each of them reflects the mood of society at a specific period and portrays scenarios that might have once been thought futuristic, but now have become reality.

20.3.1 *Snowden (2016)*

Oliver Stone directed this film about Edward Snowden's life, and how, disillusioned with the intelligence community, he left the National Security Agency (NSA) and decided to leak classified information. Snowden then becomes a traitor to some, a hero to others, and is forced to live in exile in Russia. The movie depicts Snowden's life from youth up until the time that he decides to leave the agency and reveal the US surveillance program to the press. Although this work is a dramatized depiction and not a documentary, the film gives a realistic view of Snowden's life, as well as his work and social environments.

20.3.2 *Blackhat (2015)*

This movie follows a talented hacker released from a 15-year prison sentence to help solve a case involving parts of code he wrote when he was young. This code has appeared in malware that triggered a terrorist attack in a nuclear power plant in China. The main theme of this movie is in fact all about cybersecurity, as it involves a terrorist attack on a nuclear power plant by passing malware through PDF files by e-mail. It is a very realistic scenario for those who are familiar with Stuxnet, a malicious computer worm, first identified in 2010 and responsible for causing substantial damage to Iran's nuclear program. Although it includes this hacking case in the plot, this movie by Michael Mann is largely a regular action film.

20.3.3 *The Imitation Game (2014)*

Most people in the cybersecurity field are familiar with the history of modern encryption, Alan Turing, and the Enigma machine. The Enigma machine, invented by German engineer Arthur Scherbius, was used from the early 1920s to protect commercial, diplomatic, and military communication. Then during the Second World War, the Germans took over the Enigma machine, improved it, and used it to code messages during the war. In 1939, Britain received information about the Germans' use of the machine and Alan Turing, Gordon Welchman, and others worked on cracking the Enigma code. This Oscar-winning movie, directed by Morten Tyldum, depicts the real-life story of cryptanalyst Alan Turing and how he and his team broke German coded messages during the Second World War. There have been numerous articles and books written about the Enigma code breach. However, this movie depicts and summarizes Turing's work and life well, although it is a Hollywood depiction of a true story, and therefore, not all events portrayed are entirely historically accurate.

20.3.4 *Identity Thief (2013)*

Directed by Seth Gordon and written by Craig Mazin, this movie begins by portraying how one can become the victim of identity theft. It follows the story of a man from Denver who is tracking down a woman from Florida who has stolen his identity to finance a luxurious lifestyle, destroying his credit. Although the initial scenes show the consequences and impact of identity theft on a person's life, the subsequent parts of the movie are simply a comical chase between the two characters.

20.3.5 *Catch Me if You Can (2002)*

The story of Frank Abagnale Jr. inspired director Steven Spielberg to make this interesting and entertaining movie that depicts the life of a con artist who poses and passes using many illegitimate identities. The movie shows the social engineering and behaviors that can be used to approach victims of identity theft. The case of Abagnale is interesting, in that it provides valuable discussion material for its portrayals of social engineering and identity theft, as he claimed to have had at least eight identities and escaped police custody twice. After serving time in prison, he began working for the federal government and the Federal Bureau of Investigation.

20.3.6 *Antitrust (2001)*

This movie follows a computer programmer whose dream job at a Portland-based firm turns nightmarish when he discovers his boss has a secret and ruthless means of dispatching antitrust problems. Made in 2001, the film is still watchable in 2017, although a lot of its depicted technology (CD, DVD, and flash memory sticks) is now obsolete. From a cybersecurity point of view, the different techniques used to hack into each side of the system are still valid, and the film demonstrates the vulnerability of the systems despite all its security procedures and technologies.

20.3.7 *Office Space (1999)*

This light, entertaining comedy is the story of three company workers who hate their jobs and decide to rebel against their greedy boss. Although it is not about cybersecurity per se, the film shows how employees of a company can gradually become sources of insider threats. The movie exaggerates its characters for comedic effect, but has fundamental truths in what the director Mike Judge tries to convey to the viewer. The behavior of managers and how employees are treated pushes the employees to neglect protecting their organization and even themselves ultimately becoming a threat to the security of the organization.

20.3.8 *The Talented Mr. Ripley (1999)*

Directed by Anthony Minghella and based on the novel written by Patricia Highsmith, this movie is an entertaining drama that illustrates how it is sometimes very easy to convince people with very little information. The movie is not about cybersecurity, but is a good resource to show how social engineers with good observational skills of human behavior can easily go so far as to pretend, convincingly, to be someone else.

20.3.9 *Hackers (1995)*

A young boy is arrested by the US Secret Service for writing a computer virus and is banned from using a computer until his 18th birthday. Director Iain Softley and writer Rafael Moreu incorporated many of the more common cyberattack scenarios into this movie. From breaking into a supercomputer and dumpster diving, and phone tapping to changing the record on databases, the movie portrays a good overview of the new era of cybersecurity. Also, to some degree, it illustrates how the hacker community operates.

20.3.10 *The Net (1995)*

Computer programmer Angela Bennett realizes that her colleagues at her new freelance job are suspiciously dying. The story becomes more complicated following a trip to Mexico when she realizes that her identity has been erased. Directed by Irwin Winkler and written by John Brancato and Michael Ferris, this is another early movie on cybersecurity packed with a complex plot line.

20.3.11 *Sneakers (1992)*

This thriller is all about computers, cryptography, government espionage, and the secret service, as well as deception, betrayal, and friendships. Made in 1992, this movie is almost as relevant now as it was in the past. All events in the movie are security related, and it includes surveillance, social engineering, encryption, and tailgating. Directed by Phil Alden Robinson, with actors such as Robert Redford and Sidney Poitier, this movie still is a very entertaining film.

20.4 Documentaries

There are many documentary films on cybersecurity. The following are a few that are worth watching.

20.4.1 *Citizenfour* (2014)

Edward Snowden is a former contractor at the CIA and US NSA. While working for the NSA, he downloaded NSA documents onto a thumbnail drive and then contacted journalists MacAskill and Greenwald from *The Guardian*. The three arranged to meet in a hotel in Hong Kong with documentary filmmaker Laura Poitras, who filmed the encounter and their discussion. As shown in this documentary, the documents that Snowden brought to the meeting revealed in detail how the NSA is using Presidential Policy Directive 20, a top-secret document issued in October 2012, to tap fiber-optic cables, intercept telephone landing points, and bug communication on a global scale. An Academy Award winner for Best Documentary and a long list of other awards, this film shows the potential dangers and breadth of governmental surveillance. The viewer feels the anxiousness, fear, and suspense, like being with these people in the meeting room in real time. No matter if one supports what Snowden did or not, seeing this film is an eye-opener for surveillance programs and how people's cybersecurity and privacy can be compromised in a surveillance state. Without question, viewing this film is a must from a human factors point of view, as it illustrates the state of mind of those involved in surveillance and how the privacy of citizens is not only compromised but they are also under watch Orwellian-style.

20.4.2 "State of Surveillance" (2016)

In this episode of *Vice* on HBO, cofounder Shane Smith conducts an in-depth interview with NSA whistleblower Edward Snowden along with the presentation of a series of the stories supporting Snowden's findings. This documentary reveals how easy it is to run a surveillance program and collect unimaginable information about anybody just through their own smart devices. The case is made for how the US government managed to run a massive surveillance program, making this a must-see for anyone interested in surveillance programs.

20.4.3 *America's Surveillance State* (2014)

This six-part documentary series is directed by Danny Schechter, who also directed *America Before the Bubble Bursts* (2006), *WMD: Weapons of Mass Deception* (2004). and *Beyond "JFK": The Question of Conspiracy* (1992). In this series, he dissects current US surveillance strategies and reviews how, in the digital world with mass surveillance tools and technologies, a citizen's privacy is fundamentally compromised. He concludes that human society must adjust to the reality that whatever one does or says is watched can be recorded at all times.

20.4.4 *The Hacker Wars* (2014)

This documentary by Vivien Lesnik Weisman presents profiles of Internet hackers who break into government and big business computer networks. It also explores the hacktivist militant group that includes Andrew "weev" Aurenheimer, Prodigy, Jeremy Hammond, and journalist Barrett Brown. The documentary includes comments from

NSA whistleblower Thomas Drake, Anonymous' attorney Jay Leiderman, and journalist Glenn Greenwald. It provides a good inside look into some militant hackers' lives and their perspectives on their actions.

20.4.5 "Defeating the Hackers" (2013–2014)

This is the third episode of the 12-episode BBC documentary series, *Horizon*. It explores the world of hackers that are out to steal money and identities and wreak havoc with people's online lives and the scientists who are joining forces to help defeat them. It includes two men who uncovered the world's first cyberweapon, the pioneers of what is called ultraparanoid computing, and the computer expert who worked out how to hack into cash machines.

20.5 Informative media

There are several informative videos available as YouTube videos from different sources. In this section, more reliable and informative ones are reviewed.

20.5.1 "Kevin Mitnick" (2015)

This talk is at Live Hack at CeBIT Global Conferences 2015. Kevin David was arrested in 1995 and then spent five years in prison for various computer-related and communications-related crimes. He is now a computer security consultant and author. This is an interesting talk, although it is now somewhat dated.

20.5.2 "Glenn Greenwald and Edward Snowden live on stage at #CGC15" (2015)

This is a recording of Glenn Greenwald and Edward Snowden live on stage at the CeBIT Global Conferences 2015.

20.5.3 Zero Days—Security Leaks for Sale (2015)

Uploaded by VPRO, this is a documentary television series by the Dutch public broadcasting organization. This Zero Days video explores online safety. A foreboding voice informs us near the opening that "...our entire power supply can be cut off, ... our systems can be taken over, ... hospitals deprived of power, would cease to function, ... and, ... it's not if, it's when." The documentary then elaborates on vulnerabilities that can be exploited by hackers. There are many other videos available in the "Zero Days" series on YouTube.

20.5.4 Anonymous—The Hacker Wars (2015)

This documentary gives an inside look at a hacker community and their activities and perspectives. It includes Andrew Aurenheimer, hacker hero Jeremy Hammond, and journalist Barrett Brown.

20.5.5 Science of Surveillance (2015)

This video takes viewers into the disturbing world of surveillance technology and scans the latest and future technologies used to spy on the private lives of citizens. This is more like propaganda for showing the good side of surveillance and spy agency advertising.

20.5.6 Anonymous Documentary—Inside a Hacker's World full documentary (2014)

This is a full documentary about Anonymous, an international network of activist and hacktivist entities that was founded in 2003. More information about the group is available on their website http://anonofficial.com/.

20.5.7 TED Talks

- "Governments don't understand cyberwarfare. We need hackers" by Rodrigo Bijou (2015)
- "How to avoid surveillance … with the phone in your pocket" by Christopher Soghoian (2015)
- "Government surveillance—This is just the beginning" by Christopher Soghoian (2013)
- "Your phone company is watching" by Malte Spitz (2012)
- "Hire the hackers!" by Misha Glenny (2011)
- "Cracking Stuxnet, a 21st-century cyberweapon" by Ralph Langner (2011)

20.6 Podcasts

There are many free and subscription-based podcasts that are available to download or listen to online. They address a range of cybersecurity issues from beginners to advanced levels. The following are just a few selected ones.

20.6.1 Security now

Security Now is a program for both news and tech junkies, providing analysis and detailed technical discussion of security headlines.

20.6.2 Defensive security

Defensive Security is a weekly information security podcast which reviews recent high-profile security breaches, data breaches, malware infections, and intrusions to identify lessons that we can learn and apply to the organizations we protect.

20.6.3 Exploring information security

The Exploring Information Security podcast interviews a different professional each week exploring topics, ideas, and disciplines within information security. Prepare to learn, explore, and grow your security mind-set.

20.6.4 The social-engineer podcast

The Social-Engineer podcast is an amalgamation of indie music, deep interviews with security experts, and a topical roundtable discussion about pretexting strategies companies are most likely to encounter.

20.6.5 Tenable

Tenable covers the latest in security news, vulnerabilities, and Tenable's Nessus and Security Center products.

20.6.6 OWASP security

This podcast focuses on all aspects of web application security. Many of the episodes are short interviews with experts in this field. This podcast helps readers learn about or keep on top of web application security topics.

References

America's Surveillance State (2014)
Director: Danny Schechter
Writers: Michael German, Roy Singham, Thomas Drake Himself, Sam Antar, Himself, and Thomas S. Blanton
Stars: Michael German, Roy Singham, and Sam Antar
http://www.imdb.com/title/tt4284058/?ref_=fn_al_tt_1=
Available on Youtube: https://www.youtube.com/watch?v=NI1NdigJHP0

"Anonymous—The hacker wars" (2015)
Directors: Vivien Lesnik Weisman
Writers: Vivien Lesnik Weisman and Meredith Raithel Perry
https://www.youtube.com/watch?v=ku9edEKvGuY
http://www.imdb.com/title/tt4047350/

"Anonymous documentary—Inside a hacker's world full documentary" (2014)
https://www.youtube.com/watch?v=gcAhsl2wo8I

Antitrust (2001)
Director: Peter Howitt
Writer: Howard Franklin
Stars: Ryan Phillippe, Tim Robbins, and Rachael Leigh Cook
http://www.imdb.com/title/tt0218817

Blackhat (2015)
Director: Michael Mann
Writer: Morgan Davis Foehl
Stars: Chris Hemsworth, Viola Davis, and Wei Tang
http://www.imdb.com/title/tt2717822/?ref_=fn_al_tt_1

Catch Me If You Can (2002)
Director: Steven Spielberg
Writers: Frank Abagnale Jr. (book) and Jeff Nathanson (screenplay)
Stars: Leonardo DiCaprio, Tom Hanks, and Christopher Walken
http://www.imdb.com/title/tt0264464/?ref_=fn_al_tt_1

Citizenfour (2014)
Director: Laura Poitras
Stars: Edward Snowden, Glenn Greenwald, and William Binney
https://citizenfourfilm.com

"Cracking Stuxnet, a 21st-century cyber weapon" (2011)
TED Talk by Ralph Langner
https://www.ted.com/talks/ralph_langner_cracking_stuxnet_a_21st_century_cyberweapon

"Defeating the Hackers" (2013–2014)
Director: Kate Dart
Episode 3 of 12
http://www.bbc.co.uk/programmes/b0391z20
https://www.youtube.com/watch?v=HQJMg6FdcvQ&t=1620s

"Glenn Greenwald and Edward Snowden live on stage at #CGC15" (2015)
https://www.youtube.com/watch?v=C3JVWVqtgLQ&t=1699s

"Governments do not understand cyber warfare. We need hackers" (2015)
TED Talk by Rodrigo Bijou
https://www.ted.com/talks/rodrigo_bijou_governments_don_t_understand_cyber_warfare_we_need_hackers

"Government surveillance—This is just the beginning" (2013)
TED Talk by Christopher Soghoian
https://www.ted.com/talks/christopher_soghoian_government_surveillance_this_is_just_the_beginning

Hackers (1995)
Director: Iain Softley
Writer: Rafael Moreu
Stars: Jonny Lee Miller, Angelina Jolie, and Jesse Bradford
http://www.imdb.com/title/tt0113243/?ref_=nv_sr_1

"Hire the Hackers!" (2011)
TED Talk by Misha Glenny
https://www.ted.com/talks/misha_glenny_hire_the_hackers

"How to avoid surveillance … with the phone in your pocket" (2015)
TED Talk by Christopher Soghoian
https://www.ted.com/talks/christopher_soghoian_a_brief_history_of_phone_wiretapping_and_how_to_avoid_it

Identity Thief (2013)
Director: Seth Gordon
Writers: Craig Mazin (screenplay) and Jerry Eeten (story)
Stars: Jason Bateman, Melissa McCarthy, and John Cho
http://www.imdb.com/title/tt2024432/?ref_=fn_al_tt_1

"Kevin Mitnick" (2015)
Live Hack at CeBIT Global Conferences 2015 https://www.youtube.com/watch?v=NtzZBTjKngw&t=1974s

Nineteen Eighty-Four (1984)
Director: Michael Radford
Writers: George Orwell (book) and Michael Radford (screenplay)
Stars: John Hurt, Richard Burton, and Suzanna Hamilton
http://www.imdb.com/title/tt0087803/?ref_=fn_al_tt_8

Office Space (1999)
Director: Mike Judge
Writer: Mike Judge
Stars: Ron Livingston, Jennifer Aniston, and David Herman
http://www.imdb.com/title/tt0151804/

"Science of Surveillance" (2015)
Produced by National Geographic Explorer science of surveillance
https://www.youtube.com/watch?v=XXIonD93H4Y

Sebastian (1968)
Director: David Greene
Writers: Leo Marks (book) and Gerald Vaughan-Hughes (screenplay)
http://www.imdb.com/title/tt0063570/

Sneakers (1992)
Director: Phil Alden Robinson
Writers: Phil Alden Robinson and Lawrence Lasker
Stars: Robert Redford, Dan Aykroyd, and Sidney Poitier
http://www.imdb.com/title/tt0105435/?ref_=ttmd_ph_tt1

Snowden (2016)
Director: Oliver Stone
Writers: Kieran Fitzgerald and Oliver Stone
Stars: Joseph Gordon-Levitt and Melissa Shailene Woodley
http://www.imdb.com/title/tt3774114/

State of Surveillance (2016)
Produced by Vice News
Stars: Ben Anderson, Andreas Bakke Foss, and Jason Leopold
http://www.imdb.com/title/tt5764906/?ref_=fn_al_tt_1
Available on YouTube: https://www.youtube.com/watch?v=ucRWyGKBVzo

Stolen Identity (1953)
Director: Gunther von Fritsch
Writers: Alexander Lernet-Holenia (book) and Robert Hill (screenplay)
Stars: Donald Buka, Joan Camden, and Francis Lederer
http://www.imdb.com/title/tt0046372/?ref_=fn_al_tt_2

The Great Impersonation (1935)
Director: Alan Crosland
Writers: E. Phillips Oppenheim (book) and Eve Greene (screenplay)
Stars: Edmund Lowe, Valerie Hobson, and Wera Engels
http://www.imdb.com/title/tt0026438/?ref_=fn_al_tt_1

The Hacker Wars (2014)
Director: Vivien Lesnik Weisman
Writers: Vivien Lesnik Weisman and Meredith Raithel Perry
Stars: Andrew "weev" Auernheimer, Andrew Blake, and Barrett Brown
http://www.imdb.com/title/tt4047350/fullcredits?ref_=tt_ov_st_sm
Available on Youtube: https://www.youtube.com/watch?v=ku9edEKvGuY

The Imitation Game (2014)
Director: Morten Tyldum
Writers: Andrew Hodges (book) and Graham Moore (screenplay)
Stars: Benedict Cumberbatch, Keira Knightley, Allen Leech, Rory Kinnear, and Mark Strong
http://theimitationgamemovie.com

The Net (1995)
Director: Irwin Winkler
Writers: John Brancato and Michael Ferris
Stars: Sandra Bullock, Jeremy Northam, and Dennis Miller
http://www.imdb.com/title/tt0113957/?ref_=fn_al_tt_1

The Talented Mr. Ripley (1999)
Director: Anthony Minghella
Writers: Patricia Highsmith (book) and Anthony Minghella (screenplay)
Stars: Matt Damon, Gwyneth Paltrow, and Jude Law
http://www.imdb.com/title/tt0134119/?ref_=fn_al_tt_1

Three Days of the Condor (1975)
Director: Sydney Pollack
Writers: James Grady (book) and Lorenzo Semple Jr. (screenplay)
Stars: Robert Redford, Faye Dunaway, and Cliff Robertson
http://www.imdb.com/title/tt0073802/

"Your phone company is watching" (2012)
TED Talk by Malte Spitz
https://www.ted.com/talks/malte_spitz_your_phone_company_is_watching

"Zero days—has security leaked for sale" (2015)
VPRO series by Dutch public broadcasting organization
https://www.youtube.com/watch?v=XpYTE8-PlZA&t=1354s

Podcasts

Security Now
https://www.grc.com/securitynow.htm

Defensive Security
https://defensivesecurity.org/category/podcast/

Exploring Information Security
http://www.timothydeblock.com/eis/

The Social-Engineer Podcast
https://www.social-engineer.org/category/podcast/

Tenable
http://www.tenable.com/podcast

OWASP Security
https://www.owasp.org/index.php/OWASP_Podcast

Glossary

ACTivating InnoVative IoT smart living environments for AGEing well (ACTIVAG) A European multicentric large-scale pilot on smart living environments. The main objective is to build the first European Internet of things (IoT) ecosystem across nine deployment sites in seven European countries, reusing and scaling up underlying open and proprietary IoT platforms, technologies, and standards and integrating new interfaces needed to provide interoperability across these heterogeneous platforms.

Active and healthy ageing (AHA) The potential to increase functional dependency and loss of autonomy by 2 years without impacting life span.

Activity tree The range of the types of activities that an employee (potential insider) might take part in as part of their expected daily workload.

Advanced persistent threats (APT) A network attack in which an unauthorized person gains access to a network and stays there undetected for a long period of time.

Agence Nationale de la Sécurité des Systèmes d'information (ANSSI) A French national agency that protects government systems against cyberattacks. It also operates the French government's public key infrastructure and root certificate authority called IGC/A.

Alliance for Internet of Things Innovation (AIOTI) The European Commission created AIOTI in 2015. Its goal is to facilitate the communication and interaction among IoT players in Europe.

Ambient assisted living (AAL) Embedding intelligent objects in the environments of assisted and independent living situations (mostly for elderly residents). This is a relatively new information and communication technology (ICT) trend.

American Institute of Certified Public Accountants (AICPA) system and organization control (SOC) A set of SOC service offerings that allow system-level controls of a service organization or the entity-level controls of other organization to be audited by certified public accountants.

American Recovery and Reinvestment Act of 2009 (ARRA) A stimulus package enacted by the 111th US Congress and signed into law by President Barack Obama in February 2009. Developed in response to the great recession, the primary objective of the ARRA was to save existing jobs and create new ones as soon as possible (according to Wikipedia).

Anomaly detection methodology A methodology that can be used for external threats by using the three steps of (1) data collection, (2) feature extraction, and (3) internal threat detection using *k*-nearest neighbor algorithms.

Anti-cybersquatting Consumer Protection Act The Anti-cybersquatting Consumer Protection Act (ACPA), 15 USC § 1125(d), is a US law enacted in 1999 that established a cause of action for registering, trafficking in, or using a domain name confusingly similar to, or dilutive of, a trademark or personal name (according to Wikipedia).

Antimoney laundering (AML) A set of procedures, laws, and regulations designed to stop the practice of generating income through illegal actions.

Article 29 Working Party (WP29) The Article 29 Working Party (Art. 29 WP) includes a representative from the data protection authority of each European Union (EU) member state, the European Commission, and the European Data Protection Supervisor.

Association of Certified Fraud Examiners An antifraud organization and provider of antifraud training and education.

Attack attribution Determination of the identity or location of an attacker or attackers' intermediary.

Attack trees Conceptual diagrams showing how an asset, or target, might be attacked. Attack trees have been used in a variety of applications. In the field of information technology, they have been used to describe threats on computer systems and possible attacks to realize those threats (according to Wikipedia).

Audit trails Cybersecurity audit trails that allow detection and response to cybersecurity events must be implemented and records must be maintained for at least three years.

Authentication authorization accounting (AAA) A term including the framework for intelligently controlling access to computer resources, enforcing policies, auditing usage, and providing the information necessary to bill for services.

Authentication The process or action of verifying the identity of a user or process.

Authority principle A multilayered and widely accepted system of authority confers an immense advantage upon society. It allows the development of sophisticated structures for the production of resources, trade, defense, expansion, and social control that would otherwise be impossible.

Automatic certificate management environment (ACME) Currently an Internet Engineering Task Force (IETF) draft that was designed to simplify the process by which certificates were issued while attempting to validate the legitimacy of the requestor.

Behavioral biometrics A term encompassing the many biometric patterns of detected human activity. This activity can include mouse movements, movements of a hand holding a smartphone, keystroke dynamics, and details of touch screen gestures.

Big data analysis Big data analytics is the process of examining large and varied datasets to uncover hidden patterns, unknown correlations, market trends, customer preferences, and other useful information.

Biometric cryptosystem The method of generating a biometric key and helper date from an enrollment sample and random bits, used in revocable biometrics.

Biometric fusion The process of recognizing a subject that relies upon data from biometric samples that pertain to multiple biometric modalities.

Biometric identification A method of comparing a biometric sample from a subject against all the samples and data in a database of enrollment samples with the intention to identify the subject among the enrollment samples.

Biometric key A randomized bit string, used in revocable biometrics, that can be consistently regenerated from varying but genuine biometric samples, in conjunction with helper data from which it is infeasible to derive any useful biometric information.

Biometric matching A process that determines if two samples have been taken from the same subject. The samples consist of a sample or data derived from an enrollment sample and a biometric sample taken from a subject.
Biometric modality A method of recognizing individuals by measuring a particular biometric trait using a variety of biometric techniques.
Biometric recognition A method that incorporates both biometric verification and biometric identification to validate a biometric sample provided by the subject against one or more enrollment samples.
Biometric sample A measurement of a biometric trait taken by a sensor or input device.
Biometric template An encoding of relevant features of a biometric sample used in biometric matching.
Biometric trait A measurable quality of the human body that can be used to recognize a subject.
Biometric verification A method of confirming and validating the accuracy of a biometric sample provided by a subject. The subject sample is compared against a single enrollment sample, or the data derived from it, from an earlier sample provided from the subject in order to ensure that the subject has produced both samples.
Bitcoin (BTC) A type of digital currency that is created and held electronically. Bitcoin is not printed and is not a tangible form of currency. It is not controlled or regulated, and people and, increasingly, businesses using software that solves mathematical problems on computers all over the world produce it.
Bonjour protocol Apple's implementation of zero-configuration networking (zeroconf), a group of technologies that includes service discovery, address assignment, and hostname resolution. Bonjour locates devices such as printers, other computers, and services that those devices offer on a local network using multicast domain name system service records (according to Wikipedia).
Bring your own device (BYOD) Also called bring your own technology (BYOT), bring your own phone (BYOP), and bring your own personal computer (BYOPC); refers to the policy of permitting employees to bring personally owned devices (laptops, tablets, and smartphones) to their workplace and to use those devices to access privileged company information and applications. The phenomenon is commonly referred to as information technology (IT) consumerization (according to Wikipedia).
Building management system A computer-based control system in buildings that monitors and controls the building\s electrical and mechanical equipment, such as ventilation, lighting, power systems, fire systems, and security systems. This term is also known as a building automation system (BAS).
CAN-SPAM Act A law that sets the rules for commercial e-mail, establishes requirements for commercial messages, gives recipients the right to have you stop e-mailing them, and spells out tough penalties for violations (https://www.ftc.gov).
Canadian Office of the Privacy Commissioner (OPC) The Privacy Commissioner of Canada is an agent of parliament whose mission is to protect and promote privacy rights. The OPC oversees compliance with the Privacy Act, which covers the personal information-handling practices of federal government departments and agencies, and the Personal Information Protection and Electronic Documents Act (PIPEDA), Canada's federal private sector privacy law. (https://www.priv.gc.c).
Cardholder data environments (CDE) A computer system or networked group of IT systems that processes, stores, and/or transmits cardholder data or sensitive

payment authentication data. A CDE also includes any component that directly connects to or supports this network.

Center for the Protection of National Infrastructure (CPNI) The UK government authority which provides protective security advice to businesses and organizations across the national infrastructure. Their advice aims at reducing the vulnerability of the national infrastructure to terrorism and other threats, keeping the United Kingdom's essential services (delivered by the communications, emergency services, energy, finance, food, government, health, transport, and water sectors) safer (according to Wikipedia).

Command and control (C2) Remote communication between the malware and attacker servers for transmitting information, receiving updates, and executing commands.

Computer emergency readiness team (CERT) By collaborating with high-level government organizations, such as the Federal Bureau of Investigation (FBI) and other law enforcement, the intelligence community, the Department of Defense, and the Department of Homeland Security; CERT develops products, methods, and tools to help organizations assess their security-related practices and to conduct forensic examinations, analyze vulnerabilities, and monitor large-scale networks.

Certificate database A database that maintains a record of certificates that have been issued or revoked for audit purposes.

Certificate management system (CMS) Uses centrally defined policies that govern the issuance, distribution, and life cycle management of certificates.

Certificate revocation lists (CRLs) Certificates that have been issued but do not need to be trusted any longer (for a variety of reasons such as key compromise and entities that have left the organization) are revoked by the issuing authority [certification authority (CA)] and put on a "blacklist" called a CRL that can be used by relying entities to check on the status of known/unknown parties in a transaction.

Certificate transparency (CT) Essentially a framework for monitoring and auditing the issuance of certificates in near real time. Google has helped promote this by requiring that certificates have publicly accessible issuance records in order to be treated as the most trustworthy within the Chrome browser—currently indicated by a green lock icon in the address bar of the browser, which can be interpreted by Internet users as Google-provided confidence in the validity of the site that they intend to visit.

Certification authority (CA) Issues digital certificates. CAs can be public (trusted by anyone on the Internet) or private [trusted only by specific organization(s)] for the purposes of internal transactions and are the root of trust.

Certification authority authorization (CAA) A standard designed to help protect websites by preventing the issuance of rogue or unauthorized secure sockets layer (SSL)/transport layer security (TLS) digital certificates.

ChaCha/Poly1305 A cryptographic message authentication code created by Daniel J. Bernstein. It can be used to verify the data integrity and the authenticity of a message. It has been standardized in RFC 7539 (according to Wikipedia).

Charitable contributions fraud The act of using deception to get money from people who believe they are making donations to charities. Often, a person or a group of people will make material representations that they are a charity or part of a charity and ask prospective donors for contributions to the nonexistent charity (according to Wikipedia).

Cloud computing An IT paradigm, a model for enabling ubiquitous access to shared pools of configurable resources (such as computer networks, servers, storage, applications, and services), which can be rapidly provisioned with minimal management effort, often over the Internet (according to Wikipedia).

Cloud security alliance (CSA) A not-for-profit organization with a mission to "promote the use of best practices for providing security assurance within Cloud Computing, and to provide education on the uses of Cloud Computing to help secure all other forms of computing." The CSA has over 80,000 individual members worldwide (according to Wikipedia).

CloudLock A cloud security company focused on providing enterprise class security solutions for data in the cloud.

Code signing The notion of signing application code to prove its integrity and trustworthiness has assumed particular significance in the last 20 years because of the explosive growth of malicious software (commonly referred to as "malware").

Code signing keys Code signing is the process of digitally signing executables and scripts to confirm the software author and guarantee that the code has not been altered or corrupted since it was signed. The process employs the use of a cryptographic hash to validate authenticity and integrity (according to Wikipedia).

Command and control (C2) phase The influence an attacker has over a compromised computer system that they control. For example, a valid usage of the term is to say that attackers use "command and control infrastructure" to issue "command and control instructions" to their victims (according to Wikipedia).

Commercial general liability (CGL) An insurance policy issued to business organizations to protect them against liability claims for bodily injury and property damage, arising out of premises, operations, products, and completed operations, and advertising and personal injury liability.

Common data repository (CDR) The CDR contains more than 11 million anonymized records and is the result of an extensive, multilevel, and heterogeneous data collection which incorporates data from hosts (i.e., logins, password updates), applications [i.e., secure shell (SSH), e-mail, and web server logs], networks (i.e., Stealthwatch, honeynet, Snort intrusion detection system, and e-mail sensors), and physical data (i.e., card access records).

Confidentiality integrity availability (CIA) Known as the CIA triad, this is a model that guides policies for information security within an organization.

Consistency and commitment principle Once we have made a choice or taken a stand, we will encounter personal and interpersonal pressures to behave consistently with that commitment.

Control flow integrity (CFI) A general term for computer security techniques which prevent a wide variety of malware attacks from redirecting the flow of execution of a program. Associated techniques include code-pointer separation, code-pointer integrity, stack canaries, shadow stacks, and vtable pointer verification (according to Wikipedia).

Corporate insider threat detection (CITD) By combining behavioral actions with technical activities, the CITD can assess the threats posed by individuals. It encompasses user and role-based profiling and is designed for large-scale repositories and activity logs.

Counterfeit check fraud Counterfeit checks can come in many forms, from cashier's checks and money orders to corporate and personal checks. Although the account, bank, and routing numbers on the counterfeit check may be real, the check can still be fake and it can be printed with the names and addresses of legitimate financial institutions.

Cross site scripting (XSS) XSS attacks inject malicious scripts into benign and trusted websites. XSS attacks occur when an attacker uses a web application to send malicious code, generally in the form of a browser side script, to a different end user.

Cryptographic hash function A hash function which takes an input (or "message") and returns a fixed-size alphanumeric string. The string is called the "hash value," "message digest," "digital fingerprint," "digest," or "checksum". It is extremely easy to calculate a hash for any given data (according to Wikipedia).

Cryptographic key The core part of cryptographic operations and is a string of bits used by a cryptographic algorithm to transform plain text into cipher text, or vice versa. The key remains private and ensures secure communication, and these systems can include pairs of operations, such as encryption and decryption.

Cryptographic protocol A security protocol (cryptographic protocol or encryption protocol) is an abstract or concrete protocol that performs a security-related function and applies cryptographic methods, often as sequences of cryptographic primitives. A protocol describes how the algorithms should be used (according to Wikipedia).

Customer proprietary network information Customer proprietary network information in the United States is information that telecommunications services such as local, long distance, and wireless telephone companies acquire about their subscribers. It includes not only what services they use but also their amount and type of usage.

Cyberdenial and deception Use of decoys to collect data on the behaviors and induce poor decision-making of the adversaries. The most established technical implementation is honeypot, which serves as a decoy network that is intended to be attacked and often closely monitored to obtain valuable attackers' information and produce and early warning call slowing down external cyberattack.

Cyberhygiene The process of improving an security of an entire system through a reduction of attack surface. This behavior promotes safety and security.

Cyberinsurance An insurance product used to protect businesses and individual users from Internet-based risks and, more generally, from risks relating to IT infrastructure and activities. Risks of this nature are typically excluded from traditional commercial general liability policies or at least are not specifically defined in traditional insurance products (according to Wikipedia).

Cyberkill chain A systematic process of staging a cyberattack involving seven phases reconnaissance, weaponization, delivery, exploitation, installation, command and control (C2), and actions on objectives.

Cyberphysical system (CPS) A system or mechanism controlled or monitored by computer-based algorithms that are tightly integrated with the Internet and its users.

Cybersecurity audit data The events that are used to determine outlier behavior or policy violations and are associated with the organizational/demographic data of an employee.

Cybersecurity Enhancement Act of 2014 An ongoing, voluntary public–private partnership to improve cybersecurity and to strengthen cybersecurity research and development, workforce development and education, public awareness and preparedness, and for other purposes (https://www.congress.gov).

Cybersecurity framework (CSF) A framework created through collaboration between industry and government and consisting of standards, guidelines, and practices to promote the protection of critical infrastructure. The prioritized, flexible, repeatable, and cost-effective approach of the framework helps owners and operators of critical infrastructure manage cybersecurity-related risk (https://www.nist.gov).

Cybersecurity intelligence index Provides a high-level overview of the major threats to businesses worldwide over the past year.

Cybersecurity Strategy of the European Union 2013 The intention of this strategy is to ensure the effective and strong protection and promotion of citizens' rights to make the EU online environment the safest in the world. The strategy outlines the EU's vision in this domain, clarifies the roles and responsibilities, and proposes specific activities at the EU level. It is jointly adopted by the Commission and the High Representative (https://ec.europa.eu).

Cybersupply chain risk management (C-SCRM) The process of identifying, assessing, and mitigating the risks associated with the distributed and interconnected nature of IT/OT product and service supply chains.

Cyber Unified Coordination Group (UCG) The White House National Security Council (NSC) is the principal forum used by the president of the United States for the consideration of national security and foreign policy matters with senior national security advisors and cabinet officials and is part of the executive office of the president of the United States (according to https://csrc.nist.gov).

Cyclic redundancy check (CRC) code An error-detecting code commonly used in digital networks and storage devices to detect accidental changes to raw data. Blocks of data entering these systems get a short check value attached, based on the remainder of a polynomial division of their contents (according to Wikipedia).

Dark Triad (Machiavellianism, narcissism, and psychopathy) A subject in psychology that focuses on three personality traits narcissism, Machiavellianism, and psychopathy. Use of the term *dark* implies that people possessing these traits have malevolent qualities (according to Wikipedia).

Damage potential, reproducibility, exploitability, affected users, discoverability (DREAD) Previously used at Microsoft and currently used by OpenStack and other companies, DREAD is part of a system that assesses risk and computer security threats.

Data aggregation Process of gathering and expressing the information in a summary form, for purposes such as statistical analysis.

Data anonymization A type of information sanitization whose intent is privacy protection. It is the process of either encrypting or removing personally identifiable information (PII) from datasets, so that the people whom the data describe remain anonymous (according to Wikipedia).

Data breach hypothetical(s) The data breach hypothetical tabletop exercise requires a bit of imagination to posit some basic facts Who or what caused this imaginary breach? How did we find out about it? What was the reaction of the media, the government, and the individuals affected? Ultimately, the goal of the exercise is to create a checklist of items to review systems and procedures to avoid such a scenario from actually occurring.

Data deidentification/deidentified data Deidentification is the process used to prevent a person's identity from being connected with information. Common uses of deidentification include human subject research for the sake of privacy for research participants. Common strategies for deidentifying datasets include deleting or masking personal identifiers, such as name and social security number, and suppressing or generalizing quasi-identifiers, such as date of birth and zip code (according to Wikipedia).

Data minimization The idea that companies should collect, use, disclose, and store only the minimum data necessary to perform a task and that they should limit the data they collect and retain and dispose the data once it is no longer needed. By doing this, companies can reduce the amount of data that can be misused or leaked.

Data protection directive of the European Commission/General Data Protection Regulation (GDPR) The General Data Protection Regulation (GDPR) (Regulation (EU) 2016/679) is a regulation by which the European Parliament, the Council of the European Union, and the European Commission intend to strengthen and unify data protection for all individuals within the EU. It also addresses the export of personal data outside the EU (according to Wikipedia).

Data retention The policies of persistent data and records management addressing the retention and secure disposal of nonpublic data.

Data-in-motion A stream of data moving through any kind of network. It is one of the two major states of data, the other being data-at-rest. It can be considered the opposite of data-at-rest as it represents data which is being transferred or moved, while data-at-rest is data which is static and is not moving anywhere (https://www.techopedia.com/).

Decision-making When presented with a situation that has some ambiguity or risk, this term refers to the process of selecting a course of action as long as the decision takes more than 1 s to make.

Deep belief network (DBN) In machine learning, a generative graphical model or, alternatively, a class of deep neural network, composed of multiple layers of latent variables ("hidden units"), with connections between the layers but not between units within each layer.

Deep packet inspection (DPI) Also called complete packet inspection and information extraction or IX; a form of computer network packet filtering that examines the data part (and possibly the header) of a packet as it passes an inspection point, searching for protocol noncompliance, viruses, spam, intrusions, or defined criteria to decide whether the packet may pass or if it needs to be routed to a different destination or for the purpose of collecting statistical information that functions at the application layer of the open systems interconnection model (according to Wikipedia).

Demilitarized zone (DMZ) A physical or logical sub-network that separates an internal local area network from other untrusted networks, usually the Internet.

Denial-of-service (DoS) attack In computing, a cyberattack where the perpetrator seeks to make a machine or network resource unavailable to its intended users by temporarily or indefinitely disrupting services of a host connected to the Internet. DoS is typically accomplished by flooding the targeted machine or resource with superfluous requests in an attempt to overload systems and prevent some or all legitimate requests from being fulfilled (according to Wikipedia).

Departmental silos The image of a grain or water silo tower standing tall and alone in the field is a vivid one, and it accurately represents the idea that large companies have different departments for each functionality, and the departments often do not communicate with each other often or well. There may be different legal departments even for privacy and for security issues, or security operations may be housed in IT while privacy compliance is settled in with the regulatory group.

Derived or delegated trust model The issuing authority (CA) is what is trusted by the relying party in essence any (server or client) certificate that chains up under the trusted CA is considered trustworthy. This allows for certificate reissuance or new identities to be established without having to redefine the trust relationship (as long as the issuing CA continues to operate within its defined parameters).

Developing and Growing the Internet of Things Act (DIGIT) The bill would bring together a working group of federal entities that would consult with private sector stakeholders to provide recommendations to Congress.

Devil's ivy A vulnerability in a piece of code called gSOAP widely used in physical security products. It is potentially allowing faraway attackers to fully disable or take over thousands of models of Internet-connected devices from security cameras to sensors to access card readers.

DigiNotar A Dutch certificate authority owned by VASCO Data Security International, Inc. On September 3, 2011, after it had become clear that a security breach had resulted in the fraudulent issuing of certificates, the Dutch government took over the operational management of DigiNotar's systems. That same month, the company was declared bankrupt (according to Wikipedia).

Digital certificate Also known as a public key certificate or identity certificate; an electronic "password" that allows an organization or person to securely exchange data over the Internet using public key infrastructure (PKI).

Digital Millennium Copyright Act (DMCA) A US copyright law that implements two 1996 treaties of the World Intellectual Property Organization. It criminalizes the production and dissemination of technology, devices, or services intended to circumvent measures that control access to copyrighted works (commonly known as digital rights management or DRM) (according to Wikipedia).

Digital signature algorithm (DSA) A federal information processing standard (FIPS) for digital signatures. In August 1991, the National Institute of Standards and Technology (NIST) proposed DSA for use in their digital signature standard and adopted it as FIPS 186 in 1993. Four revisions to the initial specification have been released FIPS 186-1 in 1996, FIPS 186-2 in 2000, FIPS 186-3 in 2009, and FIPS 186-4 in 2013 (according to Wikipedia).

Digital signature algorithm (MD5) A widely used hash function producing a 128-bit hash value. Although MD5 was initially designed to be used as a cryptographic hash function, it has been found to suffer from extensive vulnerabilities. It can still be used as a checksum to verify data integrity, but only against unintentional corruption (according to Wikipedia).

Direct trust model The public key certificate of the entity is directly trusted by the relying party. Any changes to the certificate (upon renewal, reissuance, etc.) will require that trust be reestablished, manually.

Disaster recovery plan (DRP) A documented process or set of procedures to recover and protect a business IT infrastructure in the event of a disaster.

Distributed architecture When components are presented on different platforms and several components can cooperate with one another over a communication network in order to achieve a specific objective or goal.

Distributed hash table (DHT) A decentralized distributed system that provides a lookup service similar to a hash table.

Distributed network protocol (DNP3) A set of communications protocols used between components in process automation systems.

Distrust A state of miscalibrated trust in which the trustor does not sufficiently trust the trustee.

DMZ network Demilitarized zone network.

Document signing European entities, both government and business, have adopted standards such as XML DSig, XAdES, PAdES, and CAdES to standardize the document signing process and allow for interoperability within and between organizations.

Domain name system (DNS) A hierarchical and decentralized naming system for computers, services, or other resources connected to the Internet or a private network. The DNS associates various information with the domain names that are used for each of the participating entities.

Domain name system-based authentication of named entities (DANE) A protocol that allows X.509 certificates, commonly used for TLS, to be bound to DNS names using DNS security extensions (DNSSEC). It is proposed in RFC 6698 as a way to authenticate TLS client and server entities without a CA (according to Wikipedia).

Domain name system security extension (DNSSEC) Internet protocol (IP) networks include secure information provided by the DNS. This security comes from the DNSSEC, a suite of IETF specifications.

Electronic Frontier Foundation (EFF) A nonprofit organization defending civil liberties in the digital world. It is an international nonprofit digital rights group that is funded by industry heavyweights such as Cisco, Akamai, and Mozilla and launched a free CA in 2016—Let'sEncrypt (the web).

Electronic healthcare record (EHR) The systematized collection of patient and population electronically stored health information in a digital format. These records can be shared across different healthcare settings (according to Wikipedia).

Electronic personal health information (ePHI) Any protected health information that is covered under Health Insurance Portability and Accountability Act of 1996 (HIPAA) security regulations and is produced, saved, transferred or received in an electronic form.

Electronic product code (EPC) A universal identifier that gives a unique identity to a specific physical object. This identity is designed to be unique among all physical objects and all categories of physical objects in the world, for all time. (according to www.epc-rfid.info).

Electronic product code information system (EPSIC) In computer science, users can gain a shared view of physical or digital objects within a relevant business context by using the EPCIS, a global GS1 standard for creating and sharing visibility event data, both within and across enterprises.

Elliptic-curve cryptography (ECC) An approach to public key cryptography based on the algebraic structure of elliptic curves over finite fields. ECC requires smaller keys compared to non-ECC cryptography (based on plain Galois fields) to provide equivalent security (according to Wikipedia).

Encryption In cryptography, the process of encoding a message or information in such a way that only authorized parties can access it. Encryption does not itself prevent interference, but denies the intelligible content to a would-be interceptor. In an encryption scheme, the intended information or message, referred to as plaintext, is encrypted using an encryption algorithm—a cipher—generating cipher text that can only be read if decrypted (according to Wikipedia).

Encryption coverage The conversion of data into a form, called a cipher, that cannot be understood by unauthorized people.

Encryption eavesdropping [man-in-the-middle (MITM)] Also called a Janus attack; in cryptography and computer security, an attack where the attacker secretly relays and possibly alters the communication between two parties who believe they are directly communicating with each other. One example of MITM attacks is active eavesdropping, in which the attacker makes independent connections with the victims and relays messages between them to make them believe that they are talking directly to each other over a private connection, when in fact, the entire conversation is controlled by the attacker (according to Wikipedia).

End-to-end encryption with datagram transport layer security (DTLS) In this type of encryption, any data that are transferred through a WebRTC system is encrypted using the DTLS method. This type of encryption is built in to compatible web browsers (Firefox, Chrome, and Opera), so that eavesdropping or data manipulation cannot happen.

Enrollment sample A biometric sample taken from a subject that will be used for the recognition of the subject in the future.

Entrenchment Installation of sensors or unauthorized software.

Equal error rate (EER) The equalization of the false accept rate and the false reject rate resulting from the configuration of a biometric matching process.

EU FP7 Project Rerum Rerum will develop, evaluate, and try an architectural framework for dependable, reliable, and secure networks of heterogeneous smart objects supporting innovative smart city applications. The framework will be based on the concept of "security and privacy by design," addressing the most critical factors for the success of smart city applications (https://ict-rerum.eu/).

European Network and Information Security Agency (ENISA) Since its inception in 2004, ENISA actively seeks to contribute to a high level of network and information security (NIS) within the EU. Its aim is to raise awareness of NIS and to develop a culture of NIS within society in order to ensure proper functioning of the internal market (https://www.enisa.europa.eu).

EU's GDPR It will take effect in 2018 and will have a significant effect on US businesses that operate in the EU, touch personal information of EU citizens, or simply do business on an international basis. It will provide several cautionary tales for companies due to its high EU-based privacy standards, its comprehensive regulations, and its focus on monetary fines.

Expectancy A user's expectations for what, how, where, or when a stimulus will appear.

Exploit A piece of software, a chunk of data, or a sequence of commands that capitalize on a bug or vulnerability to induce unintended or unanticipated behaviors in digital systems.

False accept rate (FAR) The probability that a presented sample is accepted as genuine when it is not genuine.

False match rate (FMR) Synonymous with false accept rate.

False nonmatch rate (FNMR) Synonymous with false reject rate.

False reject rate (FRR) The probability that a presented sample is rejected when it is in fact genuine.

FBI Internet Crime Complaint Center (IC3) The IC3 register online Internet crime complaints from either the actual victim or from a third party to the complainant.

Feature set A biometric template where the relevant features are encoded as an unordered set.

Feature vector A biometric template where relevant features are encoded as an ordered sequence.

Federal Information Security Management Act of 2002 (FISMA) An act defining a framework to protect government information, operations, and assets against natural or human-made threats in the United States. FISMA was signed into law as part of the Electronic Government Act of 2002. FISMA "requires each federal agency to develop, document, and implement an agency-wide program to provide information security for the information and systems that support the operations and assets of the agency, including those provided or managed by another agency, contractor, or other sources."

Federal Risk and Authorization Management Program (FedRAMP) A US federal government-wide program for the assessment and authorization of cloud services providers. It is a federal program for assessing and authorizing cloud service providers. It provides a "do once, use many time" framework for both cloud service providers and US agency buyers. It follows FISMA guidelines and uses NIST 800-53 controls to define its cybersecurity requirements. FedRAMP applies to software-as-a-service, platform-as-a-service, and infrastructure-as-a-service providers.

Federal Trade Commission (FTC) A US agency that provides information to help consumers identify, prevent, and avoid scams and fraud and works to stop fraudulent, deceptive, and unfair business practices.

File transfer protocol (FTP) A protocol used to transfer files between computers on a network. FTP allows for files to be exchanged between computer accounts, transferred between an account and a desktop computer, or accessed online in software archives. Many FTP sites are heavily used.

Filter bubble phenomenon A filter bubble is a state of intellectual isolation that can result from personalized searches when a website algorithm selectively guesses what information a user would like to see based on information about the user, such as location, past click behavior, and search history (according to Wikipedia).

General Data Protection Regulation (GDPR) Individuals in the EU rely on the GDPR, which is used by the European Parliament, the Council of the European Union, and the European Commission in order to strengthen and unify data protection.

Genuine sample The verified and confirmed biometric sample from a subject.

Gramm–Leach–Bliley Act (GLBA) An act enacted that requires financial institutions, according to the FTC, to disclose their data security and privacy practices to their customers and to create security protocols for their customer's personal data.

Hacktivism The practice of gaining unauthorized access to a computer system and carrying out disruptive actions as a means of achieving political or social goal.

Harvesting (e.g., credentials) The attacking technique or activities of stealing legitimate user identification and passwords to gain access to target systems for illegal or malicious purposes.

Health Information Technology for Economic and Clinical Health (HITECH) Act An act enacted as part of the American Recovery and Reinvestment Act of 2009 and signed into law on February 17, 2009, to promote the adoption and meaningful use of health information technology.

Health Information Trust Alliance (HITRUST) A not-for-profit organization advocating programs instrumental in safeguarding health information and managing information risk while ensuring consumer confidence in the organizations that create, store, or exchange their information.

Health Insurance Portability and Accountability Act (HIPAA) HIPAA (Public Law 104-191) is intended to provide continuous health insurance coverage for workers who lose or change their job and to reduce the administrative burdens and cost of healthcare by standardizing the electronic transmission of administrative and financial transactions. It is an Act to amend the Internal Revenue Code of 1986 to improve portability and continuity of health insurance coverage in the group and individual markets; to combat waste, fraud, and abuse in health insurance and health care delivery; to promote the use of medical savings accounts; to improve access to long-term care services and coverage; to simplify the administration of health insurance; and for other purposes (according to Wikipedia).

Heuristic (decision-making) A shortcut that reduces the time or cognitive effort needed to make a decision. See also *satisficing*.

Homomorphic encryption A form of encryption that allows computation on ciphertexts, generating an encrypted result which, when decrypted, matches the result of operations performed on the plaintext. The purpose of homomorphic encryption is to allow computation on encrypted data (according to Wikipedia).

Honeypot A security resource whose value lies in being probed, attacked, or compromised.

Hop-to-hop In computer networking, a hop is one portion of the path between source and destination. Data packets pass through bridges, routers, and gateways as they travel between source and destination. Each time packets are passed to the next network device, a hop occurs (according to Wikipedia).

Host-based firewall Placing host-based firewalls on every virtual machine one may have in the cloud environment.

Human automation teaming With the nature of human–computer interaction changing in response to automation and its increasing capability to make decisions and interact with humans in naturalistic ways, this term encapsulates this style of interaction.

Human–machine interface (HMI) A software application for an operator or user to observe parameters of a process and to input control settings of equipment.

Identity Enables users to authenticate themselves to access sensitive content. This is typically limited to government and military personnel, typically using smart card initiatives such as the common access card or the personal identity verification program.

Identity proofing Verification and authentication of the identity of legitimate customers. Identity proofing has become more critical than ever.

IEEE 802.1X An Institute of Electrical and Electronics Engineers (IEEE) Standard for port-based network access control. It is part of the IEEE 802.1 group of networking protocols. It provides an authentication mechanism to devices wishing to attach to a local area network or a wireless local area network (according to Wikipedia).

Incident response plan A plan providing instructions for responding to a number of potential scenarios, including data breaches, DoS/distributed DoS attacks, firewall breaches, virus or malware outbreaks, or insider threats.

Industrial control systems cyberemergency response team (ICS-CERT) ICS-CERT partners with law enforcement agencies and the intelligence community to coordinate efforts among federal, state, local, and tribal governments and control systems owners, vendors, and operators. By doing so, the ICS attempts to reduce risks within and across all critical infrastructure sectors. Also, the ICS-CERT works together with the private sector and international computer emergency response teams to share control systems-related security incidents and mitigation measures (according to https://ics-cert.us-cert.gov/).

Information and communication technologies (ICTs) A more extensive term for IT that emphasizes the role of unified communications and the integration of telecommunications (phone lines and wireless signals), computers, middleware, storage, enterprise software, and audiovisual systems that allow users to access, store, transmit, and manipulate information.

Information security awareness (ISA) An evolving part of information security that focuses on raising consciousness regarding potential risks of rapidly evolving forms of information and rapidly evolving threats to that information, which target human behavior (according to Wikipedia).

Information security management system (ISMS) A set of policies and procedures for systematically managing the sensitive data of an organization. An ISMS typically addresses employee behavior and processes as well as data and technology.

Information security policy (ISP) A policy that addresses data, programs, systems, facilities, tech infrastructure, and users of technology and third-party organizations. An ISP provides employees with guidelines concerning how to ensure information security when they utilize information systems to perform their jobs (according to Wikipedia).

Information Systems Audit and Control Association (ISACA) An independent, nonprofit, global association and engages in the development; adoption; and use of globally accepted, industry-leading knowledge and practices for information systems.

InfraGard An organization locally run by FBI field offices. It has 84 chapters with more than 46,000 members nationwide helping to protect and defend critical infrastructures. The InfraGard program was designed to facilitate public–private collaboration between the private sector and the government.

Insider threat prediction model (ITPM) A model that initially estimates the potential impact of an incident, as well as the suspected insider's role, the hardware and software tools they are capable of using, their historical behavior, etc.

Integration of over-the-air (OTA) The integration of OTA programming mechanisms as part of the IoT ecosystem allows bugs to be solved or new functionalities to be added in deployed devices. Moreover, OTA is essential in order to ensure that existing infrastructures will be able to address future security flaws without requiring the substitution of massive amounts of old units or manual flashing procedures.

International Association of Privacy Professionals (IAPP) Known to be the world's largest and most comprehensive global information privacy community, the IAPP is a resource for professionals who want to develop and advance their careers by helping their organization successfully manage these risks and protect their data.

Internet Engineering Task Force (IETF) An international community of network designers, operators, vendors, and researchers concerned with the evolution of the Internet architecture and the smooth operation of the Internet.

Internet protocol version 6 (IPv6) The communications protocol that provides an identification and location system for computers on networks and routes traffic across the Internet.

Internet of things (IoT) Physical devices; vehicles (also referred to as "connected devices" and "smart devices"); buildings; and other items embedded with electronics, software, sensors, actuators, and network connectivity are able to collect and exchange data using the IoT.

Internet of things architectures (IoT-A) By outlining guidelines and principles for the technical design of protocols, interfaces, and algorithms, the Iot-A aims to develop an architectural reference model for the interoperability of IoT systems.

Internet of Things European Research Cluster (IERC) With the intention of defining a common vision of IoT technology and development research challenges in the European view of global development, the IERC brings together various EU-funded projects to accomplish this aim.

IP spoofing (also known as IP address forgery) The creation of IP packets with a false source IP address, for the purpose of hiding the identity of the sender or impersonating another computing system. A technique used to gain unauthorized access to computers, whereby the intruder sends messages to a computer with an IP address indicating that the message is coming from a trusted host. To engage in IP spoofing, a hacker must first use a variety of techniques to find an IP address of a trusted host and then modify the packet headers so that it appears that the packets are coming from that host.

Iris code A feature vector derived from an iris image.

ISO/IEC (International Organization for Standardization/International Electrotechnical Commission) 27000 series A set of standards that provide background, common terminology, principles, techniques, and guidance on information security management systems (ISMSs). It describes what is required to implement an ISMS, including understanding organizational context, required leadership support, risk assessment, and continuous improvement. The main document in the ISO/IEC 27000 series is ISO/IEC 27001, officially titled "Information technology—Security techniques—Information security management systems—Requirements," was updated and rereleased in 2013. It describes what is required to implement an ISMS including understanding organizational context, required leadership support, risk assessment, and continuous improvement.

Issuer distinguished name Who issued the certificate—depending on whether the issuer is a known or unknown entity, this can be used to determine the level of trust placed in the owner of the certificate.

Invisible Internet Project (I2P) By using anonymous web surfing, blogging, chatting, and file transfers, the I2P allows applications to send messages to each other pseudonymously and securely. This is a garlic routing which uses overlay networks and darknet.

***k*-Nearest neighbor (*k*-NN) algorithm** In pattern recognition, a non-parametric method used for classification and regression. In both cases, the input consists of the *k* closest training examples in the feature space (according to Wikipedia).

Key escrow/archival server Used to store copies of private keys corresponding to entities to audit/inspect communications between human and machine entities or for disaster recovery purposes.

Key usage and extended key usage An attribute that controls what the certificate can be used for (to authenticate digital identities, to encrypt messages, for smart card authentication, etc.).

Liking principle We like to say yes to people whom we like and know on a personal level. The salesperson who tries to create a sense of friendship with potential customers primarily uses the liking principle.

LINDDUN A privacy threat analysis methodology that supports analysts in eliciting privacy requirements but does not explicitly provide risk analysis support. It provides a list of privacy solutions to mitigate elicited threats so that the threats can then be translated into privacy requirements.

Liveness detection Sometimes generally used as a synonym for presentation attack detection; the method of confirming if a biometric sample is live, meaning it has been presented in real time or has been provided by a live body.

Low-power and lossy networks (LLNs) Composition of many embedded devices with limited power, memory, and processing resources interconnected by a variety of links, such as low-power Wi-Fi.

Malicious insider A trusted entity that is given the power to violate one or more rules in a given security policy. An insider can thus be defined with regard to two primitive actions (1) violation of a security policy using legitimate access and (2) violation of an access control policy by obtaining unauthorized access.

Malicious insider attack—fraud Intending to (1) steal information that leads to an identity crime (i.e., credit card fraud or identity theft) or (2) obtain or accomplish modification, addition, or deletion of the data of an organization (not their systems or programs) for personal gain.

Malicious insider attack—IT sabotage Intending to cause harm to a specific individual or organization by an insider's knowledge and capabilities with IT.

Malicious insider attack—miscellaneous Any situation of insider attacks that do not fall under the categories of fraud, IT sabotage, or intellectual property theft.

Malicious insider attack—theft of intellectual property Intending to steal intellectual property from an organization using knowledge of IT by either an insider or an outsider taking part in industrial espionage.

Malicious software (malware) Software designed to infiltrate and damage a computer system without the user's consent. The term covers all different types of threats to your computer safety such as viruses, spyware, worms, Trojans, adware, and rootkits.

Malware infections Malicious software, known more commonly in its abbreviated form as malware, refers to a variety of types of intrusive or hostile software, including worms, Trojan horses, spyware, viruses, ransomware, scareware, adware, and other malicious programs. Software deficiencies or problems that cause unintended harm do not fall into this category. Instead, malware is very much defined by its malicious intent and can take the form of scripts, active content, executable code, and other software.

Manipulation of cyberassets Changing file permissions; suppressing or altering information content.

Masquerading attack One of the most serious types of computer abuse; an attack where one user impersonates another. It cannot be detected by authentication and access control since the original user's credentials are presented. The best existing solution is to detect departures from normal user behavior.

Mental models The ability to generate descriptions of system purpose and form, explanations of system functioning and observed system states, and predictions of future states.

Microsegmentation A security technology that breaks the data center into logical elements and manages them with high-level IT security policies.

Microsoft security development life cycle (SDL) A software development process that helps developers build more secure software and address security compliance requirements while reducing development cost.

Minutia A feature of a fingerprint template specifying the position and orientation of an end or bifurcation of a friction ridge.

Mirai Malware that infects consumer IoT devices such as IP cameras and home routers and utilizes them as part of a "botnet" to conduct large-scale network attacks. The source code for Mirai is freely accessible and has been adapted and used to disrupt a number of highly networked websites including those belonging to Twitter, Netflix, Airbnb, Reddit, and GitHub.

Mobile device management (MDM) A service that can secure iOS, Android, and other devices.

Mobile health (mHealth) The delivery of healthcare services via mobile communication devices.

Money mule An unwitting or witting person who transfers money acquired illegally (e.g., stolen), through a courier service, or electronically, on behalf of others. Typically, the mule is paid for services with a small part of the money transferred.

Multifactor authentication (MFA) Covered entities must implement MFA or similar controls for access to systems and nonpublic data. In particular, access to internal networks from external networks is called out.

Multifactor authentication tool A tool that adds an extra layer of protection on top of the username and password. With MFA enabled, when a user signs in to a cloud computing services website, they will be prompted for their username and password (the first factor—what they know) as well as for an authentication code from their AWS MFA device (the second factor—what they have).

National Conference of State Legislatures (NCSL) A nongovernmental organization established in 1975 that has since been a champion for state legislatures. By giving states the tools, information, and resources to craft the best solutions to difficult problems, the NCSL has helped states remain strong and independent from unwarranted actions in Congress and has saved states more than $1 billion.

National Cyber Investigative Joint Task Force (NCIJTF) A US task force providing coordination among law enforcement and intelligence communities. The NCIJTF is composed of over 20 partnering agencies and has representatives who are colocated and work jointly to accomplish mission of the organization from a whole-of-government perspective.

National Cybersecurity and Communications Integration Center (NCCIC) A center serving as a 24/7 cybermonitoring, incident response, and management center and as a national point of cyberincident and communications incident integration.

National Security Council (NSC) The White House NSC is the principal forum used by the president of the United States for consideration of national security and foreign policy matters with senior national security advisors and cabinet officials and is part of the executive office of the president of the United States (according to Wikipedia).

Network intrusion monitoring Observation and recording of network data including packet signature, time stamps, origin and destination IP addresses of network attempts, types of network event, and frequency of attempts/events.

NIS Directive A directive outlining security requirements and incident notification rules for digital signal processing that are different from those that apply to the operator of essential services (OESs) and digital service providers (DSPs). It was adopted in July 2016, and EU member states have until May 2018 to translate it into national laws and an additional six months to identify the OESs to which it applies (https://www.itgovernance.eu).

NIST cybersecurity framework A framework created through collaboration between industry and government and consisting of standards, guidelines, and practices to promote the protection of critical infrastructure. The prioritized, flexible, repeatable, and cost-effective approach of the framework helps owners and operators of critical infrastructure to manage cybersecurity-related risk.

Nonpublic Information (NPI) The privacy rule protects a consumer's "nonpublic personal information" (NPI). NPI is any "personally identifiable financial information" that a financial institution collects about an individual in connection with providing a financial product or service, unless that information is otherwise "publicly available."

Object name system (ONS) An automated networking service similar to the DNS that points computers to sites on the World Wide Web. The ONS leverages the DNS to obtain information about a product and related services from the EPC.

OCEAN (openness, conscientiousness, extraversion, agreeableness, neuroticism) A theory using descriptors of common language and suggests five broad dimensions commonly used to describe the human personality and psyche. The big five personality traits, also known as the five-factor model, are based on common language descriptors of personality. When factor analysis (a statistical technique) is applied to personality survey data, some words used to describe aspects of personality are often applied to the same person. For example, someone described as "conscientious" is more likely to be described as "always prepared" rather than "messy." This theory is therefore based on the association between words but not on neuropsychological experiments.

Office of Cyber Infrastructure and Analysis (OCIA) A US agency whose mission is to support efforts to protect the critical infrastructure of the nation through an integrated analytical approach evaluating the potential consequences of disruption from physical or cyberthreats and incidents.

Office of Cybersecurity and Communications (CS&C) The US agency responsible for enhancing the security, resilience, and reliability of the cyberinfrastructure and communications infrastructure of the nation. CS&C works to prevent or minimize disruptions to critical information infrastructure in order to protect the public, the economy, and government services.

Office of Infrastructure and Protection A US agency that leads and coordinates national programs and policies on critical infrastructure security and resilience. The office conducts and facilitates vulnerability and consequence assessments to help critical infrastructure owners and operators and state, local, tribal, and territorial partners understand and address risks to critical infrastructure.

Online pharmaceutical fraud Pharmaceutical fraud involves activities that result in false claims to insurers or programs such as Medicare in the United States or equivalent state programs for financial gain to a pharmaceutical company. There are several different schemes used to defraud the healthcare, system which are particular to the pharmaceutical industry (according to Wikipedia).

Onion router (Tor) Free software and an open network that helps one defend against traffic analysis, a form of network surveillance that threatens personal freedom and privacy, confidential business activities and relationships, and state security.

Open Interconnect Consortium (OIC) Developers, manufacturers, and end users rely on a standard and open source project called OIC in order to obtain "just works" interconnectivity.

Open-source intelligence (OSINT) Data collected from publicly available sources to be used in an intelligence context. In the intelligence community, the term *open* refers to overt, publicly available sources (as opposed to covert or clandestine sources) (according to Wikipedia).

Open Web Application Security Project (OWASP) A 501(c)(3) not-for-profit charitable organization focused on improving the security of software.

OWASP software assurance maturity model (SAMM) A open framework that helps organizations formulate and implement a strategy for software security that is tailored to the specific risks facing the organization.

Packet replay attacks A replay attack (also known as playback attack) is a form of network attack in which a valid data transmission is maliciously or fraudulently repeated or delayed. This is carried out either by the originator or by an adversary who intercepts the data and retransmits it, possibly as part of a masquerade attack by IP packet substitution (according to Wikipedia).

Payment Card Industry Data Security Standard (PCI DSS) A global organization founded by some of the largest payment card processing companies—American Express, Discover Financial Services, JCB International, MasterCard, and Visa Inc.—the PCI DSS defines operational and technical requirements for the protection of payment card account data.

The PCI DSS is an information security standard for organizations that handle branded credit cards from the major card schemes.

The PCI Standard is mandated by the card brands and administered by the PCI Security Standards Council (according to Wikipedia).

Peer to peer (P2P) P2P computing or networking is a distributed application architecture and a decentralized communications model in which each party has the same capabilities and either party can initiate a communication session.

Penetration testing and vulnerability assessments This regulation requires monitoring and testing to evaluate the effectiveness of the cybersecurity program. It states that monitoring should be continuous or penetration testing is required at least annually and vulnerability assessments are required at least biannually.

Perimeter-based access control Regulation of data/communication traffics to prevent unauthenticated data or restrict authenticated data into a selected part of the network. For example, firewalls can setup a DMZ to buffer corporate network into the trusted supervisory control and data acquisition (SCADA) one.

Personal health information (PHI) or personal health record (PHR) An electronic, universally available, lifelong resource of health information needed by individuals to make health decisions.

Personally identifiable information (PII) PII, or sensitive personal information as used in information security and privacy laws, is information that can be used on its own or with other information to identify, contact, or locate a single person or to identify an individual in context.

Phishing Attempt to obtain sensitive information such as usernames, passwords, and credit card details (and, indirectly, money), often for malicious reasons, by disguising as a trustworthy entity in an electronic communication.

Phishing e-mail The fraudulent practice of sending e-mails purporting to be from reputable companies in order to induce individuals to reveal personal information, such as passwords and credit card numbers.

PKI trust models A trust model is a collection of rules that informs application on how to decide the legitimacy of a digital certificate. There are two types of trust models widely used. For PKI to work, the capabilities of CAs must be readily available to users. The model that has been shown to this point is the simple trust model. PKI was designed to allow all these trust models to be created. They can be fairly granular from a control perspective (https://www.scribd.com).

Planned obsolescence Planned obsolescence, or built-in obsolescence, in industrial design and economics is a policy of planning or designing a product with an artificially limited useful life, so it will become obsolete (that is, unfashionable or no longer functional) after a certain period of time (according to Wikipedia).

Presentation attack When an adversary presents a biometric sample with the objective of impersonating a subject and the subject's genuine sample. In this case, the adversary's sample does not come from their body but could be from an artifact, a disguised adversary, or a digitally altered sample that has come from the adversary.

Presentation attack detection An attempt at determining if a presentation attack is taking place at the moment that a biometric sample is being presented.

Privacy Act of 1974 The Privacy Act of 1974, 5 USC § 552a, establishes a code of fair information practices that governs the collection, maintenance, use, and dissemination of information about individuals that is maintained in systems of records by federal agencies (https://www.justice.gov).

Privacy breach The loss of, unauthorized access to, or disclosure of personal information. Some of the most common privacy breaches happen when personal information is stolen, lost, or mistakenly shared.

Privacy-by-default It means that once a product or service has been released to the public, the strictest privacy settings should apply by default, without any manual input from the end user. Also, any personal data provided by the user to enable the optimal use of a product should only be kept for the amount of time necessary to provide the product or service.

Privacy by design An approach to systems engineering which takes privacy into account throughout the whole engineering process. The concept is an example of value sensitive design, i.e., to take human values into account in a well-defined manner throughout the whole process and may have been derived from this (according to Wikipedia).

Privacy-enhancing technology (PET) A standardized term that refers to specific methods that act in accordance with the laws of data protection. PETs allow online users to protect the privacy of their PII provided to and handled by services or applications (according to Wikipedia). These technologies refer to the specific methods that are in accordance with the laws of data protection.

Privacy impact assessment (PIA) A type of impact assessment conducted by an organization (typically, a government agency or corporation with access to a large amount of sensitive, private data about individuals in or flowing through its system). The organization audits its own processes and sees how these processes affect or might compromise the privacy of the individuals whose data it holds, collects, or processes (according to Wikipedia).

Private information retrieval (PIR) Users can retrieve an item from a server in possession of a database without revealing which item is retrieved by using a type of cryptography called PIR protocol.

Programmable logic controller (PLC) Digital computers ruggedized and adapted for the control of industrial processes, such as assembly lines and power generation, that requires high reliability control and fault diagnosis.

Propensity to trust The potential ability to trust an entity before having had any experience with it.

Psychological profile (PPs) A profile dynamically constructed from behavioral patterns.

Public key A key that records the public key part of the entity's key pair that can be used to send encrypted messages to the entity.

Public key infrastructure (PKI) The ecosystem that controls the issuance, storage, and distribution of digital certificates.

Pubic key pinning (PKP) The PKP extension for HTML5 is a security feature. It tells a web client to associate a specific cryptographic public key with a certain web server to decrease the risk of MITM attacks with forged certificates.

Quality of service (QoS) The method of measuring or outlining the overall performance of a service, such as a telephony, computer network, or cloud computing service—especially the performance seen by the network users.

Radio-frequency identification device (RFID) By using electromagnetic fields, RFID can automatically identify and track tags (which contain electronically stored information) attached to objects.

Reciprocity or obligation to repay principle According to anthropologists, the rule of reciprocity is apparent in all human societies.

Regional cyber forensic laboratories (RCFLs) RCFL examiners combine the talents and experience of federal, state, and local law enforcement agencies. A midsize RCFL consists of 15 people 12 of the staff members are examiners and 3 staff members support the RCFL. Digital forensics is the application of science and engineering to the recovery of digital evidence in a legally acceptable method (https://www.rcfl.gov/about).

Registration authority (RA) Responsible for the verification of identities prior to the issuance of certificates.

Reliable, resilient, and secure IoT for smart city applications (Rerum) By increasing the trustworthiness of IoT, Rerum enhances and improves devices and middleware functionalities and provides an overall security, privacy, and trust framework to address citizens' requirements for advanced, resilient, reliable, and secure smart city applications that respect their privacy.

Remote terminal unit (RTU) An electronic device transmitting telemetry data from physical equipment to supervisory control servers or computers systems.

Revocable biometrics A general term that refers to any use of a biometric key that has been regenerated from a biometric sample and helper data. The helper data and biometric key are randomized and can be revoked and regenerated using different random bits. This method can be used to recognize a subject by directly or indirectly inferring that a biometric key regenerated from a presented sample is identical to a biometric key originally generated from an enrollment sample, instead of matching the presented sample to the enrollment sample or data derived from the enrollment sample such as a biometric template.

Right to data portability in Article 20 It states that the data subject shall have the right to receive the personal data concerning him or her, which he or she has provided to a controller, in a structured, commonly used and machine-readable format and have the right to transmit those data to another controller without hindrance from the controller to which the personal data have been provided (https://www.privacy-regulation.eu).

Right to erasure ("right to be forgotten") in Article 17 The data subject shall have the right to obtain from the controller the erasure of personal data concerning him or her without undue delay, and the controller shall have the obligation to erase personal data without undue delay under certain grounds (https://www.privacy-regulation.eu).

RSA (Rivest–Shamir–Adleman) RSA is one of the first practical public key cryptosystems and is widely used for secure data transmission. In such a cryptosystem, the encryption key is public, and it is different from the decryption key which is kept secret (private). In RSA, this asymmetry is based on the practical difficulty of the factorization of the product of two large prime numbers, the "factoring problem" (according to Wikipedia).

SANS Institute A cooperative research and education organization.

Sarbanes–Oxley Act (SOX) An act passed in 2002 by the US Congress that mandated strict reforms to improve financial disclosures from corporations and to protect investors from risks of fraudulent accounting activities by corporations.

Satisficing Making a decision that is sufficient even if not optimal.

Scarcity Wanting what may not be available.

Schema theory A theory stating that a person can only make meaning of a concept if it relates to knowledge that the person already possesses. Federick Bartlett first introduced it.

Schonlau dataset Fifteen thousand truncated UNIX commands for each user, 70 users; 100 commands as one block. Each block is treated as a "document" and randomly chooses 50 users as victims. Each user's first 5000 commands are clean, and the rest have randomly inserted dirty blocks from the other 20 users.

Script kiddie In programming and hacking culture, a script kiddie or skiddie is an unskilled individual who uses scripts or programs developed by others to attack computer systems and networks and deface websites. It is generally assumed that most script kiddies are juveniles who lack the ability to write sophisticated programs or exploits on their own and that their objective is to try to impress their friends or gain credit in computer-enthusiast communities (according to Wikipedia).

Secure file transfer protocol, the SSH file transfer protocol (SFTP) Also known as the secure file transfer protocol, SFTP enables secure file transfers between networked hosts. SFTP provides remote file system management functionality, allowing applications to resume interrupted file transfers, list the contents of remote directories, and delete remote files, which is different from secure copy protocol.

Secure shell (SSH) A cryptographic network protocol for operating network services securely over an unsecured network. The best-known example application is for remote login to computer systems by users. SSH provides a secure channel over an unsecured network in a client–server architecture, connecting an SSH client application with an SSH server (according to Wikipedia).

Secure sockets layer (SSL) SSL certificates have a key pair a public and a private key. These keys work together to establish an encrypted connection.

Security/privacy impact assessments (PIAs) PIAs and, in the EU, data PIAs, are, in a nutshell, a data audit of your company. Where are the data and how are they stored, transmitted, and processed? It is common to evaluate these issues internally, and assign red, yellow, or green colors to each existing practice to evaluate whether they are dangerous, questionable, or just fine as is.

Security vulnerabilities In computer security, a weakness that allows an attacker to reduce the information assurance of a system. Vulnerability is the intersection of three elements a system susceptibility or flaw, attacker access to the flaw, and attacker capability to exploit the flaw (according to Wikipedia).

Sensitive authentication data (SAD) Security-related information including, but not limited to, card validation codes/values used to authenticate cardholders and/or authorize payment card transactions. SAD must not be stored after authorization.

Session key Sometimes called symmetric keys (since the same key can be used for both encryption and decryption; an encryption and decryption key that is randomly generated and ensures the security of a communications session between a user and another computer or between two computers.

Service level agreement (SLA) An official commitment that prevails between a service provider and a client. Particular aspects of the service—quality, availability, and responsibilities—are agreed between the service provider and the service user (according to Wikipedia).

Shoulder surfing In computer security, a type of social engineering technique used to obtain information such as personal identification numbers, passwords, and other confidential data by looking over the victim's shoulder (according to Wikipedia).

Signal-to-interference noise ratio (SINR) A quantity that is used in telecommunication engineering and information theory, the SINR gives theoretical upper bounds on channel capacity (or the rate of information transfer) networks and other wireless communication systems.

Smart building management By integrating the mechanical, security, and electrical systems in a facility, a smart building management system is able to enhance a particular IT infrastructure.

Smart contract (SC) Instead of using legal language in print or on paper, SCs can be written and transmitted in computer code, which can be stored on the Blockchain.

Smart objects for intelligent building management (SOrBeT) SOrBet (FP7 MC-IAPP, project no. 612361) is a Marie Curie Industry–Academia Partnerships and Pathways (IAPP) project. It is started in January 2014 and is funded by the EU. It aims toward highly distributed, self-organizing, self-managing, wirelessly communicating smart objects, enabling the robust management of energy-efficient smart buildings.

Social proof Confirmation of the validity of something by finding out if other people also think it is valid.

Sociotechnical system A complex operational environment characterized by diverse actors participating with risk and dynamism.

SP 800-63 SP 800-63 provides an overview of general identity frameworks, using authenticators, credentials, and assertions together in a digital system, and a risk-based process of selecting assurance levels. SP 800-63 contains both normative and informative material (http://nvlpubs.nist.gov).

Spoofing Taking part in a presentation attack with the intent to impersonate a subject.

SQL injection A code injection technique used to attack data-driven applications in which nefarious SQL statements are inserted into an entry field for execution (e.g., to dump the database contents to the attacker). SQL injection must exploit a security vulnerability in the software of an application, for example, when user input is either incorrectly filtered for string of literal escape characters embedded in SQL statements or strongly typed and unexpectedly executed. SQL injection is mostly known as an attack vector for websites but can be used to attack any type of SQL database (according to Wikipedia).

Standard Internet protocol (SIP) The session initiation protocol is a communications protocol for signaling and controlling multimedia communication sessions in applications of Internet telephony for voice and video calls, in private IP telephone systems, as well as in instant messaging over IP networks (according to Wikipedia).

States (psychological) Different mental moods or feelings within individuals.

STRIDE Used in partnership with a model of a target system that can be constructed in parallel, STRIDE was initially created as part of the process of threat modeling. STRIDE helps reason and finds threats to a system. Microsoft originally developed it for dealing with computer security threats.

Structural anomaly (SA) SA uses graph analysis, dynamic tracking, and machine learning to detect anomalies.

StuxNet attack Stuxnet is a malicious computer worm, first uncovered in 2010 by Kaspersky Lab. Thought to have been in development since at least 2005, Stuxnet targets SCADA systems and was responsible for causing substantial damage to Iran's nuclear program. Although neither country has admitted responsibility, since 2012, the worm is frequently described as a jointly built American/Israeli cyberweapon (according to Wikipedia).

Subject distinguished name Who the certificate was issued to—this could be a human or a machine identity.

Supervisory control and data acquisition (SCADA) A control architecture adopted by most industrial control systems that enable human and automation in monitoring and controlling industrial processes. SCADA systems usually contain three segments of ICT—field devices, the SCADA network, and the corporate network.

Symmetric keys Symmetric keys are preferred over asymmetric keys as they offer better encryption performance, yet have the requirement of both parties in a transaction needing access to the (same) symmetric key. It is for this reason that symmetric keys are typically used to secure data-at-rest—disk encryption, file encryption, database encryption, etc.

System and organization controls (SOC) SOC for cybersecurity is a risk framework that establishes common criteria and guidelines for communicating about cybersecurity risk management program of an organization. It is a reporting framework through which organizations can communicate relevant useful information about the effectiveness of their cybersecurity risk management program.

Traits (psychological) The characteristics of an individual that is generally stable between task situations such as age, personality, and propensity to trust.

Transparency The descriptive quality of an interface pertaining to its abilities to afford an operator's comprehension about an intelligent agent's intent, performance, future plans, and reasoning process.

Transport layer security (TLS) A protocol proposed as a stronger alternative to SSL in 1999 and is now required by most modern applications for encryption. Like with SSL, TLS has gone through multiple versions with TLS 1.3 being the latest version of the protocol.

Triple A Authentication, authorization, and accountability Authentication is the "process of verifying a claim that a system entity or system resource has a certain attribute value." Authorization is the "process for granting approval to a system entity to access a system resource." Authorization controls who can do what to which objects, while authentication involves identifying who is seeking the access, being often a specific part of the authorization process. Accountability enables "the detection of actions to be traced to the potentially responsible entity."

Trust in automation Trust in a nonhuman agent.

Trustee The entity that is the recipient of a person's trust.

Trustor The person who trusts another entity.

Truth in Lending Act The Truth in Lending Act (TILA) of 1968 is US federal law designed to promote the informed use of consumer credit, by requiring disclosures about its terms and cost to standardize the manner in which costs associated with borrowing are calculated and disclosed (according to Wikipedia).

US Department of Health and Human Services (HHS) The US department that protects the health of all Americans and provides essential human services, especially for those least able to help themselves.

US National Institute of Standards and Technology (NIST) A US agency that implements practical cybersecurity and privacy through outreach and effective application of standards and best practices necessary for the United States to adopt cybersecurity capabilities.

Unintentional Insider Threat (UIT) An unintentional insider threat is defined as a current or former employee, contractor, or business partner who has or had authorized access to an organization's network, system, or data and who, through their action/inaction without malicious intent cause harm or substantially increase the probability of future serious harm to the confidentiality, integrity, or availability of the organization's information or information systems.

User datagram protocol (UDP) In computer networking, one of the core members of the IP suite. The protocol was designed to send messages, in this case referred to as datagrams, to other hosts on an IP network.

US Family Educational Rights and Privacy Act (FERPA) FERPA (20 USC § 1232g; 34 CFR Part 99) is a federal law that protects the privacy of student education records. The law applies to all schools that receive funds under an applicable program of the US Department of Education.

Validity Governs when a certificate was issued and when it expires to ensure that keys are periodically regenerated (much like passwords) to ensure that they do not become susceptible to cracking attempts.

Virtual desktop infrastructure (VDI) Virtualization technology that hosts a desktop operating system on a centralized server in a data center. VDI is a variation on the client–server computing model, sometimes referred to as server-based computing.

Virtual identity An interface that represents the user in a virtual world such as a chat room, video game, or virtual common space.

Virtual private network (VPN) The technology that creates a safe and encrypted connection over a less secure network, such as the Internet.

WannaCry malware exploit The WannaCry ransomware attack was a May 2017 worldwide cyberattack by the WannaCry ransomware cryptoworm, which targeted computers running the Microsoft Windows operating system by encrypting data and demanding ransom payments in the Bitcoin cryptocurrency (according to Wikipedia).

Wireless sensor network (WSN) By cooperatively passing their data through the network to other locations, WSNs, sometimes called wireless sensor and actuator networks, are spatially distributed autonomous sensors that can monitor environmental or physical conditions such as pressure, sound, and temperature.

World Trade Organization (WTO) An intergovernmental organization that regulates international trade. The WTO deals with regulation of trade between participating countries by providing a framework for negotiating trade agreements and a dispute resolution process aimed at enforcing participants' adherence to WTO agreements, which are signed by representatives of member governments.

X509 A standard that is widely deployed and understood when it comes to defining the format of digital certificates.

Zero-day exploit A vulnerability previously undisclosed to the public exploited by attackers for system access and control.

Zero-effort attack When an adversary presents a biometric sample from their own body with the objective of impersonating a subject and the subject's genuine sample.

Index

Page numbers followed by f, t, and b indicate figures, tables, and boxes, respectively.

D

K

L

M

U